19

Springer Series in Chemical Physics

Edited by Robert Gomer

Springer Series in Chemical Physics

Editors: V. I. Goldanskii R. Gomer F. P. Schäfer J. P. Toennies

Volume 1 **Atomic Spectra and Radiative Transitions** By I. I. Sobelman

Volume 2 **Surface Crystallography by LEED** Theory, Computation and Structural Results By M. A. Van Hove, S. Y. Tong

Volume 3 **Advances in Laser Chemistry** Editor: A. H. Zewail

Volume 4 **Picosecond Phenomena** Editors: C. V. Shank, E. P. Ippen, S. L. Shapiro

Volume 5 **Laser Spectroscopy** Basic Concepts and Instrumentation By W. Demtröder

Volume 6 **Laser-Induced Processes in Molecules** Physics and Chemistry Editors: K. L. Kompa, S. D. Smith

Volume 7 **Excitation of Atoms and Broadening of Spectral Lines** By I. I. Sobelman, L. A. Vainshtein, E. A. Yukov

Volume 8 **Spin Exchange** Principles and Applications in Chemistry and Biology By Yu. N. Molin, K. M. Salikhov, K. I. Zamaraev

Volume 9 **Secondary Ion Mass Spectrometry SIMS II** Editors: A. Benninghoven, C. A. Evans, Jr., R. A. Powell, R. Shimizu, H. A. Storms

Volume 10 **Lasers and Chemical Change** By A. Ben-Shaul, Y. Haas, K. L. Kompa, R. D. Levine

Volume 11 **Liquid Crystals of One- and Two-Dimensional Order** Editors: W. Helfrich, G. Heppke

Volume 12 **Gasdynamic Laser** By S. A. Losev

Volume 13 **Atomic Many-Body Theory** By I. Lindgren, J. Morrison

Volume 14 **Picosecond Phenomena II** Editors: R. Hochstrasser, W. Kaiser, C. V. Shank

Volume 15 **Vibrational Spectroscopy of Adsorbates** Editor: R. F. Willis

Volume 16 **Spectroscopy of Molecular Excitons** By V. L. Broude, E. I. Rashba, E. F. Sheka

Volume 17 **Inelastic Particle-Surface Collisions** Editors: E. Taglauer, W. Heiland

Volume 18 **Modelling of Chemical Reaction Systems** Editors: K. H. Ebert, P. Deuflhard, W. Jäger

Volume 19 **Secondary Ion Mass Spectrometry SIMS III** Editors: A. Benninghoven, J. Giber, J. László, M. Riedel, H. W. Werner

Volume 20 **Chemistry and Physics of Solid Surfaces IV** Editors: R. Vanselow, R. Howe

Secondary Ion Mass Spectrometry

SIMS III

Proceedings of the Third International Conference,
Technical University, Budapest, Hungary,
August 30 - September 5, 1981

Editors:
A. Benninghoven J. Giber J. László
M. Riedel H.W. Werner

With 289 Figures

Springer-Verlag Berlin Heidelberg New York 1982

Professor Dr. A. Benninghoven
Physikalisches Institut der Universität Münster
D-4400 Münster, Fed. Rep. of Germany

Professor Dr. J. Giber,
J. László
Physical Institute of the Technical University
H-1111 Budapest, Hungary

Dr. M. Riedel
Department of Physical Chemistry and Radiology, Eötvös University
H-1111 Budapest, Hungary

Dr. H.W. Werner
Philibs Research Laboratories, Eindhoven
The Netherlands

International Organizing Committee: A. Benninghoven (Chairman), V.T. Cherepin, C.A. Evans, Jr., K.F.J. Heinrich, J. Okano, G. Slodzian, H. W. Werner

Local Organizing Committee: J. Giber, J. László, M. Riedel, A. Sólyom, V. Mesterházy
Physical Institute of the Technical University Budapest, Budafoki ut 8,
H-1111 Budapest, Hungary

Sponsored by: Hungarian Academy of Sciences, Ministry of Culture and Education of Hungary, Technical University of Budapest, European Physical Society, Lorand Eötvös Physical Society

---

Series Editors

Professor Vitalii I. Goldanskii
Institute of Chemical Physics
Academy of Sciences
Vorobyevskoye Chaussee 2-b
Moscow V-334, USSR

Professor Robert Gomer
The James Franck Institute
The University of Chicago
5640 Ellis Avenue
Chicago, IL 60637, USA

Professor Dr. Fritz Peter Schäfer
Max-Planck-Institut für
Biophysikalische Chemie
D-3400 Göttingen-Nikolausberg
Fed. Rep. of Germany

Professor Dr. J. Peter Toennies
Max-Planck-Institut für Strömungsforschung
Böttingerstraße 6–8
D-3400 Göttingen
Fed. Rep. of Germany

ISBN 978-3-642-88154-1 ISBN 978-3-642-88152-7 (eBook)
DOI 10.1007/978-3-642-88152-7

Softcover reprint of the hardcover 1st edition 1982

2153/3130-543210

Preface

Following the biannual meetings in Münster (1977) and Stanford (1979) the Third International Conference on Secondary Ion Mass Spectroscopy was held in Budapest from August 31 to September 5, 1981. The Conference was attended by about 250 participants. The success of the 1981 Conference in Budapest was especially due to the excellent preparation and organization by the Local Organizing Committee. We would also like to acknowledge the generous hospitality and cooperation of the Hungarian Academy of Sciences. Japan was chosen to be the location for the next conference in 1983.

SIMS conferences are devoted to two main issues: improving the application of SIMS in different and especially new fields, and understanding the ion formation process. Needless to say, there is a very strong interaction between these two issues.

The major reason for the rapid increase in SIMS activities in the last few years is the fact that SIMS is a powerful tool for bulk, thin-film, and surface analysis. Today it is extensively and successfully applied in such different fields as depth profiling and imaging of semiconductor devices, in isotope analysis of minerals, in imaging biological tissues, in the study of catalysts and catalytic reactions, in oxide-layer analysis on metals in drug detection, and in the analysis of body fluids.

Yet in spite of these successful applications of SIMS, the ion formation process remains the central question. Up to now we still know very little about the fundamental processes which result in the emission of a charged atomic or molecular particle from an ion-bombarded surface. Thus the present situation may be summarized by the assessmant that we have succeeded in developing an excellent and powerful analytical technique, but based upon processes that are not sufficiently understood.

This volume contains almost all of the invited and contributed papers presented at the Conference. The contributions are arranged following the sections of the Conference program. All papers included in this volume were reviewed at the Conference. The final responsibility for the manuscripts, however, is with the authors only. For technical editorial work we thank Dr. I. Bârsony, Mr. Z. Soha and Mrs. Z. Pâkh.

The editiors hope that this volume will help those who could not participate in the meeting become familiar with the recent developments and activities in the field of secondary ion mass spectrometry.

Budapest, Hungary
December, 1981

A. Benninghoven
J. Giber · J. László
M. Riedel · H.W. Werner

Contents

Part I. **Instrumentation**

*Instrumental Aspects of Spatially 3-Dimensional SIMS Analysis
By F.G. Rüdenauer 2

*Some Problems of Construction Implied by Requirements of Up-To-Date SIMS Instrumentation. By R.L. Gerlach 22

Description and Applications of a New Design Cs^+ Ion Source on the COALA Ion Microprobe for Negative Ion SIMS
By B.L. Bentz and H. Liebl 30

Operational Data of a Simple Microfocus Gun Using an EHD-Type Indium Ion Source. By M.J. Higatsberger, P. Pollinger, H. Studnicka, and F.G. Rüdenauer 38

First Results on a Scanning Ion Microprobe Equipped with an EHD-Type Indium Primary Ion Source. By F.G. Rüdenauer, P. Pollinger, H. Studnicka, H. Gnaser, W. Steiger, and M.J. Higatsberger 43

Simple Double-Channel SIMS Instrument
By V.T. Cherepin, I.N. Dubinsky, and Ya.Ya. Dyad'kin 49

Principles and Applications of a Dual Primary Ion Source and Mass Filter for an Ion Microanalyser. By J.J. Le Goux and H.N. Migeon .. 52

A Quadrupole Mass Spectrometer with Energy Filtering for SIMS Studies
By R.-L. Inglebert and J.-F. Hennequin 57

Development and Operation of Special SIMS-Equipment for Use in Iron and Steel Analysis. By J. Dittmann 61

*Design Concept of a New Secondary Ion Optics System for Use with Quadrupole Mass Spectrometers
By R. Jede, O. Ganschow, and A. Benninghoven 66

Improved Analysis of Insulators in an ARL IMMA Using Positive Primary Ions and an Electron Gun. By J.D. Brown and D.J. Gras 72

Performance and Use of Dissector Ion Microanalyzer
By V.T. Cherepin and V.L. Ol'khovsky 77

---

*Invited paper

Distortion of Secondary Ion Extraction Due to Sample Surface Irregularities. By W. Bedrich, B. Koch, H. Mai, U. Seidenkranz, H. Syhre, and R. Voigtmann 81

A Combined Direct Imaging Laser Ionization Secondary Ionization Mass Spectrometer. By B.K. Furman and C.A. Evans, Jr. 88

Advances in Ion Probes A-DIDA. By H. Frenzel and J.L. Maul 94

A Novel Ion Etching Unit Applicable for Depth Profiling with SIMS and IIR. By K.H. Guenther, E. Hauser, G. Hobi, P.G. Wierer, and E. Brandstaetter 97

Improvements and Applications of the Riber MIQ 156 By F. Simondet and D. Kubicki 102

Part II. **Fundamentals I. Ion Formation**

*Molecule Formation in Oxide Sputtering. By H. Oechsner 106

*Dependence of Ionization Yields Upon Elemental Composition; Isotopic Variations. By G. Slodzian 115

Measurements of the Energy Distributions of Positive Secondary Ions in the Energy Range from 0 to About 500 eV By C. Pahlke, H. Düsterhöft, and U. Müller-Jahreis 124

Ion Dose Effects in Static SIMS By W. Speckmann, S. Prigge, and E. Bauer 128

Current Density Effects on Secondary Ion Emission from Multicomponent Targets. By A. Barcz, M. Domanski, and B. Wojtowicz-Natanson 134

Isotope Effect in Secondary Ion Emission By J.C. Lorin, A. Havette, and G. Slodzian 140

Caesiated Surfaces and Negative Ion Emission By M. Bernheim and G. Slodzian 151

Secondary Ion Mass Spectrometry of Organic Compounds; A Comparison with Other Methods (EI, CI, FI, FD, FAB). By K.D. Klöppel 159

Secondary Ion Emission from NbV-Alloys By J. Schou, G. Flentje, W.O. Hofer, and U. Linke 165

Part III. **Fundamentals II. Depth Profiling**

*Depth Profiling by SIMS: Depth Resolution, Dynamic Range and Sensitivity. By C.W. Magee, R.E. Honig, and C.A. Evans, Jr. 172

*Disturbing Effects in Sputter Profiling. By S. Hofmann 186

*The Theory of Concentration Depth Profiling by Sputter Etching By W.O. Hofer and U. Littmark 201

Surface Topography Development During SIMS Investigations and Using it to Get Additional Information on Polycrystalline and Heterogeneous Solids. By W. Hauffe 206

Sputtering of Metals with 20 keV O_2^+; Characteristic Etch Patterns, Sputtered Atom Yields and Secondary Ion Mass Spectra By K. Tsunoyama, T. Suzuki, Y. Ohashi, and M. Konishi 211

Depth Resolution of Ion Bombardment Technique Applied to NiPd, PiPt, PtPd Thin Layer Systems. By J. Giber, D. Marton, J. Lászlő, and I. Mizsei 216

The Influence of Ion Bombardment on the Results of AES-Depth Profiling on Silicides. By Th. Wirth 222

A Study of Secondary Ion Energy Distributions During Sputtering of MIS Layer Structures. By G.Ph. Romanova, P.I. Didenko, A.A. Yefremov, V.G. Litovchenko, and R.I. Marchenko 227

Structural Effects in SIMS at the Depth Profiling of Boron Implanted Silicon. By V.T. Cherepin, A.K. Kosyachkov, A.D. Krasyuk, and M.A. Vasiliev 233

Comparison of Compositional Thin Film Depth Profiling Obtained by SIMS, IIR and AES. By E. Hauser, G. Hobi, K.H. Guenther, and E. Brandstaetter 238

Part IV. **Quantification**

*Quantification of SIMS. By G.H. Morrison 244

Quantitative Chemical Analysis of Standard Iron Alloys by SIMS Technique. By Z. Jurela 257

Application of the LTE Model to Quantifying the Secondary Ion Spectra of Steels. By J. Suba and A. Stopka 264

Modification of the MISR Method with the Use of Implantation of Standard Elements. By J. Giber, A. Sólyom, L. Bori, and J. Gyulai 269

Use of Ionic Implantation for Quantification of SIMS Analysis in Metals and Oxides – Application to Corrosion Studies By J.C. Pivin, D. Loison, C. Roques-Carmes, J. Chaumont, A.M. Huber, and G. Morillot 274

Secondary Ion Emission from Binary and Ternary Amorphous Alloys By H. Gnaser, M. Riedel, J. Marton, and F.G. Rüdenauer 282

Experimental Procedures for Quantitative Analysis of Silicate Minerals. By A. Havette and G. Slodzian 288

SIMS Isotopic Measurements at High Mass Resolution By E. Zinner and M. Grasserbauer 292

Computer Peak Identification and Evaluation of SIMS Spectra By J. Antal, S. Kugler, and M. Riedel 297

Part V. Application I. Depth Profiling

Depth Profiling of Copper Atoms Gettered in Ion-Damaged GaP
By M. Griepentrog, H. Kerkow, H. Klose, and U. Müller-Jahreis 302

The Optimisation of SIMS for the Analysis of Semiconductor Materials
By J.B. Clegg 308

Impurity Redistribution in GaAs Epilayers. By A.M. Huber, G. Morillot, P. Merenda, M. Bonnet, and G. Bessonneau 314

Quantitative Distribution Analysis of B, As and P in Si for Process Simulation. By M. Grasserbauer, G. Stingeder, E. Guerrero, H. Pötzl, R. Tielert, and H. Ryssel 321

High Spatial Resolution SIMS Depth Profiling of Cr Dopant in CdSe Thin Film Transistors
By J.D. Brown, F.R. Shepherd, and W.D. Westwood 330

SIMS Investigation of p-n Junction Quality in Ion Implanted cw Laser Annealed Silicon
By M. Maier, D. Bimberg, H. Baumgart, and F. Phillipp 336

Profiles of Implanted or Diffused Dopants (Be, Zn, Cr, Se) in Indium Phosphide. By M. Gauneau, A. Rupert, and P.N. Favennec 342

Applications of SIMS in Studies of Slow Diffusion and Isotope Diffusion. By A. Lodding, H. Odelius, and U. Södervall 351

Rapid Diffusion and Gettering Studies of Bulk Oxygen in Silicon by Cs/SIMS. By C.A. Evans, Jr., B.K. Furman, and T.J. Magee 357

Water Diffusion in Fused Silica and Iron-Making Slag
By M. Kobayashi, K.S. Goto, and M. Someno 361

Combined SIMS-AES-XPS Investigation of the Composition and Interface Structure of Anodic Oxide Layers on $Cd_{0.2}Hg_{0.8}Te$ (CMT). By U. Kaiser, O. Ganschow, J. Neelsen, H.M. Nitz, L. Wiedmann, and A. Benninghoven 365

Part VI. Application II. Surface Studies, Ion Microscopy

The Chemical Composition of Oxide Films on Aluminium and Its Influence on Surface Properties Studied by SIMS, XPS and AES. By M. Textor .. 372

Study of the Adsorption of Water on Titania by Secondary Ion Mass Spectrometry. By J. Marien and E. De Pauw 377

SIMS Studies on Oxygen Adsorption on Aluminium and Its Alloys
By D. Marton and Á. Csanády 383

Oxidation and Segregation at the Surfaces of Different Aluminium Foils and Sheets. By Á. Csanády, D. Marton, and T. Turmezey 388

Oxygen Adsorption on Polycrystalline Pt_3Pb at Elevated Temperatures. A SIMS Study. By W. Unger, L. Bori, and D. Marton 394

SIMS Investigations on TiFe Nitrided by NH_3-Pretreatment
By G. Kirch and H. Züchner 398

*SIMS/TDMS Studies of Hydrocarbon Interaction with Nickel
By M. Schemmer, P. Beckmann, D. Greifendorf, and A. Benninghoven .. 405

SIMS Investigation of Adsorption and Chemical Modification of C_2H_4 and C_2H_2 on a Polycrystalline Ni-Surface. By H. Kaarmann, B. Leidenberger, H. Hoinkes, and H. Wilsch 411

*Secondary Ion Emission from UHV-Deposited Amino Acid Overlayers on Clean Metal Surfaces. By W. Lange, M. Jirikowsky, D. Holtkamp, and A. Benninghoven 416

*SIMS Investigation of Adsorption of O_2, H_2O, CO, CO_2, CH_2O, and CH_3OH and Coadsorption of O_2 with CH_2O and CH_3OH on Polycrystalline Silver Surfaces. By L. Wiedmann, N.L. Wang, R. Jede, L.D. An, O. Ganschow, and A. Benninghoven 421

Distribution of Ni, Co, Ga, and Cu in Iron Meteorities
By J. Okano and H. Nishimura 426

Metallurgical Applications of Ionic Microscopy. By E. Darque-Ceretti, R. Dennebouy, J.C. Pivin, and C. Roques-Carmes 431

*Secondary Ion Mass Spectrometry of Organic Compounds
By A. Benninghoven 438

Index of Contributors 443

Part I

Instrumentation

Instrumental Aspects of Spatially 3-Dimensional SIMS Analysis

F.G. Rüdenauer

Austrian Research Center Seibersdorf,
A-1082 Wien, Austria

Limits of Mass Resolution, Transmission and Lateral Resolution

It has become clear from detailed ion optical analyses [1,2,3] that matching analyzer acceptance with secondary ion beam emittance is essential to obtain high transmission at given mass resolution, independent of the particular imaging principle used (scanning probe, microscope, image dissecting ion probe). Close-to-optimum matching can be obtained using a "transfer optic" system (consisting of an immersion lens immediately at the target and one or more image transfer lenses between immersion lens and mass spectrometer) as described by SLODZIAN [1,4]. Future scanning probes and microscopes therefore will have to incorporate such a system, taylored to the particular mode of analysis and primary beam illustration. When comparing in particular the ion microscope and the scanning ion microprobe mode of operation for their relative merits, it is only fair to assume the same type of double focusing spectrometer for both systems. We consider an instrument which is partially corrected for second-order aberrations [5], the most important remaining aberrations being second-order chromatic aberration and third-order aperture aberration. Without taking into account detailed ion optical properties of particular instrument designs, LIEBL [6] gives an estimate for the maximum tolerated divergence angle and the slit width s' for such an instrument in dependence on the required mass resolution R (= $M/\Delta M$)

$$\alpha_x = (40\ R)^{-1/3}$$

$$s' = L/20\ R \qquad (1)$$

$$L = 8\ r_m$$

where L is the total ion path length and r_m the magnetic deflection radius. Owing to the second-order chromatic aberration coefficient $A_{\delta\delta}$, the maximum allowed initial energy $\Delta\Phi_{ch}$ of the secondary ions is determined by

$$\Delta\Phi_{ch} = 0.45\ V\ (R.A_{\delta\delta})^{-1/2} \qquad (2)$$

where V is the final energy of the secondary ions and it has been assumed that the chromatic aberration contribution to the width of the slit image, $r_m A_{\delta\delta}\ (\Phi/V)^2$, is about half the slit width s'.

SLODZIAN [1] and LIEBL [2,31] showed that in a scanning ion microprobe transmission can be drastically increased and in principal can be made independent of the size of the field of view when the "dynamic emittance matching (DEM)" technique is applied. DEM has only recently been incorporated in a scanning ion probe and should be integrated in any "fourth-generation" ion microprobe instrument [32].

The ion optics of the immersion lens (transfer optics) mass spectrometer system has been treated in great detail by SLODZIAN [1]. He arrives at a formula for the number dN of detected ions from a sputtered sample depth dz and sample area δ_o which can be rewritten as

$$dN^+ = \pi\delta_o dz\left[\int_0^{\Phi_{om}} \left(\frac{dN}{d\Phi}\right)d\Phi + \Phi_{om}\int_{\Phi_{om}}^{\Delta\Phi_{ch}} \left(\frac{dN}{d\Phi}\right)d\Phi\right] \tag{3}$$

where $dN/d\Phi$ is the energy distribution of the sputtered ions averaged over the solid angle, Φ the initial energy of the sputtered ions and Φ_{om} the maximum emission energy the ions can have in order to be transmitted independent on their emission angle. Φ_{om} in turn is a function of spectrometer acceptance and final beam energy V. Introducing $\bar{N}^+$, the total number of ions emitted from the small volume $\delta_o dz$, the useful yield can be calculated from (3) [8]

$$\tau_n = \alpha^+\left[\int_0^{\Phi_{om}} \frac{1}{\bar{N}^+}\left(\frac{dN^+}{d\Phi}\right)d\Phi + \Phi_{om}\int_{\Phi_{om}}^{\Delta\Phi_{ch}} \frac{1}{\Phi}\cdot\frac{1}{\bar{N}^+} + \left(\frac{dN^+}{d\Phi}\right)d\Phi\right]. \tag{4}$$

Since $\tau_u \cong \alpha^+\cdot T$, the term in the brackets is the transmission of the total ion optical system. For a numerical evaluation of (4) we have to know the integrated relative energy distribution function $T_\Phi(\Phi)$

$$T_\Phi(\Phi) = (1/\bar{N}^+)\int_0^{\Phi} d\Phi(dN^+/d\Phi)\ , \tag{5}$$

i.e. the number of ions emitted with energies below Φ. We assume a GRIES distribution [9]

$$\frac{dN^+(\Phi)}{d\Phi} \cong \frac{E(E+B)^{1/2}}{(E+B)^3}\ , \tag{6}$$

(B - surface binding energy) the integral of which can be calculated numerically but which can also be approximated by power functions. From data published by LIEBL [6] we obtain the approximations in Table 1.

From (4) we can derive the transmission T

$$T \cong \frac{\Phi_{om}}{\Delta\Phi_{ch}}\cdot T_\Phi(\Delta\Phi_{ch}) + \Phi_{om}\int_{\Phi_{om}}^{\Delta\Phi_{ch}} \frac{1}{\Phi^2}\,T_\Phi(\Phi)d\Phi\ . \tag{7}$$

Table 1 Approximations to $T_{\Phi}(\Phi)$ (fraction of ions emitted with energies below Φ)

Φ [eV]	$T_{\Phi}(\Phi)$
$200 \geq \Phi \geq 10$	$6.6\cdot 10^{2}\sqrt{\Phi}$
$10 \geq \Phi \geq 2$	$1.4\cdot 10^{-2}\ \Phi$
$2 \geq \Phi \geq 0$	$7\cdot 10^{-3}\Phi^{2}$

From Eq.(2) and using data for a "standard" medium-sized mass spectrometer (see below) we note that for all mass resolution values of practical interest in SIMS, $\Delta\Phi_{ch}$ is at least of the order of 50 eV so that the first approximation in Table 1 for $T_{\Phi}(\Phi)$ is appropriate. We can therefore derive from (2) and (7)

$$T \cong 9.8\cdot 10^{-2} R^{1/4} A_{\delta\delta}{}^{1/4} \Phi_{om}/\sqrt{V} + \Phi_{om} \int_{\Phi_{om}}^{\Delta\Phi_{ch}} \frac{1}{\Phi^{2}} T_{\Phi}(\Phi)\, d\Phi. \qquad (8)$$

At this point we have to split up our considerations between the ion microscope and the scanning ion probe (with DEM).

Scanning Ion Probe

The bombarded area is very small. The diameter δ_V of the aberration disc which is "seen" by the mass spectrometer (following the DEM stage) depends on the maximum transversal energy Φ_o^* accepted to form the aberration disc. For illuminated spot sizes in excess of 10 μ in general, the emittance E(V) of the secondary ion beam is not determined by the diameter of the illuminated spot but by that of the aberration spot

$$E(V) \cong (2\delta_V \cdot \sqrt{\Phi_o^*/V})^2 = 4D^2(\Phi_o^*/V)^4 . \qquad (9)$$

For all ions with $\Phi < \Phi_o^*$ to be transmitted, the emittance (9) has to be smaller than the mass spectrometer acceptance A_{MS}, which can be calculated from (1)

$$A_{MS} = (2s'\ \alpha_x)^2 = 8.6\cdot 10^{-4} R^{8/3} L^2. \qquad (10)$$

This "emittance matching" condition leads to an expression for Φ_o^*

$$\Phi_o^* = 0.12(L/D)^{1/2}\ V\cdot R^{2/3}. \qquad (11)$$

For a numerical evaluation we want to introduce a "standard spectrometer" with operating parameters as given in Table 2. This standard instrument should also be partially aberration corrected, as indicated above.

With these values we can calculate Φ_o^* for a range of mass resolutions and note that it is always in a range where the first approximation in Table 1 for the integrated energy distribution

Table 2 Opearating parameters of a medium-sized "standard" SIMS mass analyzer

Total ion path length (L)	100 cm
Sample extractor distance (D)	0.5 cm
Acceleration Voltage (V)	5000 V
2.O. Chrom. Aberr. const. ($A_{\delta\delta}$)	0.1

function is appropriate. Inserting now (11) into the expression for the transmission [18], we obtain an approximate formula for the transmission of a scanning probe:

$$T_{SP} \cong 4.6\ 10^{-2}(L/D)^{1/4}V^{1/2}R^{-1/3} - 1.2\ 10^{-2}A_{\delta\delta}^{1/4}V^{1/2}(L/D)^{1/2}R^{-5/12} \quad (12)$$

which, for the values from Table 2, we can reformulate

$$T_{SP} \cong 12.1\ R^{-1/3} - 6.75\ R^{-5/12}. \quad (13)$$

A numerical evaluation of this formula is shown in Fig. 1.

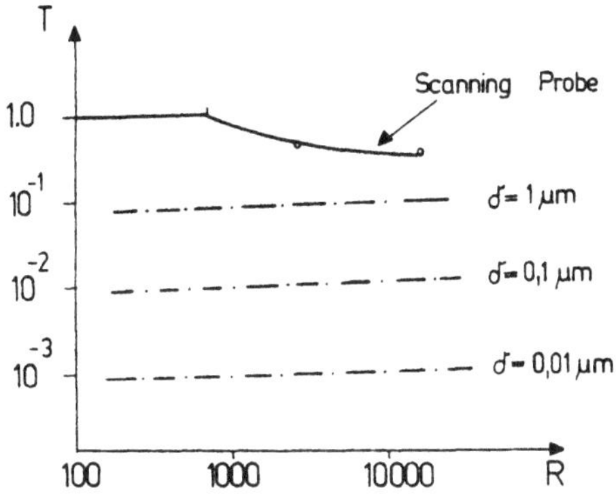

Fig. 1

Ion Microscope

In the ion microscope, local resolution is determined by the diameter δ_V of the aberration spot in the virtual image plane [1]

$$\delta_V \cong D\cdot\Phi_{om}/V\ . \quad (14)$$

Another limitation to transmission is set up by the diameter d_f of the field of view which has to be fully accomodated by the acceptance of the mass spectromater (in contrast to the scanning probe with DEM!). The emittance E(V) of the secondary ion beam can be calculated in the plane of the virtual image, considering that the maximum beam divergence is given, as $\bar{\alpha} = (\Phi_{om}/V)^{1/2}$, so that with (14)

$$E(V) = (2\bar{\alpha}d_f)^2 = 4d_f^2\delta_v/D \; . \tag{15}$$

When optimum matching with mass spectrometer acceptance (10) is again assumed, we obtain a relationship between the diameter of the field of view which can be observed at a given mass resolution and local resolution

$$d_f \leq 1.5\cdot 10^{-2}\, L\sqrt{D}\, R^{-4/3}\delta_v^{-1/2} \; . \tag{16}$$

For standard instrument parameters from Table 2 we calculate the following values.

Table 3

δ_v [µm]	d_f [µm]			Φ_{om} [eV]
	R = 250	R = 2500	R = 15000	
10	200	9.3	0.85	10
1	635	30	2.7	1
0.1	2000	93	8.5	0.1
0.01	6350	300	27	0.01

The transmission again is calculated from Eq.(11) with Φ_{om} now taken from (14). From the previous table we note that Φ_{om} varies in a range where all three approximations to $T_\Phi(\Phi)$ in Table 1 have to be used. When we are limiting ourselves to the submicron range, the first approximation from Table 1 applies for the first term in (8) and in the second term the integrand varies in a range where all three appoximations apply, leading to

$$T_{MS} \cong (9.8\cdot 10^{-2}A_{\delta\delta}{}^{1/4} - 0.196A_{\delta\delta})V^{1/2}R^{1/4}\delta_v/D + \\ + 7.82\cdot 10^{-2}\delta_v\cdot V/D \; . \tag{17}$$

Using again numerical values for the "standard instrument" an estimation of (17) is

$$T_{MS} \cong 782\;\delta_v(1 + 0.1\;R^{1/4}) \tag{18}$$

which has a negligible dependence on mass resolution R (see Table 4 and Fig. 1).

Table 5 shows a collection of experimental data obtained on different instruments from a boron implant in silicon. Lines 4 and 6 of that table show that the particular ion microscope and scanning probe are about equivalent in transmission, at least for the rather large field of view indicated. It appears that the microscope has been developed closer to its intrinsic limitations, particularly when the τ_u value for Si in line 7 is considered: when we assume $\alpha = 0.1$, a

Table 4 Relation between mass resolution R, local resolution $1/\delta_V$, diameter of field of view d_f, and transmission T in an ion microscope

R	δ_V [cm]	T	d_f [cm]
250	10^{-4}	0.11	$4.5\ 10^{-2}$
	10^{-5}	0.011	$1.4\ 10^{-1}$
	10^{-6}	0.0011	$4.5\ 10^{-1}$
2500	10^{-4}	0.13	$2.0\ 10^{-3}$
	10^{-5}	0.013	$6.3\ 10^{-3}$
	10^{-6}	0.0013	$2.0\ 10^{-2}$
15000	10^{-4}	0.17	$1.9\ 10^{-4}$
	10^{-5}	0.017	$6.0\ 10^{-4}$
	10^{-6}	0.0017	$1.9\ 10^{-3}$

Table 5 Data obtained on different instrument types

Instrum.	Element	Prim. gas	Prim. current I_p[μA]	Oxygen yet	Bombarded area	Accepted area	Secondary current I_s [cps]	I_s/I_p [cps/nA]	Pract. sensitivity S_p[cps/nA]	Useful yield
ATOMIKA[a)]	B	O_2^+	1.2	-	$(750\mu)^2$	$(235\mu)^2$	$9.2\cdot10^4$	77	$3.8\cdot10^4$	-
RIBER[b)]	B	O_2^+	0.030	-	$(400\mu)^2$	$(100\mu)^2$	$2.1\cdot10^3$	70	$3.5\cdot10^4$	$2.4\cdot10^{-6}$
IMS 300[c)]	B	O_2^+	3.5	-	$(120\mu)^2$	$(300\mu)^2$	$8.3\cdot10^4$	24	$1.9\cdot10^5$	$6.3\cdot10^{-6}$
IMS 300[d)]	B	O_2^+	2.0	+	$(1250\mu)^2$	$(50\mu)^2$	$6.3\cdot10^4$	32	$1.0\cdot10^7$	$7.4\cdot10^{-4}$
IMMA[e)]	B	O^-	?	-	?	?	?	?	$4.5\cdot10^5$	
IMMA[e)]	Si	O_2^+	?	-	?	?	?	?	$5.5\cdot10^5$	
IMS 3f[f)]	Si	O_2^+	?	+	?	?	?	?	$2.3\cdot10^8$	$1.6\cdot10^{-2}$

a) 12 keV, O_2^+, 2° angle of incidence, [23]

b) 10 keV, O_2^+, 0°, H. Gnaser, private communication 1981

c),d) 5.5 keV O_2^+, 64°, [23]

e) 13.5 keV $^{16}O^-$; ca. 150μA/cm^2 [24]

f) 5.5 keV O_2^+, 60°, local resol. 5 μ[4]

*) B - implanted in Si, $c(B) \cong 2\cdot10^{-3}$ (10^{20} cm^{-3})

value of $\tau_u = 1.6\ 10^{-2}$ would mean, according to (5), that a transmission of T = 0.1 is obtained; this value is close to the limit of an ion microscope operated at low mass resolution and lateral resolution (see Fig. 2). Note also that there is about an order-of-magnitude performance difference between good quadrupole instruments (lines 1 and 2 in Table 5) and the double focusing instruments, operated at low mass resolution.

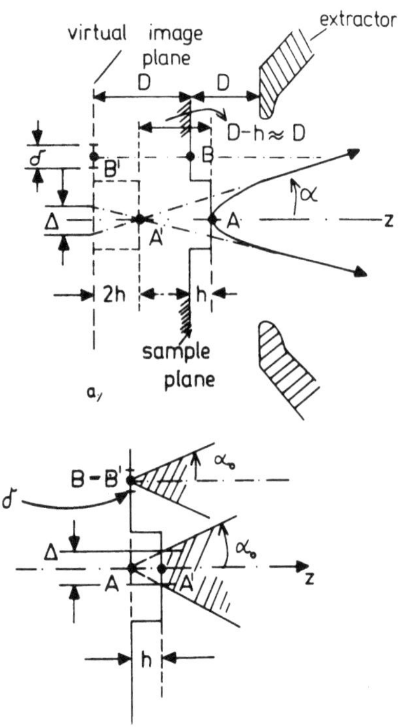

Fig. 2

Useful Resolution

When three-dimensional local microanalysis is performed with an ion probe or ion microscope, the limit to the lateral and depth resolution is set up by the destructive nature of the sputtering process [12]. When the local resolution is required to be equal in all spatial coordinates the useful resolution δ_u can be derived from the expression of CASTAING and SLODZIAN as

$$\delta_u = 10^{4/3}(np^2\alpha^+cT)^{-1/3} \cong 0.55\cdot 10^{-6}(p^2\alpha^+cT) \qquad (19)$$

where n is the total number of atoms/cm^3, p the precision (in %) of the determination of the local ion count, α^+ the degree of ionization of the secondary ions, c the fractional atomic concentration and T the transmission of the instrument. Using again the "standard mass spectrometer" and making the following coarse estimate (within ± 25 %) for the transmission of scanning probe and microscope (see (13) and (18)): $T_{SP} \geq 0.4$; and $T_{MS} \geq 0.82\cdot 10^3\cdot\delta_V$, we obtain expressions for the useful resolution of these two instrument types:

$$(\delta_u)_{SP} = 0.75\cdot 10^{-6}(p^2\alpha^+c)^{-1/3}$$
$$(\delta_u)_{MS} = 3.8\cdot 10^{6}(p^2\alpha^+c)^{-1/4}. \qquad (20)$$

The ratio of (three-dimensional) useful resolutions is therefore given as

$$(\delta_{MS}/\delta_{SP})_u \cong 5.1(p^2\alpha^+c)^{1/12} \tag{21}$$

which is numerically evaluated for 2 different values of the degree of ionization α^+ at various concentrations c.

Table 6

c	$(\delta_{MS}/\delta_{SP})_u$		
	$\alpha^+ = 0.1$	$\alpha^+ = 0.01$	$(\delta_{sp})_u$
1	7	5.8	17 Å
10^{-1}	5.8	4.8	36 Å
10^{-2}	4.8	4.0	78 Å
10^{-3}	4.0	3.3	170 Å
10^{-4}	3.3	2.7	360 Å
10^{-5}	2.7	2.2	780 Å
10^{-6}	2.2	1.8	1700 Å

Depth of Focus

We limit our considerations to a very simple case, viz. a flat step of height h on top of a flat planar surface (see Fig. 2). The immersion lens of an ion microscope produces a "stepped virtual image which lies in a plane at a distance -2D in front of the extraction electrode for the outer regions of the sample (B,B' in Fig. 2a) and at a distance of -2(D-h) for the central step. If the microscope is adjusted to give a sharp image of the outer section in the plane of point B'=B, then the image of point A is defocused in this plane, showing a disc of confusion with diameter Δ

$$\Delta = 4\alpha_o h \tag{22}$$

where α_o is the limiting angle of the ions leaving the virtual image A':

$$\alpha_o = \Phi_o/V. \tag{23}$$

Introducing the local resolution δ from (14), we obtain

$$\delta = \Phi_o \cdot D/V = \alpha_o^2 \cdot D. \tag{24}$$

The defocusing of the virtual image by the surface step is negligible as long as Δ is considerably smaller than the intrinsic lateral resolution δ of the immersion lens ($\Delta \ll \delta$), leading to

$$h/\delta \ll (D/\delta)^{1/2}/4 \tag{25}$$

for the maximum tolerated step height h, measured in units of lateral resolution δ. Obviously, this "relative depth of focus" increases with increasing resolution $1/\delta$ of the immersion lens.

In the case of the scanning ion microprobe, let us assume that the primary beam is focused to the plane of B (Fig. 2b). In the plane of A it will therefore be defocused with a diameter Δ, determined by the angle of convergence α_o of the primary beam and the step height h:

$$\Delta = 2\alpha_o \cdot h$$

when we are interested in microanalysis with sub micron lateral resolution, chromatic aberration in the primary beam sets the limit to obtainable beam diameters [8]. In this case, a relation between δ and the optimum divergence angle α_o can be derived:

$$\alpha_o \cong \delta/2 \cdot \sqrt{2} C_c(\delta V/V)$$

where C_c is the chromatic aberration constant of the last focusing lens, and $\delta V/V$ the relative energy spread in the primary beam. The defocusing Δ can be tolerated as long as $\Delta << \delta$; this condition finally leads to

$$h/\delta << \sqrt{2} C_c(\delta V/V)/\delta. \quad (26)$$

Table 7 gives a comparison of calculated values for the relative depth of focus in microscopes and scanning probes for different values of lateral resolution.

Table 7 Depth of focus in Ion Microscopes and Scanning Probes
$D = 0.5$ cm, $(\delta V/V = 10^{-3})$, $C_c = 4$ cm

Lateral resolution Å	microscope		scanning probe	
	h/δ	h [μm]	h/δ	h [μm]
10^4	11	11	57	57
10^3	35	3.5	570	57
10^2	112	1.1	$5.7\ 10^3$	57
10	354	0.35	$5.7\ 10^4$	57

Image Recording Time

We assume that in the microscope a scanning ion primary beam is employed to obtain a flat crater bottom; the current which can be focused into this spot determines the average current density in the rastered area and therefore the sputtering speed and signal strength; the diameter of that beam has not to be made extremely small, but should be of the order of 1/10 of the diameter of the field of view to avoid excessive crater-wall effects. Let us assume, that the current which can be focused in the primary beam spot is limited by spherical aberration (such as in the Cs^+ source above 100 Å), so that we obtain

$$I_p(d) \cong k \cdot d^{8/3}. \tag{27}$$

Assuming this assumption to be valid in the scanning probe as well, we can calculate the time needed to accumulate enough counts/pixel to determine the local current with a precision of p%. Then it is easy to calculate the image recording time (for a full image) in scanning probe and microscope:

$$\tau_{SP} \cong \frac{10^{-25} p^2}{Y_{tot} \cdot \alpha^+ \cdot c} \cdot \frac{(d_f/\delta)^2}{\delta^{8/3}}, \tag{28}$$

$$\tau_{MS} \cong \frac{1.3 \cdot 10^{-26} p^2}{Y_{tot} \cdot \alpha^+ \cdot c} \cdot \frac{(d_f/\delta)^2}{\delta \cdot d_f^{8/3}}. \tag{29}$$

The ratio of τ and τ_{MS} can be written as

$$(\tau_{SP}/\tau_{MS}) \cong 7.7 \left(\frac{d_f}{\delta}\right)^{8/3} \cdot \delta. \tag{30}$$

Table 8 gives numerical values for this ratio, assuming that the field of view consists of 10^4 resolution elements of diameter δ (100x100 image).

Table 8 Ratio of image recording times in scanning probe and microscope for a 100x100 pixel image

Resolution [Å]	τ_{SP}/τ_{MS}	τ_{MS} (Y_{tot}=2, c=1, =0.1)	
10^4	170	$1.3 \cdot 10^{-5}$	s
10^3	17	$6 \cdot 10^{-2}$	s
10^2	1.7	4.7	min
10	0.17	361	h

Primary Ion Beam

Limitations to the obtainable spot diameter are set up by the brightness of the ion beam β, its relative energy spread $\delta V/V$ (chromatic limit), the beam divergence (spherical aberration limit) and the wave nature of a moving particle (diffraction limit); at high beam currents the interaction between beam particles (space charge) also plays a role [8,13,14]. The optimum ion current which can be focused into spot sizes of 1000 Å and 100 Å for different source types is shown in Table 9.

Table 9 Characteristic data for optimized microfocus ion guns

Source type	$\beta(A \cdot cm^{-2} \cdot sr^{-1})$	I (1000 Å)	I (100 Å)
Duoplasmatron	$2 \cdot 10^2$	$4 \cdot 10^{-14}$	$2 \cdot 10^{-18}$
Surf. Ionis. (Cs^+)	$1 \cdot 10^3$	$1 \cdot 10^{-11}$	$1 \cdot 10^{-14}$
EHD (Ga^+)	$1 \cdot 10^6$	$1 \cdot 10^{-10}$	$1 \cdot 10^{-15}$

Obviously the EHD source has an advantage in the 1000 Å range whereas in the 100 Å range the surface ionization source would be preferable. Both types of sources have been developed to a stage where first experiments on secondary ion mass spectrometers have been performed [15,16]. Another point of importance for the choice of an ion source is the secondary ion yield. In the case of the surface ionization source operated with Cs^+ practical sensitivities have long been known [11]. No data have been published so far for the rather exotic type of elements with which EHD sources are operated. New data on yields of pure elements and compounds can be found in another paper of these proceedings [16]. Here we want to show some new results comparing oxygen and In primary ion bombardment (Fig. 3). The measurements were performed in the same experimental arrangement as in [16]. Obviously, sensitivities measured at low sample chamber pressure ($\approx 2 \cdot 10^{-8}$ Torr) are for many elements only insignificantly higher for oxygen bombardment than for In bombardment. Introducing oxygen gas ($p \cong 5 \cdot 10^{-6}$ Torr) causes a yield enhancement similar to that observed for rare

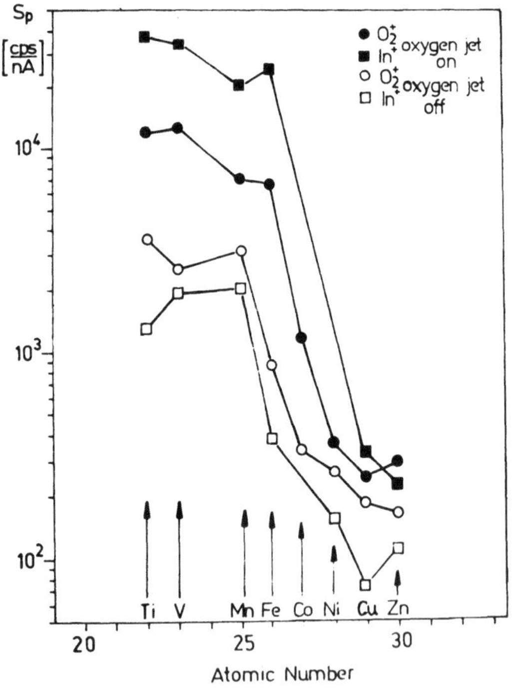

Fig. 3

gas bombardment by GIBER [17] and by many other authors. Now the positive ion yield for indium exceeds that for oxygen; a possible explanation is that the formation of an oxide layer dominates all other chemical effects on the secondary ion yields; higher practical sensitivites under In bombardment then may be explained by higher sputtering yields for the heavier element.

Quantitation

From a practical point of view, the choice of the quantitation model used to convert ion intensity data into element concentration data will be influenced by three factors: (a) the amount of a priori information needed before the experimental data can be gathered at all; (b) the amount of experimental data which have to be gathered in a particular experiment in order to allow complete quantitation of a sample; (c) the precision and accuracy of the result.

A survey of the most frequently used quantitation algorithms where these requirements have been considered is given in [18]. Increased accuracy can generally be obtained only with increased a priori knowledge or with increased experimental data (more peaks, more depth points, etc.). In two- and particularly in three-dimensional analysis the latter fact is particularly critical because of the large amount of individual analyses to be performed for each pixel or voxel; each additional element to be measured consumes sample volume and increases analysis time intolerably. Therefore, the "standard addition technique" (SAD) [19] and the internal indicator (MISR) [20] technique (in the underlined operating version) appear to be most promising. Both imply the implantation of an element into the sample before analysis: the SAD technique uses an implant of the element to be determined and quantitation proceeds via the depth profile of that element; in the MISR technique (IIN) a reference element would be implanted (acting as internal standard) and 3 ion peaks would have to be measured: unknown, reference implant and a reference element cluster or molecule. For a three-dimensional analysis extending to considerable sputtering depths many consecutive implantation operations at different sputter depths will probably be required. Off-line implantation will be very cumbersome so that a future instrument probably will have to incorporate in situ implantation facilities. The role of low-energy ion implantation appears to be promising for that purpose.

From what we have seen in the preceding, we are able to sketch down how a fourth-generation SIMS instrument, capable of spatially three-dimensional quantitative SIMS analysis in the submicron local definition range, will probably look (see Fig. 4):

1) The mass analyzer will be a double focusing instrument: entrance aperture (11), divergence aperture (13); sector fields (14), exit slit (15);
2) The secondary ion optics will consist of an immersion lens (9) and a transfer optics (10);

Fig. 4

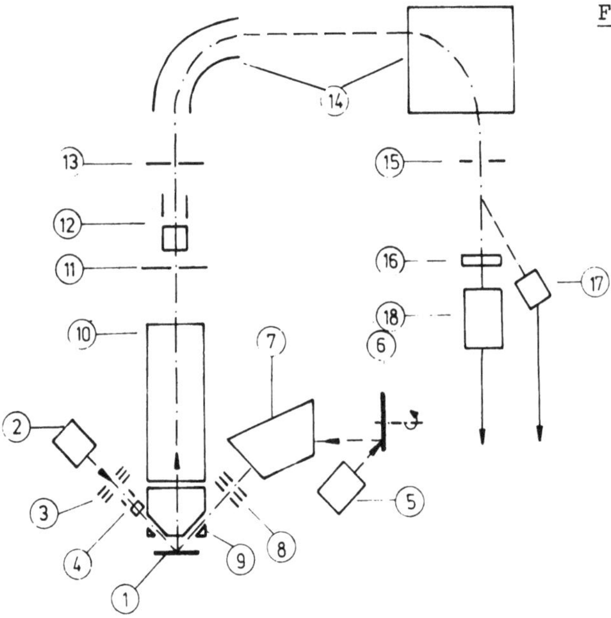

3) It will have to be operated in both the scanning probe mode (for ultimate sensitivity) and the microscope mode (for fast image acquisition) when sample consumption is not a problem;
4) In the microscope mode ion images will be projected to a channel plate (16); a digital TV camera (18) will perform fast image acquisition, digitization and storage in a computer-compatible medium;
5) In the scanning probe mode the DEM feature (4)+(12) will be included and a multiplier will be used for ion detection (17), feeding its pulses directly into the computer;
6) The primary ion source will probably be an EHD source (2); due to the purity of the EHD-beam mass separation will probably not be required; a Cs^+ source, probably also of the EHD type will also have to be included, either alternately or simultaneously;
7) On-line low-energy ion implantation will very likely be a feature to allow three-dimensional quantitation of the sample; ion implantation may either be performed by a second EHD gun or, for increased element coverage (in particular when using the standard addition technique) by a multielement implantation facility; such a facility may consist of a target wheel (6), being bombarded by a high-intensity douplamatron (5); secondary ions from the targets can be mass analyzed (7), focused (8) and finally implanted into the sample (1);
8) Comfortable computer facilities for instrument control, data acquisition and data evaluation will be prerequisite; the most stringent requirements in this respect will have to be put to the image acquisition and display sections, which probably will have to work in an interactive mode with operator and SIMS instruments.

Three-Dimensional Elemental SIMS Analysis

Owing to the surface sensitivity of the SIMS method and to the inherent sputter removal of surface layers during a SIMS analysis, any of the lateral imaging SIMS techniques can be easily extended to perform spatially three-dimensional analysis. When elemental mapping is performed at certain well-defined time intervals during the sputtering process, the two-dimensional elemental maps can be "stacked" to yield a three-dimensional distribution of a certain element(s) in a volume near the surface of the solid. The main problems with the implementation of this analytical mode are of a technical rather than a principal nature and are related to the large amount of image data to be handled and to the difficulties of efficiently displaying three-dimensional information. It therefore appears that significant progress in three-dimensional characterization and analysis of solids is only possible when data acquisition and display are fully computerized and ion or electron intensity data can be converted to digital form directly in the data-acquisition process [21].

Space-resolved characterization and analysis of solids can be performed in four "primary" modes (Fig. 5): point, line, image and volume analyses. All of these primary modes require knowledge of the lateral position and extent of the sampled analytical area. Knowledge of the depth scale (with respect to the original sample surface) is then available through measurement of local primary current dose and calculation of local sputtering rate.

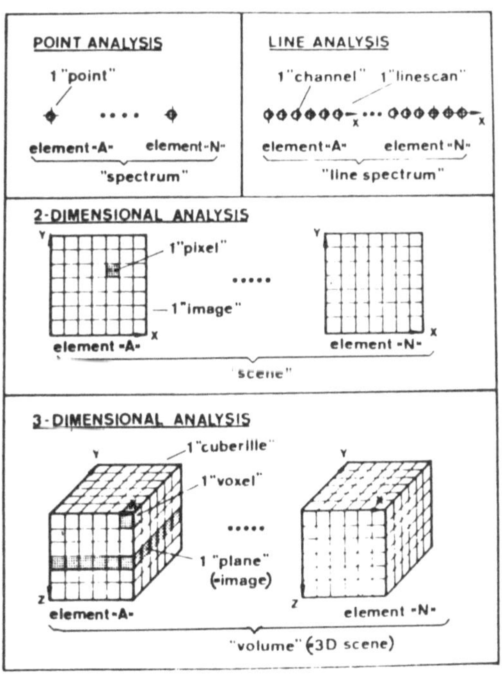

Fig. 5

Acquisition of Three-Dimensional Compositional Information

When a "three-dimensional scene" ("volume") is considered as a stack of two-dimensional images (see Fig. 5) the computerized acquisition of three-dimensional information appears to be straight-forward, particularly in the scanning probe, although recently digital video techniques have been developed [22,23] which possibly will allow maintaining the speed advantage of the microscope in the imaging mode through the digitization and data storage stage. In the scanning probe (a) the primary beam is raster-scanned across the surface, thereby continuously sputtering away consecutive surface layers of the sample; (b) one image frame each for all elements of interest is recorded and stored on magnetic disc; (c) this "mass cycle" is repeated until the sample volume of interest has been sputtered through.

The secondary-ion signals stored during this sputtering process now contain information on the three-dimensional distribution of the recorded elements in the sample. When the original sample surface is flat and the sputtering yield (and therefore the erosion speed) is locally constant, the locations of the "analytical points" corresponding to the individual voxel information stored in the computer (arranged in the "data grid") are directly arranged in an "isomeric analytical grid". This grid can be imagined as a three-dimensional grid, fixed to the sample, with equal voxel dimensions in the directions parallel to the sample surface and constant voxel dimension in the depth co-ordinate (see Fig. 6). The "data grid" in which

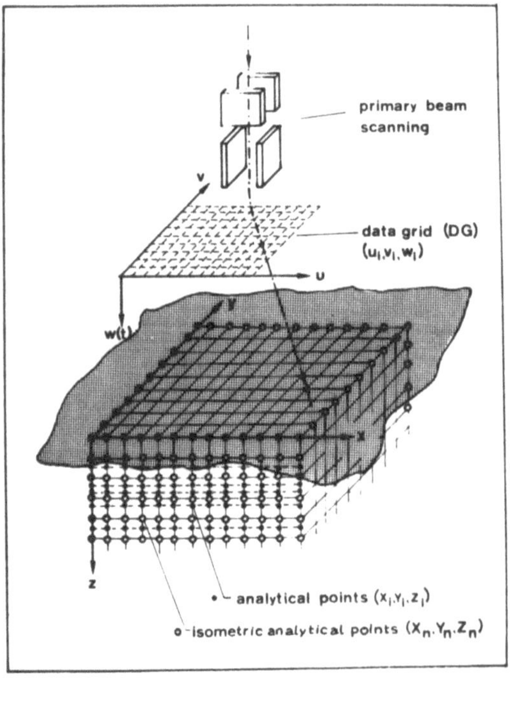

Fig. 6

the information is stored in the computer can therefore be immediately displayed and gives spatially undistorted information on the distribution of elements in the sample.

In the general case of a sample with topographically structured original surface and a locally variable sputtering yield, the situation becomes very complicated; in this case, the surface topography may change during the sputtering process, owing to local differences in sample constitution as well as "shading effects" shielding part of the surface from the primar ion beam; the analytical points will therefore be situated irregularly in (x,y,z) space. The measured analytical data therefore have to be interpolated ("registered") to the isomeric analytical grid (X_n, Y_n, Z_n) [12]. Such a registration, however, has, to the author's knowledge, not yet been applied in an actual analysis, and it has been common practice to display the analytical information as it is contained in the "data grid", i.e., in equidistant planes, respresenting successive scanning frames, with the same metric as the scanning pattern of the microprobe or digitizing equipment, respectively [24-27].

In principle, a three-dimensional scene stored in the computer contains all information obtainable on this sample by the SIMS technique. Thus, depth profiles can be obtained from the primary three-dimensional data by integrating voxel counts across individual depth planes; "line scans" between two arbitrary points within the sputtered volume can be obtained by selecting the appropriate voxels; and two-dimensional distributions of elements in any desired sample cross-section can be obtained by considering voxels intersecting only this particular plane.

Thus, for example, it is possible to look "sideways" through the sample and display the element distribution in a plane perpendicular to the sample surface ("transaxial projection"); the conventional way of displaying two-dimensional element distributions in planes parallel to the surface is called a "coaxial projection" (see Fig. 7). The method of displaying three-dimensional distributions by means of transaxial projections may be of great interest in the analysis of microelectronic devices, because the lateral distribution and the depth distribution of an element are simultaneously visible [21].

Figures 8 and 9 show examples of such a three-dimensional analysis [28]. In the Fig. 8 a) and b) are coaxial projections of elements Al, Si and Sn at sputtering depths of 820 and 2460 Å, respectiveley; the images display a number of Si/Al-contaminating grains on a tin PCB contact; c) shows transaxial projections along line 65 of the coaxial images for the same elements displaying lateral and depth distribution of elements in a thin surface volume extending from 0 (top line) to about 4100 Å (bottom line) sputter depth.

In the Fig. 9 images of Si at 30 Å, 80 Å and 120 Å sputter depth in a microelectronic circuit and transaxial projection covering the range from the surface (top) to about 150 Å depth

Fig. 7

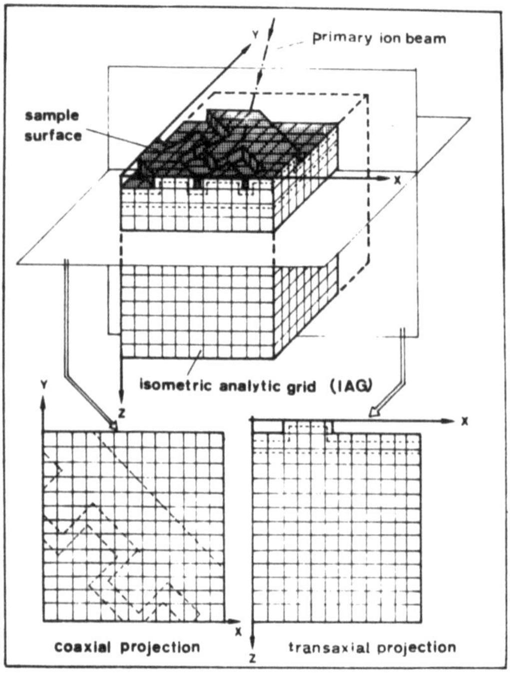

are presented. Obviously, lateral and depth structure is contained in the Si distribution in Fig. 9 which would have been hard to recognize from lateral distribution images alone. The vertical "streakiness" in the transaxial images comes from the fact that a considerable degree of interpolation had to be performed in the depth coordinate because of the limited number of coaxial ion images which can be stored on available disc space.

Note that the images in Figs. 8 and 9 are "ion micrographs", i.e. they contain the uncorrected distribution of secondary-ion signal intensity. To obtain a three-dimensional distribution of true elemental concentration a voxel by voxel application of a quantitation routine is necessary. Such routines are available for correction of two-dimensional ion micrographs [18,25,29] and can be easily extended to three-dimensional "scenes". Difficulties arise, however, with regard to internal-standard elements, which are required by most of the quantitation routines and by problems of computation time. The fastest quantitation routine available is probably the MISR technique [30] with CPU times of the order of 20 ms/voxel; a $(64)^3$ voxel scene would therefore require about 9 hr for correction on a typical minicomputer with integer arithmetic; by using floating-point arithmetic, this figure can be reduced to about 2 hr; with current array-processing techniques, a three-dimensional scene can probably be corrected in about 7 min [18]. Limits of local resolution and of sensitivity in the three-dimensional imaging mode are discussed in [21].

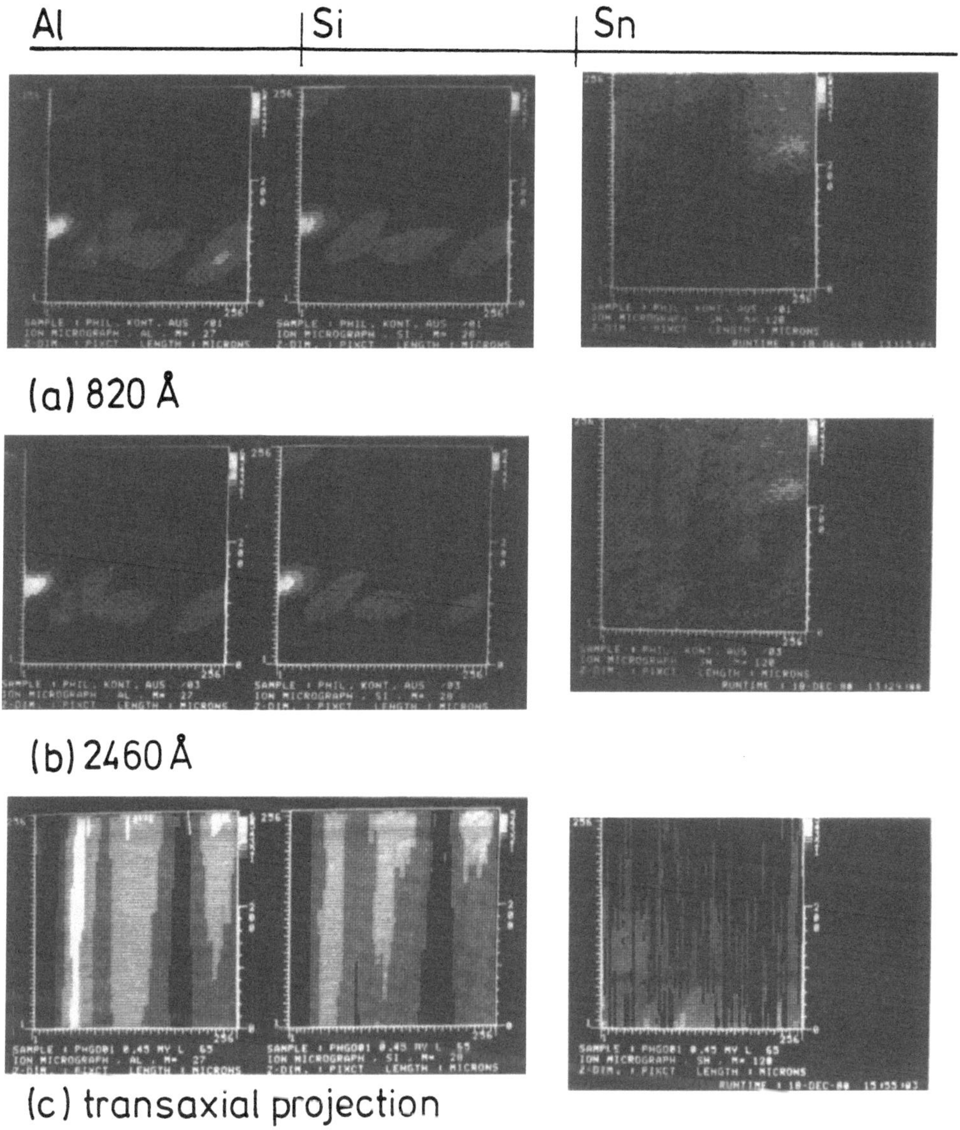

Fig. 8

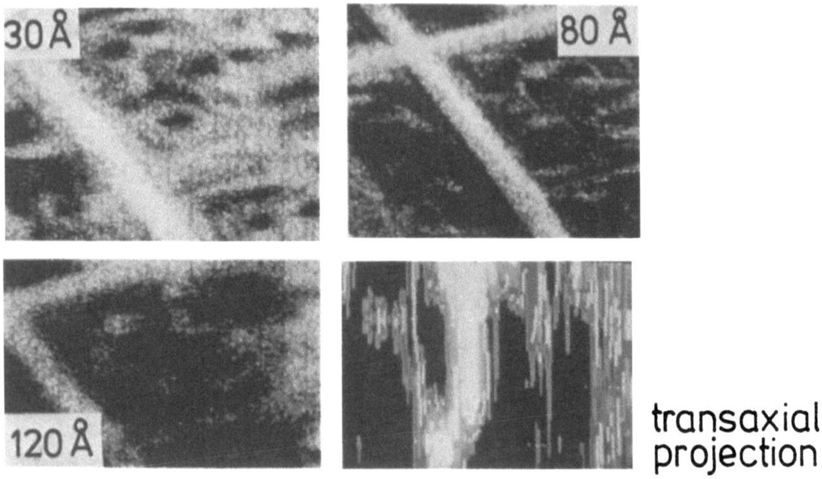

Fig. 9

Acknowledgements

The author gratefully acknowledges the work of Dr. W. Steiger and Dr. I.H. Gnaser which is partially reflected in this article.

References

1. G. Slodzian, Nat. Bur. Stand. U.S. Spec. Publ. 427 (1975)
2. H. Liebl, in Low Energy Ion Beams 1977 , K.G. Stephens et al., eds. Conf. Ser. 38, The Inst. of Phys. (1978)
3. F.G. Rüdenauer, Int. J. Mass Spectrom. Ion Phys. 6, 309 (1971)
4. G. Slodzian, in Applied Charged Particle Optics , A. Septier, ed., Academic Press (1980)
5. S. Taya, PhD Thesis, Osaka Univ., Japan (March 1978)
6. H. Liebl, Adv. Mass Spectrom. 7A (1978)
7. H.W. Werner, Dev. Appl. Spectrosc. 7A (1969)
8. A. Benninghoven, F.G. Rüdenauer and H.W. Werner, in preparation
9. W.H. Gries and F.G. Rüdenauer, Int. J. Mass Spectrom. Ion Phys. 18, 111 (1975)
10. P.R. Boudewijk and H.W. Werner, private communication (1981)
11. H.A. Storms, K.F. Brown and J.D. Stein, Anal. Chem. 49, 2023 (1977)
12. R. Castaing and G. Slodzian, J. Microsc. 1, 395 (1962)
13. H. Liebl, Messtechnik 12, 358 (1972)
14. H. Liebl, J. Phys. E 8, 797 (1975)
15. B.L. Bentz and H. Liebl, these proceedings
16. F.G. Rüdenauer, these proceedings
17. J. Giber, Thin Solid Films 32, 295 (1976)

18. P. Braun, F. Viehböck and F.G. Rüdenauer, Adv. Electron. Electron Phys. in print
19. D.P. Leta and G.H. Morrison, Anal. Chem. 52, 277 (1980)
20. M.A. Rudat and G.H. Morrison, Anal. Chem. 51, 1179 (1979)
21. F.G. Rüdenauer and W. Steiger, Mikrochim. Acta (Wien), Suppl. 9, in print (1981)
22. B.K. Furman and G.H. Morrison, Anal. Chem. 52, 2305 (1980)
23. D.M.Drummer and G.H. Morrison, Anal. Chem. 52, 2147 (1980)
24. J.H. Schilling and P.A. Büger, Int. J. Mass Spectrom. Ion Phys. 27, 283 (1978)
25. W. Steiger and F.G. Rüdenauer, Anal. Chem. 51, 2107 (1979)
26. F.G. Rüdenauer and W. Steiger, Proc. 7th Int. Vac. Congr. (1977) p.2535
27. H.W. Werner, Acta Electron. 19, 56 (1976)
28. F.G. Rüdenauer and W. Steiger, unpublished work
29. D.M. Drummer and G.H. Morrison, Anal. Chem. 52, 2117 (1980)
30. J.D. Fassett, J.R. Roth and G.H. Morrison, Anal. Chem. 49, 2322 (1977)
31. H.J. Liebl, US Patent No. 3517191, filed 1965
32. R.J. Colton, J.E. Campena, T.M. Benlak, J.J. Decorpo and J.R. Wyak, Rev. Sci. Instr. 51, 1685 (1980)

Some Problems of Construction Implied by Requirements of Up-To-Date SIMS Instrumentation

R.L. Gerlach

Perkin-Elmer, Physical Electronics Division,
Eden Prairie, MN 55344, USA

Recent advances in various instrument components and methods for SIMS bring us to the threshold of rapid growth in the application of SIMS. In spite of the present difficulties in quantitative analysis, the extreme sensitivity and promise of sub-micron imaging capability foreshadow its rapid growth in surface and thin film analyses. Here we will examine some of these new methods and components and predict the potential that might be realized in a few years for quadrupole microbeam SIMS instrumentation.

Some system requirements for static SIMS/spectral analysis, depth profiling, and imaging are listed in Table 1. One important requirement, high sensitivity, is primarily governed by the transmission η of the SIMS analyzer, the ionization factor γ, and the volume of material analyzed [1,2] according to the equation

$$C_L = \frac{n_o}{(\frac{\pi}{4} d^2 Z)\rho\gamma\eta} \qquad (1)$$

where C_L = detection limit (atomic concentration),
n_o = ion count limit,
d = diameter of area analyzed,
Z = depth sputtered, and
ρ = bulk atom density of the target.

The detection limit C_L is plotted in Fig. 1 versus d, the diameter of the analysis, for $\rho \equiv 6 \cdot 10^{22}$ atoms/cm^3 and $n_o \equiv 10$ counts. For microprobe-type instruments, d is just the primary beam diameter, while for ion emission microscopes, d corresponds to the diameter of the imaged area. State-of-the-art ion microscopes approach 10 % trasmission [2] whereas quadrupole based systems approach 1 %. Through the appropriate use of a Cs or oxygen beam on the surface [3], the ionization factors for most ion species lie between 10^{-1} and 10^{-4}. If we analyze a typical atom with an ionization factor of $3 \cdot 10^{-3}$ using a quadrupole detector with a transmission of $5 \cdot 10^{-3}$, plus a 100 µm diameter primary beam and a 100 Å sputter depth, then $\gamma\eta = 1.5 \cdot 10^{-5}$ and $C_L = 1.4 \cdot 10^{-7}$. Thus ppm sensitivity is possible with SIMS on 100 µm areas, and there is the potential for 10 % or better detection sensitivity at ~.1 µm diameter analysis area. The ion gun has been brought into this discussion

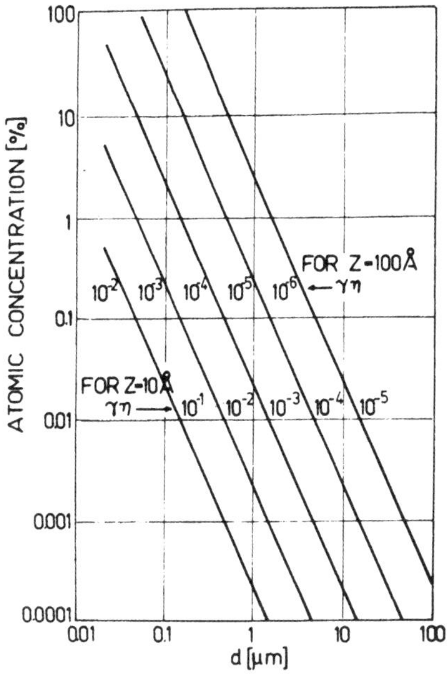

Fig. 1 SIMS detection limits as a function of depth Z and diameter d of the area analyzed. The sample is assumed to have $6 \cdot 10^{22}$ atom/cm^3 density and the detection limit is assumed to be 10 ion counts. γ is the ionization factor and η the instrument transmission

of detection limits only with regard to the beam diameter (in the case of the ion microporbe), and the influence of the primary ion species upon the ionization factor.

Table 1 SIMS system requirements

I. Static-SIMS/spectral analysis

A. Rapid sample introduction
B. Clean UHV
C. Clean ion beam with low neutrals and clusters
D. High sensitivity to all elements and radicals with low bacground
E. +/- SIMS
F. High mass capability
B. Large area rastering of primary beam
H. Well defined analysis area
I. Automated setup, analysis and data manipulation
J. Correlation to other techniques such as AES, ESCA
K. Reliable
L. Ease of operation
M. Moderate cost

II. Profiling (additional requirements)

A. High sputter rate
B. Good sputter depth resolution

III. Imaging (additional requirements)

A. High spatial resolution
B. Mechanically stable optics
C. Rapid image capability

The transmission of the mass analyzer determines the sensitivity of the SIMS instrument. For quadrupole based systems, the optics usually include an extraction or transfer lens, an electrostatic energy filter, a quadrupole, and a detector [4-6]. The primary purpose of the extraction lens is to efficiently remove ions emitted with large angular dispersion from a localized point on the sample, and transfer them with low angular dispersion to the quadrupole entrance aperture. The admittance characteristics of the energy filter/quad combination must be properly matched by the exit characteristics of the extraction lens in order to obtain optimum transmission/resolution characteristics [7].

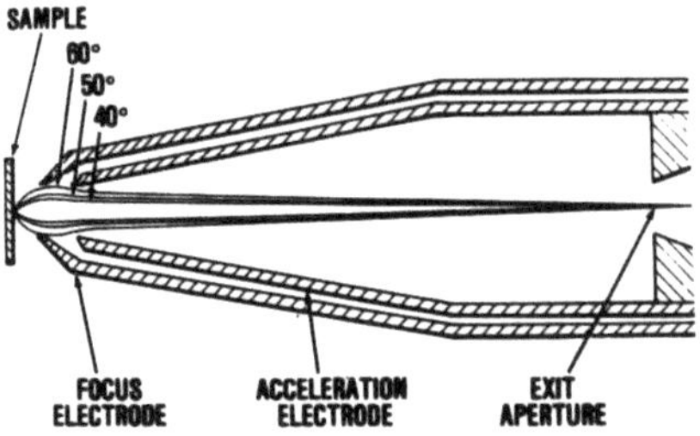

Fig. 2 Ion extraction lens similar to that developed by MAGEE, showing ion trajectories using the praxial approximation. Initial ion energy is 10 eV and final energy is 500 eV

An efficient extraction lens has been developed by MAGEE and HONIG [4], and a similar one employed by Physical Electronics on the SIMS II is shown in Fig. 2. Also shown in this figure are praxial trajectories for ions originating from a point on the target with various emission angles. The computer trajectories do not include spherical aberration which would cause the rays to just fill the exit aperture for 60° take-off angle from the sample with respect to the extractor axis. Furthermore, the high angle rays close to the focus electrode are expected to deviate strongly from the computer trajectories in a manner not included by the praxial approximation. Nevertheless, it is believed that this extraction lens will transmit extracted ions approaching 60°. The angular spread of the transmitted ions emitted from the extractor is near 10° (half angle) after deceleration to 10 eV, which is compatible with quadrupoles [8,9]. If we assume a cosine distribution of secondary ions, the transmission of this lens approaches 75 % for secondary ions that are accelerated to 500 eV in the extraction lens. The conical shaped focus lens also accurately defines the area from which ions are extracted and is geometrically compatible with multianalysis systems that include, for example , AES and/or ESCA.

The characteristics of the ion energy filter are dictated by the energy spread, ΔE, allowed by the quadrupole; ΔE is typically 1 to 6 eV in currently available quadrupole instruments [4,10]. Using the theoretical secondary ion energy distribution [7] calculated by GRIES, about 9 % of the secondary ions are transmitted for $\Delta E = 6$ eV. If the quad [8,9] and ion detector have a combined transmission of 25 %, then the overall transmission using the extraction lens of Fig. 1 and $\Delta E = 6$ eV is 1.7 %. If we allow $\Delta E = 10$ eV into the quad, overall transmission increases to 3 %. Although the mass resolution will not be very high ($M/\Delta M = 50$, e.g.) [8,9] the high transmission characteristics would be quite useful for SIMS imaging.

Spatial resolution and data acquisition speed in ion microprobes are determined by the ion gun characteristics. Ion sources may be divided into two broad categories, 1) large diameter sources, such as the duoplasmatron, electron impact and surface ionization sources, and 2) the microscopic or field emission sources [11,12]. The former generally require demagnification optics to form a microbeam, whereas the latter must be magnified because the effective source diameter is only about 150 Å. In either case, the optics operate in a spherical aberration limited condition for large beam diameters, and usually operate in the chromatic limit for small beam diameters. As we shall see, the beam diameter versus current characteristics of ion optical systems operating in the demagnification mode are vastly different from those operating in the magnification mode.

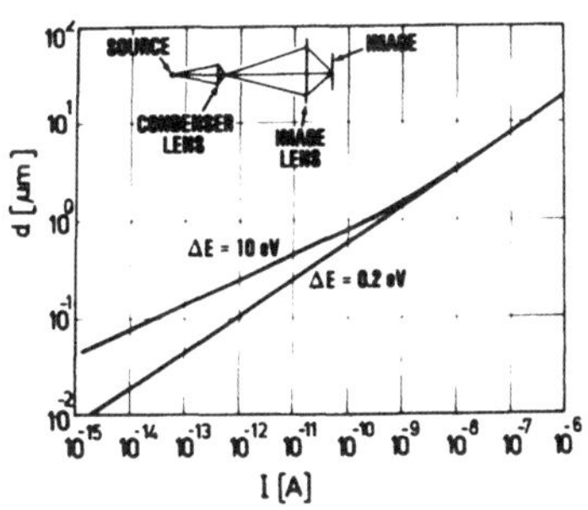

Fig. 3 Theoretical beam diameter versus ion current for an ion gun employing demagnification optics such as with a duoplasmatron or surface ionization device. The curves assume the optimum aperture setting for each beam diameter, $B = 100$ A/cm^2/strd, $E = 10^4$ eV, $C_c = 5$ cm and $C_s = 30$ cm

Demagnification optics are typically composed of one or more condenser lenses and objective lens. A simple example employing one condenser lens is shown schematically in the upper portion of Fig. 3. The optimized beam currents in the chromatic and spherical limits are given by [12]

$$I_c = \frac{\pi^2 \beta}{16 C_{c2}^2} \left(\frac{E}{\Delta E}\right)^2 d^4 \qquad (2)$$

$$I_s = \frac{3\pi^2 \beta}{16 C_{s2}^{2/3}} d^{8/3} \qquad (3)$$

where β = source brightness
C_{c2} = chromatic coefficient
C_{s2} = spherical aberration coefficient,

where C_{c2} and C_{s2} refer to the image or objective lens. Eqs (2) and (3) assume that the optimum objective aperture sizes are used for all beam diameters and that aberrations of all other optical components are negligible compared to those of the objective lens.

Consider, for example, a source requiring demagnification optics with a brightness of 100 A/cm^2/strd and E = 10^4 eV, and assume that the objective lens has aberration coefficients C_{c2} = 5 cm and C_{s2} = 30 cm. The resulting currents calculated from Eqs (2) and (4) are plotted in Fig. 3 for E = 0.2 and 10 eV. Note the very rapid (d^4) decrease in beam current with decreasing beam diameter d in the chromatic limit, which makes very small beam diameters impractial to use. Also, according to Eq.(2), a factor of 10 increase in ΔE corresponds to a factor of 100 loss in current for the same beam diameter in the chromatic limit, which shows the importance of small energy spread for microbeam applications. In order to determine the practicality of the beam currents in Fig. 3, one must compute the time τ(s) required to sputter various depths Z according to the equation

$$\tau = \frac{2\rho}{JY} \qquad (4)$$

where J = ion current density, and

Y = total sputter yield (atoms/ion). τ is plotted in Fig. 4 versus d for the ion gun optics of Fig. 3, where Y is taken to be 2 atoms/ion [13]. The time *per point* to sputter a given depth decreases dramatically with increasing beam diameter, because the current density has a d^2 dependence in the chromatic limit and a $d^{2/3}$ dependence in the spherical limit. For imaging a 100x100 array, it is practical to take a maximum of about 10^{-1} s per point or 10^3 s frame time. If one wants to achieve 0.1 μm image resolution, a small energy spread ($\Delta E \sim 0.2$ eV) of the source is imperative for a brightness of 100 A/cm^2/strd. The surface ionization Cs sources have this small energy spread. On the other hand, field emission sources have a brightness large enough so that large ΔE can be tolerated, although magnification optics must be used (see below).

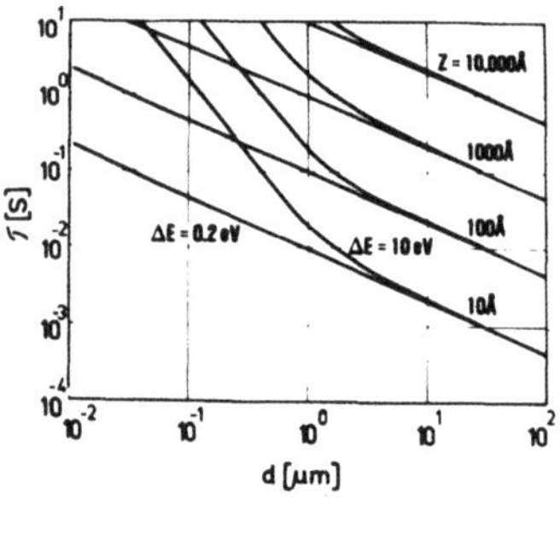

Fig. 4 Time required to sputter the depths indicated on the curves as a function of beam diameter for the ion gun depicted in Fig. 3. A sample density of $6\cdot10^{22}$ atoms/cm^3 and a sputter yield of 2 are assumed

Profiling is generally accomplished by sputtering (and raster gating) in a 10x10 primary ion beam scan array, which is practical up to about 1 s per point. Such a profile would take

10^2 s per frame or 10^4 s for a 100 "frame" profile. Here a "frame" refers to the minimum time to plot a point in a depth profile. One concludes from Fig. 4 that a 0.1 μm spatial resolution "microprofile" could be performed up to ~3 μm depth with ~300 Å depth resolution.

Thus we conclude that surface ionization sources, in particular with ΔE = 0.2 eV, could be satisfactorily used in a quadrupole based SIMS system to achieve images and micro-profiles to 0.1 μm spatial resolution. However, these maps and profiles would require some tens of minutes or hours. On the other hand, the ~10 eV energy spread and 100 $A/cm^2/strd$ brightness from duoplasmatron sources is not practical for forming 0.1 μm diameter beams for SIMS applications.

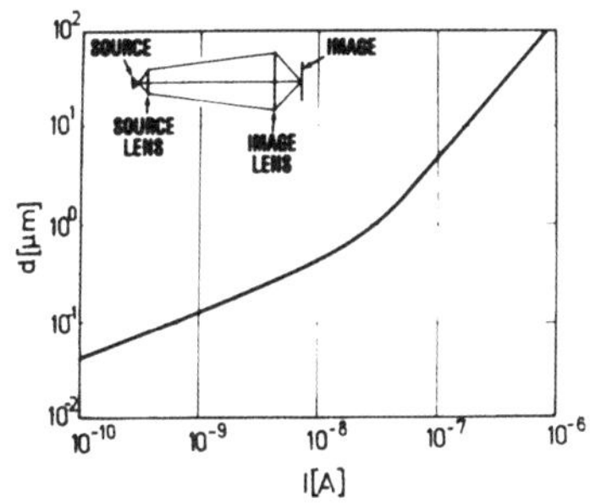

Fig. 5 Theoretical beam diameter versus ion current for an ion gun employing magnification optics with a field ion source. The source is assumed to emit dI/dΩ = 20 μA/strd with ΔE = 5 eV at 10 keV beam energy. The lenses have C_{c1} = 4 cm, C_{c2} = 5 cm, C_{s1} = 7 cm and C_{s2} = 30 cm, where the subscripts 1 and 2 refer to source and image lenses, respectively

Field emission sources [14,17], which are characterized by very high brightness but very small source diameter, require magnification optics [12], such as shown at the top of Fig. 5. The equations for the chromatic and spherical limited currents for optimized aperture conditions are

$$I_c = \frac{\Pi (E/\Delta E)^2 (dI/d\Omega)}{2C_{c1}C_{c2}} d^2 \tag{5}$$

$$I_s = \frac{3^{1/4}\Pi (dI/d\Omega)}{C_{s1}^{1/2}\, C_{s2}^{1/6}} d^{2/3} \tag{6}$$

where dI/dΩ is the angular intensity of the source and the subscripts 1 and 2 refer to the source and image lenses, respectively.

A Ga liquid field ion source has been measured to have dI/dΩ = 20 μA/strd with ΔE = 5 eV [14]. If we assume C_{c1} = 4 cm and C_{s1} = 7 cm [14] and let E = 10^4 eV, the gun performance is as shown in Fig. 5. For the smallest beam diameters, the source diameter (~150 Å) becomes significant and has been included. Nearly a nanoamp is predicted in a beam of 0.1 μm diameter. The beam diameter dependence on current is vastly different from that shown in Fig. 3 for other types of sources and the field ion emitter gives much larger currents for beams less than about 10 μm in diameter. This performance "crossover" is, of course, dependent on the particular optical design parameters assumed.

Using Eq.(4), we plot in Fig. 6 the time per point to sputter a given depth as a function of beam diameter. Here we have used a sputter yield Y = 3 because Ga is a heavy atom. The spherical limited region is peculiar in that the current density actually *decreases* with increasing beam diameter. In the chromatic limited region, the sputter rate is a remarkable ~10 Å/40 μs = 0.25 Å/μs, and thus images will require only a short time. For example, a 100 Å sputter depth image at 0.1 μm image resolution with 100x100 points will require ~4 s. From Fig. 1 we see that such an image could have ~1 % detection limits for some elements, even with a quadrupole SIMS detector with ~1 % transmission. One must be cautious in drawing too many conclusions from Fig. 6, however. The Ga field ion emitter is relatively reliable and simple but emits cluster ions [18] as well as single atom ions. The effects of these clusters on SIMS are unknown at present. Ion yields for Ga bombardment are unknown as well. Cs field ion emitters are more complex, possibly less reliable and possibly less bright, but are nevertheless important for detection of many elements, particularly electronegative elements [3].

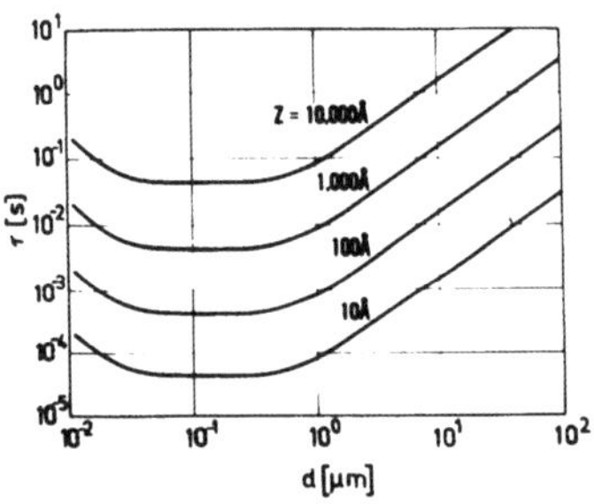

Fig. 6 Time required to sputter the depths indicated on the curves as a function of beam diameter for the ion gun depicted in Fig. 5 employing a field ion emitter. A sample density of $6 \cdot 10^{22}$ atoms/cm^3 and a sputter yield of 3 are assumed

To summarize, an optimized quadrupole SIMS system should approach ~3 % ion transmission in microbeam applications where the ions originate from a sufficiently small area and where low mass resolution is tolerable. Field ionization sources deliver greater ion currents only for moderate to small beam diameters (<10 μm for the example presented here). If field ionization sources are made to be reliable and trouble free, very rapid SIMS imaging capability will become practical at ~0.1 μm resolution with ~1 % typical detection limits.

The contributions of coworker Dr. David Welkie and the critical review of Brad Phillips are gratefully acknowledged.

References

1. H. Liebl, Scanning 3, 79 (1980)
2. R.J. Blattner and C.A. Evans, Jr., in Crystal Growth: A Tutorial Approach , W. Bardsley, D.T.J. Hurle, J.B. Mullin (eds), North-Holland Publ. Co. (1979) p. 269
3. H.A. Storms, K.F. Brown and J.P. Stein, Anal. Chem. 49, 2023 (1977)
4. C.W. Magee, W.L. Harrington and R.E. Honig, Rev. Sci. Instr. 49, 477 (1978)
5. F.G. Rüdenauer, W. Steiger and U. Kraus, Microchimica Acta Suppl. 8, 51 (1979)
6. K. Wittmaack, Adv. Mass Spectrom 7A, 758 (1978)
7. H. Liebl, Adv. Mass Sepctrom. 7A, 751 (1978)
8. J.F. Hennequin and R.L. Inglebert, Int. J. of Mass Spectrom. and Ion Phys. 26, 131 (1978)
9. J.F. Hennequin and R.L. Inglebert, Revue de Physique Appliquee 14, 275 (1979)
10. B.F. Phillips and R.L. Gerlach, in Microbeam Analysis - 1980 , David B. Wittry (ed), San Francisco Press (1980) p. 85
11. H. Liebl, In Secondary Ion Mass Spectrometry SIMS II, ed. by A. Benninghoven, C.A. Evans, Jr., R.A. Powell, R. Shimizu, and H.A. Storms, Springer Series in Chemical Physics, Vol. 9 (Springer, Berlin, Heidelberg, New York 1979) p. 176
12. Lee H. Veneklasen, in Physical Aspects of Electron Microscopy , B. Siegel and D.R. Beaman (eds), John Wiley & Sons (1975) p. 315
13. H.H. Andersen and H.L. Bay, Sputter-Yield Measurements, Ch. 4, Institute of Physics, Univ. of Aarhus, Denmark (1980)
14. J. Orloff and L.W. Swanson, Proceedings of the 16th Electron, Ion and Photon Beam Technology Symposium (1981), to be published
15. R. Clampitt and D.K. Jefferies, Nucl. Instrum. and Meth. 149, 739 (1978)
16. R.L. Seliger, J.W. Ward, V. Wang and R.L. Kubena, Appl. Phys. Lett. 34, 310 (1979)
17. J.F. Mahoney, A.Y. Yahiku, H.L. Dalex, R.D. Moore and J. Perel, Jour. Appl. Phys. 40, 5101 (1969)
18. A. Wagner, T. Venkatesan, P.M. Petroff and D. Barr, Proceedings of the 16th Electron, Ion and Photon Beam Technology Symposium (9181), to be published

Description and Applications fo a New Design Cs^+ Ion Source on the COALA Ion Microprobe for Negative Ion SIMS

B.L. Bentz and H. Liebl

Max-Planck-Institut für Plasmaphysik, EURATOM Association,
D-8046 Garching/München, Fed. Rep. of Germany

Introduction

In any mass spectrometric system, one of the key instrumentation components is the ion source, or stated similarly, the method of ionization. Apart from influencing the level at which the sought-for constituents can be quantitatively ionized, the ionization process should also preserve the native chemical integrity of the sample, that is, without inducing decomposition, compound formation, or falsifying sample structure. In the SIMS method, the practicing spectroscopist can often select a primary ion projectile species which affords advantages when sputtering into certain materials or when profiling specific elements. For example, negative primary ions can reduce the accumulation of electrostatic charge on the surface of poor-conducting samples [1]; molecular ions were used in one study to obtain information relative to damage density effects in the collision cascade [2]; and bombardment by O and Cs promotes ion yield enhancements for positive secondary ions [3] and negative secondary ions [4], respectively. The latter aspect - Cs bombardment SIMS - is of active interest in several research groups [5-7], including this laboratory [8]. We have recently designed and brought into operation a compact, microbeam Cs^+ source for SIMS [9]. This brief report is intended to describe the source, and present first results of its coupling to a scanning ion microprobe mass analyzer developed at IPP.

Ion Source

A schema of the first model of three Cs^+ ion sources developed in our laboratory is shown in Fig. 1. The source is 9.0 cm in length, 4.6 cm in width, and mounts on a 2.75" diameter vacuum flange. Positive ions are produced with low energy spread by thermal surface ionization, a mechanism used by other SIMS investigators in sources incorporating a rear-fed, heated porous W plut emitter [5-7]. Our source differs constructionally from the above mentioned one, in that a non-porous, compact Re polycrystalline surface, resistively heated and maintained at 900 $^{\circ}C$, is employed as the emitting surface. We developed this alternative geometry because the stability and durability of the pores may become questionable at high temperatures [10], in addition to the fact that the current density available at the planar plug - and therefore the source brightness - is

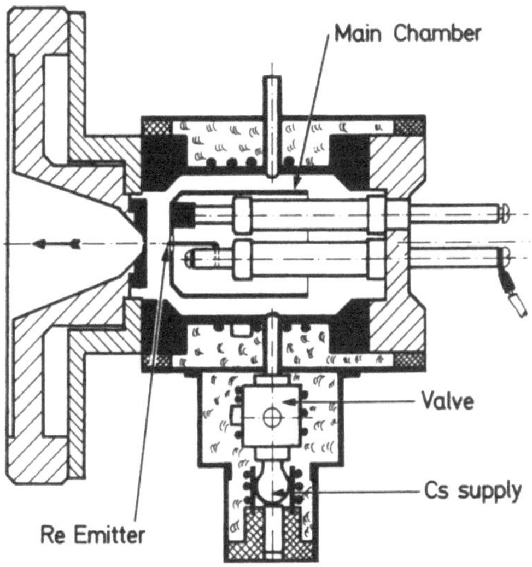

Fig. 1 Ion source generating $^{133}Cs^+$ by surface ionization

limited by space charge [11]. When forming micro-ion probes, this point can play an important role in determining the beam current available for focusing into a small spot. The Re wire is fashioned like a hairpin filament, protruding through a shield cap at filament potential, so that a very high field strength prevails at the tip preventing space charge limitation of the emission current density. The limiting parameter is the maximal Cs vapor density which can be maintained in the source chamber before voltage breakdown sets in. Since the accelerating electric field in front of the tip is near-concentric around the center of curvature of the tip, the ions emitted from the surface seemingly originate from an area around the center which is much smaller than the tip dimensions. This source is intrinsically a microsource with high brightness and small energy spread. Cs vapor is introduced into the thermostated main chamber by heating a glass reservoir containing a charge up to 1 gram of solid Cs metal; the loading is accomplished in a glove box containing argon. A heated Nupro valve isolates the supply ampule from the chamber. Ions exit the source through a 200 μm diameter orifice drilled in a grounded Ta electrode which is separated from the HV emitter by a distance of ~5 mm.

In terms of current production, the source at 5 keV accelerating potential easily delivers several tenths of microamperes through the 200 μm diameter orifice. The beam current-temperature characteristic indicated that, above threshold (critical) temperature, the ion signal decreased slowly with increasing temperature [12]. The virtual source size of 20 μm is considerably smaller than values found in a SIMS duoplasmatron-type in source. Compared to beams from a porous W plug said to operate at 1100-1200 °C [5,6], the ion beams from this source should have a thermal energy spread a factor of 1.2 lower. Beam stability is ±1 % over 1 hour. Measured beam brightness at T(Cs) = 125 °C is 33 A/cm^2 sr at 5 keV. The measured beam brightness value is probably biased low compared to the source brightness due to lens aberrations of the imaging system. Measured brightness values for Cs^+ planar plut sources are not available for comparison, although one report comments the brightness to be comparable with a duoplasmatron supplying O_2^+ (beam energy unspecified, [13]).

A low and high gain (factor 1.000 difference) mass scan of a 5 keV positive ion beam is shown in Fig. 2. Neither ions of the Re thermoemitter proper, background gases in the source chamber, outgassing products, nor impurity ions from the Cs supply are observed. The only signal arises from singly charged atomic cesium ions. This is *not* the case with a reported high brightness Cs^+ liquid metal field ion source [14], where cesium cluster ions and cesium oxide species were observed. The high beam purity of our source would be advantageous, both mechanically and financially, for SIMS instruments without primary ion magnetic mass separation.

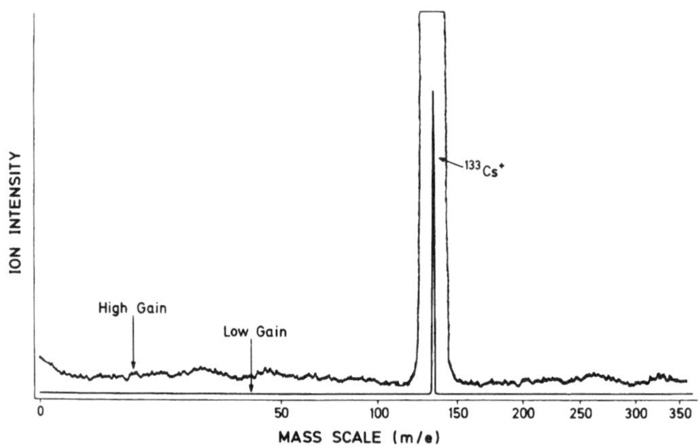

Fig. 2 Ion composition of mass analyzed beam extracted from the source. Note absence of "molecular" species and hydrogenic impurities

COALA Ion Microprobe

A schema of the ion microprobe to which the source is coupled is shown in Fig. 3. Ions leave the source as a diverging beam, but an Einzel lens positioned downstream renders the beam parallel for entrance into a 180° uniform magnetic field used for primary beam purification. A mass separating slit is placed in the middle of the half circle and mass selected ions leave the field again as a parallel beam. This configuration accomplishes mass separation without introducing any lateral probe spot elongation owing to the energy spread of the primary ions [15]. Another advantage is that the ion acceleration voltage and the magnetic field need not be extremely constant. The surface ionization Cs^+ source does not require primary beam mass analysis, but when switching over to duoplasmatron operation (with Ar^+, N_2^+, or $O^{\pm}$ for (+)ion SIMS), beam purification is needed.

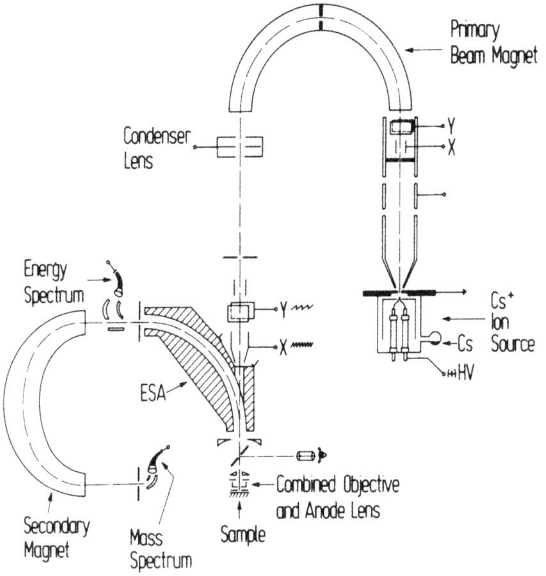

Fig.3
Schematic layout of ion microprobe

After passing the condenser lens and a set of electrostatic deflection plates used for beam adjustment and rastering, the primary beam approaches the last component of the primary ion column, the so-called COALA lens (Combined Objective and Anode Lens Analyser, [16]). This is a three electrode lens arrangement, unique in that by a suitable choice of excitation potentials and primary beam energy, the *same* set of electrodes simultaneously focuses the primary Cs^+ beam onto the sample and accelerates the negatively charged secondary ions backwards out of the lens. The two beams of opposite charge polarity thus pass this combined lens coaxially and in opposite directions. This lens design has strong advantages for ion beam microfocusing including (a) an objective lens with short focal length (~6 mm) and small spherical aberration is realized, thereby allowing a given probe spot size a relatively higher probe current; and (b) a high field strength prevails at the sample surface and therefore better lateral resolution or lower emittance of the secondary beam is provided. The lens assembly incorporates a Schwarzschild-type microscopic objective for viewing the sample *in situ*. Light is delivered to the specimen through a series of bores and by reflection from highly polished integral electrode surfaces. The sample is viewed perpendicular to the surface at a magnification of ~150.

The sample is housed in a chamber pumped by a 450 l/s turbo pump fitted with a coolable sorption trap, a 1600 cm^2 LN_2 cold wall, and Ti sublimation pump. An additional LN_2 shroud surrounds the sample to reduce H_2O and hydrocarbon partial pressures. A high sensitivity quadrupole mass filter monitors residual gas composition. In order to facilitate sample holder exchange without venting the sample chamber, the instrument includes a turbo pumped vacuum load-lock device having a base pressure in the 10^{-7} Torr range.

The secondary negative ion beam is separated from the primary beam by a 90° spherical condenser having a bore machined in the outer plate. A continuous dynode electron multiplier is placed at the exit of the condenser field, and is used as a pulse counting energy monitor for the total sputtered negative ion population. The energy spectrometer output is useful as a first stage check that the secondary ion optics are adjusted properly for optimum pick-up before commencing work with the secondary magnet. The spherical condenser and the following 180° secondary magnetic field form a stigmatic imaging double-focusing mass spectrometer, similar to the one described earlier [15]. With an energy stop of 3 mm width and 1000 V acceleration voltage, the transmitted energy band width is 25 eV. A Galileo model 4830 electron multiplier, operated in the pulse counting mode, serves for detection of the mass-analysed negative secondary ions.

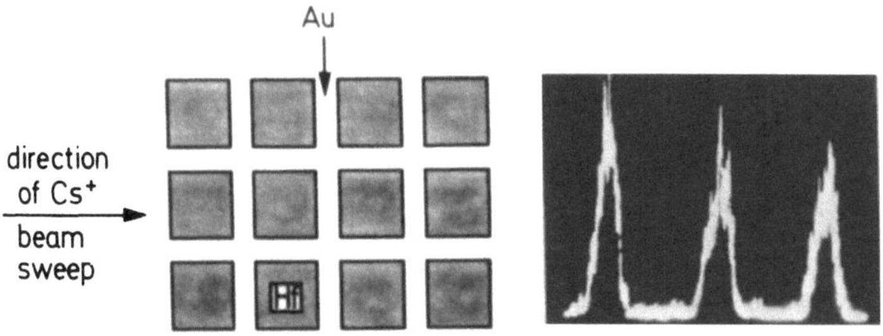

Fig. 4 Line scan along a Au/Hf grid deposit having 6.2 µm wide Au bars. $^{197}Au^-$ trace. The 6 keV Cs^+ beam was swept across 3 such bars

The instrument is equipped with an electronic aperture unit which controls gated counting of the secondary ion signal when rastering. To give an idea of the Cs^+ primary beam size achievable with the microprobe, Fig. 4 shows results of an X-line scan across a Au/Hf grid sample having Au bar width of 6.2 µm and Hf interspace width of 19.2 µm. The detected mass-analysed secondary ion was $^{197}Au^-$, and the primary beam analogue deflection circuitry was adjusted to sweep the Cs^+ beam across three Au bars. A Cs^+ primary beam with diameter in the few microns range was achieved.

Results

Initial studies involved sputtering a set of metal samples (Al, Si, Mo, Cu, Cr) and recording survey negative ion mass spectra from which relative intensities of negative ions could be compared. A static Cs^+ beam of 20 µm diameter bombarded the sample at normal incidence so that current densities of 0.1-1.0 mA/cm² were maintained. The spectra showed that 6 keV bombardment produced negative ion metal clusters and/or metal oxide formation. From a PAPYEX carbon matrix, C_n^- clusters up to n = 9 were observed, while Al showed homonuclear clusters up to n = 4. For cases where metals such as Mo and Cr are found strongly oxidized, the detection of metal oxide species is more sensistive than the metal M. No MCs^- species were observed. The study also proved useful because instrument sensitivity using Cs^+ bombardment could be evaluated; when sputtering into a Si wafer, a count rate of $1.1 \cdot 10^5$ counts/s·nA of incident Cs^+ was measured for $^{28}Si^-$. This compares within a factor of 4-10 to other Cs^+/SIMS machines described in [5,6].

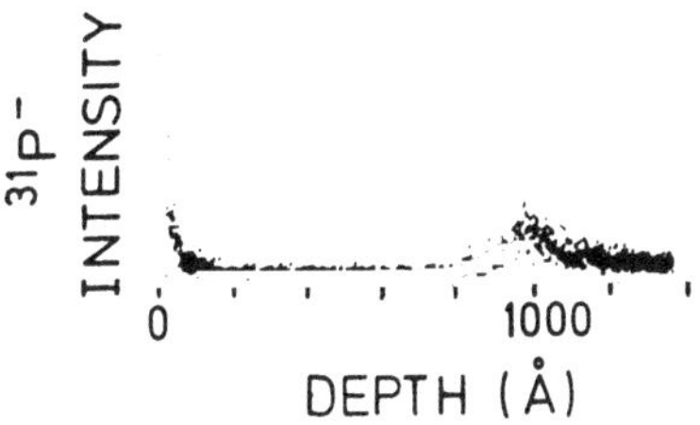

Fig.5 Depth profile showing the ability of COALA to resolve thin layers beneath much thicker covering films (6 keV Cs^+, P implant in Ta_2O_5, FWHM ≈ 125 Å)

To evaluate the depth resolution achievable with the instrument, a tantalum oxide sample was studied which had a phosphorus-enriched zone nominally 50 Å thick inserted 1000 Å beneath the surface. An area of 19 μm x 19 μm was raster scanned at 100 lines/frame (time per full frame = 3.1 s) using a 6 keV Cs^+ beam impinging 90° to the sample plane. Fig. 5 presents the phosphorus depth profile. As the mean projected range for a Cs particle, at the aforementioned energy and incident angle, is about 55 Å in this matrix, the peak of the ion damage distribution occurred at a depth less than the location of the P insert. A value of 125 Å was found for the apparent full width at half maximum (FWHM) of the phosphorus layer. This value compares well with results from other instruments using O_2^+ and detecting $^{31}P^+$ [17].

The low background at m/e=2 offered by negative ion spectroscopy was taken advantage of to profile a deuterium implant in silicon (see Fig. 6). Although a depth scale has not been assigned here, a value of 1690 Å for the average depth (mean projected range) of the implantation distribution was recently computed using the program TRIM [18]. Further work needs to be done to compare measured and calculated profiles, in view of

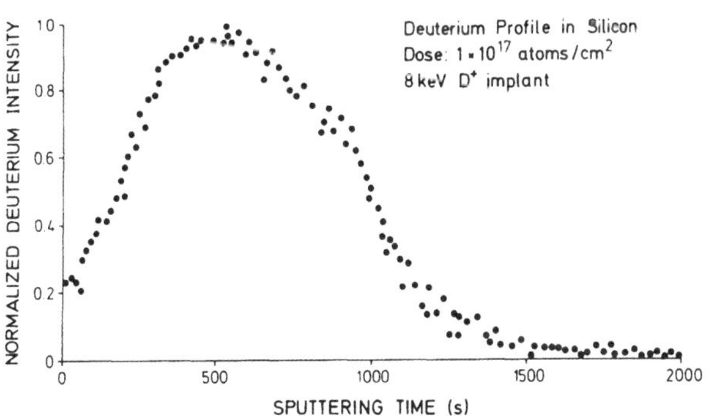

Fig.6 Profile in-depth following $^2D^-$

the implications for plasma-wall interactions in fusion plasma machines.

Summary

The Cs^+ source described in this paper can be used advantageously in SIMS for reasons including

(1) the source provides no heavy gas load into the primary ion column and sample chamber as does a gas discharge source, thus easing pumping requirements and maintenance of UHV conditions. The Cs leaving the source acts as an excellent getter for residual gas components.

(2) heavy ions have a shorter range in a solid than do lighter ions of the same energy, thus causing less extensive atomic mixing. For example, Cs^+ has a projected range in solid materials of about 60 % of that of O_2^+ of the same energy.

(3) negative ion yields of certain elements are enhanced [4,7]. When depth profiling, this can render a reduction in sample consumption for a given ion count.

(4) when profiling certain dopants in silicon, higher sensitivities are often obtained by detecting molecular ions rather than atomic. Studying a B implant in Si, we found a factor ~10 greater sensitivity for the $^{11}B^{28}Si^-$ species to $^{11}B^-$.

Herbert Weiss, Ingenieur (grad.) is acknowledged for his helpful comments and excellent technical assistance.

References

1. C.A. Andersen, H.J. Roden and C.F. Robinson, J. Appl. Phys. 40, 3419 (1969)
2. K. Wittmaack, Surface Sci. 90, 557 (1979)
3. C.A. Andersen, Int. J. Mass Spectrom. Ion Phys. 2, 61 (1969)
4. V.E. Krohn, Jr., J. Appl. Phys. 33, 3523 (1962)
5. C.W. Magee, J. Electrochem. Soc. 126, 660 (1979)
6. P. Williams, R.K. Lewis, C.A. Evans and P.R. Hanley, Anal. Chem. 49, 1399 (1977)
7. H.A. Storms, K.F. Brown and J.D. Stein, Anal. Chem. 49, 1399 (1977)
8. H. Liebl, In Secondary Ion Mass Spectrometry SIMS II, ed. by A. Benninghoven, C.A. Evans, Jr., R.A. Powell, R. Shimizu, H.A. Storms, Springer Series in Chemical Physics, Vol. 9 (Springer Berlin,Heidelberg,New York 1979) p. 176
9. B.L. Bentz and H. Liebl, paper presented at the AGMS Mass Spectrometric Isotope and Element Analysis Conference, Amsterdam, Netherlands, April 8-10, 1981
10. E.Ya. Zandberg and N.I. Ionov, Surface Ionization (Israel Program for Scientific Translation, Jerusalem 1971) Chapter 10
11. L. Loeb, Rev. Sci. Instrum. 4, 486 (1933)

12. R.G. Wilson and G.R. Brewer, Ion Beams with Applications to Ion Implantation (Wiley, New York 1973) Chapter 2 IPP 9 (1980)
13. H.A. Storms, K.F. Brown and J.D. Stein, Report on Second Japan-US Joint Seminar on SIMS, Takarazuka, Japan, October 23-27, 1978, p. 267
14. R. Clampitt and D. Jefferies, Nucl. Instr. Meth. 149, 739 (1978)
15. H. Liebl, Int. J. Mass Spectrom. Ion Phys. 6, 401 (1971)
16. H. Liebl, Vacuum 22, 619 (1972)
17. C.A. Evans, B.N. Colby, G.L. Kearns and W. Singer, Anal. Chem. 45, 398 A (1973)
18. W. Eckstein, Max-Planck-Institut für Plasmaphysik, Report IPP 9(33, Oktober 1980

Operational Data of a Simple Microfocus Gun Using an EHD-Type Indium Ion Source

M.J. Higatsberger, P. Pollinger, H. Studnicka, and F.G. Rüdenauer[1]

Institute of Experimental Physics, University of Vienna, A-1090 Wien, Austria

When a liquid conductor (metal) is exposed to a high electrostatic field the liquid surface, under the action of surface tension and electrostatic polarization force, assumes an equilibrium shape, viz. a cone with a total apex angle of 98.6° ("Taylor cone" [1,2]). When the metal is positively polarized with respect to its surroundings, intense positive ion emission from the cone apex is observed [3,4]. The mechanism of ion emission is not yet fully understood; according to GOMER [5], ion emission appears to be initiated by field desorption from the liquid apex and, at higher emission currents, changes over to field ionization of thermally evaporated atoms in front of the cone apex. We can identify several critical parameters for ion emission: the "cone forming voltage" ΔV is to be applied between liquid metal and extractor electrode before a Taylor cone can develop; if both the microscopic cone apex and the extractor are assumed to be of ideal shape (according to the "sphere-on-cone", SOC-model) ΔV is given by [5]

$$\Delta V_{threshold} = 4.53 \cdot 10^{5} \, (\gamma R_{o})^{1/2} \qquad (V)$$

where γ is the surface tension of the metal (N/m) and R_o the apex-extractor distance (m). The electrostatic field at the microscopic cone apex has to exceed the value

$$F_{vap} = 6.95 \cdot 10^{8} \, (E_{vap} + E_{i} - \Phi - 0.63)^{2} \qquad (V/m)$$

before ions can be field desorbed [3]; here E_{vap} is the evaporation voltage of a metal atom, E_i the ionization voltage and Φ the voltage of the work function of the metal, all given in volt. Once ion emission is initiated, the emission current I^+ seems to be space charge limited and follows a characteristic voltage dependence [3],

$$I^{+} = k \cdot (\Delta V / \Delta V_{threshold})^{3/2} - 1 \qquad \text{for} \quad \Delta V \geq \Delta V_{threshold}$$

where ΔV is the tip/extractor voltage difference and k a constant, somewhat depending on geometry.

These "electrohydrodynamic" (EHD) ion sources have been used in mass spectrometry for semiquantitative analysis of liquid metals [6,7], for ion lithography [8] and recently also as a primary ion source in SIMS [9]. In all these applications the

[1] Austrian Research Center Seibersdorf, A-1082 Wien, Austria

extremely small source size and high brightness of the EHD source potentially allows focusing high i on currents onto spot sizes of the order of a few 100 Å [8,10].

We have designed a simple liquid metal source for In ions, to be used as a primary ion gun in a scanning ion microprobe [11]. It was of prime importance in this application to obtain highly focused beam current (≈10 nA) rather than extremely small beam diameters, in order to obtain secondary ion currents strong enough to be of use in the microanalysis of solids. The source itself is shown schematically in Fig. 1.

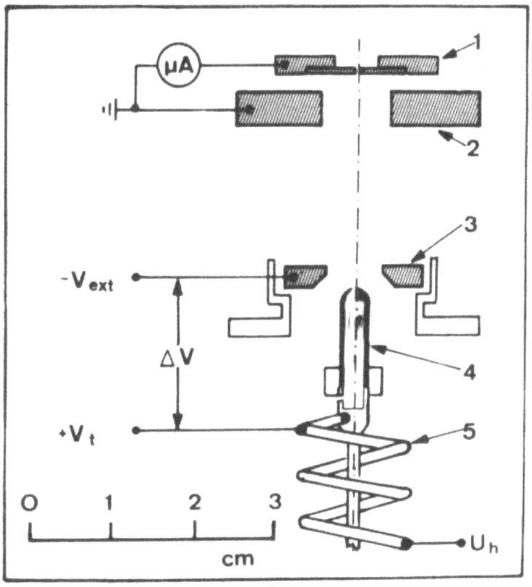

Fig. 1 Liquid metal EHD ion source of capillary type (schematic) (1... ground electrode; 2... grounded ring; 3... extractor; 4... In-reservoir; 5... heater filament)

The reservoir is filled with liquid In under atmosphere. In the filling process a sharp In tip is produced by pulling out molten In with a stainless stell wire (0.1 mm diameter) from the emitter side of the reservoir; thereby overall wetting of the bore rim is achieved too. The source, prepared as described, is now mounted in a 2 lens electrostatic microfocus ion gun (Fig.2) and evacuated to a vacuum of better than $7 \cdot 10^{-4}$ Pa. Then the extraction potentials are applied and the heater filament surrounding the source body is switched on. The originally irregularly shaped in protrusion melts under field action, immediately forming the Taylor cone, and ion emission starts. The reverse sequence is followed for shutdown of the ions source: the field is kept on and the heater is switched off. The Taylor cone solidifies and, after demounting the source, can be observed in a light microscope (Fig. 3). The source may be operated in voltage or in current stabilized mode; in the latter mode, the source current is regulated via the extractor potential, in the former mode tip and extractor voltages are kept constant. Since no serious deterioration of operating characteristics was observed, the

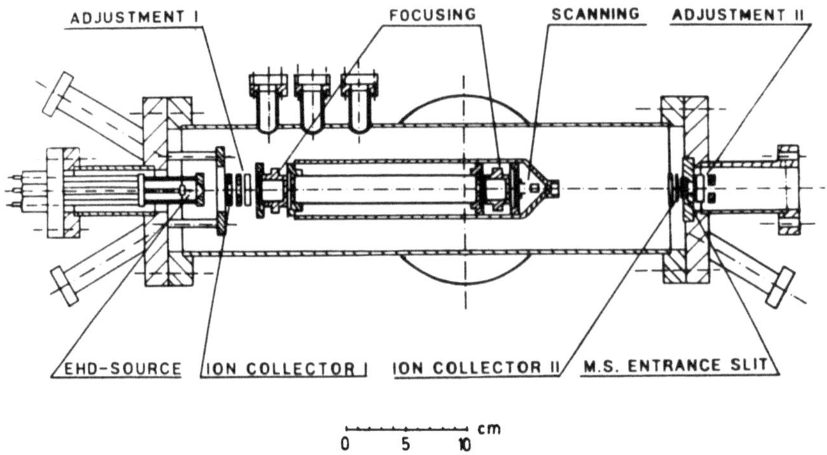

Fig. 2 Microfocus ion gun equipped with EHD ion source (schematic)

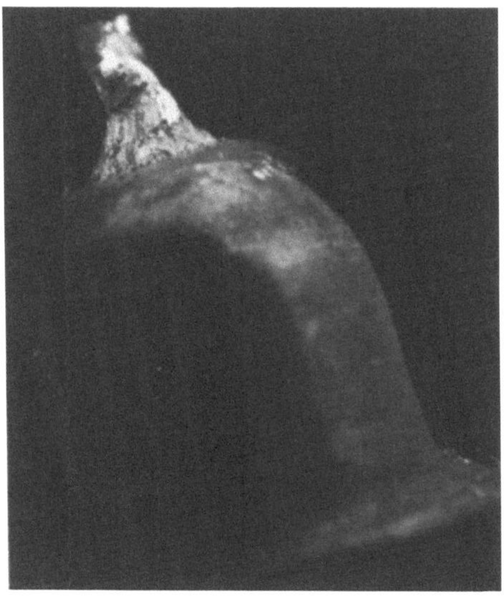

Fig. 3 Taylor cone, solidified under electrostatic field from liquid In, on ion source shown in Fig. 1

voltage stabilized mode is usually chosen because total beam energy and the focusing action of the extraction field remain constant. Commercially available power supplies with short and long term stability of 10^{-4} were used to run the source. Fig. 4 shows the current to the ground electrode (see Fig. 2), and Fig. 5 the angular intensity (current/differential solid angle) in the center of the extracted beam, both as a function of the voltage difference between tip and extractor. The angular dependence of the angular intensity is shown in Fig. 6. For the measurement of the focused beam diameter the beam was scanned across a Ni grid of 63 μm mesh width and 10 μm wire thickness;

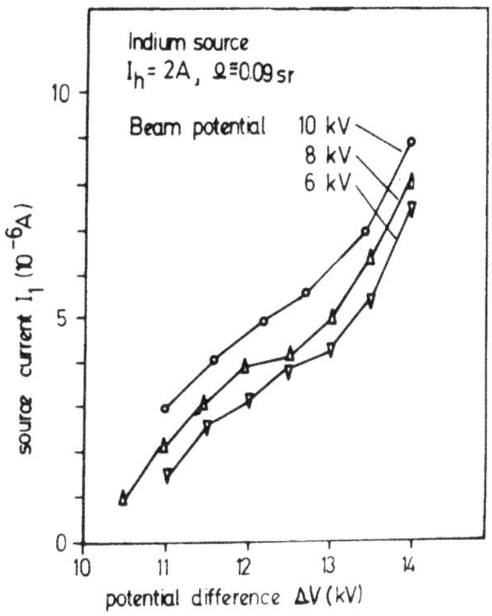

Fig.4 Current/voltage characteristic of In source

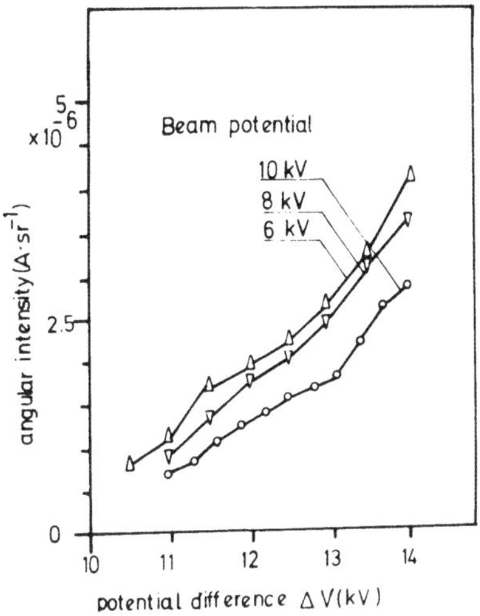

Fig. 5 Angular intensity at center of extracted In beam

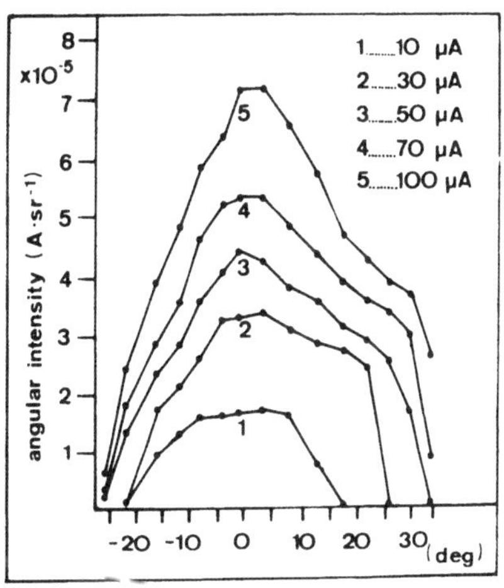

Fig. 6 Angular distribution of ion current emitted from In source; total source current is parameter

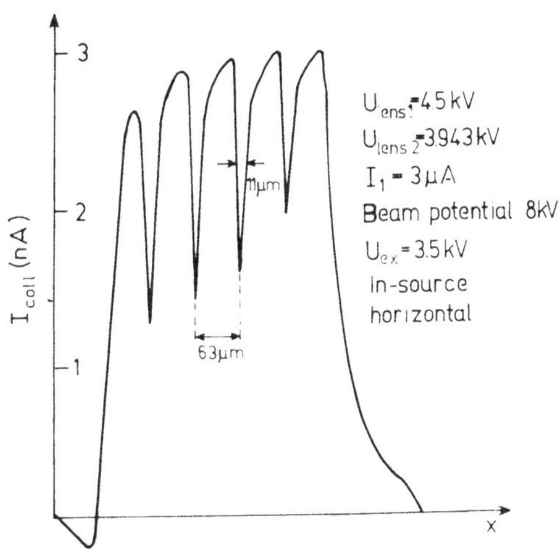

Fig. 7 Line scan of focused In beam across 63 μm grid; registration time 10 s, multiple traces superimposed

the current passing the mesh was registered in a Faraday cup arrangement. An 11 μm beam diameter was deduced from these measurements (Fig. 7). When mounted in the ion microprobe, a shorter working distance was used for the ion gun (≈35 mm) and spot sizes of the order of 5 μm were obtained at beam currents of 10 nA [11]. The beam divergence was limited by a diaphragm in the ground electrode (see Fig. 2) to a value of $\alpha = 1\cdot10^{2}$ rad. It can be shown that, due to the large energy spread of EHD sources [12], the spot diameter is generally limited by chromatic aberration [10],

$$d_{spot} \approx d_{ch} = 2 \cdot C_{ch} \cdot \alpha \cdot (\delta V/V)$$

where $\delta V/V$ is the relative energy spread in the beam. Estimating a value of C_{ch} = 100 mm for the chromatic aberration constant of the objective lens in our column, an energy spread of ΔV = 50 eV, and using the values $\alpha = 1 \cdot 10^{-2}$ rad, V_t = 8 keV (final beam energy), we calculate a spot diameter of 12.5 µm, in good agreement with the observations.

The source operates in a vertical or horizontal orientation with capillary forces alone being responsible for the metal transport to the Taylor cone. Once stable operation has been obtained, the source may usually be shut down and restarted without difficulties. Source lifetime generally is limited by the capacity of the In reservoir (≈90 h at 2 µA total source current); an ionization efficiency for the charge material between 14 % and 23 % has been determined by the weight-loss technique. Preliminary measurements of the mass spectrum of the In beam have shown that the ionized fraction mainly consists of singly charged In^+ ions with small (order of $10^{-4} \cdot I\{In^+\}$) contributions from In^{2+}, In_2^+, In_3^+. Also, impurities of the In charge materials were identified (Cu, $1:10^4$; Te, $1:10^6$), in approximate agreement with the specified purity of In (99.99%).

References

1. G.I. Taylor and A. McEwan, J. Fluid Mech. 22, 1 (1965)
2. G. Taylor and M.D. Van Dyke, Proc. R. Soc. London A 313, 453 (1969)
3. J.F. Mahoney, A.Y. Yahiku, H.L. Daley, R.D. Moore and J. Perel, J. Appl. Phys. 40, 5101 (1969)
4. V.E. Krohn, J. Appl. Phys. 45, 1144 (1974)
5. R. Gomer, Appl. Phys. 19, 365 (1979)
6. C.A. Evans, Jr. and C.D. Hendricks, Rev. Sci. Instrum. 43, 1527 (1972)
7. D.S. Simons, B.N. Colby and C.A. Evans, Jr., Int. J. Mass Spectrom. Ion Phys. 15, 291 (1974)
8. R.L. Seliger, R.L. Kubena, R.D. Olney, J.W. Ward and V. Wang, J. Vac. Sci. Tech. 16, 1610 (1979)
9. P.D. Prewett and D.K. Jefferies, Inst. Phys. Conf. Ser. 54, 316 (1980)
10. H.J. Liebl, In Secondary Ion Mass Spectrometry; SIMS II, ed. by A. Benninghoven, C.A. Evans, Jr., R.A. Powell, R. Shimuzu, H.A. Storms, Springer Series in Chemical Physics, Vol. 9, (Springer, Berlin,Heidelberg,New York 1979)
11. F.G. Rüdenauer, P. Pollinger, H. Studnicka, W. Steiger, H. Gnaser and M.J. Higatsberger, these Proceedings
12. R. Ohana, PhD Thesis, Univ. Paris-Sud/Orsay (1980)

First Results on a Scanning Ion Microprobe Equipped with an EHD-Type Indium Primary Ion Source

F.G. Rüdenauer[1], P. Pollinger[2], H. Studnicka[2], H. Gnaser[3], W. Steiger[1], and M.J. Higatsberger[2]

Electrohydrodynamic ion sources have recently received increased attention in secondary ion mass spectrometry, owing to their potential of producing extremely small primary ion beams at relatively high focused ion currents. KROHN and RINGO were the first to suggest the use of such a source as primary ion gun in an ion microprobe [1,2,3] and have reported on ion gun tests using Ga^+ and Cs^+ ions [4]. The only paper so far actually describing experiments with an EHD source coupled to a secondary ion mass spectrometer was published by PREWETT and JEFFERIES [5]; these authors obtained 0.5 μm diameter primary beams of Ga^+ at currents of 0.2 nA and were able to record secondary electron and mass separated secondary ion distribution maps from microelectronic circuits. LIEBL [6] has shown that spot currents (at spot sizes below 1 μm) are limited by chromatic aberration and calculated a limiting focusable current of $1.6 \cdot 10^{-10}$ A at a spot size of 200 Å (ca. 50 A/cm^2). Values approaching these theoretical limits have actually been obtained with a Ga^+ ion gun designed for applications in ion lithography [7,8].

Although it thus appears possible to obtain spot sizes approaching the intrinsic lateral resolution limit of SIMS (determined by the extension of a collision cascade in the sample, i.e. approximately a few 100 Å), it remains to be shown that EHD sources can be usefully applied for microanalytical purposes in SIMS, in particular, that (a) the rather exotic metal ions which have been produced in EHD sources (Cs, Ga, In, Sn, Au, etc. [9]) produce reasonably high ion yields of positive and negative secondary ions for most elements; for Cs^+ and negative secondary ions this has already been demonstrated [10]; (b) that the primary metal ion does not produce high yields of secondary molecular or cluster species interfering with the analysis; and (c) that the short time and long time stability of current and beam position is sufficient to permit recording of mass spectra, depth profiles and secondary ion images without excessive noise, systematic drift and loss of image resolution.

[1] Austrian Research Center Seibersdorf, A-1082 Wien, Austria

[2] Inst. of Exper. Physics, Univ. Vienna, A-1090 Wien, Austria

[3] Inst. of Gen. Physics, Techn. Univ. Vienna, A-1040 Wien, Austria

In our work we have used a 2-lens electrostatic microfocus In^+ EHD ion gun described in another paper of these proceedings [11], coupled to a quadrupole ion microprobe previously described [12,13]. Instrument control and data acquisition is fully computerized using a PDP 11/34 minicomputer; ion images can be transferred to a video refresh memory (512 x 512 x 4 bit), also controlled by the PDP 11/34 [14] and read out on a B/W or color TV monitor using pseudocolor display techniques. The sample can be bombarded by the standard mass separated O_2^+ gun ($\leq$ 10 keV) and the In^+ EHD gun ($\leq$ 10 keV) in the same manipulator position; angle of incidence for both beams is 45°, the secondary ion take off angle is also 45° (Fig. 1). To obtain "apparent elemental secondary ion yields" (= detected cps/nA primary current) and "practical sensitivities" (= detected cps/nA primary current/fractional atomic concentration) for various secondary ion species, an In^+ beam of 11 nA and 9 keV was focused to about 10 µm diameter. The targets were presputtered in a raster of 200x280 μm^2 until a steady secondary ion current of the matrix species was obtained; the raster was reduced to 100 x 140 μm^2 and the mass spectra were recorded. Except for the measurements on ^{11}B implanted into Si, the quadrupole mass filter was set to "standard operating conditions (SOC)", ensuring approximately mass-independent transmission over a large mass range [15]; when this condition is relaxed, yields can be increased by more than one order of magnitude by tuning the quadrupole individually for each mass range.

Table 1 shows practical sensitivities (S_p) for some elements under 9 keV In^+ bombardment in comparison with the practical sensitivities of the same elements under 10 keV O_2^+ bombardment.

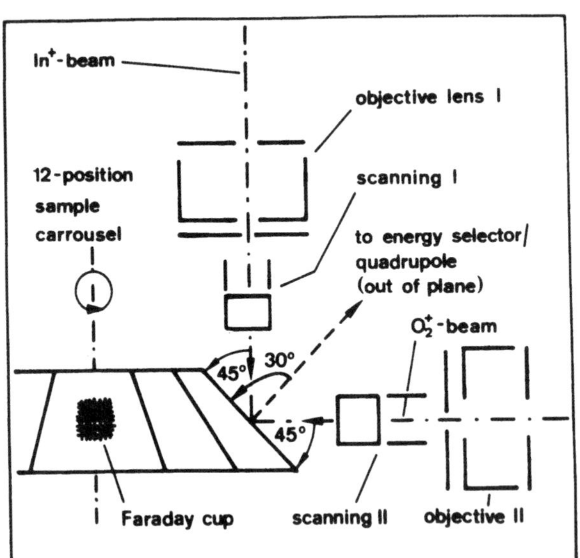

Fig. 1 Target geometry of ion microprobe equipped with dual primary ion guns

Table 1 Practical sensitivities (S_p) for some elements under 9 keV In^+ bombardment

Element X	Sample**	P_{O_2} [Torr]	$S_p(X^+)$ [cps/nA]		$Y(In^+)$ [cps/nA]	$Y(InX^+)$ [cps/nA]
			9 keV In^+	10 keV O_2^+		
Al	pure El.	$4x10^{-7}$	$6.9x10^3$	$(6.7 \pm 1.2)x10^3$	$4.4x10^2$	8
Ti	-"-	-"-	$2.6x10^3$	$(4.1 \pm 2.7)x10^4$	$2.8x10^2$	1
Fe	-"-	-"-	$1.8x10^2$	$(1.1 \pm 0.1)x10^3$	$2.3x10^2$	0.1
Ni	-"-	-"-	$2.3x10^2$	$(1.7 \pm 0.2)x10^2$	$3.7x10^3$	0.1
Ta	-"-	-"-	$1.8x10^2$	$(1.8 \pm 0.6)x10^2$	$5.0x10^2$	0.1
B	$Fe_{80}B_{17}W_3$ Met.Glass	-"-	$4.4x10^2$	$3.1x10^3$	n.d.	n.d.
Fe	-"-	-"-	$3.3x10^2$	$2.5x10^3$		n.d.
W	-"-	-"-	$3.8x10^2$	$1.7x10^3$		n.d.
Mn	st.steel*	-"-	$1.7x10^1$	$2.2x10^1$	$6.6x10^2$	n.d.
Cr	-"-	-"-	$3.3x10^2$	$9.2x10^2$		n.d.
Fe	-"-	-"-	$1.8x10^2$	$6.5x10^2$		n.d.
Ni	-'-	-"-	8.0	$3.7x10^1$		n.d.
B	B in Si	$6x10^{-6}$	$6.0x10^2$	$5.9x10^3$	$8.8x10^2$	n.d.
Si	-"-	-"-	$1.1x10^3$	$1.3x10^4$		n.d.

*data given are yields (cps/nA) rather than practical sensitivities because steel composition is not specified

**standard operating conditions, except for ^{11}B implanted into Si, where boron yield is optimized and oxygen jet is on

n.d. ... not determined

Obviously, for the pure elements yields under O_2^+ and In^+ bombardment are of comparable magnitude whereas in the multi-element systems and Si the yields and sensitivities under In^+ bombardment are considerably reduced, even when the oxygen jet is on. Fig. 2 shows a depth profile of ^{11}B implanted into Si at 30 keV ($2x10^{16}/cm^2$), obtained with the In^+-EHD source. O_2 was leaked into the sample chamber (oxygen jet) during this measurement to increase secondary ion yields. The raster-scanned area was 200 x 280 μm, the electronic aperture was set to 1/4 of the amplitude. Assuming, that the maximum of $^{11}B^+$ appears at the projected range (R_p(30 keV) = 987 Å), artefacts in the $^{115}In^+$ and $^{29}Si^+$ signals can be identified, the explanation of which is unclear at the moment but may be related to the implantation of In (R_p (9 keV) = 81 Å) into the sample. Note, however that the "surface peak" artefact, usually found in 8 profiles measured under O_2^+ or Ar^+ bombardment is absent. The background level of the order of 10 cts/frame may be due to a superposition of sputter redeposition, memory from previous measurements of 8 - containing samples, detector background (estimated to be ca. 5 cts/frame) and crater edge effects due to a beam "halo"; the In ion gun was not optimized for extreme microfocusing so that, e.g., scattering from electrodes and diaphragms or energy dispersion in the electrostatic beam centering condensers [11] may cause a halo considerably exceeding the beam diameter.

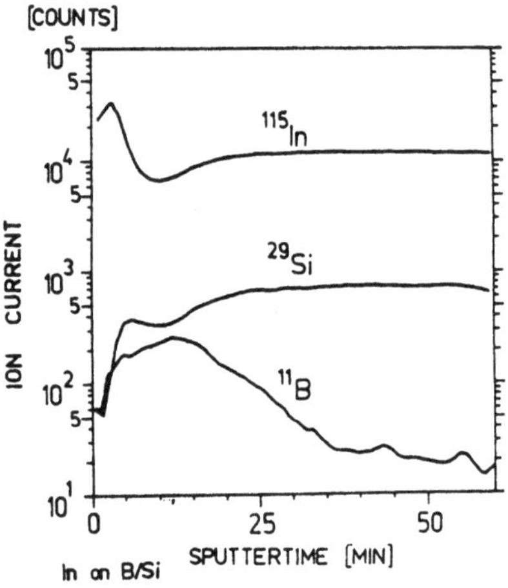

Fig. 2 Depth profile of 30 keV ^{11}B implanted into monocrystalline Si($2x10^{16}/cm^2$), obtained with the In^+-EHD primary ion source. Raster size 180 x 240 $μm^2$; primary ion current 11 nA; oxygen jet on ($6x10^{-6}$ Torr)

Figure 3 shows an example of a 3-dimensional characterization of a microelectronic circuit using the In^+-EHD primary ion gun. The target was continuously raster scanned and digital secondary ion images of the ions $^{27}Al^+$, $^{28}Si^+$, $^{115}In^+$ were stored on-line in the computer using an image raster of 256 x 256 pixels (corresponding to an imaged area of 840x1190 $μm^2$. Successive ion images ("coaxial projections" [13]) of the same element (corresponding to different sputtering depths, see Figs 3a,b,c) were "stacked" to yield 3-dimensional information on the element distribution in a near-surface volume of the

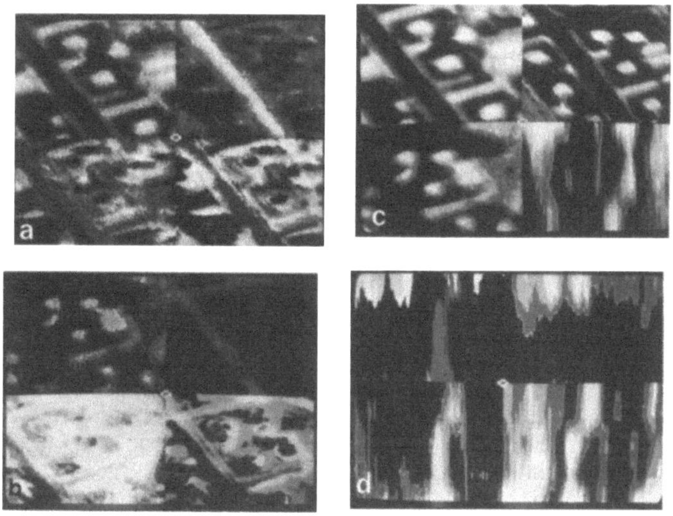

Fig. 3. 3-dimensional characterization of microelectronic circuit; scanned area 840 x 1190 μm^2, digitized into 256 x 256 pixels, linear equidistant intensity levels. 4 subimages of each fig. described in the order (1)=u.l., (2)=u.r., (3)=l.l., (4)=l.r.; l.r. images are superpositions of binary clipped images (1)-(3) with intensity level assignment useful only in pseudocolor representation.
(a): coaxial projections at depth 50 Å; (1) Al, (2) Si, (3) In;
(b) as (a), depth 250 Å;
(c): (1)-(3) coax. proj. of Al^+, 50, 150, 250 Å; (4) transaxial projection through line 120 from top of coax. proj.;
(d): transaxial projections through same line, 0-250 Å; (1) Al^+, (2) Si^+, (3) In^+

sample [13,16]. For each recorded element this 3-dimensional data field, stored in the computer, can be cut with a plane perpendicular to the sample surface and the elemental distribution in this plane can be displayed as a 2-dimensional digital image (Fig. 3d). In such a "transaxial projection" [13] the top line corresponds to the target surface, the bottom line to the maximum sputtering depth (ca. 250 Å in our case); lateral and depth distribution thus are displayed in the same image. Note that the In^+ distribution in coaxial and transaxial projections is by no means homogeneous although the local current density of the In^+ primary beam is constant. Enhanced In^+ emission mainly occurs from the Si areas of the sample (Fig. 3a); matrix effects influencing the degree of ionization and differences in the sputtering yields of Al and Si (with correspondingly different equilibrium concentrations of implanted In) may be responsible for this phenomenon. The current and positional stability of the source and gun obviously is sufficient to record images with good lateral resulution (ca. 10 µm from these images, ca. 5 µm from other parts of the same sample); the correlated noise (20-40 cycles) observable in the high intensity image areas is higher than the counting statistics and can be possibly related to vibrations from the mechanical pumps being transmitted to the liquid metal tip of the EHD source.

References

1. V.E.Krohn and G.R. Ringo, Appl. Phys. Lett. 27, 479 (1975)
2. V.E. Krohn and G.R. Ringo, Rev. Sci. Instrum. 43, 1771 (1972)
3. V.E. Krohn and G.R. Ringo, in "Secondary Ion Mass Spectrometry"; M. Someno and D.B. Wittry, eds., Riaru Kogeisha Ltd., Tokyo (1979)
4. V.E. Krohn and G.R. Ringo, Int. J. Mass Spectrom. Ion Phys. 22, 307 (1976)
5. P.D. Prewett and D.K. Jefferies, Inst. Phys. Conf. Ser. 54, 316 (1980)
6. H. Liebl, in Secondary Ion Mass Spectrometry; SIMS II, ed. by A. Benninghoven, C.A. Evans, Jr., R.A. Powell, R. Shimuzu, H.A. Storms, Springer Series in Chemical Physics, Vol. 9 (Springer, Berlin,Heidelberg,New York 1979) p. 176
7. R. Seliger, J.W. Ward, V. Wang and R.L. Kubena, Appl.Phys. Lett. 34, 310 (1979)
8. R. Seliger, R.L. Kubena, R.D. Olney, J.W. Ward and V. Vang, J. Vac. Sci. Techn. 16, 1610 (1979)
9. R. Ohana, Thesis Doct.-Ing., Univ. Paris-Sud/Orsay (1980)
10. H.A. Storms, K.F. Brown and J.D. Stein, Anal. Chem. 49, 2023 (1977)
11. M.J. Higatsberger, P. Pollinger, H. Studnicka and F.G. Rüdenauer, these Proceedings
12. F.G. Rüdenauer, W. Steiger and U. Kraus, BEDO 11, 17 (1978)
13. F.G. Rüdenauer and W. Steiger, Microchimica Acta (Wien), in print
14. W. Steiger, F.G. Rüdenauer and W. Marschik, to be published
15. H. Gnaser, M. Riedel, J. Marton and F.G. Rüdenauer, these Proceedings
16. P. Braun, F.G. Rüdenauer and F.P. Viehböck, Adv. Electron and Electron Phys. in print

Simple Double-Channel SIMS Instrument

V.T. Cherepin, I.N. Dubinsky and Ya.Ya. Dyad'kin

Institute of Metal Physics, Academy of Sciences of the UkrSSR,
Kiev 142, USSR

Secondary ion mass spectrometers serve currently as rather efficient tools for elemental and isotopic analysis of thin films, diffusion layers and other solids complex in composition and structure. Using a mass analyzer of the monopole type we have designed a compact double-channel secondary ion mass spectrometer. The objective was to produce simple, economical and technologically feasible ion optics with enhanced reliability of power supplies.

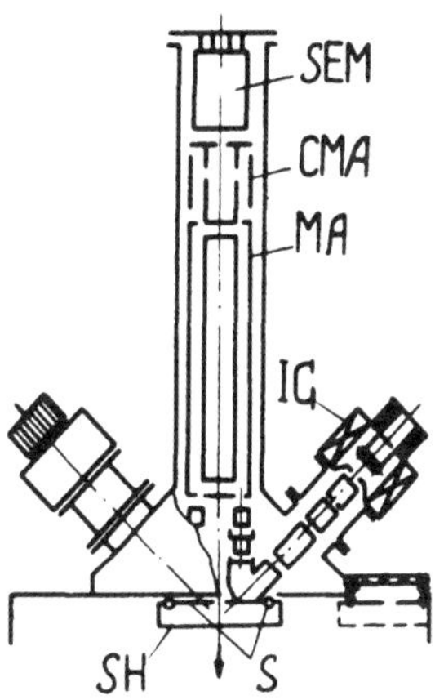

Fig. 1 Double channel SIMS. S - sample; SH - sample holder; IG - ion gun; MA - mass analyzer; CMA - cylindrical mirror analyzer; SEM - secondary electron multiplier

Secondary ions are analyzed in a double-channel monopole mass spectrometer with a grounded square electrode (Fig. 1). The proposed monopole mass spectrometer provides two identical ion-optical channels and thus makes possible the simultaneous analysis of two objects under identical conditions (e.g. the standard and the sample) or fast transition from one species of primary ions to another.

The instrument has increased resolution due to the use of a more precise ratio in the system of the field-forming electrodes "square tube-round rod" (Fig. 2). Theoretical investigation of this system gave the ratio of the round rod diameter R to the distance r_o from the angle vertex to the rod surface: $R = 1.111\ r_o$.

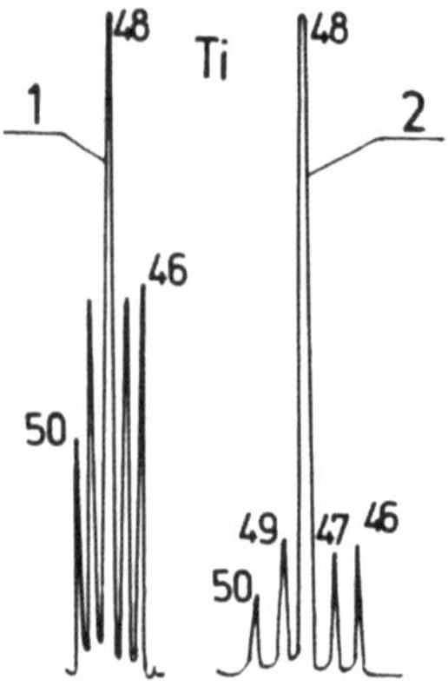

Fig. 2 Mass spectra of Ti^+ in mass analyzers with different values of R/r_o:
1 - R/r_o = 1.143;
2 - R/r_o = 1.111

The mass spectrometer is equipped with two ion sources provided with the focusing optics (Fig. 3). High voltage discharge in the magnetic field is used in the sources thus allowing one to work at 10^{-3} Pa in the discharge chamber and hence to avoid the use of an additional vacuum system for differential pumping.

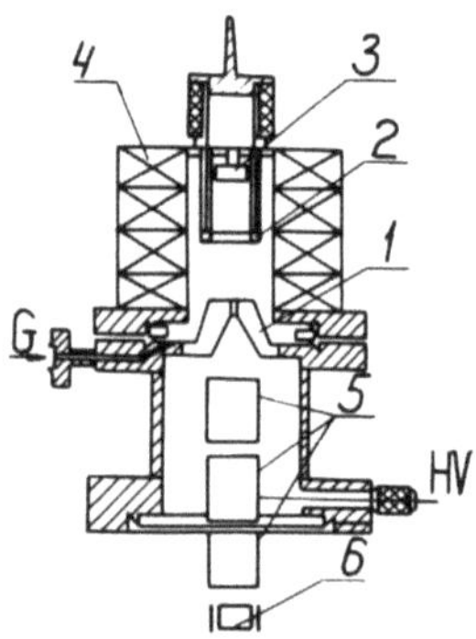

Fig. 3 Ion gun: 1 - cathode and extraction electrode; 2 - anode; 3 - cathode; 4 - magnets; 5 - focusing lens; 6 - deflecting system

The immersion lenses collect the ions to be analyzed and focus them into the entrance slits of the double-channel mass spectrometer. The ions are accelerated to improve the collection efficiency and thereafter they are decelerated to the energy required for optimal mass filter performance. The immersion lens was computer simulated for high energy ion cut off and for the focusing of low energy ions at the entrance slit of the mass analyzer. At the ion energy 10-20 eV the instrument resolution is about 500, at mass number 50. The first electrode of the immersion lens is in mechanical and electrical contact with the sample under investigation. The use of such a system makes it possible to efficiently compensate for the surface charge-up while bombarding the nonconductive samples due to effect of secondary electrons emitted from the walls of electrodes bombarded with sputtered particles and thus allows analyzing different insulators including geological samples.

The instrument is provided with the air-lock for introduction of four samples onto the specimen stage of carussel type. The design of the ion detector system ensures the shifting of the secondary electron multiplier off sight with respect to the axis of the mass analyzer, this being rather helpful in the improvement of signal-to-noise ratio and in attaining sensitivity as high as 10^{-5} at% for some elements. The detector unit is equipped either with the energy analyzer in the form of the half of the cylindrical mirror placed at the mass analyzer exit, or with the ion-electron converter designed for the use of one or two SEM. The analytical part of the device is pumped down with a turbo-molecular pump. The use of oilless pumping leads to longer device service between successive inspections. The r.f. generator of the mass spectrometer voltage is controlled with the scanning system which allows selecting four different parts of the mass spectrum.

The mass spectrum within each chosen part is registered at the preset sensitivity of the secondary ions registration channel in the dynamic range of 10^5, thus making it possible to follow the kinetics of the concentration changes for up to four components of the sample under study.

Main features of the mass spectrometer:

mass range	1-250 amu
resolution for 10 % p.h.	(1.2 - 10)xM
sensitivity	up to 10^{-5} at%
power consumption	800 W

The proposed double-channel mass spectrometer is simple enough in use and may be successfully adopted for both fundamental and experimental investigations to solve various problems in the production technology of different materials and complex structures.

Principles and Applications of a Dual Primary Ion Source and Mass Filter for an Ion Microanalyser

J.J. Le Goux and H.N. Migeon
CAMECA
F-92400 Courbevoie, France

Introduction

It is useful in some cases to filter the primary ion beam particularly when impurities are present in the primary beam. The primary ion beam from the duoplasmatron usually contains traces of non-desired ion species which may come from the feed gas or from the degassing of the source. In order to purify the primary ion beam from a duoplasmatron source, a magnetic mass filter has been designed for the CAMECA IMS 3F ion microanalyser [1]. The features are:

- A symetrical magnetic mass filter which can mount two different ion sources.
- The ability to change rapidly between the duoplasmatron and cesium ion source which allows analysis of electropositive or electronegative elements in the same area.

Description and results

The primary beam mass filter allows use of both ion sources in the same primary ion column. A schema of the system is found in Fig. 1. The ion selection from the two sources is accomplished by electrically reversing the magnetic field and fine tuning the magnet current. The mass resolution can be adjusted by the selection of the slit installed between lenses L1 and L2. The system is designed to provide a stigmatic image of the ion beam. The mass range of the magnetic analyser is from 1 to 150 amu with an ion beam energy range from 5 keV to 20 keV. Mass resolution can be adjusted up to $M/\Delta M = 150$. The spot size on the sample is adjustable from 2 microns to 200 microns with the duoplasmatron source. The filter transmission compared to the standard primary beam column used previously is 60 per cent at a mass resolution of 100. Necessary time to switch from the duoplasmatron source to the cesium source is about 30 min. Less time is needed when switching from cesium to duoplasmatron because less stabilisation time is required.

The mass spectrum of the duoplasmatron used with oxygen shows impurities in the beam (Fig. 2). The mass spectrum of the cesium ion source [2] shows only the $^{133}Cs^+$ ion peak (Fig. 3). Mass filtering of the duoplasmatron oxygen ions is important when profiling nitrogen because this element appears as an ion impurity in the primary beam.

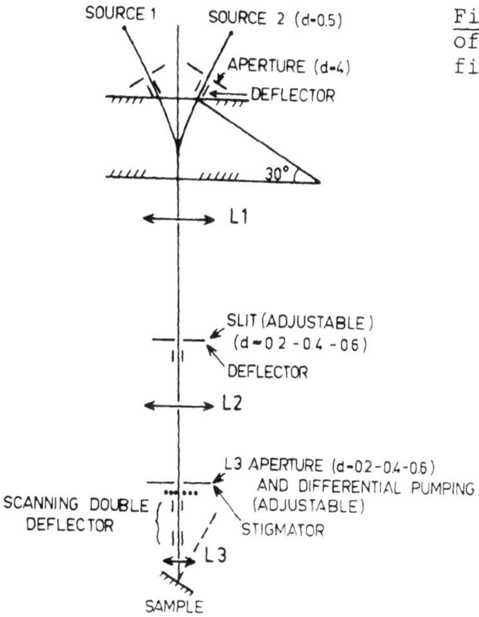

Fig. 1 Schematic diagram of the dual primary beam filter

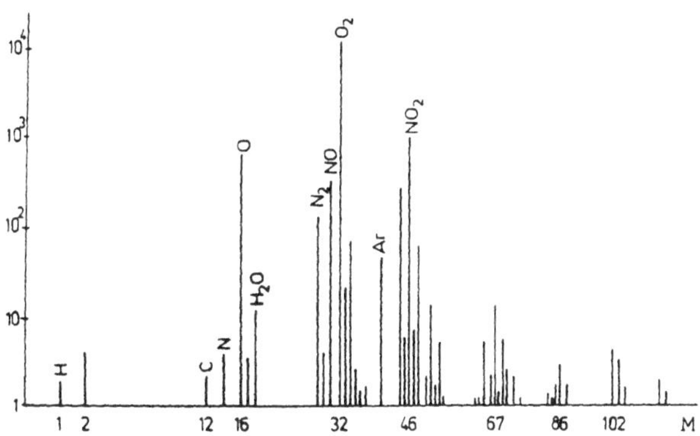

Fig. 2 Mass spectrum of primary beam from duoplasmatron FED with oxygen

The concentration profile of nitrogen implanted in a silicon oxyde layer measured with the mass filter is shown on Fig. 4. A resolving power of $M/\Delta M$ - 2000 is used to eliminate the silicon doubly charged from mass 14. Such a profile cannot be measured with an IMS 3F not equipped with the primary mass filter when bombarding with duoplasmatron oxygen ions. The same can be analyzed with the caesium source and the duoplasmatron is shown on Fig. 5. Specimen was steel containing 0.1 per cent of carbon and 0.005 per cent of oxygen treated with lithium at a high temperature. Picture of 7 Li^+ is obtained

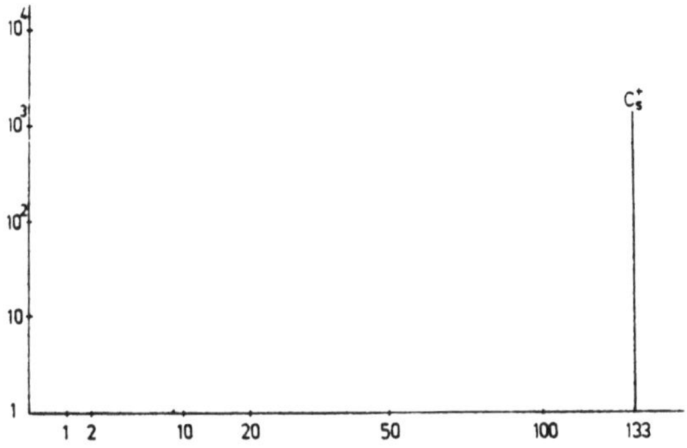

Fig. 3 Mass spectrum of primary beam from caesium source

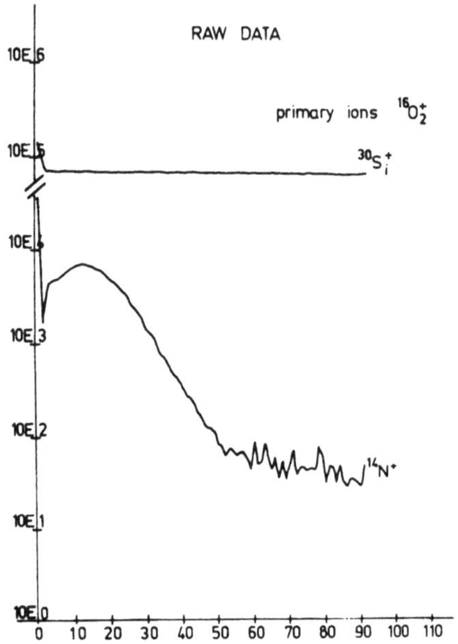

Fig. 4 Depth profile of nitrogen in silicon oxide obtained with filtered oxygen primary beam

with O_2^+ primary ions and pictures of $^{16}O^-$ and $^{12}C^-$ are obtained with $^{133}Cs^+$. Both of them correspond to the same area.

Another demonstration using the mass filter and corresponding to a boron in depth profile in silicon does not give us the expected improvement.

Up to now the dynamic range of the IMS 3F was limited to a 5.5 decade dynamic range of concentration with boron. The same dynamic range is obtained with the mass filter demonstrating that background was not produced by neutral atom bombardment. We conclude that the main problem of the limitation of dynamic range was the memory effect of the immersion lens front face. The memory effect has been found independent of the sputtering rate when primary ion current remains constant so a better signal to background ratio is obtained with increased sputtering rate in this region.

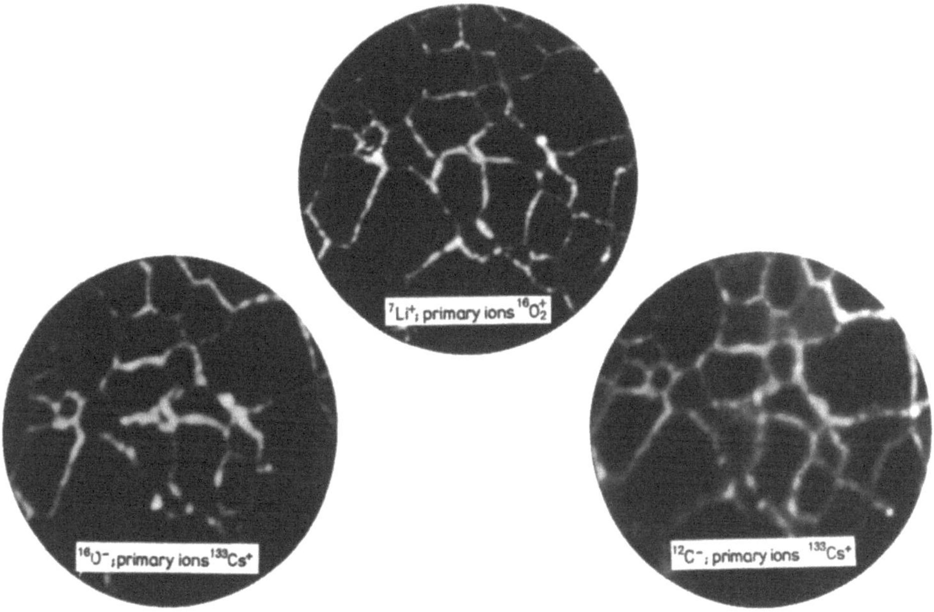

Fig. 5 Ions images from the same the duoplasmatron and the caesium source in a steel obtained successively with

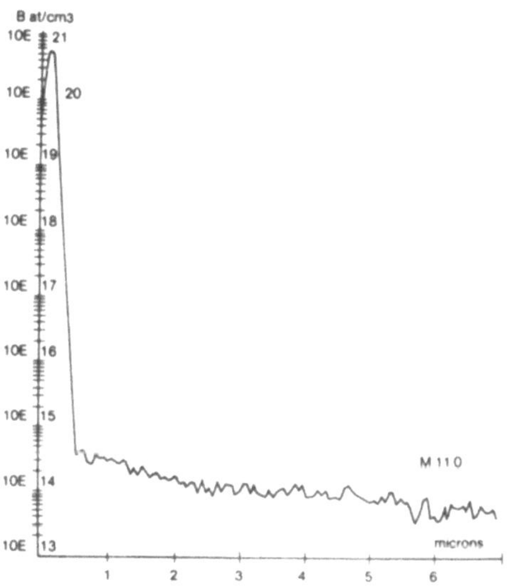

Fig. 6 Boron depth profile with increased sputtering rate in the tail region

Analysis using an increased sputtering rate in the tail region of the profile has given a 7 decade dynamic range on Fig. 6. The software used for these depth profiles normalizes the boron intensity to the silicon signal and corrects the depth scale according to the sputtering rate.

Conclusion

Mass filtering of the duoplasmatron ion beam improves the detection limit on elements corresponding to impurities present in the primary ion beam. This was seen with the nitrogen example. The ability to switch from one bombarding species to the other in order to obtain the best ionization conditions for electropositive and electronegative elements in the sample has been shown to save a considerable amount of time.

References

1. J.M. Gourgout, in Secondary Ion Mass Spectrometry II, ed. by A. Benninghoven, C.A. Evans, Jr., R.A. Powell, R. Shimuzu, H.A. Storms, Springer Series in Chemical Physics, Vol. 9 (Springer, Berlin,Heidelberg,New York 1979) p. 286-290
2. P. Williams, R.K. Lewis, C.A. Evans and P.R. Hanley, Anal. Chem. 49, 1499 (1977)

A Quadrupole Mass Spectrometer with Energy Filtering for SIMS Studies

R.-L. Inglebert and J.-F. Hennequin

C.N.R.S., Laboratoire P.M.T.M., Université Paris-Nord,
F-93430 - Villetaneuse, France

In the last few years, the quadrupole mass filter has been widely used as mass analyser in secondary ion mass spectrometry. An energy pre-filter is generally added to reduce the background intensity and to reject the higher energy ions which degrade the mass resolution. As a very high energy resolution is not useful for such studies, the retarding-accelerating energy analyser [1,2] is very interesting because of its large angular acceptance, and is now effectively used in some modern apparatus [3,4] in spite of a rather poor knowledge of the energy transmission properties of such spectrometers.

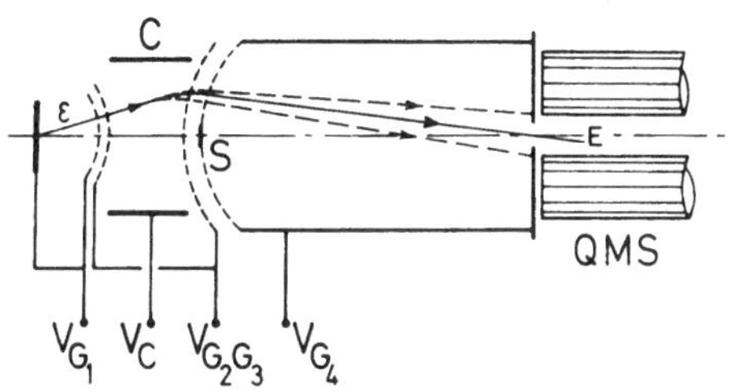

Fig. 1 Schematic view of the energy filter

Our spectrometer associates a quadrupole mass filter (RIBER model Q 156) and a pass-band energy filter with four spherical grids (Fig. 1), derived from the STAIB type energy analyser [2]. The first set of concentric grids G_1-G_2 is a retarding field arrangement and acts as a high-pass pre-filter. With an accelerating field, it also allows post-acceleration of slower ions. The second set of grids G_3-G_4 is an accelerating one, which focuses the transmitted ions into the quadrupole entrance aperture situated beyond the common centre of both grids. An energy band can thus be selected by focal separation [5]: the higher energy ions are insufficiently focused and the lower energy ions much too focused. Both ion species miss the aperture, the ions introduced near the axis being eliminated by the central screen S. The cylindrical electrode C is biased at a potential slightly higher than the common potential of G_2 and G_3; it focuses the ions emitted with a large angle from the axis and allows a slight adjustment of the band of the transmitted energies [6]. As the central screen prevents a line-of-sight path from the emitting target to the detector, the simultaneously emitted photons and excited neutral atoms cannot reach the detector and the signal-to-noise ratio is thus improved.

For an ionic species emitted with the given initial energy ε, the overall transmission $\vartheta(\varepsilon)$ of a mass spectrometer is defined as the part of such secondary ions which effectively reaches the detector [7]. It can be deduced from the elementary transmission of a narrow ion beam for each initial direction by integrating over the whole half-space in front of the sputtered sample with the assumption of a cosine law for the angular distribution of secondary ions. The elementary transmission is experimentally obtained by means of a small alkali ion source movable under vacuum in front of the entire spectrometer. The technique was previously described for the quadrupole filter alone [8] and a good agreement was found between experimental results and calculations by DAWSON's theory [9].

The overall transmission of a spectrometer can thus be theoretically predicted if the optical characteristics of the energy filter are known, as explained in more detail elsewhere [10]. In the present case, the potential distribution in the space between the grids G_2 and G_3 is obtained by numerical integration of the Laplace equation, and the ion trajectories through the energy filter are then deduced with a central electric field between G_3 and G_4. Such calculations give a value nearly equal to 1/2 for the angular magnification of the system. Taking into account the quadrupole acceptance, the secondary ions emitted from the centre of the sputtered area can be collected till an angle of nearly 25° from the axis, a value experimentally verified for a half-height mass resolution R = 55. The effective radius of the emitting area would then be lower than 0.5 mm.

The mean transmitted energy $\bar{E}$ and the band width ΔE of the energy filter depend on two parameters only: $V_C-V_{G_2G_3}$ and $V_{G_2}-V_{G_4}$. On the other hand, the overall transmission of the quadrupole mass filter shows a flat maximum when the ion velocity at the entrance into the quadrupolar field lies between 0.5 and 1 r_o/RF cycle [9,10]. The mean transmitted energy is therefore roughly determined for each ionic species (between 6 and 22 eV for Na^+ ions in our case), and the potential drop $V_{G_1}-V_{G_4}$ would be taken equal to $\bar{E}-\varepsilon$, if V_{G_4} is equal to the potential of the axis of the quadrupolar field, taken as potential origin. An energy analysis of secondary ions can thus be obtained by scanning V_{G_1} only [4], with an energy resolution ΔE constant on the whole spectrum and slightly adjustable by the value given to V_C.

Some energy transmission curves of the whole spectrometer are shown in Fig. 2, for $V_{G_2G_3}$ = 17 V, V_{G_4} = 0 V and for different values of $V_C-V_{G_2}V_{G_3}$. The experimental points are obtained with Na^+ ions emitted from an alkali ion source at a variable initial energy. The theoretical curves are calculated according to [10] and fitted at the maximum of transmission for $V_C-V_{G_2G_3}$ = 3 V, as the absolute value of the overall transmission is rather difficult to measure with enough accuracy.

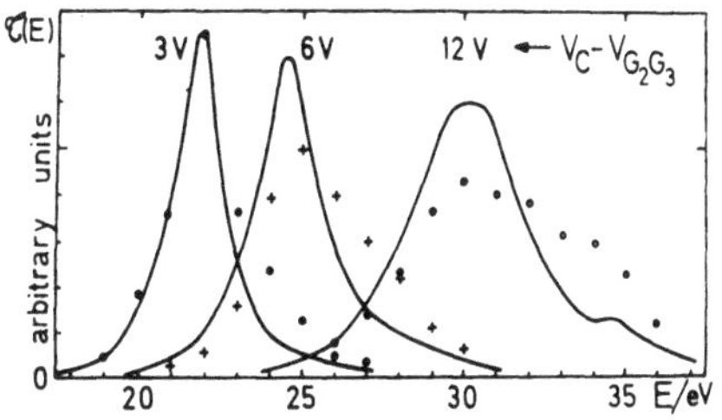

Fig. 2 Energy transmission of the spectrometer (E is the ion energy at the entrance into the quadrupolar field; ——: theoretical curve; • + O: experimental points with Na^+ ions)

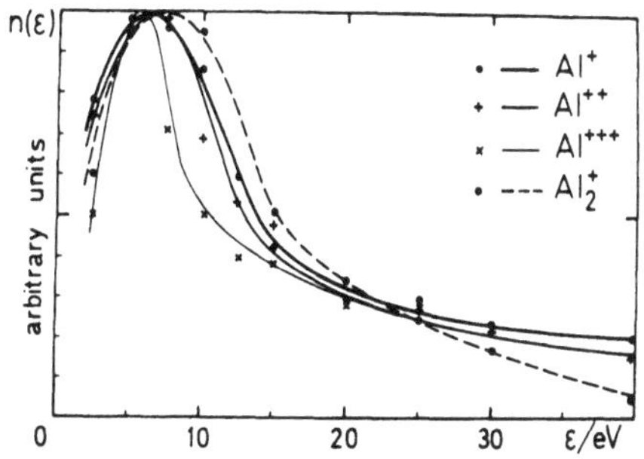

Fig. 3 Energy spectra of secondary ions sputtered from a pure aluminium sample (10 keV Ar^+ ions; $\Delta E \simeq 2.5$ eV)

The experimental transmission band is somewhat smoothed at higher energies, probably because of ion scattering on the grids. The band of transmitted initial energies can of course be translated by acting on V_{G_1}. If negative values are given to V_{G_4}, faster ions are collected and the energy band width is increased, but the retarding field near the hole at the entry into the quadrupole gives rise to an optical effect rather difficult to be theoretically taken into account.

For a first experiment with secondary ions, a pure aluminium sample was sputtered with 10 keV Ar^+ ions in ultra-high vacuum (residual gas pressure ca. 10^{-9} Torr). Fig. 3 shows some preliminary energy spectra, in good agreement with other similar studies [11, 12]. In particular, the peak position for Al_2^+ ions appears at higher energies than for Al^+ ions, but their energy spread is much smaller. Beyond some ten eV, the energy distributions are approximately proportional to ε^{-n}, with the values: $n \simeq 1.2$ for Al^+, $n \simeq 2$ for Al_2^+ and $n \simeq 0.9$ for Al^{++} and Al^{+++}

References

1. D.A. Huchital and J.D. Rigden, Appl. Phys. Lett. 16, 348 (1970); J. Appl. Phys. 43, 2291 (1972)
2. P. Staib, J. Phys. E: Sci. Instrum. 5, 484 (1972); Vacuum 22, 481 (1972)
3. P.H. Dawson and P.A. Redhead, Rev. Sci. Instrum. 48, 159 (1977)
4. A.R. Krauss and D.M. Gruen, Appl. Phys. 14, 89 (1977)
5. J.-F. Hennequin, Int. J. Mass Spectrom. Ion Phys. 19, 914 (1977)
6. P. Staib and U. Dinklage, J. Phys. E: Sci. Instrum. 10, 914 (1977)
7. R.L. Inglebert, J.-F. Henneguin: In Secondary Ion Mass Spectrometry, II, ed. by A. Benninghoven, C.A. Evans, Jr., R.A. Powell, R. Shiumuzu, H.A. Storms, Springer Series in Chemical Physics, Vol. 9 (Springer, Berlin,Heidelberg,New York 1979) p. 192, Adv. Mass Spectrom. 8, 1764 (1980)

8. J.-F. Hennequin and R.-L. Inglebert, Int. J. Mass Spectrometry Ion Phys. 16, 141 (1978); Rev. Phys. Appl. 14, 275 (1979)
9. P.H. Dawson, Int. J. Mass Spectrom. Ion Phys. 17, 423 (1975); Quadrupole Mass Spectrometry and its Applications (Elsevier, Amsterdam 1976)
10. R.-L. Inglebert and J.-F. Hennequin, Rev. Phys. Appl. 15, 1489 (1980)
11. G. Blaise, in "Material Characterization Using Ion Beams", ed. by J.-P. Thomas and A. Cachard (Plenum Press, New York 1978) p. 143
12. K. Wittmaack, Surf. Sci. 53, 626 (1975)

Development and Operation of Special SIMS-Equipment for Use in Iron and Steel Analysis

J. Dittmann

Thyssen Aktiengesellschaft vorm. August Thyssen-Hütte,
Research Laboratories,
D-4100 Duisburg, Fed. Rep. of Germany

Analysis of steel surfaces and in-depth profiling of the layer structure of different materials by SIMS often implies some compromises with the commercial instruments available today Ion microprobes with high sputtering rates and ion microscopes with similar possibilities are very successful in describing the distribution of phases and elements in micro-range, but they have no representative analysis area, most needed for technical information on products. Ion-scanning instruments with a larger area of analysis are often restricted by these conditions in fast sputtering in-depth.

To analyse technical products - especially flat products - very often both the parameters must be combined and there is no need for an ion picture of distribution. Therefore we have developed special SIMS equipment in the Research Laboratories of Thyssen for this purpose in iron and steel analysis. The instrument is a combination of a high dose primary ion gun, a target chamber with double vacuum pumping system and the SIMS-typical components including on-line computer system. Fig. 1 shows the scheme of the target chamber. The target lock-in and -out is constructed as a one-way transportation of the target. The target just transported into the chamber pushes out the sample analysed before. This can be realised without breaking vacuum in the chamber with help of the second turbo-system which operates the lock-in-out system. Transportation is realised by tooth-rack and rotation feed-through. Vacuum in the system is at $\sim 4 \cdot 10^{-5}$ Pa. High-dose primaries are produced by a freon-cooled duoplasmatron source from General Ionex with a modified lens system. We work with primary argon ions in the energy range from 2-15 keV. Gas inlet is controlled with an electronic thermo leak valve. The Einzel-lens system now can focus the beam on areas from ~ 7 mm^2 up to more than 1.5 cm^2. The normal distance from source to focal plane was altered by a distance ring. Primary ion intensity is variable up to some mA/cm^2, the result is a sputter rate between ~3 nm/min and ~1 μm/min, measured by weight difference experiments. The energy filter is a condenser [1] with modified entrance slit, to suppress background. This system is followed by an Extranuclear Elfs-quadrupole with channeltron multiplier. Fig. 2 shows the target chamber with an ion source and the first turbomolecular pump.

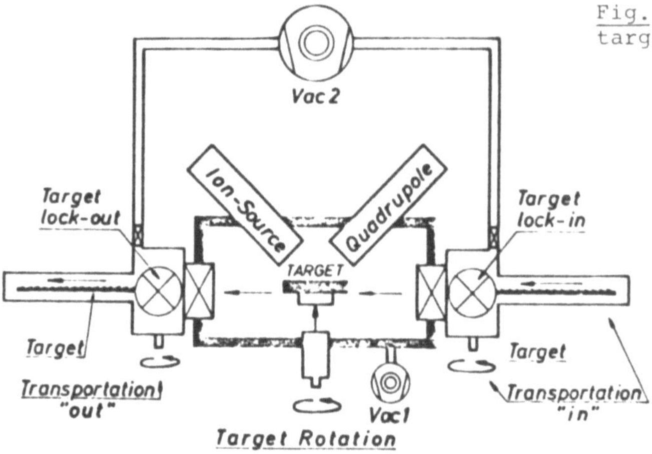

Fig. 1 Scheme of the target chamber

Fig. 2 Target chamber with transporting system, ion source and first turbomolecular pump

The target has an inclimation of 45 degrees to the primary beam and quadrupole. Directly in front of the chamber there is a magnetically coupled rotation feed-through. This allows the rotation of the target round the normal direction of target surface. Rotating the sample while bombarding is a way to minimize the result of crystallographic influence on sputtering polycrystalline material and the formation of large cones [2]. The velocity of rotation can be controlled by a stepping motor in relation to beam intensity. Besides the feed-through one can see the Faraday cup, movable with a pneumatic bellow to measure the intensity of primary ions at the target plane.

The instrument involves a DEC computer PDP 11/03 with a floppy disc. On-line coupling with a PDP 11/40 of the laboratory computer system is provided. In the momentary state all data of the ion source can be registered on-line with special interfaces. Due to the high potential of the source the data have to be acquired via an opto-electronic coupler.

With this instrument we have recorded until now nearly 25000 spectra of steel targets, most of them from rolled and coated steels. The questions for the chemical situation - elements and compounds - of surface or of the interlayers in the depth are coming from the steel plant, from rolling mills, from divisions of the research laboratories and from custromers of the material. So often the problems are failures and claims. Some examples may illustrate problems and qualitative results. The first step is in all cases a correct sampling procedure; this provides that the customer has sufficient knowledge about the analytical method. The second step is the necessity to get all the information on the special technical process to be able to differ between relevant results combined with it and results of no significance. Very important is the problem of preparation. We found it best when preparation is done or controlled by those working analytical who best know the risk of contamination from oil, dust, fingerprints, etc.

In the positive spectrum of a zinc-coated steel sheet passivated by chromate ions we found, beside a number of well-known ions from hydrogen to zinc, the group 80-85 amu. Typically such ions are only to be seen in limited areas darker in colour than the typical grey colour of the sheet. The ions are identified to come from a coincidence of ZnO- and ZnOH-ions, a result of intensive chemical reactions of the surface with the solutions of the process. In Fig. 3 one can see the relative intensities, calculated and measured. In a great number of examples we found the presence of hydroxyle-containing ions to be significant. In rolling and reactive surface treatment fluids are in contact with the metallic or oxidic surfaces often at high temperatures. With help of a computer program we calculate the relative intensities of any molecular ion group. On storage are the data of all isotopes of the periodic system. Any combination can be called up by teletype with element symbols. We now have a compilation of a thousand such diagrams.
Fig. 4 shows the ions Zn-Fe-O, which can occur in the small intermetallic alloy layer of hot dip zinc-coated steels. The

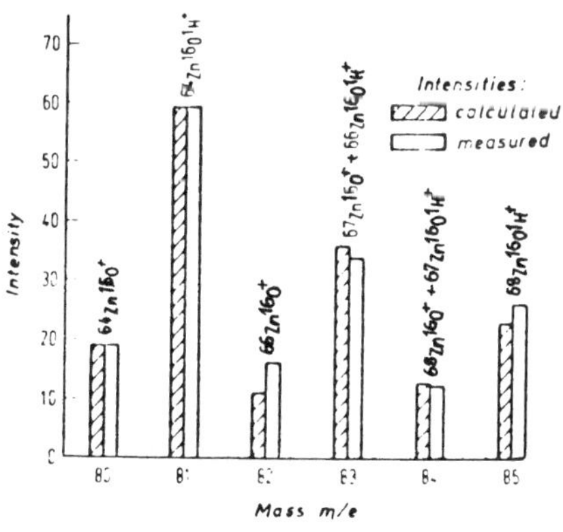

Fig. 3 Measured and calculated intensities of ZnO-ZnOH ions from a zinc-coated steel surface

```
FORM:   ZN-FE-O -

MASS:      134       135       136       137       138       139       140
RFL%:   2.8385    0.0011   46.3249    1.3241   26.7634    4.3784   17.2537
MAX%:   6.1274    0.0023  100.0000    2.8583   57.7733    8.4515   37.2448

MASS:      141       142       143       144       145       146
REL%:   0.4346    0.6634    0.0146    0.0033    0.0000    0.0000
MAX%:   0.9382    1.4320    0.0316    0.0072    0.0001    0.0000


FORM:      ZN-FE-O -

MASS  REL 0    1    2    3    4    5    6    7    8    9    0
          02468024680246802468024680246802468024680246802468 0

134   2.8 **
135   0.0 I
136  46.3 *************************
137   1.3 *
138  26.8 ***************
139   4.4 ***
140  17.3 **********
141   0.4 I.
142   0.7 *
143   0.0 I
144   0.0 I
145   0.0 I
146   0.0 I

MASS  REL 0    1    2    3    4    5    6    7    8    9    0
          02468024680246802468024680246802468024680246802468 0
```

Fig. 4
Computer diagram of ions Zn-Fe-O

identification of coincidences often implies difficulties, with reactive elements of the target forming "synthetic" secondary ions [3].

Thin interlayers in a depth of 20 μm or more often cannot be found in a cross section by ion microprobe, but can be analysed with greater analysis area and high sputtering. Fig. 5 shows the enrichment of oxygen in the depth of an aluminium coated steel. The failure at the beginning of production was a bad adhesion of aluminium. Oxygen intensities in repetitive spectra and continuous registration show the enrichment in a depth of ~20 μm, the reason of replacement.

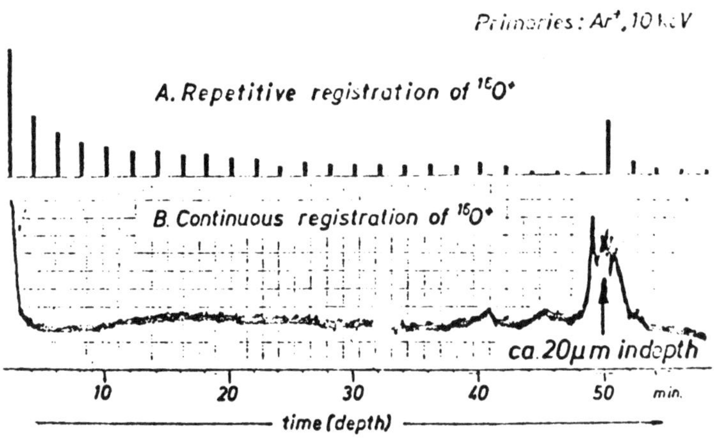

Fig. 5 Al-coated steel sheet with failure at the beginning of production: Oxygen-enrichment ~20 μm indepth

In the rough reality of industrial application, important factors for the technical usefulness are the time needed to get results and to realise an understandable form of the complex answers. A condensed result today is often much better than a very sophisticated paper with the state of the art some weeks later. We must balance between these two extremes.

References

1. K.Wittmaack, J. Maul and F. Schulz, Int. J. Mass Spectrom. Ion Phys. 11, 23 (1973)
2. J. Dittmann, Microchim.Acta, Suppl. 5, 411 (1974)
3. J. Dittmann, Microchim.Acta, Suppl. 5, 359 (1975)

Design Concept of a New Secondary Ion Optics System for Use with Quadrupole Mass Spectrometers

R. Jede, O. Ganschow, and A. Benninghoven

Physikalisches Institut, Universität Münster,
D-4400 Münster, Fed. Rep. of Germany

1. Introduction

Most commercially available SIMS optics systems for use with quadrupole mass spectrometers suffer from the problems caused by mixing the parameters governing secondary ion emission with those related to the ion injection into the mass filter. In this paper, we propose a system which overcomes these difficulties. They have been partly solved by a previous design [1] used for a dedicated SIMS instrument, which therefore has been a guideline for the present work. Because of the intended use in combination with XPS, this instrument is required to have a very large lateral acceptance of 10 mm diameter. In experiments where static conditions must be applied, this large acceptance allows reduced current density at a given sensitivity. On the other hand, in depth profiling the detection limits are improved at a given erosion rate. Additional requirements to be met have been energy filtering and reduction of the quadrupole mass discriminating effects. A detailed discussion of this system and its performance will be published elsewhere.

Figure 1 shows the different ion optical elements. The theory of operation is based upon the following considerations: The extraction field and the energy filter of CMA type have to account for the secondary ion emittance of the target at fixed primary beam position, whereas the front lens and the dynamical emittance matching (DEM) compensate for varying beam positions during raster scanning without energy dispersion and virtually without spherical aberration. The lens (4) matches the secondary ion beam to the energy filter acceptance. The injection lens (6) has to compensate the mass number dependence of the quadrupole acceptance. Finally, the ions leaving the quadrupole mass filter are collected by the immersion lens (8), pass through a 90° electrostatic deflector, where electrons are suppressed magnetically, thus giving enhanced signal-to-background ratio.

2. Extraction Field Geometry, Front Lens and DEM

In order to optimize the extraction field geometry, which primarily determines the ion optical quality of the secondary beam, different arrangements have been studied. For an estimation of their collection efficiency as a function of lateral

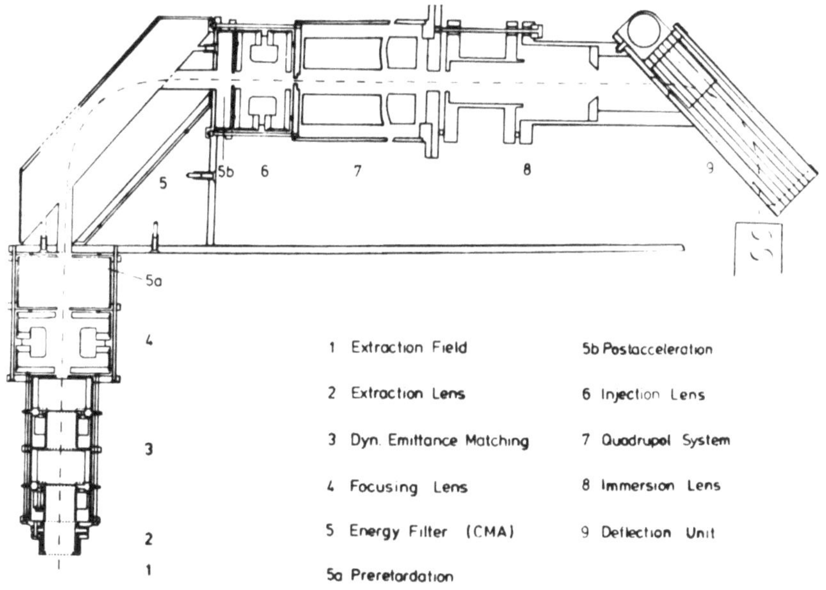

Fig. 1 Overall schema of the secondary ion optics system

acceptance and secondary ion energy, a cosine angular distribution of the emitted secondary ions has been assumed. Despite the fact that other geometries would have allowed for perpendicular incidence of the primary beam, which is preferable from the point of view of shadowing and roughening of the surfaces, the parallel plate condenser has been chosen. As compared to other arrangements, the main advantages which it offers to the intended system are the independence of secondary ion collection efficiency from the position of the emitting area, and the one-to-one correspondence between the location of the entrance to the optics and the position of the emitting spot. These factors are specifically important with regard to the DEM.

A grid is used as an entrance plate so that memory effects during prolonged sputtering, which are common to most systems, are greatly reduced. Furthermore, charged background particles created at the grid can be easily discriminated in the energy filter, since their potential of origin differs from the target potential. The same argument of course applies to all subsequent grids, so that the dynamic ratio does not deteriorate. Typical parameters of operation are a distrance from the sample to the extraction grid of 15 mm, and an extraction potential of 100-300 V

Because of restrictions in the available diameter, special care has to be taken to ensure that the acceptance of the front lens is matched to the desired 10 mm diameter field of view at the sample surface. This implies that the front lens must handle a secondary ion beam of typically 15 mm diameter and 16° full divergence angle at the entrance grid. This is achieved by a lens which consists of the entrance and exit grids and an intermediate cylinder. The lens as a whole acts like an immersion lens. It has been numerically optimized with respect to the diameter-to-length ratio and the sequence of potentials. It renders a secondary beam which is inclined to the axis and slightly convergent in itself. The collection efficiency of the front lens has been calculated at optimum operation parameters for a cosine angular distribution of secondary ions. At 200 V extraction potential and 15 mm distance from sample to optics, it is in the range of 0.6-0.4 for 2-20 eV ions. Thus, energy filtering remains feasible. Since the acceptance to diameter ratio of the front lens is fairly large, one cannot expect its action to be Gaussian at all. In this situation, LIEBL's proposal [2] of a dynamic emittance matching system (consisting of a single quadrupole deflection condenser centered at the point of intersection of the central secondary ion trajector and the optical axis) is not likely to work successfully.

Therefore a different DEM system has been designed, which consists of two orthogonal pairs of deflection condensers, as indicated in Fig. 2a. The deflection voltages are fed to the second condenser of each pair with reversed polarity. Thus for each direction, the first condenser bends the secondary ion trajectories towards the axis, and the second one back onto the axis. The condensers for one direction act as drift spaces for the orthogonal direction. Because the secondary beam is injected

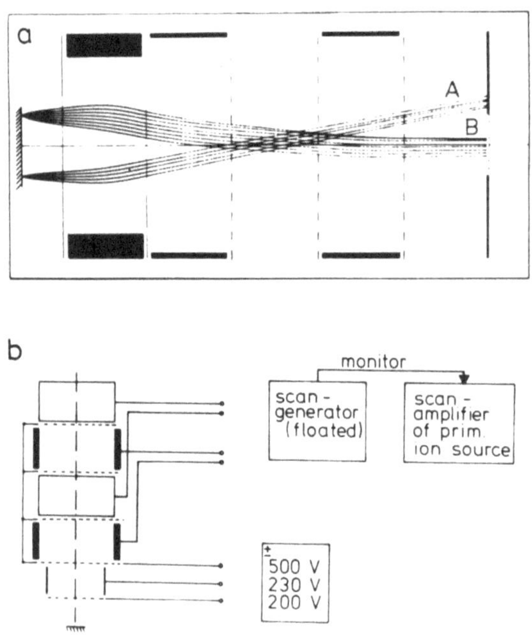

Fig. 2 Principle of operation of the dynamic emittance matching (DEM). (a) Ion trajectories originating from an off-axis emission spot (A) without, (B) with DEM; (b) Potential scheme for the DEM

through the exit grid of the front lens, which is typically kept at 500 V, the deflection voltage supply is floated to this potential, as is indicated in Fig. 2b. In order to avoid fringe field distortions, in particular energy dispersion, the different condensers are separated by grids which are electrically connected to the exit grid of the front lens. The geometrical transmissivity of each grid is 0.94; additionally they are aligned. The deflection voltage amplitudes need not be larger than 150 V. Their polarity is changed according to secondary ion polarity.

The raster scan generator of the secondary ion optics controls acts as a master to the primary beam deflection control, thus giving synchronous operation. In order to achieve optimum coincidence of the primary and secondary scanning fields, some additional features have been incorporated. The primary beam scan center can be adjusted independently for each extraction field (i.e. secondary ion) polarity. For each polarity, the scan amplitude ratio can be adjusted according to flange position and deflection sensitivity. Additionally, the primary beam scanning area, which is slightly trapezoidal, can be rotated by appropriate adjustments of the primary deflection voltage supply. The entire system gives a slightly convergent secondary beam with a typical diameter of 8 mm at the entrance of lens (4) (Fig. 1).

3. Energy Filtering

The energy filter is made as a section of a cylindrical mirror analyzer, which has an outer diameter of 100 mm and is operated in the $\vartheta = 42.3^{o}$ second order focussing condition. In the region where neutrals and high energetic ions and electrons would hit the outer cylinder wall, it is replaced by grids in order to reduce the background intensity. The fringe fields are corrected by semicircular wires kept at the appropriate potentials. A pre-retardation facility is provided (5a) for adjusting the energy resolution. Therefore the focussing lens (4) is required to ensure constant illumination. Additionally, a variable aperture is provided at the entrance to the post-acceleration stage (5b). Because the theory of CMA operation is well known [3-6], further details are not discussed here.

4. Quadrupole Mass Filter Acceptance Matching by Mass Dependent Ion Injection

After passing the post-acceleration stage, the secondary beam may have an energy of up to 500 eV (limited by the maximum allowable field axis potential). Lens (6) is then used for matching the secondary beam to the quadrupole acceptance. As is well known from the theory of operation of quadrupole mass filters [7], their acceptance fulfils

$$r_{max} \sim M^{-1/2}, \qquad \dot{r}_{max} \sim M^{1/2}$$

if operated in the Δm = const. mode at fixed ion energy. In order to achieve this, the lens potential is swept with ion mass as shown in Fig. 3. Because of the particular lens geometry chosen [8-9], the function U(m) required for acceptance matching behaves approximately like $m^{1/2}$, if the distances to the source

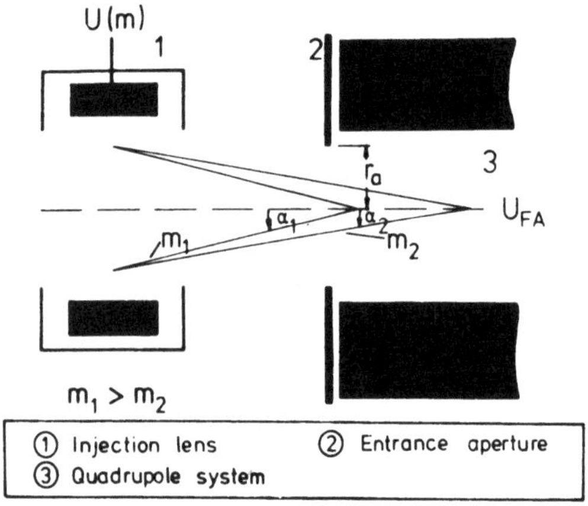

Fig.3 Schema of the mass-dependent ion injection into the mass filter

point and the quadrupole entrance aperture are suitably chosen. Nevertheless the concept requires to have a quadrupole field as large as possible. So the Extranuclear Model 324/9 mass filter with 19 mm rod diameter and ELFS is used. The retardation of the beam to an energy level allowable from the point of view of mass resolution is achieved by a suitable setting of the post-acceleration to quadrupole field voltage drop. This concept reduces quadrupole fringe field problems substantially.

5. Extraction from the Quadrupole Mass Filter

Because of the time and mass dependence of the electrostatic field of the mass filter and the additional influence of the entrance and exit fringe fields, it is difficult to get an estimate of its defocusing action. Therefore, as shown in Fig. 4, an immersion lens has been incorporated which is operated such as to focus the secondary ions at approximately one third of the trajectory length inside the parallel plate condenser. By numerical analysis it has been found that this is the most favourable geometry when accounting for the drift space action of the condenser in the direction parallel to the plates.

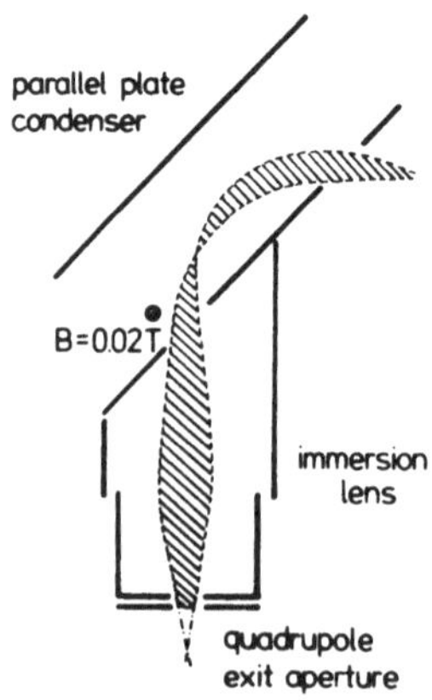

Fig. 4 Schema of an ion trajectories of the extraction system

In order to remove background particles originating from the mass filter, the outer plate has been designed as a grid. A permanent magnet has been added with its field direction perpendicular to the condenser field in order to remove the electron background, which cannot be fully eliminated by the previous optical elements in the case of negative SIMS [10]. The focussing properties of the extraction arrangement have been checked by using secondary electron multipliers with different active areas. No difference in the transmission could be observed. As compared experimentally to the conventional setup which simply uses an off-axis multiplier, the new extraction and deflection system gives an increase in signal strength by a factor of 50 and additionally in signal-to-background ratio by a factor of 2 [11].

References

1. C.W. Maggee, W.L. Harrington and R.E. Honig, Rev. Sci. Instrum. 49, 4 (1978)
2. H. Liebl, Low Energy Ion Beams 1977, Inst. Phys. Conf. Ser. No.38, p.266 (1978)
3. V.V. Zashkvara, M.I. Korsunskii and O.S. Kosmachev, Sov. Phys. Tech. Phys. 11, 96 (1966)
4. H.Z. Sar-El, Rev. Sci. Instrum. 38, 1210 (1967)
5. P.W. Palmberg, G.K. Bohn and J.C. Tracy, Appl. Phys. Letters 15, 254 (1969)
6. J.E. Draper and C. Lee, Rev. Sci. Instrum. 48, 852 (1977)
7. W. Paul, H.P. Reinhard and U.V. Zahn, Z. Phys. 152, 143 (1958)
8. W. Lippert and W. Pohlit, Optik 9, 456 (1952); 10, 477 (1953); 11, 181 (1954)
9. F. Heise, Optik 4, 203 (1949)
10. A. Benninghoven, C. Plog and N. Treitz, Int. J. Mass. Spectrom. Ion Phys. 13, 415 (1974)
11. R. Jede, Thesis, Münster 1981

Improved Analysis of Insulators in an ARL IMMA Using Positive Primary Ions and an Electron Gun

J.D. Brown

Faculty of Engineering Science and Centre for Interdisciplinary Studies in Chemical Physics. The University of Western Ontario, London, Canada

D.J. Gras

Metaalinstitut TNO, Apeldoorn, The Netherlands

Introduction

When analyzing insulating specimens by secondary ion mass spectrometry (SIMS) using positive primary ions as the bombarding species, a surface potential develops which partially or completely inhibits the extraction of ions into the secondary mass spectrometer. The result is that depth profiles, if measured intensities are observed at all, can be severely distorted, particularly when changes in composition lead to changes in conductivity of the specimen. Therefore surface charging limits the reliability of SIMS in analyzing insulators.

A number of solutions to the surface charging phenomena have been proposed. ANDERSEN, RODEN and ROBINSON [1] have recommended the use of $^{16}O^-$ for analysis of insulators such as minerals in combination with very small specimens or a thin conducting surface layer. However, the intensity of $^{16}O^-$ from a typical duoplasmatron is much less than for positive ions, and in analysis of $^{18}O/^{16}O$ ratios, where specimens are oxidized in ^{18}O enriched oxygen to study oxidation mechanism [2], such a solution is not possible. WERNER and MORGAN [3], using a CAMECA IMS 300, have recommended the use of metallic diaphragm to overcome the effects of charging under most circumstances. However, the use of such a diaphragm limits the area available for analysis.

A number of authors [4-7] have used a low energy electron gun to supply electrons coincident with the ion beam to neutralize charge. These guns have been installed in a variety of instruments. As it has been pointed out by several authors [3,5,6], complete charge neutralization is not always possible but use of such a gun can improve the reliability of SIMS data in many circumstances. This paper describes the installation of an electron gun in an ARL ion microprobe mass analyzer (IMMA).

Installation

The electron gun (Fig. 1) is a miniature gun from an oscilloscope tube modified by Philips Research Laboratories for our purpose. The gun has focussing and stigmator electrodes as well as deflection plates for electrostatic deflection which can be seen projecting from the glass envelope in Fig. 1. The gun was mounted in an aluminium bracket suspended from the lower support arm for the primary beam column (Fig. 2). The various voltages required for the gun were brought through an existing 21 pin

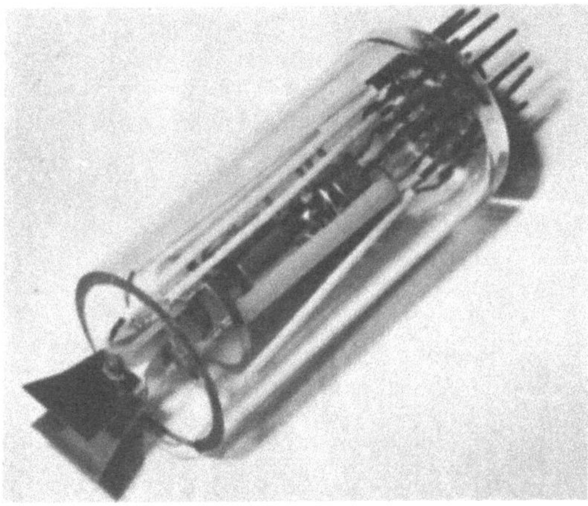

Fig. 1 The electron gun

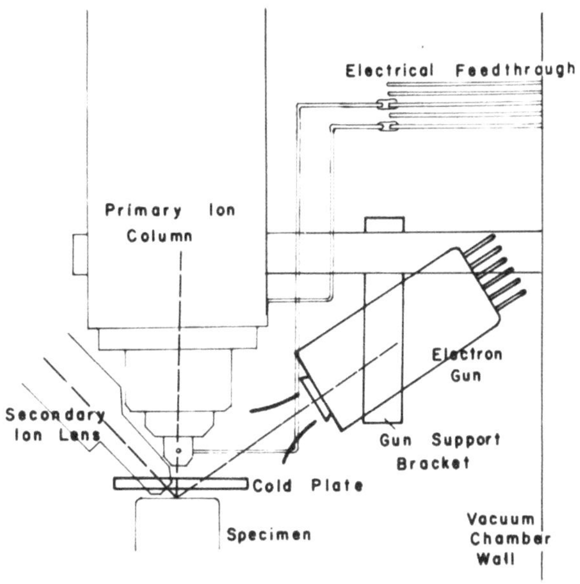

Fig. 2 Diagram of the IMMA showing the position of the electron gun relative to the primary column and specimen

feed-through which fortunately had a sufficient number of unused pins. Connections from the feed-through to the electron gun pass through the machined opening of the support arm. The gun was positioned in its bracket on the arm to ensure rough coincidence of the electron and ion beam on the sample. A few of the leads to the primary column had to be repositioned to make room for the gun. Fig. 3 shows the electron gun properly mounted just before the vacuum system was reclosed. The electrons strike the sample at an angle of 33.5 degrees.

Fig. 3 A view of the electron gun mounted on its support arm

Fig.4 The cold plate modified to extend the slot for the electrons

The only other modification to the instrument was to the cold plate which is positioned just above the specimen (Fig. 2). This cold plate has a slot in it to fit around the extraction lens for the secondary ions and to permit the primary ions to strike the specimen. To allow the electrons to reach the sample this slot was extended by about 1.5 cm as seen in Fig. 4. The cold plate was removed from the instrument for machining through the opening for the primary column so that the entire installation required only that one indium metal seal be broken and then remade.

Operation

The filament of the electron gun, when part of an oscilloscope tube, is run at roughly ground potential with the fluorescent screen at a high positive potential. For use in the ion microprobe, a negative dc potential had to be applied to the filament and all the other electrode potentials appropriately adjusted. This was accomplished using a 3 kV DC power supply, a floating filament transformer and a number of potentiometers. The deflection plates were driven with a ± 300 volt supply and the necessary potentiometers. The result is that the gun could be operated successfully and focussed from less than 1 kV to at least 3 kV. The minimum spot size is about 0.5 mm and the deflection plates can move the spot at least ± 1 cm from the central position. Alignment and focussing was achieved by viewing the spot on a zinc sulfide fluorescent screen through the light optics of the IMMA. The gun current can be varied from a few nA to more than 1 μA as measured by the sample current monitor.

Several phenomena in the use of the gun have been observed. If oxygen is used as the primary gas, a gradual decrease in the emission from the filament of the gun occurs. This must be compensated by an increase in the filament temperature. The higher emission of the gun is restored by overnight pumping or by using the gun with argon or nitrogen primary ion beams.

A major problem in the use of the gun has proven to be stray magnetic fields which cause a deflection of the electron beam when the secondary magnet field strength is changed. In the moment when the secondary magnet is moved from one mass to another, a rapid deflection of the electron beam of several hundreds of μm occurs. The beam returns to almost but not quite the same position when the magnet settles to its new value. A calculation shows that such a deflection of 2 kV electrons over a path of 15 cm would require a field of only a few gauss. The deflection will be removed by appropriate magnetic shielding or will be compensated by applying the necessary voltage to the deflection plates.

The gun has been used successfully to neutralize charge on the NBS glasses from which good spectra were obtained provided a grid was placed in contact with the surface of the specimen. As with other authors, the use of small specimens and a conducting coating in addition to the electron gun has yielded good data from many but not all insulating specimens.

Conclusions

An electron gun has been successfully installed in an ARL IMMA as an aid to charge neutralization. The gun performs well at energies between 1 and 3 keV. Deflection of the electron beam by magnetic fields during secondary mass changes will require that either magnetic shielding be installed or that the electron beam deflection plates be used to compensate for the displacement.

References

1. C.A. Andersen, H.J. Roden and C.F. Robinson, J. Appl. Phys. 40, 3419 (1969)
2. J.S. Sheasby and J.D. Brown, Oxidation of Metals 12, 405 (1978)
3. H.W. Werner and A.E. Morgan, J. Appl. Phys. 47, 1232 (1976)
4. K. Nakamura, S. Aoki, H. Tamura and H. Doi, "Proc. 6th Int. Conf. X-ray Optics and Microanalysis", ed. by G. Shinoda, K. Kohra and I. Ichinokawa, U. Tokyo Press, 1971, p. 447
5. B. Blanchard, P. Carrier, N. Hilleret, J. Marguerite and J.C. Rocco, Analusis 4, 180 (1976)
6. K. Wittmaack, J. Appl. Phys. 50, 493 (1979)
7. W. Reuter, M.L. Yu, M.A. Frisch and M.B. Small, J. Appl. Phys. 51, 850 (1980)

Performance and Use of Dissector Ion Microanalyzer

V.T. Cherepin and V.L. Ol'khovsky

Institute of Metal Physics, Acad. Sci. UkrSSR,
Kiev 142, USSR

The solution of the problem of local in-depth analysis of solids using secondary ion emission encounters difficulties related to effects produced by the walls of the crater resulting from ion etching. One of the ways to overcome these difficulties might be the formation of an intermediate ion image of the surface of the sample under study with subsequent separation of the image element of the required locality and its analysis with the aid of a mass filter. This simultaneously decreases the effect of non-uniform distribution of current density of the primary beam. Just this way was chosen in the laboratory of the Institute of Metal Physics, Acad. Sci. UkrSSR to design a dissector ion microanalyzer for the in-depth local analysis of the isotopic composition of solids and for the determination of impurity concentration profiles through the depth of the sputtered layer.

The main analytical feature of the microanalyzer described below is found in the fact that in front of the mass filter there is a device (the ion-electron converter) which enables us to visualize the unseparated ion image of the surface under study, this image being created by the immersion lens.

The unseparated ion image delivers information on the distribution of chemical elements along the surface, this being possible due to the difference in the coefficients of secondary ion emission for different elements. Information can also be obtained about the topography of the surface due to the oblique incidence of the primary beam. The surface topography image facilitates sample positioning when the area for the local analysis is to be selected.

The ion-optical scheme of the microanalyzer is shown in Fig. 1. The primary beam of positive ions (Ar, H_2, O_2) formed in the duoplasmatron with a cold cathode strikes a spot on the surface of the sample. The energy of bombardment depends on the difference between the potentials of the ion source and the sample and amounts to 6 keV when the sample potential is +4 kV. The focusing system of the primary beam forms the ion probe on the sample (0.8 mm to 2.0 mm in diam.) with an average current density from 2 mA/cm^2 to 0.15 mA/cm^2, respectively.

The immersion lens accelerates and focuses secondary ions ejected from the sample and forms an enlarged surface ion image in the plane of the cathode in the ion-electron converter. The energy of secondary ion acceleration depends on the requirement

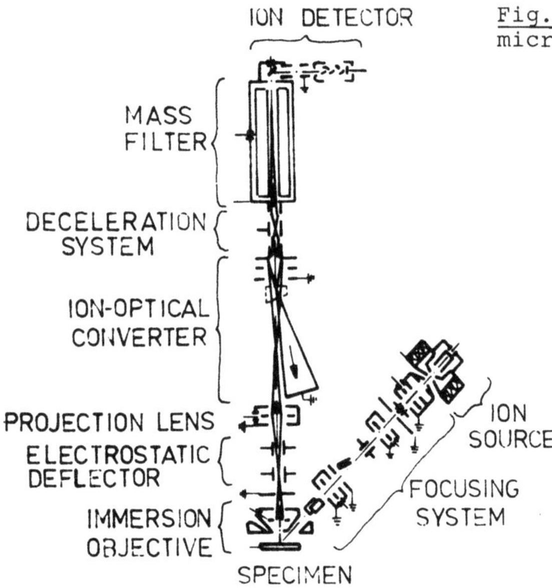

Fig.1 Ion optical scheme of microanalyzer

to reduce the effect of initial ion energy spread on ion optical lateral resolution.

Electrons ejected from the cathode are focused onto the luminescent screen and form an enlarged image of the sample surface as also of the orifice in the cathode. The latter is seen on the screen as a dark fixed spot.

By moving the sample or by deviating the secondary ion beam with the aid of a deflection system it is possible to fit the image of the cathode orifice to the image of that structural element of the surface under study which must be analyzed. The spatial resolution is defined by the size of the cathode orifice and the linear magnification of the ion image in the cathode plane. After passing through the cathode orifice the ions are slowed down in the field of a three-electrode decelerating system to the energy required for proper action of the monopole mass analyzer. The energy of ions entering the monopole is determined by the difference between the potentials of the object and the monopole and amounts to 100 eV.

By varying the optical power of the projecting lens one can alter the image magnification and thus obtain different spatial resolution of the analysis with the cathode orifice size remaining constant. At a definite optical power of the projecting lens the full ion beam is focused into the cathode orifice. This implies that the mass analyzer obtains a complete image of the surface area up to 600 μm in diam. The optical resolution of the unseparated ion image is 2-3 μm at a field of view of the analysis from 10 to 600 μm. Thus by image convolution the average composition from an area 600 μm in diam. can be obtained, while by image scanning or by shifting the object concentration cross-sections or concentration profiles with a specified spatial resolution may be obtained.

The microanalyzer consists of three racks: the analytical rack, the power rack and the registration rack. The analytical rack is a high-vacuum system with an analytical column, the object chamber and the air-lock chamber, the latter serving for the introduction of the casette containing the sample of 80 mm maximum diam. and 10 mm thickness. The specimen chamber contains an XY specimen stage with a programmed control; the specimen can be transferred for a maximum distance of 100 mm along each coordinate in 5 μm steps. The vacuum system provides a vacuum of 10^{-7} Torr with the ion source in operation using the diffusion pumps H-2 and H-05.

The microanalyzer is equipped with a monopole mass spectrometer kept at high potential (up to 4 kV) with respect to ground. The mass range of the instrument is 1-250 amu. Mass resolution (on the level of 10 % of the peak height of the M mass) is not worse than M.

The ion detector is a SEM in an off-axis position with respect to the mass analyzer axis. The r.f. generator and the registration channel are controlled by the scanning system which allows selecting four sections in the mass spectrum and five different sensitivities for each section within a dynamical range of up to 10^5.

One can judge the mass spectral resolution and sensitivity from the mass spectra of pure Ti, Mo and W recorded in the same range of secondary ion currents (Fig. 2). It may be seen that the described system of secondary ion collection and focusing ensures high detection efficiency due to ion acceleration in the immersion lens with subsequent slowing down of ions at the mass filter entrance.

Significant advantage of this microanalyzer becomes immediately apparent in the case of local in-depth analysis. Contrary to the methods of depth resolution improvement by primary beam scanning and by the use of an electronic diaphragm, the method of the dissector microanalyzer gives an increase in depth resolution with a decrease of the area under study.

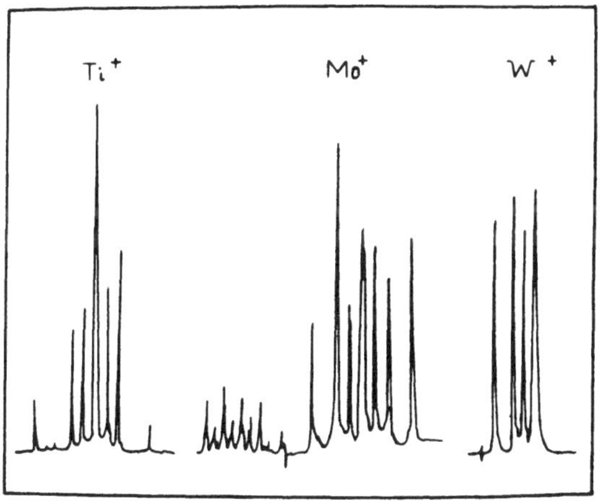

Fig.2 Mass spectra of Ti, Mo and W

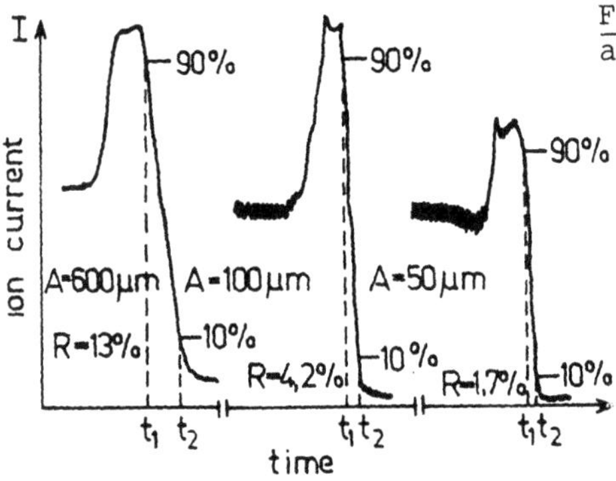

Fig.3 Effect of the analyzed area (A) on depth resolution (R)

Figure 3 shows examples of the in-depth analysis of a pure Al film of 10 000 Å thickness deposited on the "sitall" substrate. The depth resolution was estimated from the rate of the $^{27}Al^{+}$ peak fall off while approaching the substrate. The time of this falling was measured from the 90 % maximum magnitude up to 10 % and was compared with the total time of the film sputtering. It may be seen from Fig. 3 that the change in the analyzed area (A) from 600 μm to 50 μm diameter improves the depth resolution (R) from 14 % to 1.7 %.

Distortion of Secondary Ion Extraction Due to Sample Surface Irregularities

W. Bedrich, B. Koch, H. Mai[1], U. Seidenkranz[1], H. Syhre[2], and R. Voigtmann[1]
Zentrum für Forschung und Technologie Mikroelektronik, Dresden, DDR

Introduction

Hitherto numerous publications concerning the analytical application of Secondary Ion Mass Spectroscopy (SIMS) have appeared. In contrast, a rather limited number of investigations dealing with the influence of instumental effects on quantitative analysis have been published till now. One of these is the interpretation of the results of the comparative SIMS study of selected glasses [1]. NEWBURY showed that the relative sensitivity factors for the same element obtained by 22 different laboratories vary over more than one order of magnitude. Under the assumption that well characterized standards had been distributed to all cooperating laboratories the main reasons for deviations of such an extent must be supposed in different primary ion (PI) beam parameters and instrumental conditions incomparable to each other. Solely standardized conditions for creation, extraction, analysis, and detection of the secondary ions (SI) and mathematical compensation of every discriminating effect will lead to data that can be handled successfully by quantitative models.

In this respect particular attention must be paid to SI extraction within the target chamber and mass spectrometer analyzer transmission. During the last decade especially for the CAMECA-ion microscope SMI 300 some successful attempts have been published to solve these problems. Basic considerations about SI extraction and analysis have been discussed by SLODZIAN [2-4] whereas RUDAT and MORRISON [5-9] described experimental investigations of discriminating effects. WERNER and MORGAN [10] dealt in detail with the problem of charge build-up at the surface of insulating specimens and judged the effectivity of techniques for charge compensation. Unfortunately the IMMA of ARL has not found similar attention up to now. LIEBL [11] actually discussed the basic layout of the instrument but no publication about instrumental effects on output data has appeared. Therefore on the basis of an approximative description of the SI extraction optics of the IMMA the interferences of sample surface irregularities will be discussed briefly.

---

[1] Zentralinstitut für Festkörperphysik und Werkstofforschung der AdW der DDR, Dresden, DDR

[2] Zentralinstitut für Kernforschung der AdW der DDR, Rossdorf, DDR

SI Extraction Optics

The extraction optics of the IMMA consists of the specimen surface held at a potential of plus or minus 1.5 kV and the grounded pickup electrode with a 4 mm x 1.5 mm diaphragm of about elliptical cross section. Contrary to the usual cathode lens of the SMI 300 the two planes defining the acceleration space of interest have an inclination of 112° against each other. Thus in addition to the collecting action of the immersion lens the SI-trajectories leaving the specimen surface are turned through an angle of typically 45°. In the vicinity of that part of the SI beam which can enter the analyzer the accelerating field is assumed to have cylindrical geometry (i.e. no field component in z-direction). The detailed description of the SI extraction optics will be published later.

To avoid lengthy trajectory calculations the electrostatic field line distribution within the area of interest has been approximated by a uniform field near the specimen surface and a circular sector field in the remaining space.

The calculation of the SI trajectories has been carried out by repeated application of the SNELLIUS law [11] for one hundred potential steps. Fig. 1 shows the appropriate trajectory ranges for initial energies of E_o = 0.1 (dotted) and 100 eV (solid) and angles of emission $\alpha_o = -88^{\circ}...+88^{\circ}$ and $-35^{\circ}...+6^{\circ}$ resp. The trajectories for emission in normal direction have been particularly marked. Their different R_{max} values at pickup electrode location indicate the energy dispersion of the accelerating field.

This model roughly representing the extraction optics is assumed to be satisfying for the discussion of interferences to SI extraction.

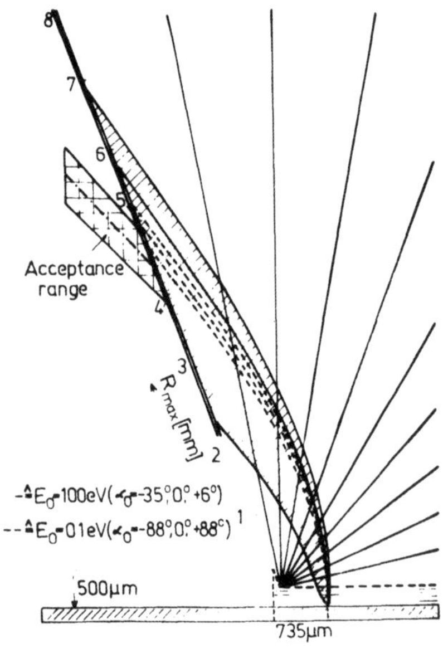

Fig. 1 SI trajectories in the target room

Interferences Due to Specimen Surface Irregularities

The physical behaviour of the specimen surface is of major importance for the analytical result. On one hand the specimen is involved in the ion optical layout of the instrument as an electrode of ideal shape and low electrical resistance, on the other hand no other part of the microanalyzer is as often subjected to modifications as the sample to be analyzed, with regard to these properties.

Serious effects on SI extraction have to be considered from:

- specimen topography (roughness and structured surface),
- misalignment of the sample, and
- charge build-up at the surface.

- Specimen topography

The micro roughness of the surface gives rise to local changes in the angle of incidence of the PI as well as in the orientation of the SI emission characteristics. Micro tips create distortions of the accelerating field in the vicinity of the SI origin. Surface structures (e.g. steps at IC-chips) can cause changes of the field shape in larger dimensions. An example for the interference caused by micro roughness is given in Fig. 2.[1] For a maximum roughness of 5 μm the $^{50}Cr^{+}$ - SI signal of a stainless steel surface shows intensity variations up to a factor of two.

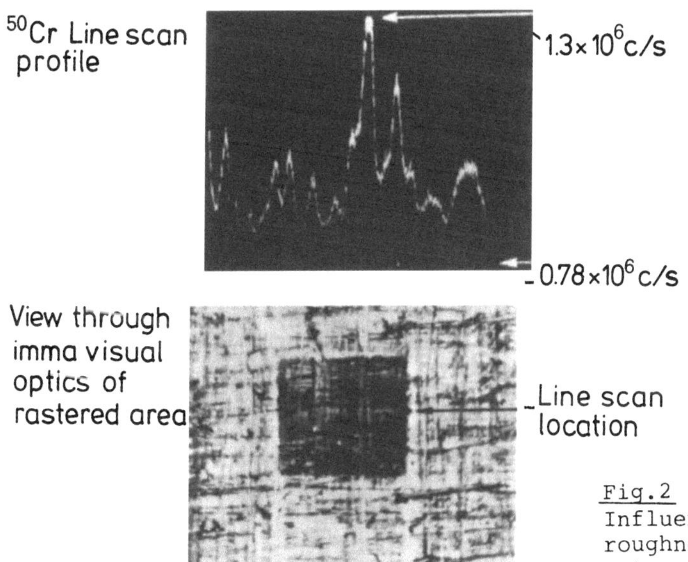

Fig.2
Influence of surface roughness on SI extraction

[1] Pictures taken by D. Comaford, ARL, Sunland, sample preparation ZFW, Dresden.

- Misalignment of the sample

A light tilt of the surface normal with respect to the PI beam direction already can lead to distortions of the accelerating field which strongly affect the intensity and the energy distribution of the SI signal. The z-axis position will give a negligible contribution because it can be adjusted with high precision (z-axis drive: 1.75 μm steps). The influence of a two degree tilt of a larger surface region on the intensity signals of several SI species is shown in Fig. 3. The specimen was made from semiconductor silicon and the measurements have been carried out after sufficient time had elapsed for sample amorphization by the PI (species: $^{16}O_2^+$, energy E_p = 20 keV, current I_p = 43 nA, beam diameter d_p = 10 μm). For three analysis spots differing in their distance from the edge between horizontal and tilted surface regions the SI angle is Θ. An increase in the intensity variation is observed for increasing distance from the horizontal surface region. Further on atomic ions are less influenced than molecular ions. For molecular ions the interference increases with increasing number of constituents. The ratio of the maximum-to-minimum value ranges from 1.17 to 3 for $^{30}Si^+$ and $^{28}Si_2\,^{16}O_2$ resp. The maximum value is achieved for that specimen position where the normal of the tilted part is directed to the pickup electrode and the minimum value if it is turned away.

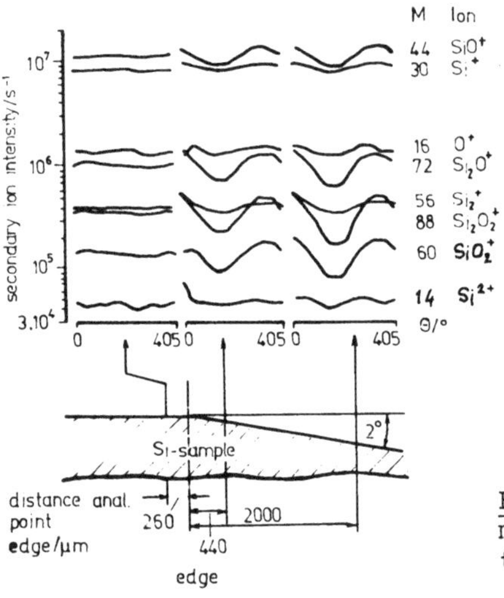

Fig. 3 Influence of specimen misalignment on SI extraction

- Charge build-up

This phenomenon observed for analyses of insulating samples by negative PI causes primarily a distortion of the extractor field. The correction of the resulting deviation of the sample potential at SI origin and the compensation of the field distortion by sample voltage optimization is only of partial success. Even by most careful adjustment it is not possible to

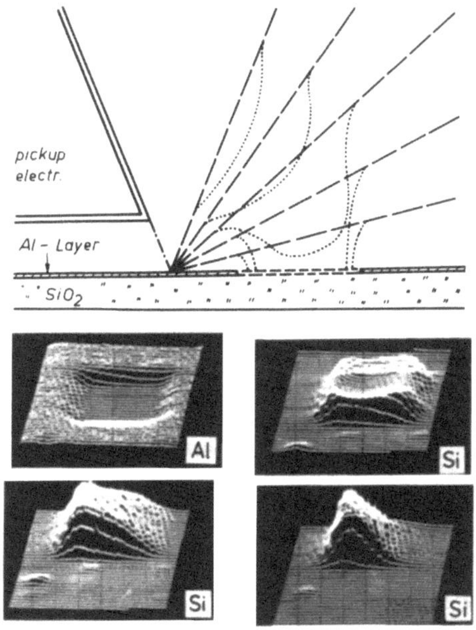

Fig. 4 Influence of charge build-up on SI extraction

attain a uniform ion extraction like in the case of a conducting specimen, Fig. 4 (equipotential lines - longer dash). It shows a scheme roughly representing the experimental arrangement for an aluminium coated silica sample. The field distortion (dotted curves) is indicated for crater potential decrease due to negative ion bombardment. The intensity mappings indicate that for $^{27}Al^+$ a uniform extraction around the crater is obtained whereas for $^{28}Si^+$ within the crater only nonuniform extraction patterns of different shape are observed for graded sample voltages. The experimental conditions are characterized by:

primary ions	$^{16}O^-$
ion energy	20 keV
current	15 nA
beam diameter	15 μm
crater size	150 x 150 μm^2
raster size for mapping	175 x 175 μm^2

The directed emission of single crystals - another effect on SI extraction - is kept out of the discussion because it is not considered as an intrinsic surface phenomenon.

Interpretation of Measurements on Tilted Sample

The experimental conditions (see Fig. 3) have been approximated by a variation of the sector angle φ of the extractor field (φ = 110°, 112° and 114°). For SI initial energies E_0 = 2, 10 and 100 eV the maximum radii of the trajectories at pickup electrode location have been calculated as a function of emission angle α_0. Fig. 5 (left hand side) shows the position of the resulting curves in respect to the acceptance range of the

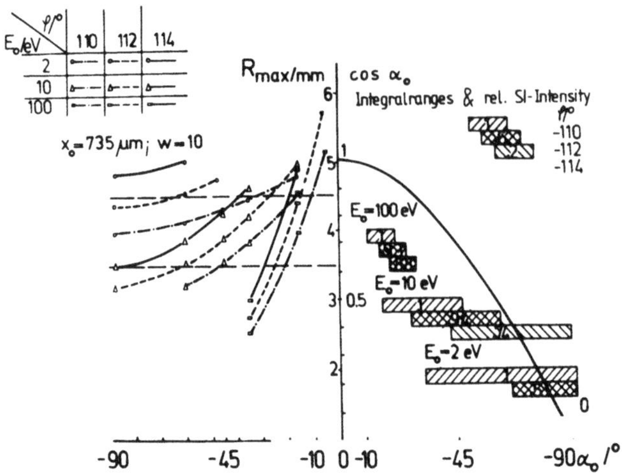

Fig. 5 Explanation of initial energy dependence of the misalignment effect on SI extraction

mass spectrometer (horizontal long dashed lines). Only ions having radii within this range are able to enter the analyzer. From the abscissa the appropriate range of the emission angle α_0 can be read. x_0 gives the distance between SI origin and the pickup electrode edge. The ratio between the potential differences of sector field and uniform field section (see Fig.1) is given by w. Particularly for ions of low initial energy for the indicated changes of φ a major variation in intensity can be concluded from the diagram whereas ions of high initial energy are less influenced. This is in qualitative accordance with the experimental where molecular ions (lower initial energies) are stronger influenced than atomic ions. The relative intensities for different initial energies have been calculated for a cosine distribution. The values given in the bars at right hand side of Fig. 5 demonstrate the appropriate intensity decrease for the assumed changes of φ. It can be seen that the intensity ratios obtained are comparable to those of the experiment.

Conclusions

The results of the experiments described indicate that even minor variations in the ideal configuration of the accelerating space i.e. changes in shape and properties of the specimen electrode can seriously influence the analytical outcome. Therefore measurements for quantitative analysis, for comparison of different samples at the same instrument or for comparative studies involving a variety of instruments will be successful only in the case that the data has been acquired under identical conditions. Thus e.g. the SI extraction process must be reproduced carefully from experiment to experiment and from instrument to instrument. Only effects which will influence all ionic species or a certain group of them (e.g. the atomic ions) to the same degree can be eliminated by referencing procedures.

References

1. D.E. Newbury, In Secondary Ion Mass Spectrometry SIMS-II, ed. by A. Benninghoven, C.A. Evans, Jr., R.A. Powell, R. Shimizu, H.A. Storms, Springer Series in Chemical Physics, Vol. 9 (Springer, Berlin,Heidelberg,New York 1979) p. 53
2. G. Slodzian, Surf. Sci. 48, 161 (1975)
3. G. Slodzian, Ann. Phys. 9, 591 (1964)
4. G. Slodzian, NBS-Publ. Nr. 427, Washington
5. M.A. Rudat and G.H. Morrison, Int. J. Mass Spectrom. and Ion Phys. 27, 249 (1978)
6. M.A. Rudat and G.H. Morrison, Int. J. Mass Spectrom. and Ion Phys. 29, 1 (1979)
7. M.A. Rudat and G.H. Morrison, Int. J. Mass Spectrom. and Ion Phys. 32, 53 (1979)
8. M.A. Rudat and G.H. Morrison, Surf. Sci. 82, 549 (1979)
9. M.A. Rudat and G.H. Morrison, Anal. Chim. Acta 112, 1 (1979)
10. H.W. Werner and A.E. Morgan, J. Appl. Phys, 47, 1232 (1976)
11. G.B. Hagen, Ann. de Phys. 6/13, 1, Heft 6-8 (1953)

A Comined Direct Imaging Laser Ionization Secondary Ionization Mass Spectrometer

B.K. Furman and C.A. Evans, Jr.

Charles Evans & Associates,
San Mateo, CA 94402, USA

The advances in materials used for a wide variety of technologies place great demands on the characterization techniques developed for a study of these materials. At the present time, there exist only two techniques capable of the microanalysis of semiconductor and metallurgical and related materials. These are the well-known high performance secondary ion mass spectrometer or ion microanalyzer and the recently developed laser ionization microprobe mass spectrometer. The first technique, SIMS, is the topic of this conference and no more details are required except to say that, for the applications of this report, we used an instrument which employed direct imaging of the ion emission from the sample surface through stigmatic optics as exemplified by the Cameca IMS-3f ion microanalyzer [1]. In this instrument, images are formed not by the rastering of a finely focussed primary ion beam, but by optics which allows the transfer of an image of the emission areas from the sample surface through the entire mass spectrometry optics for detection. The images are detected by projecting this ion image onto a channel plate wherein the ions are converted into electrons, multiplied by impacting a fluorescent screen for visual observation and recording by photographic and video methods.

The second technique, the laser ionization microprobe, as is generally configured in the commercially available instrument [2], employs a pulsed ND:YAG laser which brings about instantaneous volatilization and ionization of a particle or area of a material which is irradiated opposite to the mass spectrometer by the laser. The ions produced during the explosive detonation process are extracted into a time-of-flight mass spectrometer. The ions which have been created over an extremely short period of time, of the order of 20 nanoseconds, are accelerated to a constant kinetic energy by the imposed electrostatic voltage. According to the expression,

$$E_{final} = 1/2\ mv^2,$$

the ions separate in time according to their mass, since the velocity of the lighter ions is greater than that of the heavier ions. After time dispersion in the time-of-flight mass spectrometer, the ions are detected by specialized ion detection which allows for the recording of ion intensity as a function of time thereby producing the time-of-flight mass spectrum. At present the instrument allows for only the transmission mode of excitation so those applications include three specific sample types; (1) the dispersion of particles on a transparent substrate, such as a copper grid, (2) thin sections of biological materials, and (3) thin sections of geological materials. Recently, the laser ionization mass analyzer from Cambridge Consultants has been operated in a front side illumination or reflection mode and has been shown to produce ions from the front surface of the material.

However powerful they may be, the two techniques discussed above, high performance secondary ion mass spectrometry and laser ionization mass spectrometry, have advantages and limitations. The advantages of the SIMS technique are well known and include the ability to provide high quality depth profiles as well as rapid lateral imaging with a good quantitative capability of standards are available. However, it is limited in its ability to see organic species as distributed on the surface of the material and suffers from the difficulties in a survey analysis of a microvolume of material.

The combination of a laser and a CAMECA IMS-3f Ion Microanalyzer to provide a direct imaging laser mass analyzer (DILMA) has successfully produced the first ion images resulting from photon irradiation of various materials. This paper will present data collected from conducting and non-conducting inorganic materials. In addition, spectra from an organic thin film will be presented. This paper will emphasize combining laser excitation with the imaging capabilities of the IMS-3f Ion Microanalyzer. In addition, results utilizing the cw laser source for "in situ" laser processing will also be presented.

Experimental

Figure 1 shows a diagram of the instrumentation used for these experiments. A CAMECA IMS-3f ion microanalyzer was the backbone to the instrumentation. Two very different lasers: a Nd:YAG pulsed laser and an Ar ion, cw laser, were utilized as laser sources. A more complete description of each can be found in Table 1. The laser beam was focused by a lens with a ≈1 meter

Table 1 Laser source

Type	Nd:YAG	Ar^+
Manufacturer	Quanta-Ray	Lexel
Model	DCR-1 A	95
Mode	Pulsed ~100-200 ns un Q-switched	 cw
Beam	Focused toroid	Focused spot
Energy	<60 mJ/pulse	<4 Watts
Power density (W/cm^2)	$<10^8$	$<10^6$
λ (nm)	530	514 (multi-line)

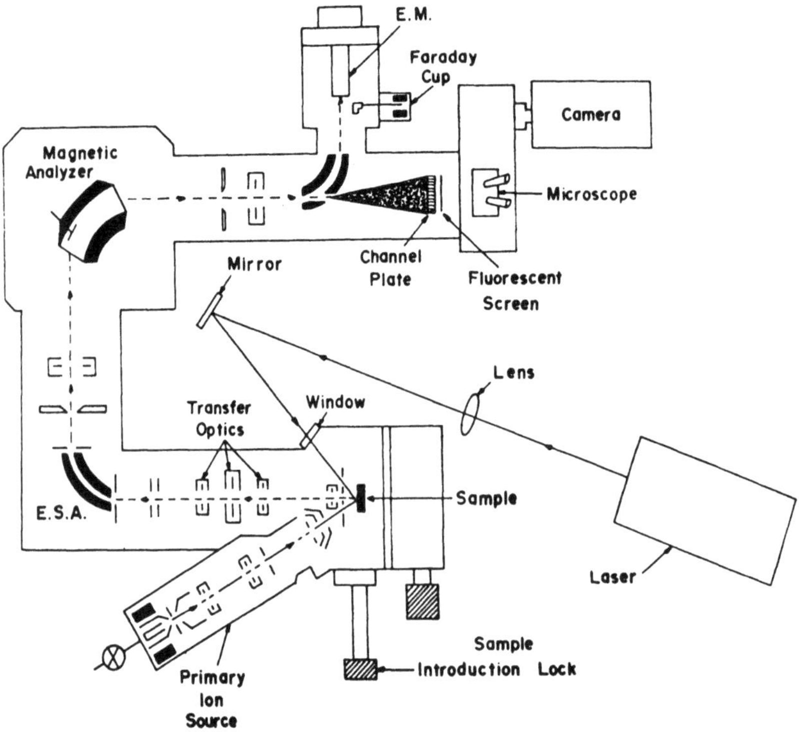

Fig. 1 Schematic drawing of CAMECA IMS-3f Ion Microanalyzer with laser

focal length to a spot size of ≈1 mm diameter. The beam was introduced into the CAMECA IMS-3f through a high purity quartz window and irradiated the sample surface at ≈45° incidence. This geometry allowed the laser beam to strike the sample surface in the same area as the conventional ion beam. Secondary ions produced by the laser were then extracted by an immersion lens in an identical manner as those produced by ion bombardment. Both positive and negative secondary ions could be extracted. Using the Nd:YAG laser, it was repeatedly pulsed at ≈10 pulses/s. Spectra were obtained by repeatedly pulsing while scanning the magnet and mowing the sample as described by HERSCH et al., who called this technique repeatable laser desorption (RLD) [3]. In all cases the laser was operated in the desorption mode at low power ($<10^7$ W/cm^2).

Results and Discussion

The intent of this study was to employ the unique excitation capabilities of either a pulsed or cw laser with the ion imaging and mass spectrometric capabilities of the Cameca IMS-3f ion microanalyzer. As has been previously shown, the laser excitation provides surface specific and molecular containing information while the Cameca IMS-3f ion microanalyzer allows for imaging of the ions produced at the sample surface. The following will summarize the use of the Nd:YAG pulsed laser.

This laser was not operated in the Q-switch mode so as to maintain a low power density to prevent damage to the optical window into the vacuum system. Therefore, the majority of the species obtained from the surface of the material were those which could be obtained in the desorption rather than in the explosive detonation mode. These included Li, Na, K, Ca, Rb, and Cs in the positive spectroscopy mode and H, C, O, and the molecular species CO in the negative spectrum. Halogens such as F and Cl were also observed when present. Ion images were obtained from the particles present on the surface of a Si wafer from a sample contaminated partially with Na and partially uncontaminated and from the surface of bulk NBS standard glass insulators. These last results were the most astounding since we found that there were no charging problems associated with the production of ions from the surface of these bulk glass materials. Generally speaking, the species found to be emitted were thought to be surface contamination either in the form of adsorbed layers or particulates rather than actual consumption of the bulk glass material.

The second set of experiments that we performed was to examine the desorption of organic molecules from the surface of the sample. As has been previously observed, the organic molecules were desorbed as positive ions produced by the combination of a cation from surface contamination such as Na and K, with the specific neutral organic molecule. Silicon, coated with a thin layer of arginine and NaCl, was examined by both ion and laser bombardment. The resulting Na^+ images of an edge are shown in Fig. 2. The mass spectrum obtained in the RLD mode is shown in Fig. 3. Molecular ion peaks of $(M+H)^+$, $(M+Na)^+$ and $(M+K)^+$ are in excellent agreement with results previously reported using laser desorption [3,4]. In contrast, spectra obtained using ion bombardment in a dynamic mode show

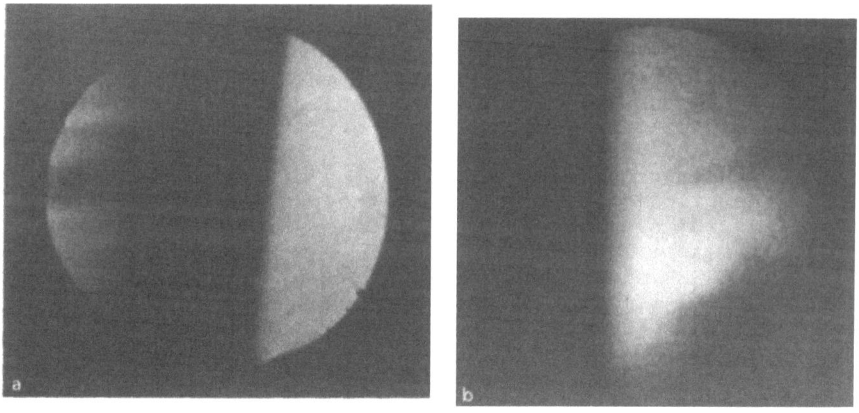

Fig. 2 Ion image of Na^+ distribution at edge of Si wafer using (a) Cs^+ primary ion bombardment (b) laser radiation only. Imaged field 150 μm

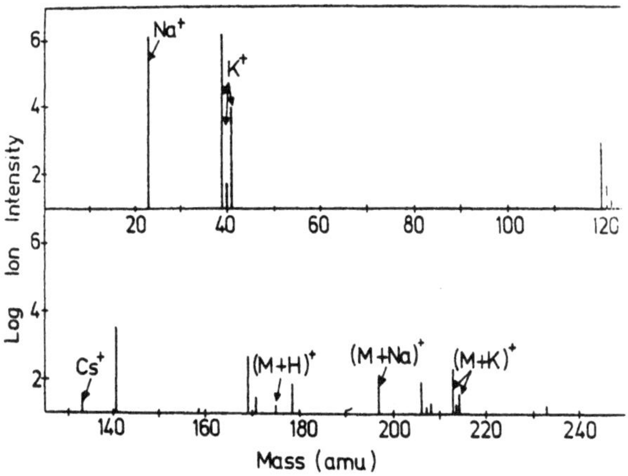

Fig. 3 Mass spectrum of arginine thin film obtained using an RLD mode of DILMA

no indication of arginine pseudo-molecular ions and are saturated with silicon molecular ion peaks.

A set of experiments was then performed using a cw argon ion laser. The first series of experiments were used to evaluate the imaging mode and we found that again, the cw laser brought about surface desorption of species on the material surface so as to produce lateral distributional information. If we examine a biological sample such as rat kidney, we find that under low power excitation, the ions tend to come from these regions as well as the more homogeneously illuminated and homogeneously contaminated regions.

In addition to the unique analytical applications of laser excitation combined with direct imaging mass spectrometry, we found that the laser provided an unique capability for actual processing of the material in the sample chamber of the ion microanalyzer. The first set of experiments was to produce heating of the surface of the sample so as to reduce the contribution of residual vacuum to an analysis such as for oxygen. We found that the irradiation of the sample surface by the cw laser produced a rapid heating of the surface which reduced the sticking coefficient or the resident time of residual vacuum species on the surface of the irradiated sample. This is illustrated in Fig. 4. The first profile labelled NO LASER shows the profile of an ion implant of oxygen into float zone silicon. In this case there was a background which had been irradiated by the laser approximately 30 s to 1 min prior to the analysis, we see the attainment of an improved background due to the decreased contribution of the residual vacuum. During that analysis the laser was turned off, the sample began to cool thereby allowing for a greater incorporation of oxygen in the analysis. This effect is found to be reversible since turning on the laser brings back a reduction in the residual vacuum contribution as the sample

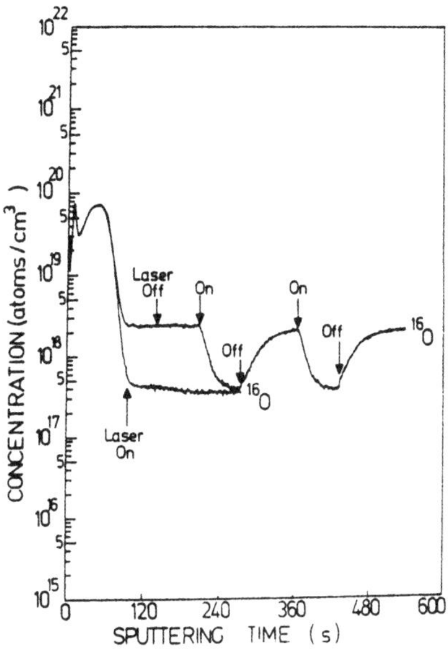

Fig. 4 In-depth distribution of oxygen ion implanted into FZ silicon with and without laser heating

surface heats up under irradiation. Such an effect was found to occur for any species contained in a vacuum which had a significant sticking coefficient on the surface of the sample.

Summary

We would like to propose that the use of a laser irradiation in combination with the specialized ion optics necessary to provide high performance secondary ion mass spectrometry provides a new, unique analytical capability as well as allows for rapid study of material related phenomena. Thus we have been able to provide a new analytical capability in the imaging and laser desorption mode of molecular ion species.

Acknowledgement

The authors are grateful to A.F. Findeis for the initial suggestions that led to this research, and to the San Francisco Laser Center (University of California at Berkeley and Stanford University) supported by the NSF grant CHE-79-16250. The partial financial support of this research by CAMECA Instruments was gratefully appreciated.

References

1. M. Lepareur, Rev. Tech. Thomson-CSF 12, 225 (1980)
2. F. Hillenkapm et al., Appl. Phys. 8, 341 (1975)
3. F. Heresch, E.R. Schmid and J.F.K. Huber, Anal. Chem. 52, 1803 (1980)
4. M.A. Prosthumus, P.G. Kistemaker and H.L.C. Meuzelaar, Anal. Chem. 50, 985 (1978)

Advances in Ion Probes A-DIDA

H. Frenzel and J.L. Maul

ATOMIKA Technische Physik GmbH, D-8000 München, Fed. Rep. of Germany

The basic concept of the ION PROBES A-DIDA was set up 10 years ago. While the basic concept proved to be highly successful, a number of advances in the A-DIDA design features could be introduced since the SIMS II conference, which considerably improved the analytical capability of the ION PROBES A-DIDA. The basic concept of the ION PROBES A-DIDA has already been described elsewhere [1,2]. Different to the horizontal configuration of the A-DIDA 2000 the new IONMICROPROBE A-DIDA 3000-30 is a vertical configuration, as commonly used in SEM design, with the primary ion beam coming from the top. The samples are mounted horizontally which allows running powder samples.

The IONMICROPROBE A-DIDA is microprocessor controlled. Microprocessor control incorporates for instance automatic vent and pump down cycling of the ion gun, analysis chamber and sample transfer system, digital mass programming on 10 channels with individual selection of mass/mass range, secondary ion energy and integration/scan time in each channel, control of digital and analog data output for mass spectra, peak intensities and depth profiles. A central mode control features push-button programming of system flunctions like mass spectrum, profile and image. To illustrat the convenience of the system operation a few examples: In mass adjust mode a 3 amu wide part of the mass spectrum around the desired mass is displayed on the CRT with a cursor to select the exact mass position (see Fig. 1). In energy adjust mode the secondary ion energy distribution of the selected mass is displayed on the CRT with a cursor to select the exact position of the "energy window" individually for each selected mass. In depth profile mode simultaneously with the depth profile plotting the ion image of the sputtered area is displayed on the CRT with a visualization of the acutal gated (analyzed) area.

It should be noted that various other advances have been made recently: The dynamic range in depth profiling has been extended to 6 orders of magnitude [3]. The sensitivity has been improved to obtain useful yields as large as 10^{-3}. Moreover detection limits down to 1 ppb have been demonstrated for various dopants in GaAs [4,5]. The sample throughput has been increased by adding an UHV sample transfer system. Pump down of the sample introduction chamber from atmospheric pressure to

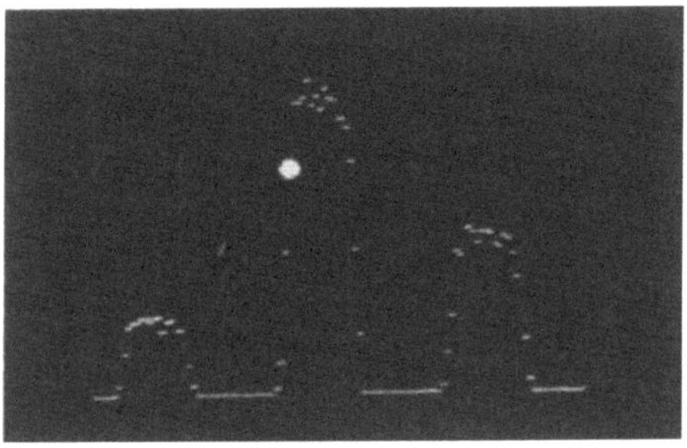

Fig.1 Secondary ion mass adjust. On the CRT: Chosen center mass$\pm$1.5 amu

$1 \cdot 10^{-6}$ Torr by a turbomolecular pump takes only 4 minutes. The transfer of the sample holder from the introduction chamber to the sample caroussel in the analysis chamber takes only seconds. A working pressure in the analysis chamber of $1 \cdot 10^{-9}$ Torr (with primary ion beam on the target) is achieved within about one minute after sample introduction. Routine analysis at residual gas pressure in the 10^{-10} Torr range is possible.

Detection of negative secondary ions from insulating material with a high dynamic range constitutes one of the more difficult SIMS problems. Negative secondary ion detection from insulators has so far been demonstrated only by Magee in a quadrupole based SIMS system [6].The measurement of negative secondary ions with an acceptable dynamic range is hampered by the electron background, caused by the secondary electrons which pass the secondary ion energy filter and are too fast to be mass separated by the quadrupole mass spectrometer. The electron induced noise is even more severe if simultaneous electron bombardment is used for charge equilibration on insulating specimens. The resulting background typically amounts to 10^4 cps for SiO_2 sample bombarded with 10 μA, 500 eV electron beam. A magnetic field is used to separate the electrons from ions by their momentum difference. The position and strength of the magnetic field has been optimized empirically. The influence of this magnetic field on the trajectories of secondary ions is compensated by an electrostatic deflector, already part of the secondary ion optics. Position and strength of the magnetic field have been chosen in a way that neither the detection of low mass secondary ions like hydrogen nor the mass peak shape is degraded. Fig. 2 shows a negative secondary ion mass spectrum from ordinary glass. In this case the background count rate is only 4 cps.

It is worthwhile mentioning that in the A-DIDA system very low primary ion energies (< 1 keV) can be used also in case of negative secondary ion detection because of the low secondary ion extraction field ($\approx$ 100 eV). Use of low primary ion energy is mandatory to minimize the distortion of the original depth profile in the investigation of shallow depth profiles and of interfaces. Depth profiling of very shallow implants down to 1 keV implantation energy have been demonstrated recently in an A-DIDA system [7].

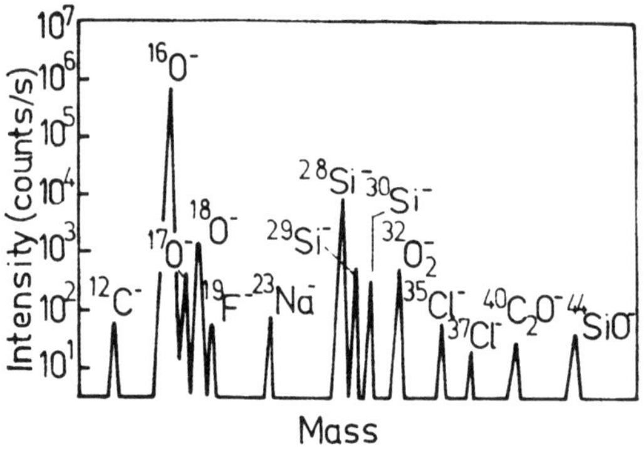

Fig. 2 Negative secondary ion mass spectrum from glass. Primary ions: 10 keV O_2^+, 100 nA. Charge equilibration: 500 eV e^-, 10 μA

References

1. K. Wittmaack, J. Maul and F. Schulz, Proc. 6th Int. Conf. Electron and Ion Beam Science and Technology (Ed.: R.Bakish) The Electrochemical Society Inc., Princeton (1974) p. 164
2. K. Wittmaack, Proc. 8th Int. Conf. X-ray Optics and Microanalysis (Ed.: D. Beaman, R. Ogilvie and D. Wittry)
3. K. Wittmaack and J.B. Clegg, Appl. Phys. Lett. 37 (1980)...
4. K. Wittmaack, J.B. Clegg and M.G. Dowsett, to be published
5. J.B. Clegg et al., J. Appl. Phys. 52 (1981)...
6. C.W. Magee and W.L. Harrington, Appl. Phys. Lett. 33, 193 (1978)
7. W. Wach and K. Wittmaack, Proc. 9th Int. Conf. on Atomic Collision in Solids, to be published in Nucl. Instr. Meth.

A Novel Ion Etching Unit Applicable for Depth Profiling with SIMS and IIR

K.H. Guenther, E. Hauser, G. Hobi, P.G. Wierer, and E. Brandstaetter[1]

BALZERS AG., FL 9496 Balzers, Fürstentum Liechtenstein

Introduction

During the past years, ion etching became a common means for removing surface layers from solid samples, for cleaning sample surfaces in UHV experiments, for depth profiling of multilayer structures, and for thinning specimens down to electron transparent thickness for investigations in transmission electron microscopy (TEM). Particularly in the case of the latter application, Balzers' Ion Etching Unit IEU 100 has been developed as a compact easy-to-operate laboratory equipment, which also enables the possibility of attaching analytical equipment.

Employing a quadrupole mass analyser for the detection of secondary ions emitted from the solid surface under primary ion bombardment, results in the basic set-up of a secondary ion mass spectrometer (SIMS). The use of a photomultiplier in combination with narrow band interference filters or a monochromator enables the detection of ion-impact induced radiation (IIR), which gives additional information about the chemical composition of the solid surface under investigation. Equipped with these two analytical accessories, the IEU 100 became a powerful, yet comparatively inexpensive surface analysis instrument with the capability of multielement depth profiling [1].

Basic configuration

The vacuum chamber of the unit, a stainless steel cross-tube of 100 mm inner diameter is directly evacuated by an automatic turbomolecular pumping stage TPH 270. The Penning type glow discharge ion source is mounted on top of the cross-tube, ususally we use inert gases for the glow discharge (argon or nitrogen) but a mixture with oxygen is also feasible. The beam of positive ions is formed by a tubular extraction lens and a permanent magnetic field. The distribution of the beam current density can be monitored by a Faraday-cup. Fig. 1 shows the variation range of the ion beam profile. The rectangular profile in Fig. 1 was obtained by the introduction of a conical diaphragm in combination with the standard tubular lens. For charge compensation on dielectric samples we used an additional

[1] on leave from Institut für Werkstoffprüfung und Metallkunde, Montanuniversität, A-8700 Leoben, Austria

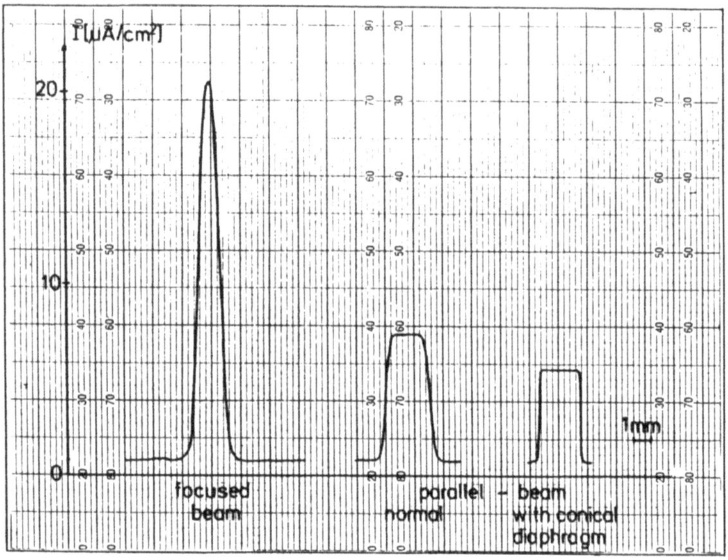

Fig. 1 Ion beam profiles

electron gun which flooded the sample surface in these experiments with electrons of 500 eV energy. The electron gun is located underneath the ion source. The specimen stage allows the motion of the sample along two axes as well as the rotation in the sample plane by externally controlled DC-motors. As the specimen stage is mounted in one of the horizontal arms of the cross-tube, it can also be inclined to the incident ion beam by manual operation in a range of 0-90°. The motion and the final position of the sample can be observed by a binocular microscope mounted in front of the sample chamber window.

Analytical Accessories

The quadrupole mass analyser QMA 150 involves a 90° ion deflection unit and an off axis SEM in order to prevent fast ions, excited neutral particles and energetic photons from reaching the conversion dynode of the SEM. The analyser is mounted in the horizontal arm of the cross-tube opposite to the specimen stage. The quadrupole mass spectrometer QMG 311, a standard product of Balzers, which enables the detection of both positive and negative ions, has been optionally equipped with the newly developed Quadrupole Data Processor QDP 101, featuring the selection of 8 mass ranges or "channels". Besides measuring a usual spectrum over a selected partial mass range in a SCAN mode, peak switching for individual masses and the integration of their intensity appropriate to a preselected scan speed is performed automatically in a POINT mode, resulting in a significant suppression of background noise. This POINT mode is most valuable for conveniently recording depth profiles of up to 8 different elements. A third mode of operation allows the recording of a complete spectrum,while other channels are simultaneously occupied with rapid peak switching. This mode is of great advantage to check for changes in the overall composition while running a depth profile with pre-set elements. Another remarkable feature of the instrument which has to be mentioned is an automatic, pseudologarithmitic display over three decades, with the advantages of a real zero point and a linear scale within the decades, as compared with a real logarithmic signal.

For the detection of spectral lines emitted by elements contained in the surface under ion bombardment a McPherson monochromator has been chosen in preference to narrow-band optical-interference filters, as they are only useful as long as the presence of a particular element is already known, e.g. for depth profiling of vacuum deposited coatings. For the surface analysis of samples of unknown composition, however, the possibility of a spectral scan in the wavelength region from 200 through 800 nm should exist. As most of the elements emit spectral lines in the UV region, a quartz lens had to be employed for imaging the emitting sample area onto the entrance slit of the monochromator. Although for transparent samples, the signal detection is generally possible from both sides of the sample, only the radiation emitted into the free space is used for analytical purposes, as the radiation transmitted through the glass substrate might be influenced by the optical properties both of the glass and of the coating under investigation.

Experimental Results and Discussion

Out of numerous practical examples a plain surface of optical glass BK 7 has been selected for the comparison of SIMS- and IIR-Spectra (Fig. 2a, 2b). Obviously both spectra reflect the qualitative composition of the glass quite well. Especially remarkable is the sensitivity of both methods for light elements. Besides Si and K (which can also be detected using an energy dispersive X-ray (EDX) analyzer e.g. with a scanning electron microscope), Na and even B are analysed distinctively. Usually, Na is below the limit of detection of an EDX-analyzer and B is not detectable by conventional Si (Li)-detectors. However, a quantitative analysis is not easily done in both cases, as both the secondary ion emission yield and the optical emission probability depend strongly on the chemical binding of the metal atoms, and may be considerably higher for oxides than for pure metals. Hence, for the analysis of glasses and oxide coatings, a certain level of residual oxygen or water vapour pressure is even desirable, in order to prevent the surface under ion bombardment to be reduced to suboxide or metallic states, which would cause lower signals. Thus, the vacuum conditions obtainable with the IEU 100, i.e. an ultimate pressure of 10^{-7} mbar, prove to be sufficient for accomplishing both SIMS and IIR surface analyses and depth profiles of oxides. Fig. 3 shows the signal traces of the spectral lines Ti (334.9 nm) and Si (288.2 nm) as a function of etching time, revealing the depth profile of an all-dielectric optical multilayer coating used as a dichroic beam splitter. A comparison of depth profiling obtained by SIMS, IIR and AES will be presented in another paper [2].

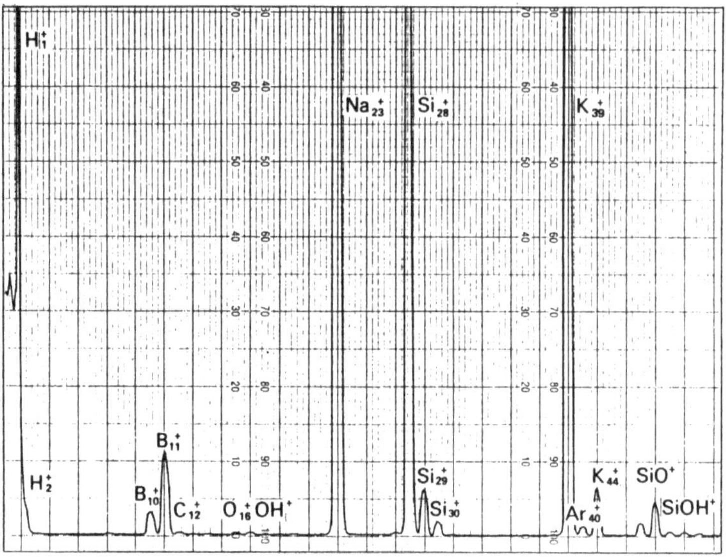

Fig. 2a SIMS spectrum of BK 7 glass

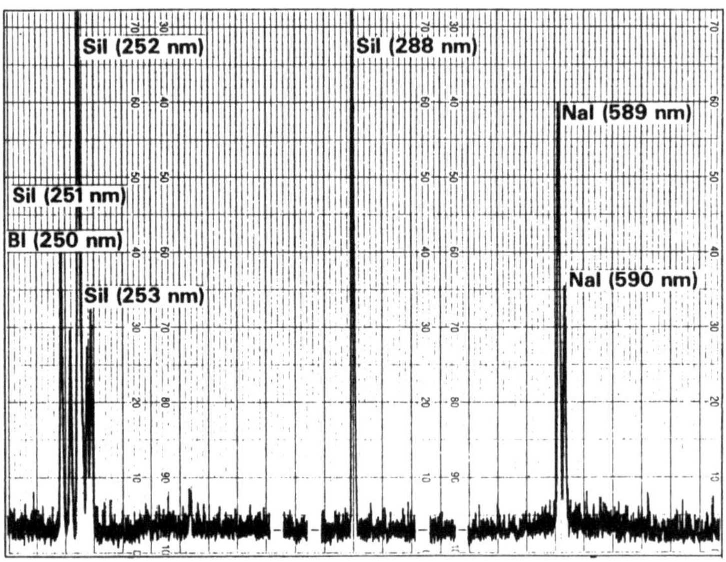

Fig. 2b IIR spectrum of BK 7 glass

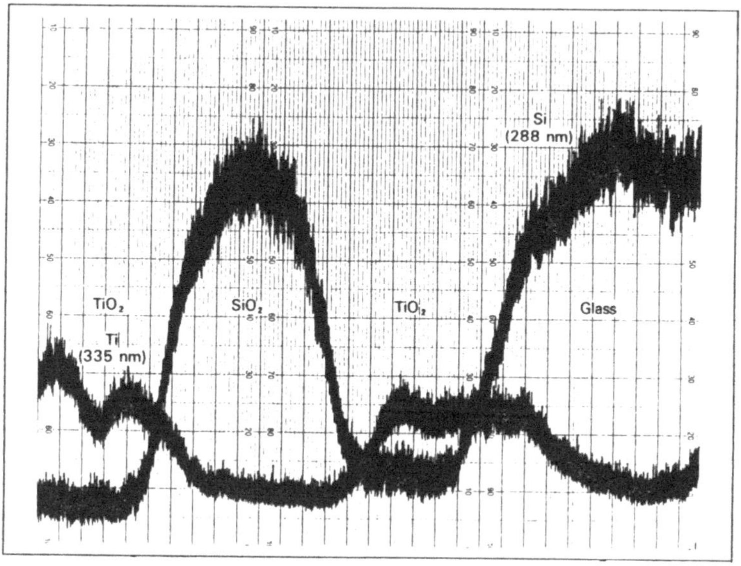

Fig. 3 IIR depth profile of dichroic beamsplitter

Conclusions

We have shown that compositional surface analysis and depth profiling of glasses and oxide coatings can be conveniently accomplished using the Balzers IEU 100 equipped with a quadrupole mass spectrometer and an optical spectrometer. We are fully aware, however, that the vacuum conditions might be insufficient for the analysis of pure or ultra-pure metal surfaces, e.g. used for adsorption studies, although we did not conduct such experiments. However, depth profiling can be done succesfully especially when oxygen is added to the sputtering gas, in order to enhance the ionization and emission probability.

References

1. H. Bach, Beitr. elektronenmikroskop. Direktabb. Oberfl. 12/1, 277 (1979) [English translation: BALZERS AG Report Nr. BB 800 012 DE 8009]
2. E. Hauser, G. Hobi, K.H. Guenther and E. Brandstaetter, This volume

Improvements and Applications of the Riber MIQ 156

F. Simondet and D. Kubicki

RIBER,
F-92503 Rueil Malmaison, France

Introduction

The techniques of in depth profile analysis have been improved considerably during the last few years. These methods include secondary ion mass spectrometry (SIMS) and surface analysis techniques combined with sputtering such as Auger electron spectroscopy (AES), electron spectroscopy for chemical analysis (ESCA) etc. We have put the emphasis on our equipment combining the three analytical methods mentioned above (LAS 3000): because comparison of the techniques indicates that, although unique in regard to certain applications, they all suffer from limitation regarding their general applicability in surface layer analysis. However a new ion gun has been designed [1] for the SIMS equipment because new needs have appeared, among these are a shorter time of analysis, a higher sensibility and an improvement in the dynamic range in depth profile.

Apparatus

The MIQ 156 SIMS equipment is an improved version of a static SIMS, it is mounted in an UHV chamber (LAS 3000). The main parts of the set-up are the following:

1. *Ion gun.* A schematic view of this ion gun is given in Fig. 1. It can be operated at energy from 0.5 to 15 keV. The primary ion beam is supplied by a duoplasmatron ion source which can be equipped either with a hot cathode (unreactive gases, e.g., noble gases, N_2,...) or with a cold cathode (reactive gases, e.g., O_2,...). A transfer lens matches the ion beam from the source to a mass selector, a 90° magnetic sector, which has resolution better than a mass. The ion gun is ended by a set of two lenses which allows the primary ion beam to be focused down to 5 µm.

2. *Mass spectrometer.* It is composed of a quadrupole mass spectrometer preceded by an energy filter (45° electrostatic sector). The electron multiplier is set off-axis of the quadrupole.

3. *Automation equipment.* It is done by a dedicated computer, PDP 11/23 which controls the quadrupole, the data acquisition and the target voltage. In case of use in depth profiling mode, an electronic gating system allows the choice of an analysed area ranging from 1 % to 100 % of the bombarded area.

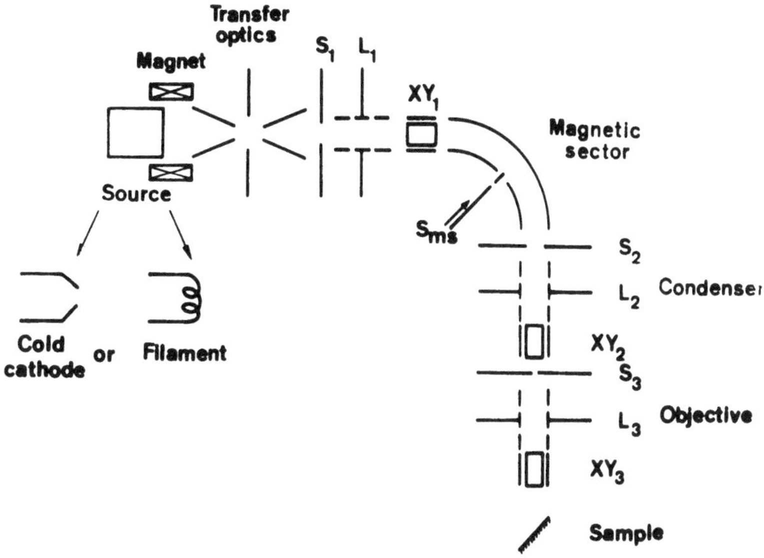

Fig. 1 Schematic view of the new ion gun, CID 15

4. *Charge neutralisation equipment.* It consists of an electron gun delivering low energy electrons (1500 eV). The sample is bombarded simultaneously with primary ions and electrons, this is used to compensate the change induced by ion bombardment during the analysis of insulating specimen.

Application

The UHV chamber (LAS 3000) equipped with ESCA and AES equipment and complete with the new ion microprobe MIQ 50 has enlarged our capabilities of analysis in various fields; among these are imaging, trace profile analysis, insulator analysis, shallow and deep layer analysis, surface analysis.

In this section we will give some examples of analysis performed with the above described SIMS equipment.

Analysis of multilayer sample. Depth profile of very thin layers required special conditions in order to reduce the depth needed to have sputtering equilibrium and to avoid mixing or knock-on effects. Amont these are use of low energy ion (3 keV) and of different types of ions (noble gases ions, N_2^+, O_2^+, ...). Fig. 2 shows a depth profile of an MBE growth superlattice. This analysis was performed with O_2^+ at 3 keV.

Analysis of semiconductor device. In Fig. 3, we give a depth profile of arsenic and boron in silicon. It is a typical transistor structure from an integrated circuit. In this type of applications low residual pressure is needed in order to get rid of interferences and in this way to improve the limit of detection of those elements. Moreover, this example shows that

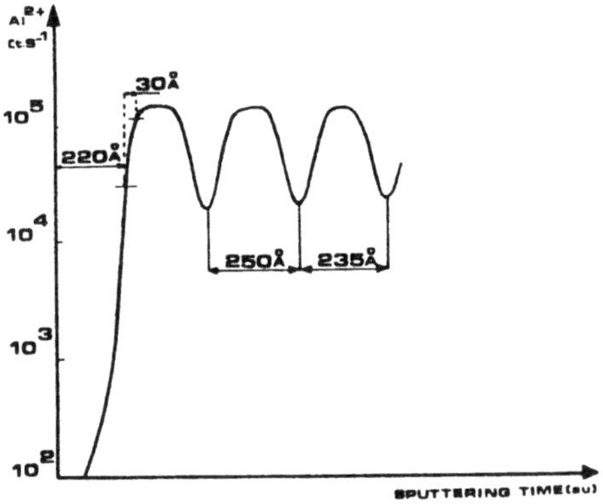

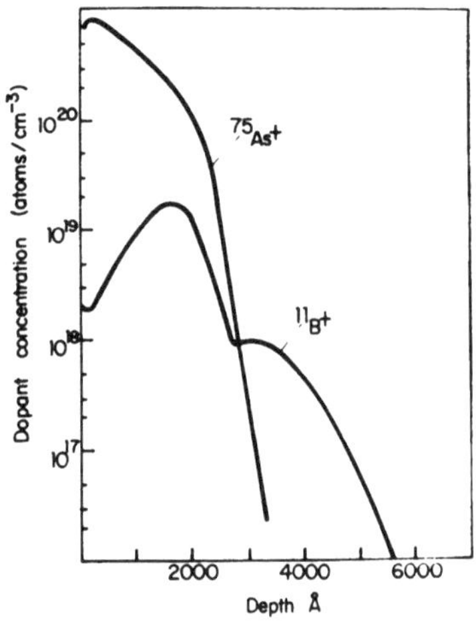

Fig.2 SIMS depth profile analysis of an MBE growth superlattice (O_2^+, 3 keV)

Fig.3 SIMS depth profile analysis of a transistor junction (N_2^+, 12 keV)

a good depth resolution is reached which allows studies of diffusion and implantation induced effects as shown in the boron depth profile.

References

1. H. Liebl, Max Planck Institut Garching, BRD Patent Pending

Part II

Fundamentals I. Ion Formation

Molecule Formation in Oxide Sputtering

H. Oechsner

Physikalisches Institut der Technischen Universität Clausthal und Sonderforschungsbereich 126 Göttingen-Clausthal
D-3392 Clausthal-Zellerfeld, Fed. Rep. of Germany

1. Introduction

The understanding of the formation process of sputtered oxide molecules, in particular of MeO (Me: metal), is important for a number of reasons, such as

- an increased insight in surface processes induced by ion bombardment,
- oxidation studies at solid surfaces by mass spectrometric methods like SIMS or SNMS (Sputtered Neutral Mass Spectrometry [1]),
- the enhancement of secondary ion yields by oxygen exposure or bombardment
- the great fraction of MeO in the flux of oxide specific sputtered particles consisting mainly of neutrals also for oxides.

For all quantitative applications of SIMS and SNMS to oxidized samples an understanding of the MeO formation process is mandatory, since the SIMS or SNMS signals $I(X^{\pm})$ or $I(X^{o})$ for a sputtered species X depend linearly on the MeO partial sputtering yield Y_X by

$$I(X^{\pm,o}) = I_p Y_X \alpha_X^{o,\pm} \eta_X^{\pm,o} , \qquad (1)$$

where I_p denotes the bombarding ion current, α_X the ionisation (or postionisation) coefficient in SIMS (or SNMS) for X, and η_X a geometry and transmission factor (Eq.(1) is valid as long as $\alpha_X^{\pm} \ll 1$ [2]). In the case of oxides, information of the oxygen surface concentration c_O^s can be obtained from $I(MeO^{\pm})$ or $I(MeO^{o})$ when the c_O^s-dependence of the partial sputtering yield Y_{MeO} has been clarified (and when for SIMS the variation of $\alpha_{MeO}^{\pm}$ with c_O^s is known, additionally).

It is the aim of the present paper to establish such a relation between Y_{MeO} and c_O^s by discussing possible formation models for sputtered MeO molecules. A quantitative check of the models will be made with experimental results from in situ SNMS and AES measurements supplying $I(MeO^{o}) \sim Y_{MeO}$ (from Eq.(1)) and c_O^s.

2. Concepts of Molecule Formation by Sputtering

Sputtered molecules have been observed very early in mass spectroscopic investigations of electrically charged particles leaving an ion bombarded solid [3]. More recently, mass spectra of sputtered neutral particles obtained by means of effective postionisation techniques (cf. to [1], e.g.) are also found to display a considerable amount of molecular species [4-6]. In particular, the neutral particles sputtered from oxide surfaces were shown to consist almost exclusively of Me and MeO and a much smaller amount of MeO_2 [7]. For the sputtering of Ta_2O_5 and Nb_2O_5 by low energy Ar^+ ions the neutral particle partial sputtering yield is highest for TaO and NbO [7,8].

Apart from the naive qualitative picture that an atomic cluster being loosely bound to the sample is sputtered away as a whole, two different quantitative models for the formation of sputtered molecules have been developed which both are based on the assumption of binary collision cascades, and which may be characterized as 1) "Atomic Combination Model" (ACM) and 2) "Direct Emission Model" (DEM).

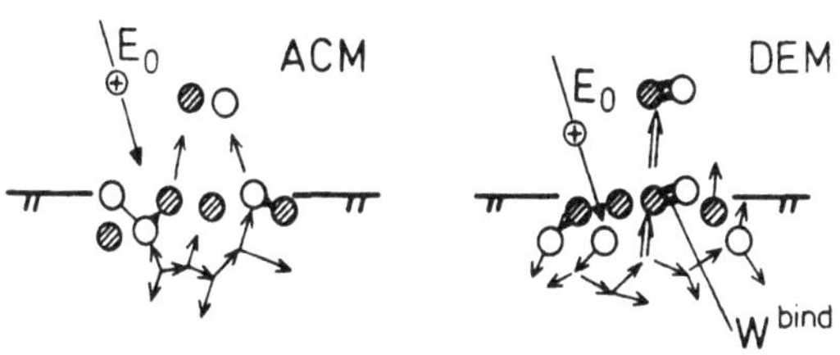

Fig. 1 Schematic representation of the atomic combination model (ACM) and the direct emission model (DEM) for the formation of sputtered molecules

The basic ideas of these two models are explained by Fig. 1:

1.) In ACM two (or more) not necessarily next neighboured surface atoms being knocked on independently in one single binary collision cascade may agglomerate while being ejected, when their emission is properly correlated in time and space, and when their center-of-mass energy is below the binding energy of the molecule to be generated. Such a mechanism being proposed first in [7] has been worked out in detail by GERHARD [9], and independently by KÖNNEN et al. [10].

From the statistics of the combination processes between single atoms sputtered from one individual collision cascade, the partial dimer and trimer yields Y_{Me_2} and Y_{Me_3} for pure metals are found to be proportional to Y_{tot}^2 or Y_{tot}^3, respectively [9], where the total sputtering yield Y_{tot} describes in the elemental case the average number of atoms ejected per single collision cascade. Such a behaviour is well confirmed by experimental results [6,11]. If ACM is expanded to the sputtering of non elemental solids containing two different atomic constituents A and B, we obtain from similar statistical arguments for the partial dimer yields

$$Y_{AB} \sim Y_A A_B; \quad Y_{A_2} \sim Y_A^2; \quad A_{B_2} \sim Y_B^2 \ . \tag{2}$$

Since for stationary atomic sputtering the partial yield Y_i for a component $i(i = A,B)$ due to mass conservation is given by

$$Y_i = c_i^b \, Y_{tot} \tag{3}$$

(c_i^b bulk concentration of i), ACM is predicting the partial molecule yields to vary as

$$Y_{AB,A_2,B_2} \sim Y_{tot}^2, \tag{4}$$

when a solid of constant bulk composition is sputtered under different bombarding conditions, e.g. at different bombarding energies. Analogous relations can be readily established for the trimer sputtering yields of non-elemental samples.

2.) The basic process for DEM is the co-emission of a surface knock-on together with one (or more) strong-bonded next neighboured atoms [8]. When a surface atom A with atomic mass M_A obtains from a binary collision cascade an outward directed momentum corresponding to an energy W^{lab} in the laboratory frame, its relative energy to a next neighboured atom B being at rest in the laboratory frame is given by

$$W^{cm} = W^{lab} \frac{M_B}{M_A + M_B} \ . \tag{5}$$

Then co-emission of the atomic pair AB is possible when again

$$W^{cm} < W^{bind}(AB), \tag{6}$$

where W^{bind} now describes the bond strength between two neighboured atoms A and B in the surface. As a necessary condition for the ejection of A and B, W^{lab} has to be greater than the total effective energy U_o, with which the AB pair is bound to the solid surface. Since W^{bind} and U_o are expected to be in the same order of magnitude, condition Eq.(6) is only fulfilled for $M_A >> M_B$, i.e. for a heavy knock-on A with a light next neighbour B at rest or with low kinetic energy in the laboratory frame.

Hence, appropriate systems for molecule formation according to DEM are presumably heavy metal oxides with $M_{Me} >> M_O$, and strong ionic Me-O bonds at the surface and in the sputtered MeO molecule. Since the number of the metal knock-ons at the surface is proportional to the metal surface concentration c_{Me}^s, the sputtered MeO intensity is given by

$$I(MeO) \sim c_{Me}^s \ f(c_O^s), \tag{7}$$

where f describes the probability for coejection of O together with the Me knock-on. If the probability function f is assumed to be proportional to the number n_O of next nearest O neighbours around a metal atom, Eq.(7) can be written as

$$I(MeO) \sim c^s_{Me} n_O p_O \qquad (8)$$

with p_O being the probability for Me-O coejection, when the energy condition Eq.(6) is fulfilled. The linear relation $f = n_O p_O$ is only valid for $p_O << 1$, which has been established experimentally [13]. Since over a wide range of c^s_O the relation $n_O = kc^s_O$ with $k \simeq 8.4$ can shown to be valid, we obtain with $c^s_M = 1-c^s_O$

$$I(MeO) \sim c^s_O - (c^s_O)^2. \qquad (9)$$

This relation predicts a parabolic dependence of I(MeO) (or of the partial sputtering yield Y_{MeO}) on the surface oxygen concentration, c^s_O and maximum I(MeO) or Y_{MeO} at $c^s_O = 0.5$.

In order to judge which of the two models is responsible for the formation of sputtered MeO, the final relations Eqs.(4) and (9) resulting from ACM or DEM, respectively, have to be compared with corresponding experimental data.

3. Experimental Test of Molecule Formation Models by SNMS

A direct experimental check of the models described in sect. II is possible with the results from recent SNMS measurements. Since in SNMS the particle formation process is decoupled from the ionisation process, α^o_x in Eq.(1) is for a certain ejected species X a constant of the apparatus [2]. Therefore, the SNMS signals $I(X^o)$ in Eq.(1) are directly proportional to the partial sputtering yields Y_x.

For the MeO case discussed here, detailed results for the partial and total sputtering yields of Ta_2O_5 and Nb_2O_5 under normal bombardment with Ar^+ ions below 600 eV are available [7,8]. Under such conditions the secondary ion fractions are always small [12]. Hence, Eq.(1) leading to $I(X^o) \sim Y_x$ for SNMS, is always well applicable (cf. the note to Eq.(1) and [2]).

According to Eq.(4), the atomic combination model is claiming Y_{MeO} to be proportional to Y^2_{tot}, when oxides of constant bulk stoichiometry like Ta_2O_5 or Nb_2O_5 are sputtered at such low temperatures that diffusion effects are negligible. Corresponding plots for the partial TaO and NbO yields obtained from the results in [7] and [8] are shown in Fig. 2a and b. Quite obviously, the behaviour of the experimental results is in both cases severely contradicting the predictions from ACM, thus ruling out that atomic combination processes dominate in the formation of sputtered TaO and NbO molecules. The same conclusion can be drawn for CeO and GdO from SNMS studies of oxidized Ce and Gd [2].

In order to examine experimentally the second possibility for MeO formation described by the direct emission model DEM, polycrystalline Ta surfaces have been exposed to different oxygen doses under simultaneous bombardment with Ar^+ ions of 2 keV at 45° incidence. After constant exposure times at dif-

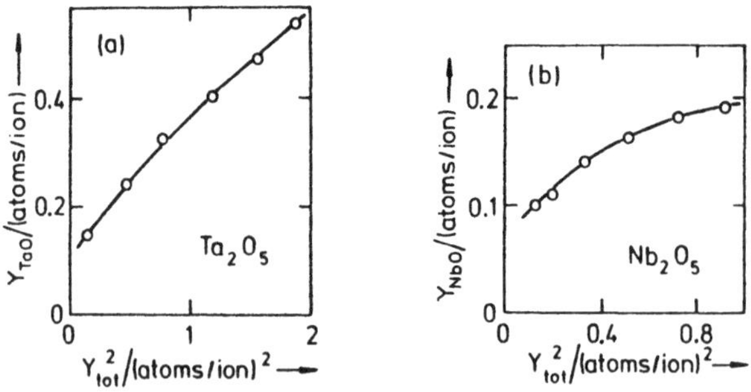

Fig. 2 Partial yields Y_{MeO} for TaO (a) and NbO (b) molecules sputtered from Ta_2O_5 and Nb_2O_5 versus the square of the corresponding total oxide sputtering yields Y_{tot}. All yields are given in atoms/ion and refer to normal bombardment with Ar^+ ions of 100 to 600 eV [7,8].

ferent oxygen pressures the produced thin surface oxide layers have been removed under continuous ion bombardment while registrating the SNMS signals of TaO and Ta.

Corresponding results for 4 different exposures are shown in Fig. 3 [13]. For the lowest oxygen exposure I(TaO°) decreases monotonously with sputtering time, while I(Ta°) increases, thus demonstrating a continuous removal of oxygen from the surface. Beginning with the second lowest oxygen dose, however, a maximum of the SNMS signal for TaO starts to develop which becomes clearly visible for the higher oxygen exposures. This behaviour of the intensity of sputtered TaO, at first increasing and passing thorugh a maximum while the oxygen

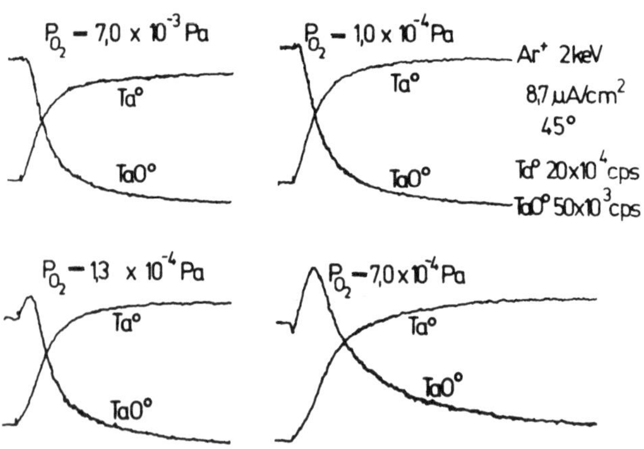

Fig. 3 Variation of SNMS signals for TaO and Ta during sputter removal of thin oxide layers (thickness 1 to several atomic layers) by Ar^+ ions of 2 keV. The oxide coverages have been produced on polycrystalline Ta surface by oxygen exposure at different partial oxygen pressures under Ar^+ bombardment. (Exposure time 180 s; experimental parameters as given in the figure)

content at the surface is continuously reduced, agrees qualitatively very well with the parabolic behaviour of I(MeO) predicted by the direct emission model according to Eq.(9).

A quantitative check of EM is combined with demonstrating the possibilities of SNMS as a quantitative surface analytical method: If Eq.(9) is valid, the surface oxygen concentration c_O^s can be determined in a self calibrating manner from the SNMS signals of MeO normalized to the maximum MeO signal $I^{max}(MeO^o)$, which should be correlated to $c_O^s = 0.5$. Then, we obtain from Eq.(9) [13]

$$c_O^s = 0.5[1 \pm (1 - I(MeO^o)/I^{max}(MeO^o))^{1/2}] \, . \qquad (10)$$

To prove Eq.(10) and simultaneously DEM quantitatively, the c_O^s values obtained from SNMS by Eq.(10) have to be compared with in-situ measurements of c_O^s by an independent surface analytical technique like Auger Electron Spectroscopy (AES). Such a comparison is made in Fig. 4 using the results from in-situ SNMS and AES investigations of oxygen covered Ta [14]. The c_O^s values determined by AES and by SNMS via Eq.(10) agree within the experimental accuracy, thus proving not only the validity of DEM for the formation of TaO, but showing also the analytical possibilities of SNMS.

Another quantitative proof for MeO formation according to DEM is obtained from the previously discussed results for the partial yields of TaO and NbO sputtered from Ta_2O_5 and Nb_2O_5 [7,8]. In Table 1 the fractions $Y^* = Y_{MeO}/Y_{tot}$ of sputtered MeO particles for normal bombardment with 600 eV Ar^+ ions are listed together with the normalized relative energies $W^* = W^{cm}/W^{lab}$ governing the DEM mechanism. For different systems with equal Me-O bond strengths at the surface we expect from DEM that the MeO intensity is proportional to the inverse of W^{cm}, and hence

$$Y^*(MeO) \cdot W^*(MeO) = \text{const.} \qquad (11)$$

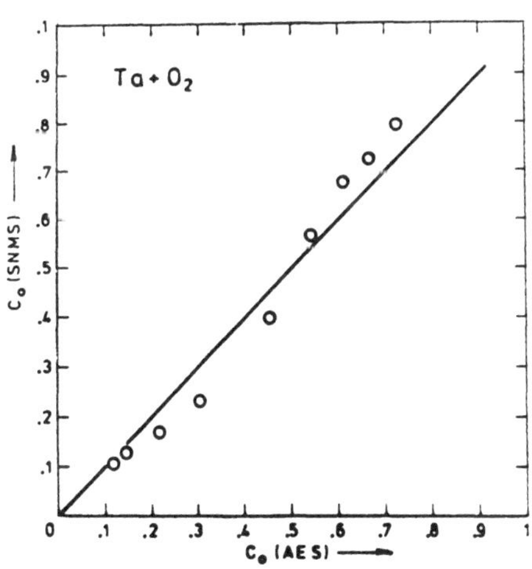

Fig. 4 Comparison of oxygen surface concentrations c_O^s determined by AES, and by SNMS according to Eq.(10)

Eq.(11) is well established by the results in Table 1, thus again proving DEM in view of the similarity of $W^{bind}(MeO)$ for Ta-O and Nb-O expected from free molecule data [15], or vice versa.

Table 1

	$Y^* = \frac{Y_{MeO}}{Y_{tot}}$	$W^* = \frac{W^{cm}}{W^{lab}}$	$Y_O^* W^*$
TaO	0.375	0.081	0.030
NbO	0.20	0.147	0.029

Whereas the direct emission model thus is shown to govern the formation of at least the majority of the MeO molecules sputtered from the oxides of heavy metals, the atomic combination model is expected to apply not only to homonuclear molecules [6,11], but also for non-elemental systems with smaller differences in atomic masses or low bond strengths between neighboured surface atoms.

As an example, the dimer formation process is discussed by means of the SNMS results for NiW and W_2 molecules sputtered from three NiW alloys with different bulk concentration ratios [16]. For stationary sputtering, the surface concentrations c^s_{Ni} and c^s_W are found to deviate significantly from the bulk concentrations due to preferential sputtering of the Ni component. For DEM, the partial sputtering yields of NiW and W_2 should be proportional to $c^s_{Ni} c^s_W$ and $(c^s_W)^2$, respectively. From ACM the yields are predicted to vary as the same expressions calculated with the bulk concentrations, if the samples are sputtered under identical bombarding parameters.

Hence, the ratios of the NiW and W_2 signals obtained by SNMS (or of Y_{NiW} and Y_{W_2}) are expected to be proportional either to the ratios of the surface concentrations or of the bulk concentrations of Ni and W for DEM or ACM, respectively.

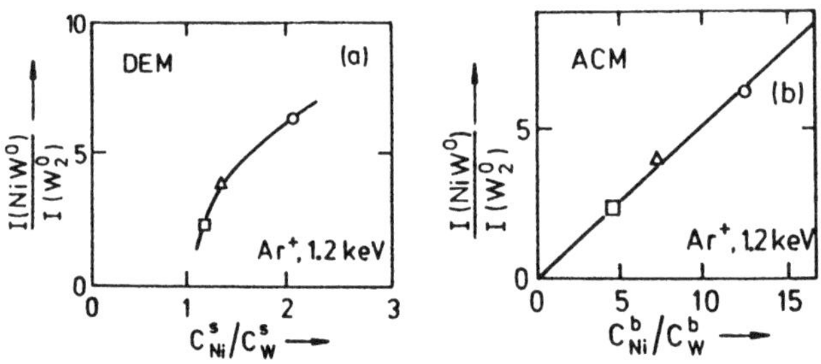

Fig. 5 Ratios of stationary intensities of sputtered NiW and W_2 molecules for three different NiW alloys under normal bombardment with 1.2 keV Ar^+ ions versus the surface (a) and the bulk (b) concentration ratios of Ni and W. The deviations between stationary surface stoichiometry and bulk composition due to preferential Ni sputtering were determined by AES, and can be taken from the different c^s_{Ni}/c^s_W and c^b_{Ni}/c^b_W ratios

Corresponding diagrams are shown in Fig. 5a and b. While no agreement with DEM is found in this case, the good proportionality between the intensity ratios and the ratio of the bulk concentrations c^b_{Ni}/c^b_W demonstrates clearly a molecule formation process according to ACM for NiW and W_2.

The formation of homonuclear molecules according to ACM has been confirmed also by computer simulations [17].

4. Conclusions

It has been shown in the present survey that, depending on the properties of the investigated sample, two different mechanisms are responsible for the formation of sputtered molecules:

a.) For strong (ionic) bonds between next neighboured atoms A and B at the ion bombarded surface and for great differences between the atomic masses of A and B, co-emission of an AB pair is possible, when the heavy species becomes knocked on in a binary collision cascade causing sputtering (Direct emission model, DEM). It has been proved experimentally that DEM dominates in the formation of MeO molecules sputtered from heavy metal oxides. In addition to the discussion of DEM in the preceding sections it should be mentioned that this moelcule formation process can also be shown to apply for MeN molecules sputtered from nitrided metal surfaces [13]. Furthermore, DEM describes the characteristic trends of the formation of sputtered MeO_2 molecules.

b.) For weak (metallic or covalent) bonds between the atomic constituents of a sample and for comparable atomic masses, sputtered molecules are dominantly generated by atomic combination during the ejection of not necessarily next neighboured but individually knocked-on surface atoms from one single collision cascade. This atomic combination model (ACM) is found to be valid at least for the majority of dimers and trimers sputtered from pure metals and from alloys [16]. ACM, however, presumably applies also to systems like Be-O with strong pair bonding but with similar atomic masses, since in such cases the energy condition of Eq.(6) may not be fulfilled.

References

1. H. Oechsner, Appl. Phys. 8, 185 (1975)
2. H. Oechsner, W. Rühe and E. Stumpe, Surface Sci. 85, 289 (1979)
3. cf. e.g. H.W. Werner, in "Electron and Ion Spectroscopy of Solids", L. Fiermans et al. eds., Plenum Press 1978, p. 324
4. J.R. Woodyard and C.B. Cooper, J. Appl. Phys. 35, 1107 (1964)
5. J.W. Coburn, E. Taglauer and E. Kay, J. Appl. Phys. 45, 1779 (1973)
6. H. Oechsner and W. Gerhard, Surface Sci. 44, 480 (1974)
7. H. Oechsner, H. Schoof and E. Stumpe, Surface Sci. 76, 343 (1978)

8. H. Oechsner, H. Schoof and E. Stumpe, Proc. 7th Int. Vac. Congr. and 3rd Int. Conf. Solid Surfaces, Vienna 1977, p. 2023
9. W. Gerhard, Thesis, Univ. of Würzburg, 1973; Z. Physik B22, 31 (1975)
10. G.P. Können, A. Tip and A.E. de Vries, Radiat. Effects 21, 269 (1974)
11. W. Gerhard and H. Oechsner, Z. Physik B22, 41 (1975)
12. A. Benninghoven, CRC Critical Rev. Solid State Sci. 6, 291 (1976)
13. H. Oechsner, E. Stumpe and A. Wucher, to be published
14. H. Oechsner and E. Stumpe, "Proc. 4th ICSS and 3rd ECOSS", Cannes 1980, Vol. II, p. 1234
15. K.P. Huber and G. Herzberg, "Constants of Diatomic Molecules", van Nostrand , 1979
16. H. Oechsner, M. Pfleging, to be published
17. D.E. Harrison, Jr. and C.B. Delaplain, J. Appl. Phys. 47, 2252 (1976)

Dependence of Ionization Yields Upon Elemental Composition; Isotopic Variations

G. Slodzian

Laboratoire de Physique des Solides, Université Paris-Sud, F-91405 Orsay, France

Owing to its great sensitivity secondary ion mass spectrometry is generally considered as a powerful technique for analyzing surfaces or microvolumes of solid samples despite many kinds of problems that remain to be solved. Among them, the ionization phenomena are directly responsible for the difficulties encountered in quantitative analysis. The subject will be restricted essentially to metallic alloys flooded with oxygen and silicate minerals bombarded with oxygen primary ions. In such experimental conditions, the elemental composition of the target controls the ionization yields which in turn prevents quantitative elemental analysis to be performed in a simple way. Moreover, it could very well be that the processes inducing the dependence of the ionization yields upon elemental composition affect isotopic abundance measurements as well.

This paper is mainly devoted to a synthetic presentation of the work done by the "Orsay Group", more extended bibliography can be found in the cited references. First, experimental results will be briefly recalled. Then a bond-breaking model will be presented as a first step towards an explanation of both matrix and isotopic effects.

Metallic Alloys

The technique of flooding the surface with oxygen while the sample is (simultaneously) bombarded with argon (or oxygen) ions has been described many times [1]. Let us just recall that as the oxygen pressure is raised the secondary ion currents are increased until a "saturation" plateau is reached. There are some strong presumptions that there is a permanent build-up of a very thin oxide film but very little is known on its true composition and thickness. Secondary ions are emitted from that assumed oxide film with much higher yields than from a bare metal bombarded with argon ions. Bombardment with oxygen ions produces similar effects but in general a saturation cannot be completed that way.

Let us take a binary alloy as a target and let [M] and [N] be the atomic concentrations. The intensities $I(M^+)$ and $I(N^+)$ of the secondary ions M^+ and N^+ can be written as:

$$I(M^+) = k\ P(M^+)\ [M], \qquad I(N^+) = k\ P(N^+)\ [N] \qquad (1)$$

where $P(M^+)$ and $P(N^+)$ are the practical ionization yields and k is a coefficient which accounts for the primary density beam and the sputtering yield; $P(M^+)$ and $P(N^+)$ are often called sensitivity coefficients. In order to eliminate k which is often a coefficient uneasy to determine, it is advisable to consider the ratio $I(M^+)/I(N^+)$ or:

$$R_N^M = \frac{P(M^+)}{P(N^+)} = \frac{I(M^+)}{I(N^+)} \times \frac{[N]}{[M]} . \qquad (2)$$

The striking experimental fact is that this relative yiels is not a constant; it varies with the elemental composition of the target. R_N^M exhibits what may be called a "matrix effect". It has been shown [2] that the R_N^M ratios fit a homographic function where [M] and [N] are the variables. In other words, $P(M^+)$ and $P(N^+)$ can be written as a linear function of the concentrations such as:

$$P(M^+) = P_M^M[M] + P_N^M[N] \qquad P(N^+) = P_M^N\,[M] + P_N^N[N] \cdot \qquad (3)$$

The coefficient P_M^M and P_N^M are the practical ionization yields of M^+ ions emitted respectively from a target of pure M and a target of N containing M as an impurity. Since only relative yields are given by the measurements reported here, only ratios like P_N^M/P_M^M are available. The experimental values of such ratios are shown in Table 1 [2] for Ni, Cr, Fe binary alloys. It can be seen that Ni^+ ions are 20 times more efficiently produced when sputtered from a chromium based alloy than from pure nickel.

Table 1

P_N^M / P_M^M		N: Ni	N: Fe	N: Cr
M	Ni	1	6.1	20
	Fe	0.23	1	2.7
	Cr	0.29	0.67	1

$P_{Cr}^{Cr}/P_{Ni}^{Ni} = 39 \pm 4$

$P_{Fe}^{Fe}/P_{Ni}^{Ni} = 8.7 \pm 0.7$

$P_{Cr}^{Cr}/P_{Fe}^{Fe} = 4.5 \pm 0.5$

The case of Cu-Al and Ni-Al alloys can be found in Table 2 [3].

Table 2

M	P_{Al}^M/P_M^M	P_M^{Al}/P_{Al}^{Al}	P_{Al}^{Al}/P_M^M
Cu	520	0.6	870
Ni	26	0.67	60

From these examples, one can be easily convinced that the dependence of the so-called sensitivity coefficients upon the elemental composition can be neglected only if one component is present at a very low level of concentration so that one term becomes negligible compared to the other in the expression of $P(M^+)$.

Let us also mention that similar results have been found for ternary alloys and mixed oxides [4]. In the latter case, it is worthwhile eliminating that oxygen flooding was necessary for reaching the saturation plateau. The most plausible reason is that differential sputtering had depleted the surface layer of oxygen. Another interesting point worth to be mentioned is that the cation valence has an influence on the ionization practical yields.

Silicate Minerals

Now we are dealing with samples which contain oxygen as a major component. Difficulties may arise from the insulating nature of the target. By vacuum depositing a metallic grid on the surface and adjusting the density of the primary beam the electrical charging-up can be often avoided. Then secondary ion currents are stable, reproducible and the ion-drift is considerably reduced.

It has been shown that the same linear approximation can provide us with a reasonable fit to the experimental results [6]. But here the problem is more complex because the number of components is high and therefore the number of coefficients which have to be determined by calibration with standards is usually much greater than for metallic alloys. The ratio of the yields of an element X and silicon is given by:

$$R_{Si}^{X} = \frac{P_{Si}^{X}[Si] + P_{Al}^{X}[Al] + \ldots}{P_{Si}^{Si}[Si] + P_{Al}^{Si}[Al] + \ldots} \tag{4}$$

where [Si], [Al], ...[X] are the cationic concentrations. Some simplifications have been proposed on an empirical basis [6]. For instance, it has been assumed that the yield of silicon is nearly independent of composition; small corrections can be needed eventually. Then $P_{Si}^{Si} \approx P_{Al}^{Si} \approx \ldots$ and the relative yields read as:

$$R_{Si}^{X} \sim K_{Si}^{X}[Si] + K_{Al}^{X}[Al] + \ldots \tag{5}$$

where $K_Y^X = P_Y^X/P_{Si}^{Si}$ are unknown coefficients that must be determined through an experimental calibration procedure. Let us consider a given element X and let us take a number n of standards equal to the number ν of components. The number ν of unknown K_Y^X is then equal to the number n of equations so that an unique set of K_Y^X coefficients can be calculated. But due to the limited accuracy with which concentrations are known in the standards and the uncertainties in the experimental values of R_{Si}^X the calculated values of K_Y^X vary greatly from one set of standards to another. A "stable" solution can be reached using a least square method when the number n of standards - that is the number of equations - is much larger than the number ν of components. Fair agreement with the reference values can be obtained this way. For instance, with a set of about twenty standards and nine different cations, a set of nine K_Y^X coefficients can be determined thereby allowing computation of relative ion yields which generally differ by less than 5 % from the measured ones.

Strictly speaking this does not prove the general validity of the linear approximation. It might very well happen that the sampling of the standards presently used does not cover a range of concentrations large enough to prevent that some relations between concentrations are accidentally favoured resulting in correlated deviations of K_Y^X that would leave unchanged the calculated R_{Si}^X. It is clear that before being accepted the linear approximation has to be checked against a much larger sampling of standards. However, as things stand, we think that it provides us with a tool reasonably good for quantitative analysis of silicates.

Quantitative Analysis

From the foregoing results it appears that one should know the composition of the target before hand to be able to calculate the relative practical yields R_{Si}^X. Then it would be a simple matter to determine the cationic concentration of X by using an expression such as:

$$[X] = \frac{I(X^+)/R_{Si}^X}{I(Si^+) + (Al^+)/R_{Si}^{Al} + \ldots} \qquad (6)$$

In other words, quantitative analysis is, in general, impossible. Fortunately, this pessimistic statement can be circumvented by using an iteration procedure [7]. First a set of concentrations is assumed, then yields are calculated and finally concentrations are determined using those yields. The new set of concentrations is fed into equations such as (5); then the new set of yields is used to calculate a new set of concentrations. The cycle is repeated over and over until the calculated concentrations reach a stable limit (7).

"Bond-breaking Model"

The simplicity of the linear approximation is a strong incentive to look for some kind of simple model that could account for it, at least on a qualitative basis. The bond-breaking model is an attempt to reach that goal.

Let us consider a M atom of the surface surrounded with oxygen atoms and let us assume that M is ejected after a collision with a N atom. Let us further assume that as M leaves the target it breaks a "bond" with the complex anion A composed of N and the neighbouring oxygen atoms, let A_N^M be the ionization probability of M during this process. Of course, M could also be ejected by a collision with a M atom, then A_M^M would be its probability to be ionized by the breaking of its bond with the complex anion A' now composed of M and the neighbouring oxygen atoms. If we assume that the flux of particles moving inside the target towards the surface and being efficient in sputtering events has the same composition as the bulk the probability for M to be ejected either by N or by M are respectively equal to [N] and [M]. The total probability is given by:

$$P(M^+) = A_N^M [M] + A_N^M [N] \quad . \tag{7}$$

After going through some averaging procedure to account for different ionization processes, energy and angular distributions and their correlated discrimination by the collecting and filtering optics, expression (7) would become identical to that assumed in the linear approximation.

The next step will be to describe the ionization process itself. However the complexity of the phenomena involved here is such that one must look for a simplified model able to lead to a qualitative understanding. The well known LANDAU-ZENER-STUECKELBERG model (LZS) for diatomic molecules seems to be a good candidate. The model is summarized in Fig. 1 where the potential curves of respectively an ionic bond corresponding to A^- and M^+ and a covalent bond between A and M have been plotted against the distance D of the two "nuclei". The curves intersect at point C. Taking into account the interaction between the two configurations leads to an anticrossing diagram classical in quantum mechanics. Now, if we imagine that the two nuclei are pulled apart there is a probability to obtain A^- and M^+ as separate ionic species. This probability is approximately given by:

$$P = \exp(- a/v_C) \tag{8}$$

where "a" is a parameter depending upon the slopes of the potential curves formerly crossing at point C and on the energy gap between the actual potential curves at point C; v_C is the velocity of one nucleus relative to the other taken at a separation distance corresponding to point C.

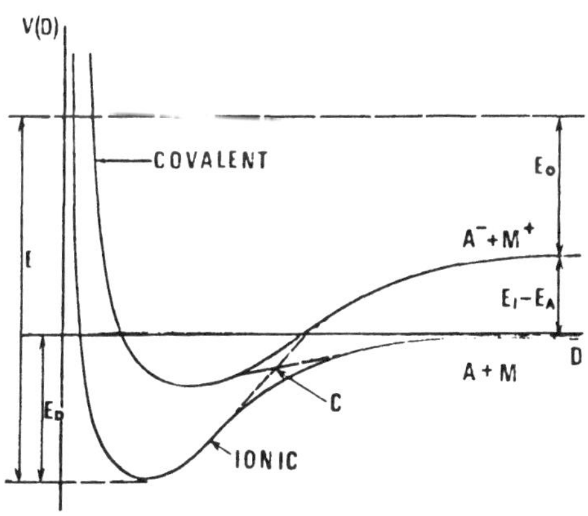

Fig. 1 Anticrossing diagram. E_I ionization potential of M, E_A electron affinity of A, E_D dissociation energy of the diatomic molecule AM, D distance between the two "nuclei" A and M. E energy received by M in the last collision

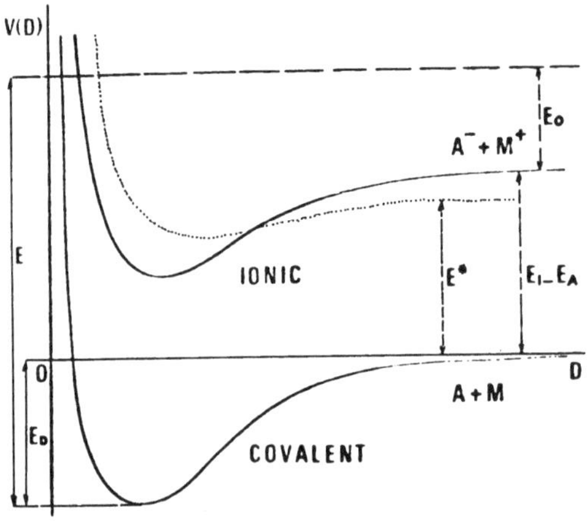

Fig. 2 Anticrossing diagram including an excited state (dotted curve) when the fundamental potential curves of ionic and covalent bonds do not intersect

Figure 2 shows another type of situations that may occur when the two potential curves do not intersect. Then an intersection point may be obtained by considering an excited state. In this case the collision between N and M could also have the effect of bringing the bond between A and M into an excited state.

When we are dealing with complex ions such as A^- the diatomic schema is obviously inadequate for a precise representation of the process but it may be good enough if we do not aim at more than a description of the most general features. For instance, the ionization probability is very sensitive to the parameter "a". Now "a" depends on the electron affinity, on the ionization potential and on the nature of the bond. For a given couple A-M excited states may produce different values of "a" and on the other hand the composition of the anion A depends on the oxygen concentration in the surface layer. The great differences being found in the P_N^M can be produced by any of the foregoing factors.

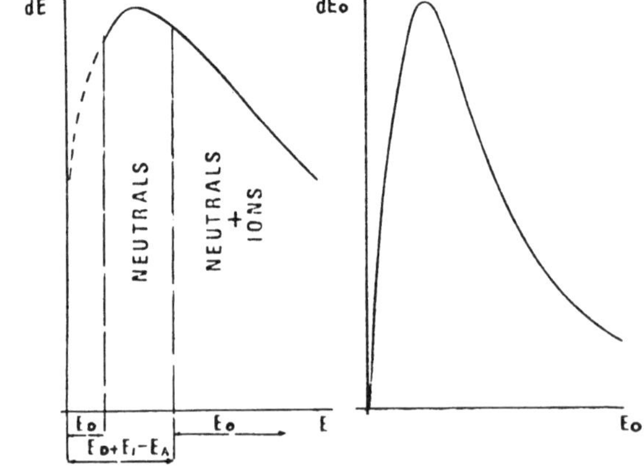

Fig. 3 Schematic diagram of the correlation between the E energy distribution in the target and the E_0 kinetic energy distribution of secondary ions

Moreover, the velocity dependence of the ionization probability could produce different shapes of the energy spectra of secondary ions and give us some clues for the understanding of isotopic effects. Concerning the first point, it should be noted that an M atom which receives an energy E will appear in the secondary ion energy spectrum with an energy $E_o = E - (E_D + E_I - E_A)$ (Fig. 3).

Isotopic Effects Upon Ionization Yields

Let M_i and M_j be two isotopes of a given element and let R^m_{ij} and R^o_{ij} be the measured ratio, $I(M_i^+)/I(M_j^+)$ and the standard abundance ratio respectively. One can define a relative deviation by $\Delta_{ij} = R^m_{ij}/R^o_{ij} - 1$ and a coefficient α such as $\Delta_{ij} = \alpha.\ (\delta M_{ij}/M_j)$ where $\delta M_{ij} = M_i - M_j$. Measurements made on terrestrial mineral samples [8,9] and on elements known to be unfractionated in geochemical processes have led to results that can be summarized as follows:

- α is a negative constant, that is the heavy isotope is emitted with an ionization probability lower than that of the light one;
- the higher the energy of secondary ions, the lower $|\Delta ij|$;
- Δij is correlated to the ionic "useful yield" τ (number of ions produced per atom sputtered of a given species), to the shape of the energy distribution curve and to the composition of the target.

Let us try to see if these features are consistent with the bond-breaking model. First the relative variation of the ionization probability (8) is given by:

$$\delta P/P = (a/v_c)(\delta v_c/v_c).$$

Two points have to be assumed. The first one is that M_i (or M_j) is set into motion by a collision with a given B atom, the second one is that the electronic exchanges occur between M_i and the complex anion as a whole. The latter is composed of B and the surrounding oxygen atoms but the overlapping orbitals are likely to be mainly centered on oxygen nulcei and M_i. Therefore v_c is the velocity of M_i (or M_j) relative to a frame of reference linked to the target. The relative variation given by $\delta v_c/v_c = K(\delta M_{ij}/M_j)$ where K is a function which depends on the dynamical conditions of the last collision (energy and mass of B, energy consumed to reach point C from the bottom of the potential curve, ...). In the foregoing conditions K is negative (as it is easy to verify in a head-on collision). If we had considered a diatomic collision between and M_i (or M_j) only, v_c would have been the relative velocity of the nuclei and K would be positive for the collision is inelastic. The expression of the relative variation of the ionization probability is given by:

$$\frac{\delta P}{P} = \frac{Ka}{v_c} \cdot \frac{\delta M_{ij}}{M_j}. \tag{9}$$

The most interesting feature is that (8) and (9) give a qualitative understanding of the experimental observations:

- negative sign of K implies that the heavier the isotope, the lower the ionization probability;
- $|K|/v_c$ is a decreasing function of the energy E_o of the secondary ions;
- at a given velocity v_c, the smaller "a" the higher the ionic yield and the lower the relative isotopic deviation;
- the ratio of the ionization probabilities of slow ions over fast ions is smaller for high values of "a". Thus, the energy spectrum looks depleted of slow ions when the isotopic relative deviation is high and the ionic yield low.

From the above considerations it appears that the ionization processes alone could account for the isotopic effects although many points have been disregarded for the sake of simplicity (such as angular effects ...). However one cannot ignore that the sputtering process itself could also be responsible for the isotopic effects. For instance it has been recently proposed [10] that the angular distributions of the sputtered isotopes are slightly different, if all secondary ions were collected it would be enough to consider the ionization probabilities. But since secondary ions are collected into a limited solid angle (which in addition varies with the initial energy of ions in our experimental set-up) the isotopic ratio could vary according to the direction and the half angle of the collection cone. It is very likely that both processes, sputtering and ionization, work together to produce isotopic effects but it presently seems difficult to evaluate the respective contribution of each one. More refined experimental results are needed.

Conclusion

The linear approximation of the ionization practical yield and the isotopic effects can be understood in the framework of a bond-breaking model. In itself that does not prove the validity of the model since many questions are still remaining. For instance, if the production of polyatomic species changed from one standard to another that would modify the ejection of monoatomic particles and induce apparent changes in the ionization yield. Other points concerning surface composition, ejection of a M particle by collision with an oxygen atom, angular effects, multiatomic versus diatomic processes, potential curves, negative ions, should be discussed.

More precise and comprehensive experimental results are needed to test the generality and the accuracy of the linear approximation itself. And more refined measurements should be made on the isotopic abundance ratios if one wishes to gain in understanding. Nonetheless, even in its simplistic form, the bond-breaking model may serve as a guide line for future developments.

References

1. M. Bernheim and G. Slodzian, Int. J. Mass. Spectrom. and Ion Phys. 12, 93 (1973)
2. J.C. Pivin, C. Roques-Carmes and G. Slodzian, Int. J. Mass. Spectrom. and Ion Phys. 26, 219 (1978)
3. C. Roques-Carmes, J.C. Pivin and G. Slodzian, In Secondary Ion Mass Spectrometry SIMS-II, ed. by A. Benninghoven, C.A. Evans, Jr., R.A. Powell, R. Shimizu, H.A. Storms, Springer Series in Chemical Physics, Vol. 9 (Springer, Berlin, Heidelberg, New York 1979) p. 160
4. J.C. Pivin, C. Roques-Carmes and G. Slodzian, J. Appl. Phys. 51, 4158 (1980)
5. G. Slodzian, R. Dennebouy and A. Havette, XVIII Colloquium Spectroscopicum Internationale, 11, 590 (1975)
6. A. Havette and G. Slodzian, C.R. Acad. Sci. Paris, 290, serie B, 51 (1980). English version J. de Phys. Lett. 41, 247 (1980)
7. A. Havette and G. Slodzian, J. Microsc. Spectrosc. Electron. 5, 131 (1980)
8. G. Slodzian, J.C. Lorin and A. Havette, C.R. Acad. Sci. Paris, 291, Series B, 121 (1980). English version J. de Phys. Lett. 44, 555 (1980)
9. A. Havette, J.C. Lorin and G. Slodzian, J. Microsc. Spectrosc. Electron. 6, 169 (1981)
10. P. Sigmund, A. Oliver, G. Falcone, IXth Int. Conf. on Atomic Collisions in Solids, Lyon (1981) to be published in Nucl. Inst. and Methods

Measurements of the Energy Distributions of Positive Secondary Ions in the Energy Range from 0 to About 500 eV

C. Pahlke, H. Düsterhöft, and U. Müller-Jahreis

Sektion Physik der Humboldt-Universität zu Berlin,
1040 Berlin, DDR

Experimental

In this paper energy distributions of singly charged ions of Bi, Sb, W, Th, Cu, Zn and Cd as well as ions of some of their oxides emitted from the pure elements are presented. The purity of the target materials was ranged from 99 % (W, Th) to 99.999 % (other elements); all the materials had a polycrystalline structure and were mechanically polished.

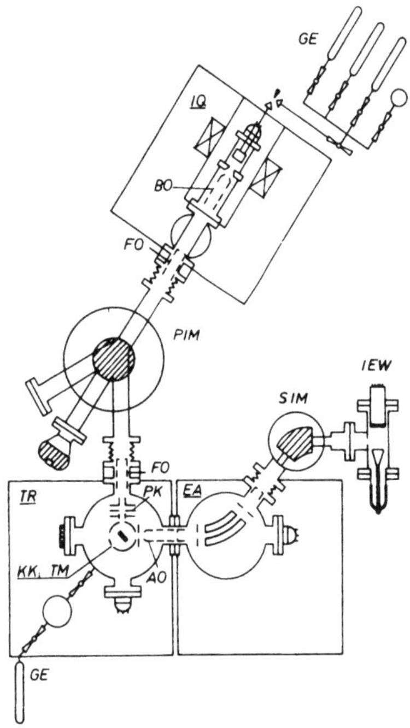

Fig. 1 Measuring arrangement. GE - gas inlets, IQ - ion source, BO, FO - acceleration and focusing system, PIM - mass analyzer of the primary ions, FO, PK - focusing system and collimator, KK, TM- target with screening box, AO - post-acceleration lens, EA - energy analyzer, SIM - secondary ion mass analyzer, IEW - ion electron converter

Figure 1 depicts the experimental set-up. The ion source (low voltage discharge type) produces Ar^+ ions, which were accelerated to 12 keV, mass selected by a magnetic mass analyzer and focused on the target to a spot of 0.88 mm^2 with a current density of 0.23 mA/cm^2. The space around the target has been carefully screened of electric fields.

Secondary ions emitted from the target at an angle of 90° with respect to the incident beam, leave the field-free zone in the vicinitiy of the target through a nickel grid (400 meshes/mm^2) and reach a post-acceleration system. Here they are accelerated up to 1200 eV and drift to the entrance slit of an electrostatic energy analyzer with a resolving power $E/\Delta E \approx 300$. The energy analyzer is followed by a magnetic mass analyzer ($M/\Delta M \approx 40$), an ion electron converter, a multiplier, and electronic devices for signal processing. To reduce systematic errors, which would result from an energy dependent transmission of the analyzing system, the measurements were carried out at constant final energy of the secondary ions (i.e. the analyzer fields were kept constant).

Some tests were performed to check the independence of the recorded distributions on the parameters of the apparatus, especially on the post-acceleration voltage. The angle of incidence of the primary Ar^+ ions was changed in the range from 20° to 70° without any influence on the energy distributions.

The measurements were performed at an enhanced O_2 partial pressure of about $6 \cdot 10^{-4}$ Pa (residual pressure $2.7 \cdot 10^{-4}$ Pa) in order to attain a coverage with oxygen under dynamic sputtering conditions. The energy distributions are presented in a normalized form.

Results

According to the shapes of the energy distributions a classification of the seven investigated elements in three groups is suggested (Figs. 2 to 4).

The first group includes the elements Bi, Sb, Cu and W (Fig. 2). Their energy distributions are characterized by a most probable energy of a few electron-volts and a high energy tail which is best fitted by a power dependence of the type E^{-n}. The most probable energies are in the order of 4 to 6 eV

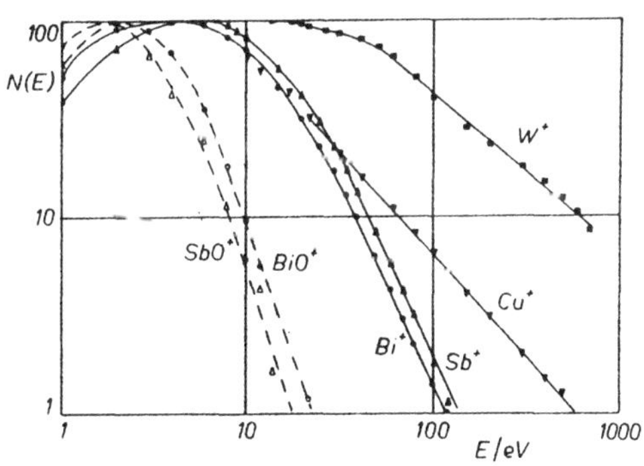

Fig.2 Normalized energy distributions of W^+, Cu^+, Bi^+, Sb^+, SbO^+ and BiO^+ emitted from the elements. O_2 pressure: $6 \cdot 10^{-4}$Pa

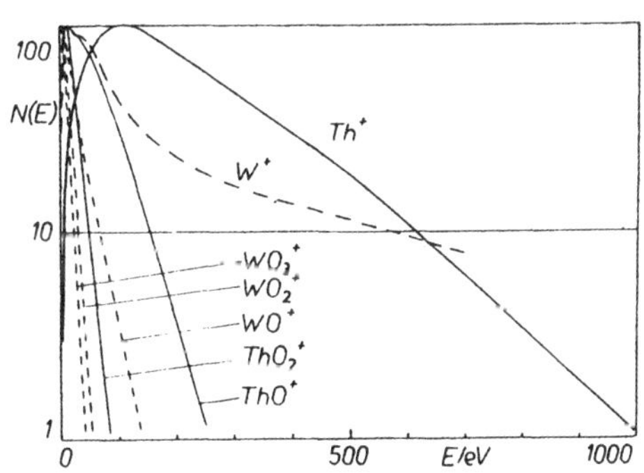

Fig.3 Normalized energy distributions of W^+ and Th^+ and of some of their oxides emitted from the elements. O_2 pressure: $6 \cdot 10^{-4}$ Pa

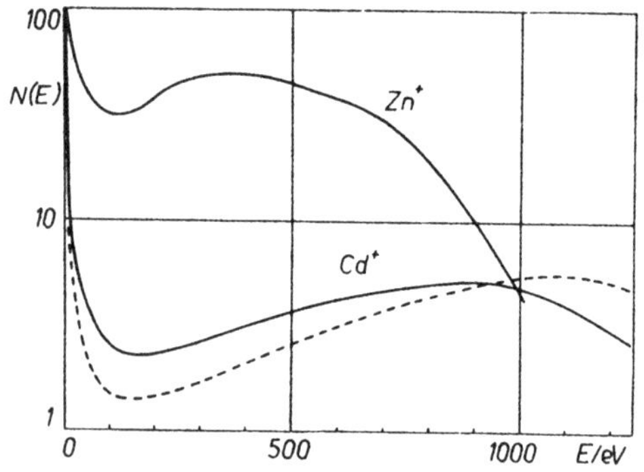

Fig. 4 Normalized energy distributions of Zn^+ and Cd^+ emitted from the elements (solid lines - O_2 pressure $6 \cdot 10^{-4}$ Pa, dashed line - oxygen pressure reduced by about one order of magnitude, resudual pressure $2.7 \cdot 10^{-4}$ Pa)

with exception of W with an energy E_{mp} of 14 eV and a shoulder in the energy distribution at about 50 eV.

The second group is only represented by thorium. In Fig. 3 thorium and tungsten are compared. In contrast to the first mentioned elements thorium exhibits a most probable energy of the order of 100 eV but a less extensive high energy part, which cannot be described by a function of the E^{-n} type but by exponential functions. On the other hand, the distributions of the oxides ThO and WO are similar. Both have low energy maxima (ThO 14 eV, WO 8 eV) and exponentially decreasing high energy parts.

The third group comprehend the elements zinc and cadmium (Fig. 4). Their energy distributions exhibit somehow unusual structure with maxima at low energies (Zn 3 to 7 eV, Cd 2 eV) and broad maxima in the high energy range (Zn 300 to 400 eV, Cd 1 keV). The position of these maxima is influenced by oxygen as demonstrated in the case of cadmium.

Discussion

The here presented results confirm that energy distributions of secondary ions often exhibit a structure which points to two or more different kinds of ionization.

The observed structure in the energy distribution of tungsten at about 50 eV coincides qualitatively with results obtained by RUDAT and MORRISON (primary ions O_2^+, 5.5 keV) [1], but not with those of KRAUSS and GRUEN (O_2^+, 3 keV) [2], who obtained a low energy peak at 34 eV and a breakpoint in the slope of the high energy part of the energy distribution at about 150 eV. Besides, the structure disappears in the case of W^+ emission from a WTh (5 %) mixture. This may be caused by the increase of the oxygen coverage on the target surface by reason of the great affinity of thorium to oxygen. The suppression of the structure in the energy distribution of tungsten by an oxygen coverage has been demonstrated [1].

The change of the slope of the energy distribution of Th at about 500 eV (see Fig. 3), occures only at enhanced oxygen pressure. At reduced oxygen pressure the energy dependence can be described by one exponential function of the type exp(-mE) with m = const. over the whole high energy range from 150 eV to 1 keV, indicating a uniform ionization process. There are some indications that the distribution of uranium [3,4] is similar to that of thorium. On the influence of oxygen both elements are characterized by their low values of the work function and iozination energy.

In the case of cadmium the ratio of maximum to minimum intensity decreases with increasing oxygen coverage (Fig. 4). This result can be compared with the energy distributions of Cd ions emitted from a CdHgTe matrix, which were obtained by WITTMAACK [5] in the energy range up to 20 eV (primary ions O_2^+, 12 keV). In [5] the primary ion current density was elevated from 0.1 to 10 mA/cm^2, the ratio of maximum intensity (at about 2 eV) to the intensity at 20 eV increased markedly. This behaviour is qualitatively explained by a post ionization process of the outgoing atoms, which is certainly dependent on the primary ion current density. In light of the here presented results, such an effect could be caused in the case of poorer vacuum conditions or at higher oxygen partial pressure by a diminuation of the oxygen coverage at an enhanced primary ion current density, too.

References

1. M.A. Rudat and G.H. Morrison, Int. J. Mass Spectrom. Ion Phys. 30, 233 (1979)
2. A.R. Krauss and D.M. Gruen, Appl. Phys. 14, 89 (1977)
3. M.A. Rudat and G.H. Morrison, Surf. Sci. 82, 549 (1979)
4. A.G. Morgan and H.W. Werner, Surf. Sci. 65, 687 (1977)
5. K. Wittmaack, Adv. Mass Spectrom. 8, 503 (1980)

Ion Dose Effects in Static SIMS

W. Speckmann, S. Prigge, and E. Bauer

Physikalisches Institut, Technische Universität Clausthal,
D-3392 Clausthal-Zellerfeld, Fed. Rep. of Germany

1. Introduction

Ion bombardment of a solid surface may lead to significant changes of the surface, i.e. a change of the chemical or structural nature due to sputtering effects. Therefore static secondary ion mass spectroscopy has been developed in order to avoid damage of the topmost layers during measurement. Static SIMS means working with mean lifetimes of one monolayer of the order of 10^4 s [1]. Nevertheless, we want to show in this paper that we find dramatic changes in the SI spectrum although the conditions for static SIMS are fulfilled.

2. Experimental

We have investigated the adsorption of sulphur on a niobium (110) single crystal surface. The adsorption behaviour has also been studied by Auger electron spectroscopy, work function change, LEED and thermal desorption measurements. The static SIMS investigations have been carried out with a primary Xe^+ ion beam of 1 keV energy and a current density of less than 5×10^{-8} A cm^{-2}. The incident angle of the primary ions has been 45°. Under these conditions less than 4.5 % of a sulphur monolayer are sputtered within 12 minutes which has been the maximum duration of a typical experiment. The normally emitted secondary ions, positive and negative, were detected with a commercial quadrupole mass analyzer. The studies have been made in an UHV-chamber with a base pressure of 3×10^{-11} Torr.

3. Results and Discussion

The contamination of the surface with oxygen, carbon and nitrogen was smaller than 2 % which is a result of our Auger calibration. From the Auger and work function measurements we conclude an adsorption behaviour of the sulphur at room temperature which is illustrated schematically in Fig. 1a. The electronegative sulphur atoms sit on the top of the first niobium layer so that the corresponding dipole moment results in an increase in the work function (Fig. 2).

At this point we should make a remark about our atom model which is, of course, a little bit simplified, especially concerning the lateral structure of the sulphur overlayer. Here

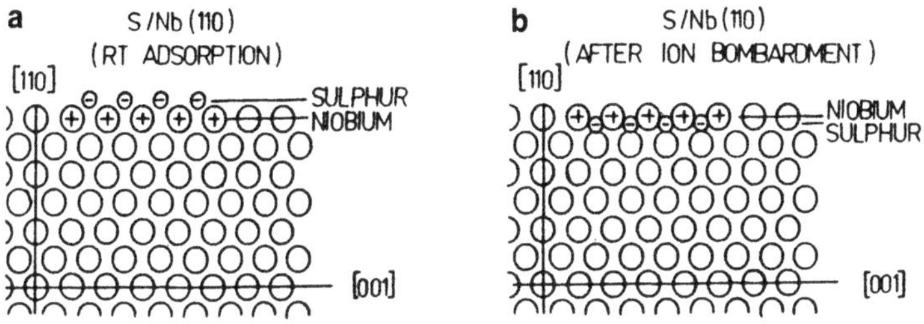

Fig. 1 A model for the position of the sulphur atoms on a niobium (110) single crystal surface a) for room temperature adsorption, b) after ion bombardment

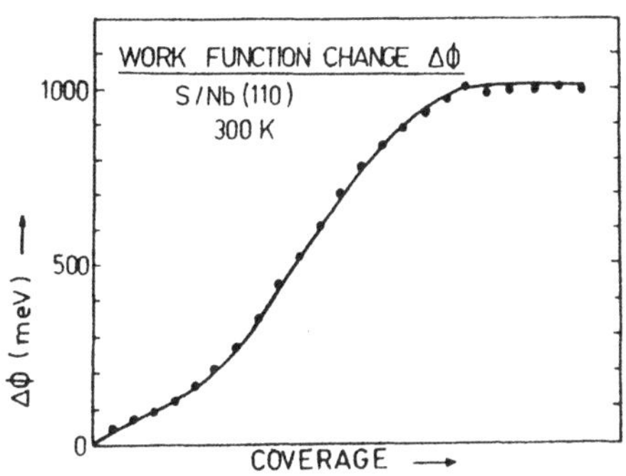

Fig. 2 Work function change for room temperature adsorption of sulphur on Nb (110)

we get changes with coverage and temperature which are neglected in our drawings. So in our model only the distribution of the sulphur normal to the surface should be taken more seriously. Because of the complex adsorption behaviour at high coverage we will concentrate only on low sulphur coverage.

So, this is the situation without bombardment. Of course, to do SIMS we need ion bombardment even in the static SIMS mode. In the following section we show what happens with the S/Nb system under ion bombardment and after that we give an explanation of our results.

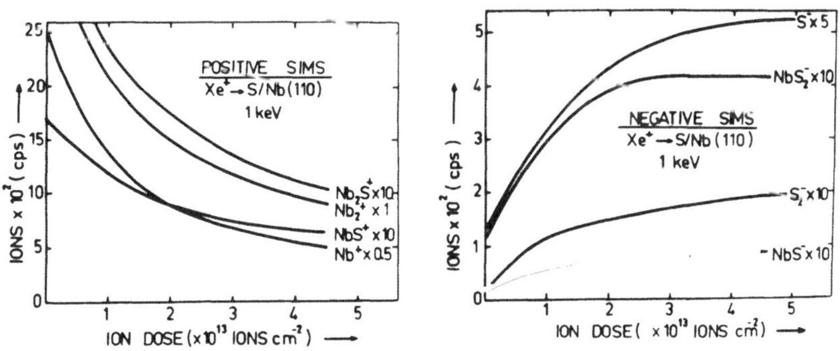

Fig.3 Change of the secondary ion yield for several ion species during Xe^+ bombardment

Figure 3 shows the intensity of the secondary ions as a function of bombarding time, or even better of bombarding Xe ion dose. In the upper part a decrease in yield of the positive ion species is shown, in the lower part an increasing negative ion yield, with ion dose. The effect itself is independent of sulphur coverage and of primary ion energy, but the effect is larger for higher coverage and higher primary ion energy.

Up to now we have the situation that the positive ion yield decreases and the negative ion yield increases upon ion bombardment. In a certain way this was found previously in 1974 by P. WILLIAMS and Ch. EVANS [3] for oxygen on different solid surfaces. Nevertheless they could not really explain this phenomenon.

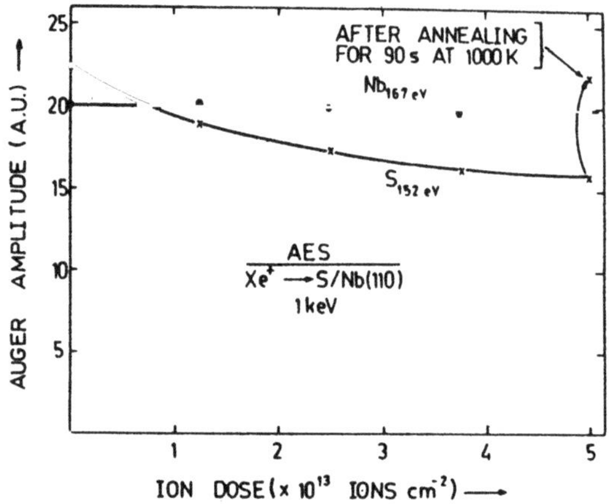

Fig. 4 Change of the Auger amplitude (peak-to-peak) of Nb (167 eV) and S (152 eV) during Xe^+ bombardment

Figure 4 shows the Auger signal of the sulphur (152 eV) and the niobium (167 eV) peak versus ion dose. From this figure one can see that the sulphur peak amplitude is reduced while the niobium peak amplitude remains constant. At the end of the bombardment procedure we have annealed the crystal for 90 s up to 1000 K and the sulphur Auger signal beocmes nearly 100 % of the value we started with before the ion bombardment. This indicates that the process induced by ion bombardment is reversible upon annealing the crystal.

Firstly we should remember that there no significant sulphur desorption occurs due to sputtering, so that the decrease in the sulphur Auger intensity is not a simple result of a decrease in the number of sulphur atoms at the surface. Therefore the sulphur-niobium system must undergo a change but with constant atom concentration. We propose an incorporation of the sulphur atoms beneath the first niobium layer due to the Xe^+ bombardment as it is shown in Fig. 1b. This would result in a shadowing of the sulphur Auger electrons by the niobium overlayer and therefore, a decrease of the sulphur Auger intensity. On the other hand the niobium Auger intensity should stay nearly constant because the small sulphur atoms have only a negligible shadowing effect for the niobium Auger electrons which we know from room temperature adsorption and low coverage.

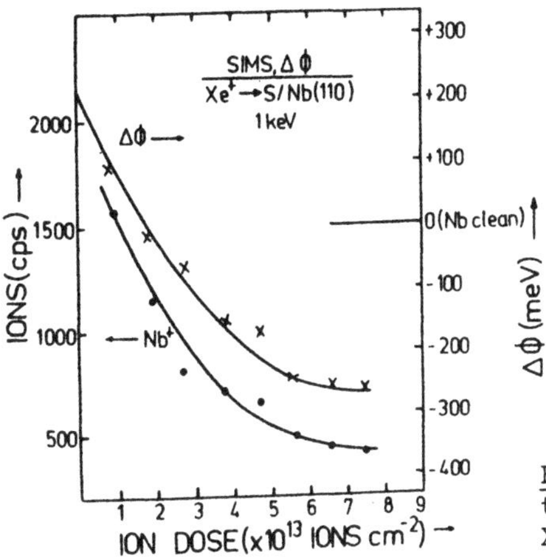

Fig. 5 Change of the work function and the Nb^+ yield during Xe^+ bombardment

Of course by incorporation or mixing of sulphur atoms into the niobium surface we produce an inversely oriented dipole moment compared to the top position of the sulphur atoms. For this reason one should expect a change of the work function, too, namely a decrease of work function with ion bombardment. This is shown in Fig. 5. Here we have the Nb^+ ion yield and the work function change as a function of ion dose. Both curves drop nearly parallel which is a further indication that the work function of the surface plays an important role in the description of secondary ion emission [2].

From the Auger results we know that the mixing effect is reversible by annealing the crystal. Our simple configuration model of the atoms then would be the same as for room temperature adsorption (Fig. 1a), i.e. the sulphur is located again on top of the first niobium layer. Therefore we get an 180° turn of the dipole moment at the surface and accordingly a work function which should be the same as for room temperature adsorption.

A sequence of some sputter and annealing cycles is shown in Fig. 6. Of course, for the clean Nb surface the work function change is zero by definition. After adsorption of sulphur the work function increases by about 350 meV. Ion bombardment with $2x10^{13}$ ions/cm^2 leads to a decrease of the work function, in this case below the value of the clean surface. Now we have induced the mixing or incorporation of the sulphur atoms into the niobium surface. After annealing for 90 s up to 1000 K we find again the same work function as for room temperature adsorption. Now the sulphur atoms have come back on the niobium surface. Another ion bombardment leads to a decrease of the work function again and a further bombardment under same conditions results in an even lower work function value. So here the work function change due to ion bombardment is more than 650 meV. One can see from the last point of this sequence that the mixing effect is still reversible upon annealing.

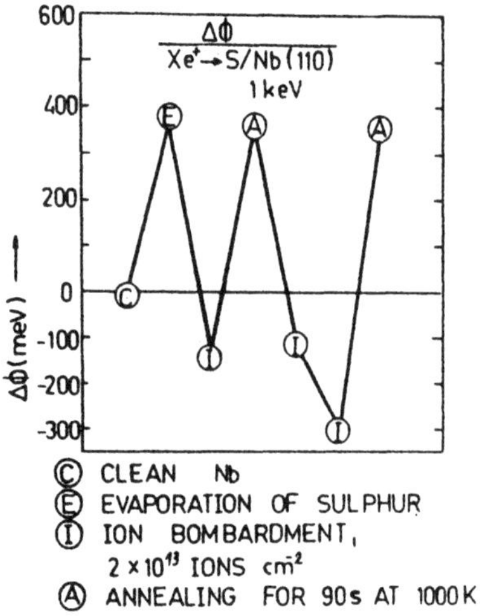

Fig.6 Change of the work function due to different surface states (see text)

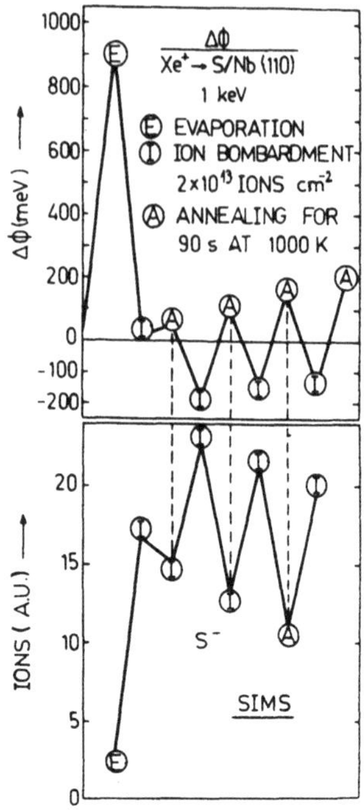

Fig. 7 Change of the work function and of the S^- yield due to different surface states (see text)

The Fig. 7 shows a similar sequence for ion bombardment and heating treatments. In the upper part we have again the work function change and in the lower part the actual S^- ion yield is drawn. The only thing we want to point out here is the fact that for a low work function the ion yield is high and for a high work function the ion yield is decreased.

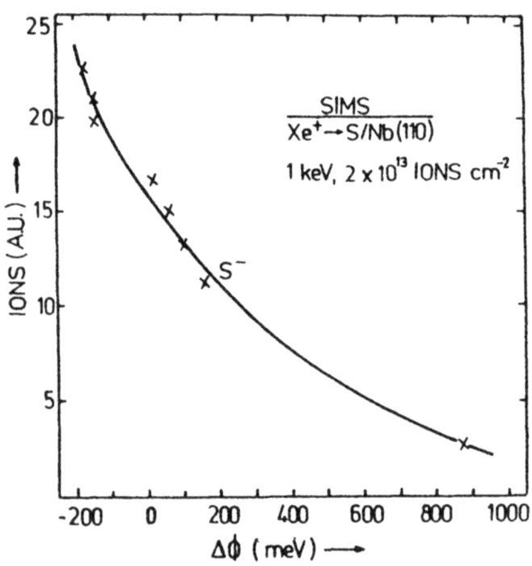

Fig.8 S^- yield vs the work function change of the surface

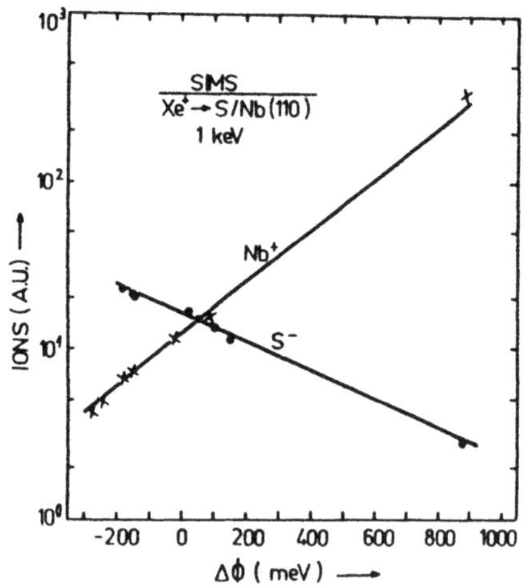

Fig.9 Secondary ion yield of Nb^+ and S^- (on a log-scale) vs the work function change of the surface

Putting these data together we get a plot of the S^- ion yield versus the work function change. This is done in Fig. 8. Here we have combined the SIMS and work function measurements independently of whether they are obtained after adsorption at 300 K or after ion bombardment or after annealing the crystal at 1000 K. One might well fit these data to a smooth curve.

From the theoretical point of view it is more useful to make a half-logarithmic plot of these data. Fig. 9 gives the results for both the Nb^+ ion yield and the S^- ion yield on a log-scale as a function of the work function. The straight lines indicate an exponential dependence of the ion yield of the work function. The different sign of the slopes is in accordance with the model of ionization and neutralization by electrons from the surface which we have observed in the case of adsorbed metals as a function of coverage [2]. A high work function of the surface corresponds to a high positive ion yield and a low negative ion yield and for a low work function of the surface the converse is true.

4. Summary

We obtained for *static SIMS* conditions:

1) mixing effect (incorporation of S atoms beneath the first niobium layer)
2) corresponding electronic changes at the surface
3) which are reversible upon annealing to 1000 K (for low sulphur coverage)
4) a reasonable dependence for the positive and negative ion yield of the work function of the surface, e.g.

$$R^+ \propto \exp(+c\Delta\phi)$$

$$R^- \propto \exp(-c'\Delta\phi) .$$

5. Conclusions

The dramatic mixing effect and electronic changes at the surface make it very difficult to get SIMS data for this and probably equivalent adsorbate/substrate systems which represent the *original* surface state.

6. References

1. A. Benninghoven, Surface Sci. 35, 427 (1973)
2. S. Prigge and E. Bauer, Adv. in Mass Spectr. 8, 543 (1981)
3. P. Williams and C.A. Evans, Surface Sci. 78, 324 (1978)

Current Density Effects on Secondary Ion Emission from Multicomponent Targets

A. Barcz, M. Domanski, and B. Wojtowicz-Natanson
Institute of Electron Technology,
02-668 Warszawa, Poland

Introduction

The use of the Secondary Ion Mass Spectroscopy for determining in-depth distribution of selected elements requires sufficiently high erosion rates [1]. On the other hand when applying this technique to study surface phenomena, sputter etching has to be minimized in order to keep the investigated area relatively undisturbed [2]. The rate at which succesive layers sputter off the material is proportional to the current density (or flux) of the incoming particles, whereas the secondary ion current, in principle, is proportional to the primary current solely. It appears, however, that the magnitude of the applied beam density can influence strongly the quantitative interpretation of the SIMS analysis. Possible effects are: (i) Temperature rise under the impact of energetic high-flux beam. Usually, in commercially available devices the power dissipated in the near-surface region is not high enough to cause appreciable macro-thermal effects. (ii) Mutual interaction of the primary and the sputtered particles above the bombarded surface leads to modifications of the ionization efficiency of the latter [3]. (iii) Ion-induced desorption is the most pronounced phenomenon directly related to the primary flux. It is well known that the presence of reactive gases on the surface can drastically intensify the ion emission [4]. In the dynamic mode, oxygen coming either from the vacuum system or intentionally introduced in the vicinitiy of the target is continuously removed by impingent ions; the resulting coverage is determined by equilibrium between the gas pressure and the primary beam density.

The object of this paper is to point out that other processes than the one mentioned above are also sensitive to the current density. Recent developments in the understanding of mechanisms governing the secondary ion formation have improved the accuracy of the absolute SIMS measurements to within a factor lower than 2 [5]. It thus becomes important to take into account all the accompanying factors affecting the relationship between the output signal and the atomic content. In this work, attention will be particularly paid to the role that ion-induced surface segregation and preferential sputtering play in promoting emission of one constituent with respect to the others from a compound material.

There is a rather general recognition that selective sputtering has no influence on SIMS bulk analysis, with the exception of transition regions i.e. before the stationary state is established [6,7]. Indeed, unlike the AES, ISS or XPS techniques where residual after-sputter surface is analysed, SIMS enables detecting sputtered species whose composition is that of the bulk, regardless the actual surface composition. Nevertheless, since the majority of the secondary ions are belived to originate from the outermost atom layer one may expect that even in a steady-state ionization behaviour will depend on the extent to which differential sputtering alters the initial surface chemistry.

Experimental

Measurements were made using a laboratory designed UHV ion microprobe (base pressure 10^{-8} Torr) equipped with a high-brightness hot cathode Pierce ion gun and Balzers quadrupole mass analyzer. Primary Ar^+ ions were not mass-separated and their energy was chosen to be 4 keV. The 2 μA beam focussed into a 170 μm diameter circular spot gives the current density of 10 mA cm^{-2} ($6 \cdot 10^{16}$ $cm^{-2}s^{-1}$). Raster scanning over an area of 2x2 mm^2 reduces this density to 50 μA cm^{-2}. Another way to decrease the beam density is to de-excite the final einzel lens which implies expanding of the beam diameter to about 500 μm while conserving the total current. A 1/4x1/4 linear electronic aperture was set to reject signals originating from crater walls. It has been reported that the species which becomes enriched at the target surface seems to eject closer to the normal direction than that which is depleted [8]. In addition, energy distribution of secondaries has been demonstrated to vary with the sample composition [9]. Therefore, the extracting optics was adjusted so that the solid angle of collection was large energy filter bandpass within 0-20 eV. The analysed ions, after passing through the quadrupole spectrometer, are directed onto an off-axis electron multiplier, then converted to pulses and amplified by a standard electronic chain, which stores information in a multichannel analyser working as a multiscaler. The counter and timer (channel advance) are both digitally controlled by the raster electronic gate.

Results

Several stoichiometric materials were investigated for the purpose of searching the effects of the primary beam density on relative intensities of secondary ions from major constituents. Focalization conditions were rapidly switched in the course of monitoring signals from selected peaks.

A_2B_6 compounds CdTe, ZnS, CdS, ZnSe, CdMnTe exhibit little or no change either in relative or in absolute values of the ion yield upon different sputter configurations. Only a small increase (~ 15 %) is observed for the Cd/S and Cd/Se peak height ratios with increasing primary flux.

Zn doped materials. The evolution of Zn^+ and As^+ signals coming from a heavily doped GaAs (~ 10^{20} Zn cm^{-3}) is shown in Fig. 1. The most striking feature is that both intensities seem to compete with each other; an increase of the Zn^+ signal

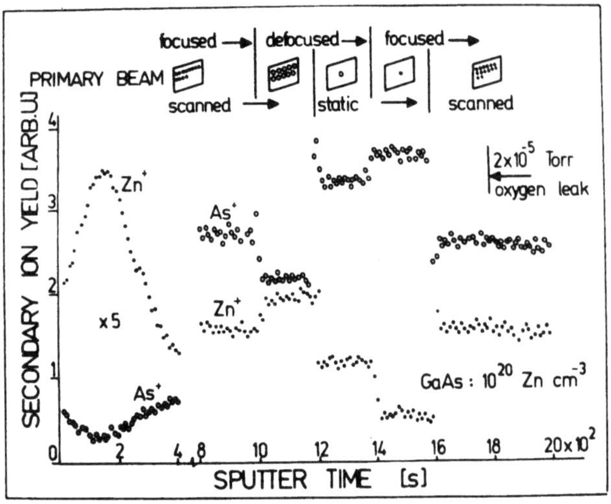

Fig. 1 Evolution of As^+ and Zn^+ secondary ion intensities from 4 keV Ar^+ bombardment of GaAs doped with Zn to a level of 10^{20} cm^{-3}. Primary ion current 2 μA. Beam diameters 170 μm and 500 μm. Scanning area 2x2 mm^2

corresponds to a decrease of the As^+ signal and vice versa. Moreover, the As^+/Zn^+ drops considerably even though the average density over the rasted area remains constant and only the instantaneous density is reduced. A strong surface enhancement of the Zn^+ signal is due probably to the zinc incorporation during diffusion process. When the target, after keeping it in the vacuum for 20 hours, was sampled in the same area for the second time, the surface peak was however again observed. Similar dependence of the Zn^+ intensity was found for gold containing 5 at% Zn, but in this case the Au^+/Zn^+ intensity ratios do not follow the pattern established for As^+/Zn^+.

Metal oxides. Clean surfaces of W and Fe were oxidized to their respective WO_2 and Fe_2O_3 oxides. Depth profiles were taken using two densities: 50 μA cm^{-2} and 10 mA cm^{-2}. In-depth distributions of W^+, O^+ and WO^+ are presented in Fig. 2. It should be noted that W^+ (10 mA cm^{-2}) is 80 % higher than W^+ (50 μA cm^{-2}) in the oxide. For Fe_2O_3 the intensity of Fe^+ (10 mA cm^{-2}) is 50 % higher than Fe^+ (50 μA cm^{-2}).

Binary alloys. Experimental points for Cu_5Ni_{95}, $Cu_{53}Ni_{47}$ and $Cu_{95}Ni_5$ bulk alloys together with pure Cu and pure Ni are plotted against the atomic composition in Fig. 3. Solid lines represent linear proportionality between the secondary ion yield and the concentration, in a semi-logarithmic scale. Application of higher flux results in decreasing the Ni^+ signal and increasing the Cu^+ signal, thereby diminishing the apparent deviation from the linearity. Assuming that this trend is not an artifact induced by different removal rates of residual oxygen, analogue data were taken in the $1{\cdot}10^{-6}$ Torr oxygen pressure (dashed curves). It is seen that here the measured values deviate in the opposite direction than those taken without oxygen leak.

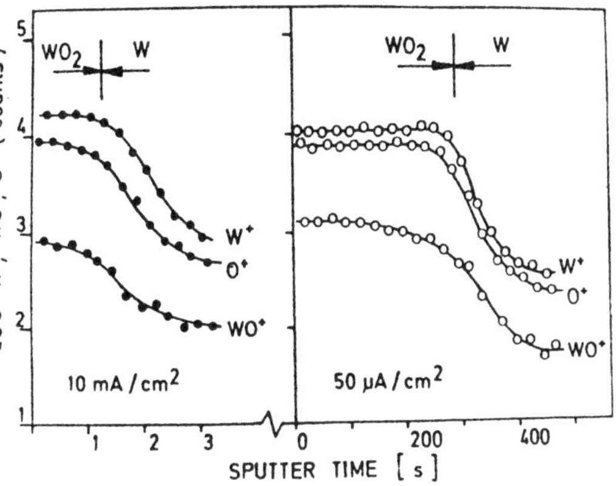

Fig. 2 Profile distribution of W^+, O^+ and WO^+ in 200 nm thick WO_2 oxide on tungsten. Beam parameters: left-static 10 mA/cm^2, right-scanned 50 $\mu A/cm^2$

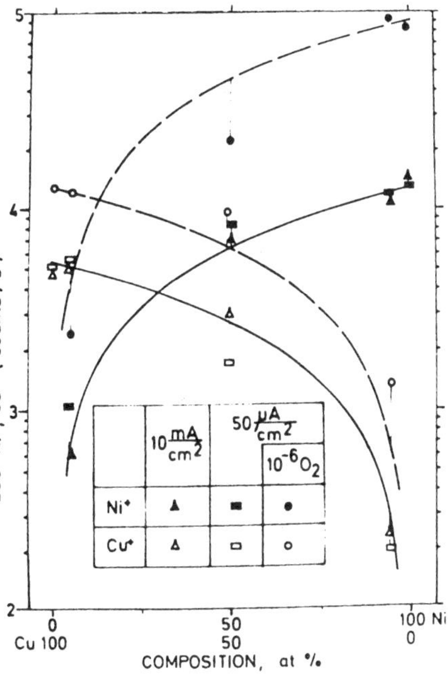

Fig. 3 Cu^+ and Ni^+ ion yields as a function of bulk composition of CuNi alloy. Solid curves correspond to linear dependence between ion signal and concentration (with no oxygen leak). Dashed curves - with $1 \cdot 10^{-6}$ Torr oxygen pressure

The result for $Ag_{90}Pd_{10}$ bulk alloy is $Ag^+/Pd^+(10\ mA\ cm^{-2})$ = = 1.3 $Ag^+/Pd^+(50\ \mu A\ cm^{-2})$.

Discussion

When a multi-component system is subjected to ion bombardment, preferential sputtering usually occurs as a consequence of different sputtering yields of the components. For the sake of simplicity we will consider a one phase binary alloy $Ag_{1-f}Pd_f$; f - atomic fraction of Pd atoms. Since the sputtering yields are: S(Ag) = 8.9 and S(Pd) = 4.8 [10], the surface is initially depleted in more volatile Ag. The sputtering rate of Ag therefore decreases. At the same time Ag atoms diffuse from the interior of the solid to the surface. Eventually, the two processes balance and reach a stationary state during which Pd and Ag sputter congruently although the altered layer persists; the bulk f^b versus surface f^s composition is given by the relation:

$$\frac{f^2}{1-f^s} = \frac{S(Ag)}{S(Pd)} \frac{f^b}{1-f^b} \quad . \tag{1}$$

The depth composition profile can be obtained by solving Fick's second differential equation [11] for a steady state:

$$f(X) = f^s + (f^s - f^b)\exp(-\frac{\dot{x}}{D}X) \tag{2}$$

where D - diffusion coefficient, $\dot{x}$ - removal velocity, X - depth. To give an idea of how the sputtering rate $\dot{x}$ can modify the composition of a near-surface region, the above function is schematically plotted in Fig. 4 for three different densities, taking $D = 10^{-13}\ cm^2s^{-1}$ [12] in $Ag_{90}Pd_{10}$.

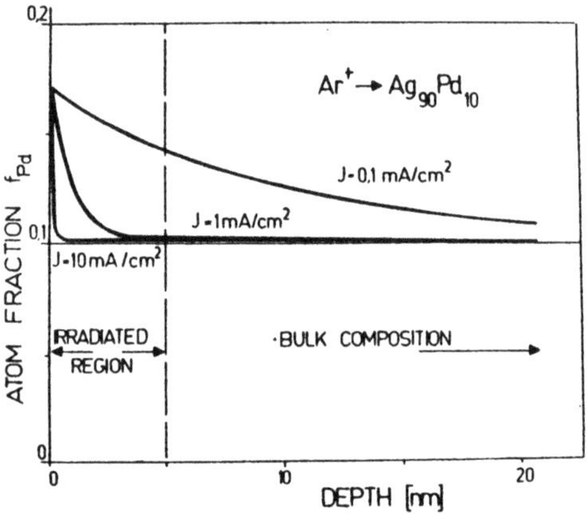

Fig. 4 Schematic plots of relative Pd concentration profiles in $Ag_{90}Pd_{10}$ alloy (Eq.2) determined by the effects of preferential sputtering, diffusion and erosion rate

Of the materials investigated in this work, the following have been reported to sputter selectively: CdS, CdSe - depletion of Cd [13], WO_2, Fe_2O_3 - loss of oxygen [14], CuNi - loss of Cu [15]. Zinc was expected to behave like an easy-to-sputter species because it is an easily diffusing and sublimating element and has been observed to outdiffuse rapidly under energetic ion impact [16].

Present measurements indicate that a rule-of-thumb conclusion may be proposed according to which when increasing the primary ion current density, the secondary ion yield is enhanced for a component being depleted at the bombarded surface. Possible explanation of this fact leads to Eq.(2)(see also Fig. 4). Even if this equation is not quite adequate in that it does not take into account different mechanisms of material transport in the bombarded layer (mixing) [17] and below (diffusion), the qualitative validity is conserved. Whatever the escape depth of secondary ions, the number of atoms with lower sputtering yield within this zone increases with increasing erosion rate, disturbing the previously established chemistry. The atoms situated closer to the surface are thus more likely to eject in ionized state. For preferentially sputtered oxides, the configuration is somewhat different because in SIMS case sputtering is accompanied by decomposition and with higher primary flux ionization of metal atoms becomes more "chemical" than for lower flux.

Most work on preferential sputtering has been done with AES using relatively low current density [15]. It is not evident whether at higher densities individual sputtering yields remain unchanged [18,19]. Another point of interest would be the investigation of selective sputtering of very diluted alloys or of doped semiconductors.

The authors would like to thank Mr. A. Trojan for his helpful technical assistance.

References

1. P. Williams and C.A. Evans, Appl. Phys. Lett. 30, 559 (1977)
2. A. Benninghoven, Surf. Sci. 35, 427 (1973)
3. K. Wittmaack, Nucl. Instr. Meth. 132, 381 (1976)
4. G. Blaise and M. Bernheim, Surf. Sci. 47, 324 (1975)
5. H.W. Werner, Surf. Interf. Anal. 2, 56 (1980)
6. G.K. Wehner, in "Methods of Surf. Anal.", ed. A.W.Czanderna, Elsevier (1975)
7. S. Hofmann, Surf. Interf. Anal. 2, 148 (1980)
8. G.K. Wehner, R.R. Olson and M.E. King, Proc. 7th Int. Vac. Congr., Vienna (1977)
9. G.Blaise and G. Slodzian, Rev. Phys. Appl. 8, 105 (1973)
10. H.J. Mathieu and D. Landolt, Le Vide, Les Couches Minces, Suppl. No.201, p. 1319
11. P. Hijer, W.T. Langendam and J.A. Lely, Philips Tech. Rev. 24, 144 (1963)
12. W. Seitz, Diffusion in Metallen (Springer, Berlin, Göttingen, Heidelberg 1955)
13. L.J. Brillson, J. Vac. Sci. Technol. 12, 249 (1975)
14. R. Kelly, Nucl. Instr. Meth. 149, 553 (1978)
15. G. Betz, Surf. Sci. 92, 283 (1980)
16. A. Barcz, unpublished
17. U. Littmark and W.O. Hofer, Nucl. Inst. Meth. 168, 329 (1980)
18. I.L. Singer, unpublished
19. A. Barcz, M. Croset, L.M. Mercandalli, Surf. Sci. 95, 511 (1980)

Isotope Effect in Secondary Ion Emission

J.C. Lorin[1], A. Havette[2], and G. Slodzian[3]

[1]Laboratoire de Mineralogie-Cristallographie, Université Pierre et Marie Curie, F-75230, Paris, France

[2]Laboratoire de Petrographie-Volcanologie, Université Paris-Sud F-91405 Orsay, France

[3]Laboratoire de Physique des Solides, Université Paris-Sud, F-91405 Orsay, France

Introduction

Isotope abundance determination on small areas of standard polished sections appears to be one of the most interesting potentialities of secondary ion mass spectrometry. Key samples for the detection in natural samples of isotope effects of nuclear or radiogenic origin are indeed often rare phases of small dimensions. Secondary ion mass spectrometry offers the unique possibility of performing isotope ratio measurements on these phases with limited chemical contamination, and with the added advantage of allowing direct correlation of isotope data with chemical data (in particular on trace elements) and with mineralogical and textural features.

Isotope abundance determination by this method has met, however, with many difficulties ever since the early investigation of the isotope composition of lithium by POSCHENRIEDER et al. [1] who reported large isotope variations in the Holbrook meteorite. These results were not confirmed by subsequent analyses of the same type performed by the BERNAS group [2], who reported however, soon after, in a more extensive study [3], variations as high as 20 % in other meteorites. Later work by GRADSZTAJN and GUEZ [4] showed that these variations were of analytical origin and the importance of mass discrimination in secondary ion mass spectrometry was recognized by these authors, who were however unable, at that time, to reach definite conclusions regarding the origin of these effects.

The recent discovery of large isotopic variations, presumed to be of nucleosynthetic origin, on a host of elements (besides rare gases) in meteoritic matter (see [5] for a review) has prompted a number of investigators to use secondary ion mass spectrometric techniques to study the fine-scale distribution of these anomalies. This has made it necessary to reappraise the question of isotope mass discrimination in secondary ion mass spectrometry and this is the subject of the present study. Since it appears that in some key meteoritic samples there exists a correlation of a fundamental, but still enigmatic nature, between linear (i.e. mass-dependent isotope fractionation) and non-linear (i.e. nuclear, etc.) isotope effects, obtaining absolute (that is unaffected by, or suitably corrected for, any discrimination of analytical origin) isotope abundance determinations in secondary ion mass spectrometry has become, in this domain, a goal of prime importance. Preliminary reports of this work have been published elsewhere [6,7].

Analytical Procedure

Isotope abundance and energy spectra determinations were performed with a CAMECA IMS 300 ion analyzer, equipped with a transfer optical system [8]. Mass resolving power (up to 5000) as well as magnetic field stability have been significantly improved upon the standard version of the instrument. Special attention has been paid in the course of analyses of insulating materials to the problem of the surface electrostatic charge-up. In order to minimize electrostatic charge-up (<<1 V), the sample surface was covered with a grid [9] and bombarded with a low intensity (c.a. $4 \cdot 10^{14}$ $cm^{-2}s^{-1}$) flux of 5.5 keV O_2^+ (or Ar^+) ions. In this way, exact charge compensation is achieved through tertiary electronic emission by the extraction electrode. In all experiments, the transfer optical system was adjusted to optimize the collection efficiency on a sample area 25 μm in diameter. Secondary ions were detected with an electron multiplier and their intensity measured in the pulse counting mode. Suitable corrections for dead-time of the counting system were applied to the ion intensity measurements. Potential interferences on analytical ions have been resolved by recourse to mass discrimination, which appears to be, at present, the safest available procedure [10,11]. Interferences which could not be resolved according to this procedure have necessitated corrections (of generally very small amplitude) by the usual peak-stripping method.

Isotope abundance ratios have been determined by hand-tuning the magnetic field to the peaks of interest. This method precludes rapid peak switching; drift with time of the secondary ion intensity was taken into account by interpolating the data with an orthogonal polynomial of 3rd degree. In the course of this work we have denoted by R_{ij} the ratio of the measured intensities of isotopes m_i and m_j and by δ the relative deviation from the standard ratio R^o_{ij}.

$$\delta = \frac{R_{ij}}{R^o_{ij}} - 1 \, .$$

Use has also been made, in order to facilitate comparison of data obtained on different elements, of the parameter α, which is chosen such that:

$$(R_{ij}/R^o_{ij}) = (m_i/m_j)^{-\alpha} \, .$$

So that, to first order:

$$\alpha = - \delta \cdot m/\Delta m \qquad \text{where} \quad m = m_i - m_j$$

Adopted reference values for the isotopic composition of investigated elements generally follow those recommended in [12], save for Mg [13], Ca [14] and Zr [15].

Results

Measurements were made on pure metals, metal alloys and oxides, and synthetic as well as natural (terrestrial) samples. In the latter, we have concentrated mainly on those elements which are known to be not or little fractionated in geochemical processes, such as Mg [13,16], Si [17] and Ca [14]. Results, corresponding to steady-state conditions in the implantation/erosion process are the following:

In all cases so far investigated the isotopic composition of emitted secondary ions (both positive and negative) was found to be biased with respect to that of the bulk unperturbed material, the bias consisting in a more or less severe heavy isotope depletion. As well demonstrated in the case of elements, such as zirconium, which have a large number of isotopes, the effect appears to be, to first order, linear with the relative mass difference between isotopes of a given element (Fig. 1), a characteristic feature of mass-dependent isotope fractionation. We have listed in Table 1 the value of the discrimination factor (0-20 eV) for positive ions of several elements emitted from different mineral targets and in Table 2 the value of the same parameter for ions emitted from pure metals, metal alloys and oxides. In Table 3 are shown some of the results obtained on negative ions.

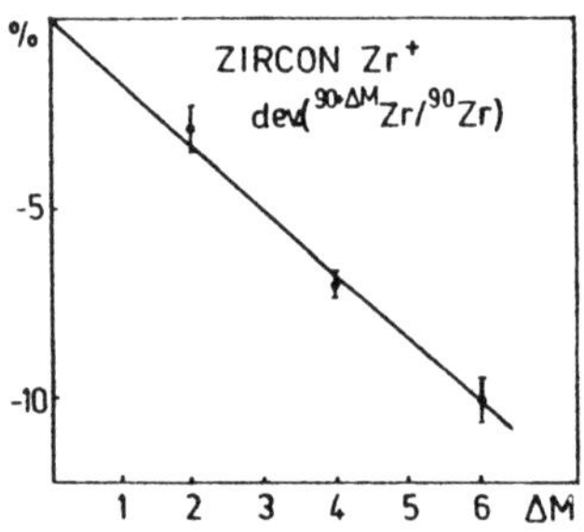

Fig. 1 Mass-dependent isotope fractionation of Zr^+ secondary ions (0-20 eV) emitted from a zircon target bombarded with a 5.5 keV O_2^+ beam. Heavier isotopes are seen to have a lower probability of being ejected/collected as ions

No appreciable change in the magnitude of the isotope effect has been observed when varying the conditions of ion extraction, for instance when modifying the size of the entrance diaphragm, tilting its position with respect to the instrument axis, or changing the cross-over magnification (Fig. 2). This provides us with no clue regarding a possible dependence of the isotope effect with the secondary ion ejection angle.

The discrimination factor is not found to be a strong function of the energy of the primary beam, at least in the energy interval 5.5 to 14.5 keV. On the other hand, the nature of of the bombarding species does appear to have some effect upon the magnitude of the isotope fractionation. Use of a primary beam of O_2^+ consistently results in an increase of the value of the discrimination factor with respect to that corresponding to Ar^+ bombardment. The increase of α(0-20 eV) may attain 40 % in the case of Mg^+ or Si^+ ions sputtered from mineral targets.

Table 1 Value of parameter α as determined under 5.5 keV O_2^+ bombardment for secondary ions of energies between 0 and 20 eV

	Mg^+	Si^+	Ca^+	Zr^+
Forsterite	0.38 ± 0.04	1.75 ± 0.07		
Sphene			0.41 ± 0.05	
Zircon		1.8 ± 0.1		1.6 ± 0.1
Melilite	0.6 ± 0.1			
Ti-rich Fassaite	0.42 ± 0.04			
Ti-poor Fassaite	0.60 ± 0.05			
Labradorite	0.4 ± 0.1	1.32 ± 0.09	0.26 ± 0.04	
Calcite			0.69 ± 0.04	
Magnesite	0.89 ± 0.06			
Apatite			0.61 ± 0.03	
Fluorite			0.81 ± 0.03	
Quartz		1.01 ± 0.04		
Obsidian glass		1.00 ± 0.03		
Hibonite	0.37 ± 0.06		0.18 ± 0.08	
Spinel	0.52 ± 0.02			
Uraninite			0.3 ± 0.1	
Scheelite			0.1 ± 0.1	

Table 2 Value of parameter d as determined under 5.5 keV Ar^+ bombardment (with the exception of Ca-Al being bombarded with 5.5 keV O_2^+ ions) for secondary ions of energies between 0 and 20 eV

Ni^+ / Nickel	0.6 ± 0.1
Si^+ / Silicium	0.77 ± 0.08
Cu^+ / Copper	0.9 ± 0.3
Cu^+ / Copper oxide	0.98 ± 0.03
Ca^+ / Ca-Al alloy	0.9 ± 0.3

Three factors essentially control the isotope effect. These factors are: nature of the secondary ion, matrix from which it is emitted and initial kinetic energy with which it is emitted.

1.) From a consideration of Table 1 it is apparent that the magnitude of the discrimination critically depends upon the nature of the emitted element. Irrespective of the

Table 3 Value of parameter α as determined under 14.5 keV Ar^+ bombardment (with the exception of chalcopyrite being bombarded with 5.5 keV O^- ions) for secondary ions of energies between 0 and 20 eV

Si^-	/ Silicium	1.3±0.1
Cu^-	/ Copper	0.1±0.5
Cu^-	/ Copper oxide	0.0±0.2
S^-	/ Chalcopyrite	2.2±0.1

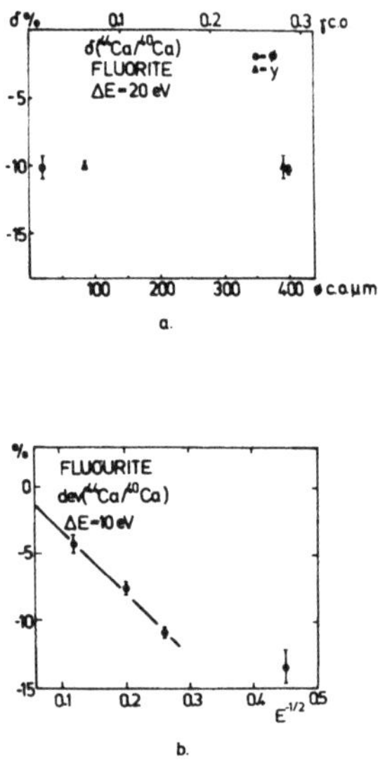

Fig. 2 (a) Magnitude of the isotope mass discrimination is left essentially unchanged when varying the diamter of the entrance diaphragm Φ or the magnification γ of the cross over. Admittedly of a restricted scope, these experiments fail to give evidence for a possible variation of the discrimination factor with the ion angle of ejection.
(b) Relative deviation with respect to the standard value of the isotope abundance of Ga^+ ions sputtered from a fluorite target exhibits a linear relationship with the inverse of the square root of the energy. This relationship breaks down at low energy, due, at least in part, to the finite energy bandwidth (10 eV) used here

matrix from which it is emitted, silicon, for instance, is always heavily fractionated, whereas calcium markedly less so.

2.) For a given element, the magnitude of the discrimination will depend upon the matrix from which it is emitted. Compare for instance, the strikingly different values of the discrimination factor for Ca^+ ions sputtered from a fluorite and a feldspar (labradorite) target, respectively (Table 1).

3.) Magnitude of the isotope discrimination is observed to show a definite dependence upon the initial kinetic energy of the secondary ion. It levels off with increasing secondary ion energy. As shown on Fig. 2b, in the case of Ca^+ ions emitted from fluorite, the effect is approximately linear with inverse ion velocity.

Other factors are found to affect to a lesser extent the discrimination of secondary ions. Of interest, because of its relevance to the so-called surface chemical effect, is the dependence observed with oxygen coverage, in the case of pure metal targets bombarded by Ar^+ ions. For instance, in the case of pure nickel, an oxygen partial pressure of 10^{-6} Torr (that is well below the saturation plateau with our density of bombardment, according to [18]) raises by about 50 % the value of the discrimination factor α(0-20 eV) over the value in the absence of oxygen flooding given in Table 2.

All these pieces of evidence conclusively demonstrate that the isotope effect we observe is essentially related to the sputtering/ionization process. Two additional facts confirm this point:

1.) The shape of the secondary ion energy spectrum is seen to depend upon both the nature of the emitted element and the matrix from which it is emitted (Fig. 3). Shape of the energy distribution, as crudely characterized by some parameter such as *most probable energy* or *full width at mean height* [19] shows a rough but definite correlation with the discrimination factor α(0-20 eV) (Fig. 4a).

2.) The pracatical ion yield, defined for convenience purpose as the number of ions collected in given experimental conditions per sputtered atom, is seen to depend also upon the nature of the emitted element and the matrix from which it is emitted. As seen on Fig. 4b, the practical ion yield displays a rough inverse correlation with the discrimination factor α(0-20 eV).

To summarize these observations, it can be stated that, as a rule, the lower the probability of an element to be emitted/collected as an ion from a given target material, the harder (i.e. the richer in high energy particles) its energy spectrum, and the larger the magnitude of the isotope fractionation. In addition, the fractionation is the smaller the higher the initial velocity of the ion as it leaves the target.

Discussion

No mention has been made so far of possible instrumental discrimination effects that could affect the results we have presented. Such discrimination effects which could arise, for instance, from the detection system are deemed to be very small in our experiments since ion intensity measurements have been effected in the pulse counting mode. It is clear, however, that high precision isotope abundance determinations will require, in the future, proper assessment of any systematic experimental bias. We wish to stress, at this point, that the isotope effect we report is observed under steady-state conditions and shows no time dependence. It follows that, in this equilibrium situation, the composition of the material leaving the target has to be identical with the undisturbed bulk sample. In principle, either the sputtering process or the ionization process might be held responsible for the isotope effect.

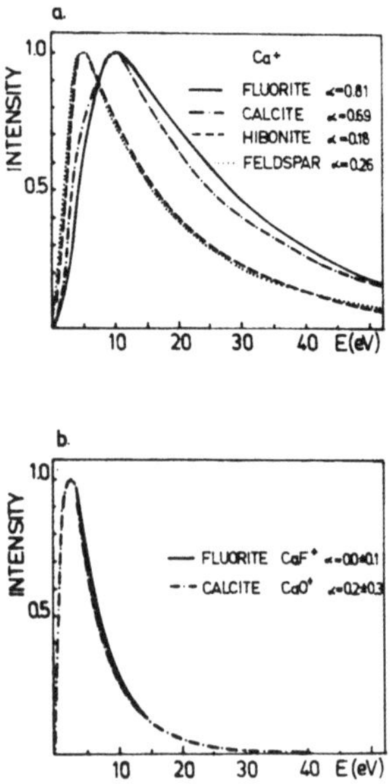

Fig.3

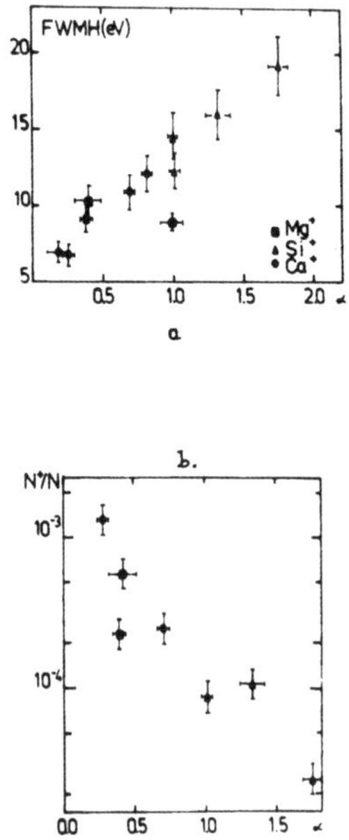

Fig.4

Fig.3 Energy spectra of secondary ions (Ca^+, CaF^+, and CaO^+) emitted from various mineral targets (energy window: 1 eV) bombarded with a 5.5 keV O_2^+ beam. The matrix effect upon the shape of the secondary ion energy spectra is conspicuous. A qualitative correlation between the width at half maximum of the energy distributions and the discrimination factor α(0-20 eV) (as defined in the text) is observed. Note that molecular ions show sharply peaked energy distributions coupled with low values of the discrimination factor

Fig.4 (a) Full width at half maximum (FWHM) of the ion energy spectrum vs discrimination factor α(0-20), for Mg^+, Si^+ and Ca^+ ions emitted from various targets. Both parameters are observed to be roughly correlated. (b) Practical ion yield (number of ions collected per sputtered atom) vs discrimination factor α (0-20). Both parameters are seen to exhibit a rough anti-correlation

An isotope effect could result from the sputtering process if the isotope composition of the sputtered material is different according to the angle of ejection. Such an effect has been reported [20,21] in low-energy (<300 eV) sputtering experiments, when the sputtering yield is less than unity, and has been attributed [20,21] to the role of reflective collisions. Such a mechanism is not expected, however, to be operative at higher bombarding energies [21] such as used in our experiments. More of concern to us, an angular isotope variation has also been reported [22] in high energy (c.a. 100 keV) sputtering experiments, and recently interpreted as the result of a screening effect due to the build-up of a thin (a few Å) altered layer at the surface of the sample [23].

Although the conditions of the experiment [22] we have mentioned differ in several important respects from those pertaining to our own, it is clear, however, that on the basis of this evidence, it cannot be ruled out, at present, that variation with ejection angle of the composition of the sputtered particles contributes to some extent to the isotope effect we observe. Experiments specifically designed to test that effect, for instance, in varying the solid angle of acceptance of the secondary ions have, it is true, given a negative result, but, in view of their limited scope, do not by themselves disprove the existence of an angular effect.

Whatever the exact contribution of the sputtering process to the isotope effect in secondary ion emission is, it seems clear, however, in view of the available evidence, that the ionization process plays a major role in this instance. Isotope discrimination effect, along with secondary ion energy spectra, provide us therefore with direct information upon the functional dependence of the ionization coefficient R^+ with the secondary ion energy. The linear relationship of the discrimination effect with the inverse velocity of the secondary ion (Fig. 2b) strongly suggests that the ionization coefficient is of the form:

$$R^+ \sim \exp(-v_o/v)$$

where v is the measured velocity and v_o some characteristic velocity.

Several ionization (or neutralization) mechanisms are expected to result in an energy dependence of this kind. In particular, mechanisms in which excitation/ionization occurs due to the potential perturbation as the atom crosses the sample/vacuum interface (e.g. [24-26]) and/or ion neutralization during escape [27] would qualify on this ground. Such mechanisms may indeed account for the isotope effect observed in secondary ions emitted from pure metal targets under Ar^+ bombardment (Table 2), and the effect of oxygen flooding on the magnitude of the discrimination is presumably due, in this case, to the competition between at least two different ionization processes. On the other hand, as regards discrimination effects observed on ions sputtered from oxides, minerals and other complex chemical compounds (Table 1), a bond-breaking mechanism [10] is likely to be the relevant process. This mechanism is expected to give the suitable energy dependence to the ionization coefficient (at least at high enough secondary ion energy), as shown in [6], and it adequately describes the salient features of the set of observations we report. In a broader context, this model satisfactorily accounts for the systematics of secondary ion yields from mineral targets [28,29] and constitutes for the present a valuable working hypothesis [30].

Applications and Implications

1.) Isotope fractionation appears to be a basic feature of secondary ion emission in sputtering experiments. Should it ever become of practical use for isotope separation, the effect is essentially of interest for extracting lighter isotopes of a given element. Adequate chemistry should be developed since the yield depends, as we have seen, upon the chemical form of the element.

2.) The isotope effect has obvious consequences on isotope abundance determination in secondary ion mass spectrometry. Isotope abundance measurements by this method have to be corrected for discrimination effects due to the secondary ion emission process by means of standards of adequate composition, valence state of elements and, maybe, structure. These requirements will be difficult to meet, especially when dealing with samples such as extraterrestrial minerals which, for some of them, result from different geochemical processes than their terrestrial counterparts. Hence the need of a more elaborate theory of mass discrimination in secondary ion emission liable to quantitative evaluation of the effect. With regard to quantitative aspects of elemental analysis in secondary ion mass spectrometry, it is clear that the isotope effect provides us with a totally new approach of the problem that deserves to be explored.

3.) It is doubtless that new insight in the physics of the ionization process will be provided by the study of this effect. We have already seen that mechanism of secondary ion production may be tested according to this scheme. Another example is afforded by the problem of formation of molecular or cluster ions, which is generally explained by two different mechanisms, fragmentation (e.g. [31]) and recombination (e.g. [32-34]). In the case of CaO^+ and CaF^+ ions emitted from a calcite and fluorite target, respectively, the value of their discrimination factor is very small (and their energy distributions very sharply peaked, see Fig. 3), in contrast to that of the corresponding atomic Ca^+ ions. This is apparently an instance where molecular ion formation does not occur through ion recombination in the main, a fact which does not contradict the synthetic model developed by OECHSNER [35] according to whom molecule formation by sputtering occurs through either direct emission or atomic combination depending upon the nature of the chemical bonds and the dynamics of the collision process.

4.) The secondary ion emission process should be considered in all astrophysical contexts where fluxes of low-energy (typically keV/amu) particles interact with solids. This is, of course, the case of planetary bodies devoid of atmospheres and strong magnetic fields, exposed to the bombardment of the solar wind. This is also the case of interstellar dust particles subject to sputtering by strong stellar winds and shock waves propagating in the interstellar medium. The ionized component of the sputtered material will be isotopically light, so that any mechanism decoupling the ionized fraction from the neutral gas (see e.g. [36]) will eventually bring about significant alterations of the initial isotopic pattern. Note that efficiency of this process is far higher than that associated with preferential sputtering, since, in the latter case, only a very small fraction of the bombarded solid will ultimately present a markedly anomalous isotopic composition.

Acknowledgements

Skilful assistance of R. Dennebouy and M. Chaintreau is gratefully acknowledged. Key experiments have been performed on a IMS 3F ion analyzer thanks to the courtesy of CAMECA Instruments and the expertise of M. Migeon.

Work was done with the financial support of C.N.R.S. through I.N.A.G. grants 37-51, 37-52 and 37-77 (A.T.P. Planetologie).

References

1. W. Poschenrieder, R. Herzog and A. Barrington, Geochim. Gosmochim Acta 29, 1193 (1965)
2. E. Gradsztajn, M. Salome, A. Yaniv and R. Bernas, Earth Planet. Sci. Lett. 3, 387 (1968)
3. R. Bernas, E. Gradsztajn and A. Yaniv, in "Meteorite Research" (ed. P.M. Millman), p. 123, D. Reidel, Dordrecht (1969)
4. E. Gradsztajn and M. Guez, Int. J. Mass Spectrom. Ion Phys. 4, 305 (1970)
5. T. Lee, Rev. Geophys. Space Phys. 17, 1591 (1979)
6. G. Slodzian, J.C. Lorin and A. Havette, J. Phys. (Paris) 41, L-555 (1980)
7. A. Havette, J.C. Lorin and G. Slodzian, J. Microsc. Spectrosc. Electron. 6, 169 (1981)
8. G. Slodzian and A. Figueras, in 8th Int. Conf. X-Ray Optics and Microanalysis (ed. D.R. Beaman, R.E. Ogilvie and D.W. Wittry) p. 659, Pendell, Midland (1980)
9. G. Slodzian, Ann. Phys. (Paris) 9, 591 (1964)
10. G. Slodzian, Surf. Sci. 48, 161 (1975)
11. S.J.B. Reed, Scanning 3, 119 (1980)
12. N.E. Holden, Brookhaven National Laboratory Nuclear Data Center, Report n°50605, 37 (1977)
13. D.N. Schramm, F. Tera and G.J. Wasserburg, Earth Planet, Sci. Lett. 10, 44 (1970)
14. W.A. Russel, D.A. Papanastassiou and T.A. Tombrello, Geochim. Cosmochim. Acta 42, 1075 (1978)
15. J.F. Minster and L.P. Ricard, Int. J. Mass Spectrom. Ion Phys. 37, 259 (1981)
16. E.J. Catanzaro and T.J. Murphy, J. Geophys. Res. 71, 1271 (1966)
17. J.H. Reynolds and J. Verhoogen, Geochim. Cosmochim. Acta 3, 224 (1953)
18. G. Blaise and M. Bernheim, Surf. Sci. 47, 324 (1975)
19. M.A. Rudat and G. Morrison, Surf. Sci. 82, 549 (1979)
20. G.K. Wehner, Appl. Phys. Lett. 30, 185 (1977)
21. R.R. Olson, M.E. King and G.K. Wehner, J. Appl. Phys. 50, 3677 (1979)
22. W.A. Russell, D.A. Papanastassiou and T.A. Tombrello, Rad. Eff. 52, 41 (1980)
23. P. Sigmund, A. Oliva and G. Falcone, paper presented at the IX Int. Conf. Atomic Collisions, Lyon, July 6-10 (1981)

24. G. Blaise and G. Slodzian, J. Phys. (Paris) 35, 243 (1974)
25. R.J. MacDonald, Surf. Sci. 43, 654 (1974)
26. J.K. Nørskov and B.I. Lundquist, Phys. Rev. B19, 5661 (1979)
27. W.F. Van der Weg and D.J. Bierman, Physica 44, 177 and 206 (1969)
28. A. Havette and G. Slodzian, J. Phys. (Paris) 41, L-247 (1980)
29. A. Havette and G. Slodzian, this Proceedings
30. G. Slodzian, this Proceedings
31. A. Benninghoven, Surf. Sci. 35, 427 (1973)
32. H. Oechsner and W. Gerhard, Surf. Sci. 44, 480 (1974)
33. G.P. Können, A. Tip and A.E. de Vries, Rad. Eff. 26, 23 (1975)
34. D.E. Harrison and C.B. Delaplain, J. Appl. Phys. 47, 2252 (1976)
35. H. Oechsner, these Proceedings
36. G. Arrhenius and H. Alfven, Earth Planet. Sci. Lett. 10, 253 (1971)

Caesiated Surfaces and Negative Ion Emission

M. Bernheim and G. Slodzian

Laboratoire de Physique des Solides, Université Paris-Sud
F-91405 Orsay, France

Introduction

The strong enhancement of the negative ion emission produced by caesium atoms is used now during analyses by secondary ion mass spectrometry more and more frequently. In most experiments, the caesium coverage results from the implantation into the target of the caesium positive ions that are producing the sputtering of the target itself [1,2,3]. After removal of a given thickness of the sample, an equilibrium is reached where the concentration of the implanted elements is controlled by the target sputtering yield. But, it is likely that the caesium concentration obtained that way is generally not sufficient to reach the saturation of the emission process. Another procedure consists of a chemisorption of neutral caesium atoms while the sample is simultaneously bombarded by a noble gas ion beam; thus the surface coverage can be adjusted by controlling the densities of either the primary ion beam or the neutral caesium jet [4,5].

Experimental Conditions

A mass-analysed primary ion beam was directed against the sample; the angle of incidence depends upon the acceleration energy of the primary ion beam and upon the voltage applied to the sample. The secondary ion collection optics was that of the direct imaging instrument [6]. A combination of magnetic and electrostatic prisms in an energy dispersive mode allowed the secondary ions to be analysed both in mass and energy [7]. Furthermore, the change of the sample work function was deduced from the shifts of the energy distributions of secondary ions. The sample could be rotated about the axis of the collecting optics without any modification of the collection efficiency and of the angle of incidence. The purpose of rotating a monocrystalline sample was to modify the sputtering yield which is known to depend on the crystal transparency seen by the primary ions (crystalline transparency lattice effect) [6,7]. In-situ information on the sample superficial concentration was obtained by measuring the intensity of the primary ions backscattered on the uppermost layer of the sample (Ion Scattering Spectrometry with 135° deflection angle) [8,9]; simultaneously, the same ion beam was producing the sample sputtering and the secondary ion emission. A neutral caesium jet obtained by heating a liquid caesium reservoir was collimated in order to obtain a caesium deposition on the center part of a large bombarded area [5].

Experimental Results

1. Determination of the caesium coverage

a) Typical variations of ion emission yields and work functions with the temperature of the caesium reservoir are shown on Fig. 1a. The maximum of Cu^- ion yield and the minimum of work function ($\Delta W = -3.1$ eV) were obtained for similar caesium fluxes (T = 228 ^{o}C and T = 242 ^{o}C, respectively); the variation of work function are very similar to the results obtained in static conditions without any sputtering (W is minimum, $\Delta W = -3.1$ eV, for coverage reaching 0.16 Cs At/Cu At) [10].

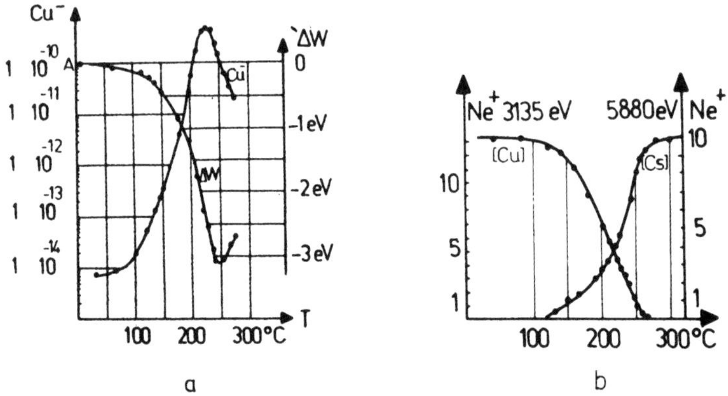

Fig.1 (a) Variation of the Cu^- emission and of the work function with temperature of the caesium neutral reservior. Measurements performed on a pure copper single crystal (energy band pass of the spectrometer 0-6 eV) with $^{20}Ne^+$ primary ions of 14 keV energy at 43° incidence angle.
(b) Ion scattering experiments performed on the same copper sample with $^{20}Ne^+$ primary ions, 10 keV energy at 45° incidence angle

b) The backscattering of the primary $^{20}Ne^+$ ion beam was used to determine the caesium coverage at different caesium fluxes; the neon ions backscattered with 3135±5 eV and 5880±9 eV correspond to binary collisions of 10 keV $^{20}Ne^+$ on copper (isotope 63) and caesium atoms respectively; their intensities can be related to the variations of superficial concentrations [11]. For instance, at 228 ^{o}C the caesium coverage reaches about half a monolayer while the copper concentration is reduced to a third (Fig. 1b).

Thus it was confirmed that, when the ion yield is maximum, the caesium coverage is about 0.16 Ca At/Cu At as was expected from work function measurements.

2. Crystalline lattice effects

a) Caesium vapour jet

A copper (100) single crystal is bombarded at 45° incidence by 9.5 keV argon ions. When the crystal is set in an opaque position, the sputtering yield is higher (by a factor 5) than the yield corresponding to transparent positions where the beam enters the crystal along the <110> transparent direction [6,7].

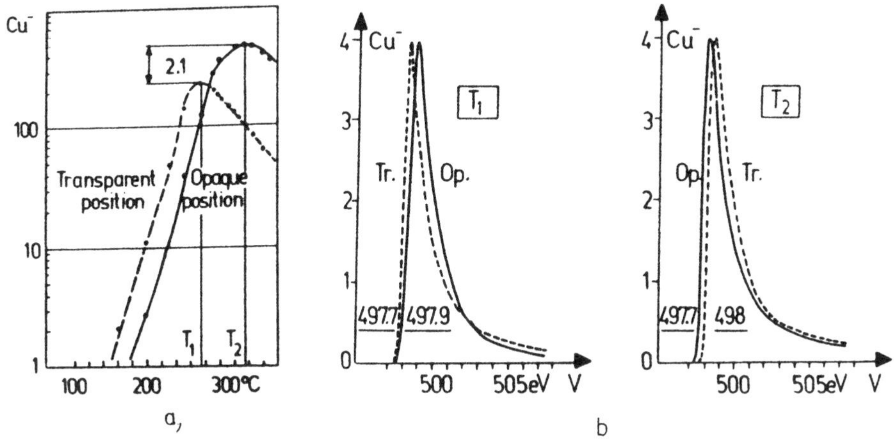

Fig.2 (a) Lattice effect on Cu^- emission with a neutral caesium supplies the (100) copper crystal bombarded with a 10 keV Ar^+ at 44 incidence angle (energy band pass of the spectrometer 0.3 eV) (b) Work function differences at T_1 and T_2 temperatures for both transparent and opaque positions (on the different recordings the Cu^- intensities are normalized to the sample peak height)

This lattice effect remains when caesium chemisorption occurs at the sample surface. For a low caesium flux, the caesium coverage is much higher in a transparent position (Fig. 2a); it can be seen that higher ionization efficiencies may overcome lower sputtering yields. For higher caesium fluxes, the situation is reversed. Furthermore, the same work function is measured at temperature T_1 and T_2 in transparent and opaque positions respectively (Fig. 2b). Thus it is very likely that at both maxima, the ionization yields are the same and therefore, the sputtering yields are in a ratio of 2.1 instead of 5 without caesium.

b) Caesium ion implantation

Let us consider now a 10.5 keV energy Cs^+ ion beam striking the same (100) Cu crystal. Once the equilibrium has been reached, the intensity of the Cu^- ions in an opaque position reaches now only a third of the intensity measured in a transparent position. In addition, Fig. 3 shows that the work function is now significantly lower in a transparent position than in an opaque setting of the target $W_1 - W_o = -0.9$ eV.

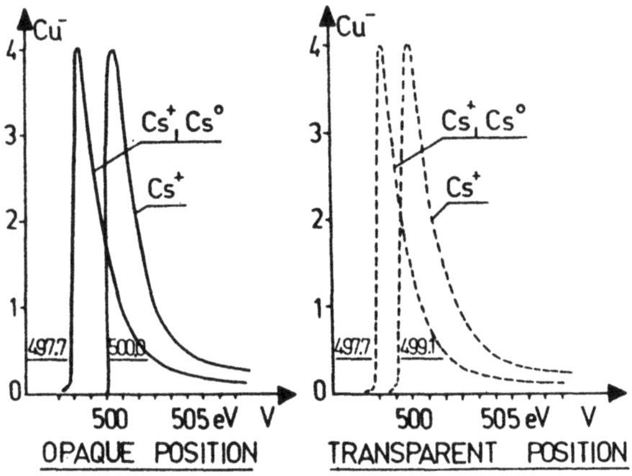

Fig. 3 Work functions differences for a Cs^+ ion bombardment (10.5 keV, 44° incidence angle); influence of an additional neutral caesium supply

c) Additional caesium vapour jet

Experiments were also performed with Cs^+ primary ions with an additional caesium vapour supply. For both positions, the temperature of the caesium reservoir was adjusted to maximize the Cu^- emission. In a transparent position, the effect of an additional neutral caesium supply was to increase the Cu^- intensity by a factor 8 whereas in an opaque position a factor 100 was reached. At the same time, the caesium jet induced a complementary lowering of the work function: $\Delta W_T = -1.4$ eV and $\Delta W_O = -2.3$ eV (Fig. 3). For both positions, the work function reached the same values that have been measured on Fig. 2b at temperature T_1 and T_2 in transparent and opaque positions, respectively.

3. Dynamics experiments

Similar experiments have already been investigated elsewhere [12]. Here the caesium primary ion beam was first implanted in an opaque position; once the equilibrium had been reached, the beam was switched off, the sample was then rapidly set in a transparent position before starting again the sputtering with Cs^+ ions (Fig. 4). During such experiments the ion beam density was kept low enough to slow down the process and lengthen the time necessary to reach the new equilibrium. In the same experiment it has been also performed from a transparent to an opaque position.

4. Useful ion yields

Copper and nickel thin films of known thickness have been used to determine the useful yield of copper and nickel targets. The useful yield is defined as the number of ions reaching the mass spectrometer detector per sputtered atom. During such experi-

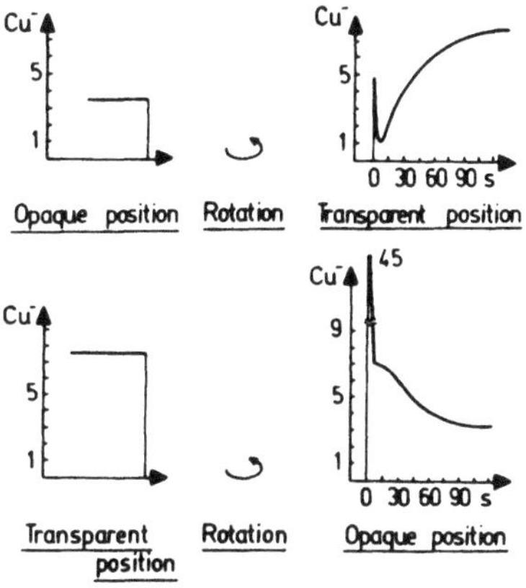

Fig. 4 Dynamics experiments, influence of the lattice effects for a Cs^+ ion beam (10.5 keV, 44° incidence) on a (100) copper crystal. (The humps observed at the beginning of the new bombardment might be due to a small neutral caesium chemisorption)

ments, the energy band pass was set at 0-4 eV and the secondary ions were accelerated to 4 keV in order to increase the collection efficiency. Table 1 gives the results where the isotopic abundancies have been corrected.

Table 1

	Useful ion yields	
	Cu^-	Ni^-
Ar^+ and Cs jet	$2.5 \pm 0.2 \cdot 10^{-2}$	$1.2 \pm 0.1 \cdot 10^{-2}$
Cs^+	$3.5 \pm 0.7 \cdot 10^{-4}$	$2 \pm 0.6 \cdot 10^{-4}$

Discussion

1.) The determination of the useful ionization yield and the influence of additional neutral caesium supply demonstrate that the caesium concentration obtained by direct implantation is not high enough to reach the saturation of the ionization processes for copper (and for nickel) neither in transparent position nor, a fortiori, in opaque position. The relation between caesium coverage Θ - determined by backscattering experiments [11] - and work function, can be applied to estimate the caesium coverage during Cs^+ bombardment. The caesium coverage in a transparent setting reaches about half the required value 0.16; the opaque situation is naturally worse (Table 2).

Table 2

Position	Work function lowering	Coverage Θ Cs At/Cu At
opaque	$\Delta W = 2.3$ eV	0.04
transparent	$\Delta W = 1.4$ eV	0.09

2.) The transparency lattice with argon ions are reduced by chemisorption of caesium atoms; but a crystalline contrast remains even for caesium saturation coverages; in contrast the iodine chemisorption on copper was much more effective (Table 3).

Table 3

	Ne^+	Ar^+	Kr^+	Xe^+	I_2^+
Cu_O^+/Cu^+	5.5	5	4.2	4	1.35
with oxygen flooding	3.5	3.4			
with iodine flooding		1			1

46° incidence, 9.5 keV primary ions (100) copper crystal

3.) The implantation conditions of Cs^+ ions in copper present some similarities to the saturation implantation of Kr^+ and Ar^+ ions into aluminium crystals reported elsewhere [7]. Since the mass ratio is such that no caesium can be backscattered at the surface of copper, most of the caesium ions are implanted into the target. Once the equilibrium has been reached, each Cs^+ ejects a mean number of atoms equal to S + 1:S copper atoms and 1 caesium atoms (S is the usual sputtering yield). Therefore, the caesium concentration at the surface can be estimated from its concentration in the sputtered material $[Cs] = (S + 1)^{-1}$. Consequently, to reach 0.16 At.Cs/Cu, the sputtering yield S should be about 5. Such a situation can only be achieved by a very low energy bombardment. For the present experimental conditions the estimated caesium coverages (Table 2) correspond to about 20 and 10 copper atoms sputtered per Cs^+ in opaque and transparent settings respectively; the ratio 2 corresponds to the sputtering yield ratio in both positions, a similar ratio was obtained for neutral caesium experiment.

4.) The caesium superficial concentrations, different in opaque and transparent positions are inducing different ionization probabilities with Cs^+ primary ions.

$$Cu_O^-/Cu_T^- = S_O P_O (S_T P_T)^{-1}$$

The experiment gives $S_O/S_T \simeq 2$ $Cu_O^-/Cu_T^- = 1/3$ so that the ionization probabilities differ by a factor 6 in opaque and transparent positions. Thus on polycrystalline sample, the useful ionization yield may be very different from one grain to another. The dispersion observed elsewhere on the measurements on metallic alloys bombarded with Cs^+ ions [3] could be attributed to the influence of the lattice effects on the relative ion yields. Indeed, it could be foreseen that such an influence would be more important for alloys of elements having large differencies in electron affinities. In the case of Cu-Ni alloys ($A.E_{Cu} = 1.23$ eV, $A.E_{Ni} = 1.15$ eV), such an effect was found to be within a factor 1.5 over ten different grains. For those alloys, the mean value of the ionization probabilities for Cs^+ bombardment is about the same as for neutral caesium supply.

5.) Another characteristic one should also point out is the more important production of polyatomic ions with the Cs^+ than when using a neutral caesium supply Ar^+ bombardment. For instance, on copper oxide sample Cu_2O, the Cu_2O^- and especially the CuO^- emissions are increased at the expense of the O^- and to a lesser degree of the Cu^- emission (Table 4). This property can be related to a primary ion mass influence.

Table 4

	O^-/Cu^-	CuO^-/Cu^-	CuO_2^-/Cu^-
Ar^+ and Cs supply	3.5	0.20	$1.8 \cdot 10^{-2}$
Cs^+	4.2	1.2	0.85
Cs^+ and Cs supply	1	0.7	0.16

Conclusion: Comparison of the Experimental Procedure

To conclude, let us summarize the main advantages and disadvantages of both experimental procedures:

- The neutral caesium supply needs a beam collimation in order to avoid a caesium accumulation on the sides of the bombarded surface. The caesium flux can easily be adjusted to reach the lowest work function and the higher ionization probability but, due to the weakness of the caesium flux, the density of the ion noble gas ion beam has to be kept low. Consequently a very good vacuum is needed in the target chamber to eliminate the adsorption of residual gas at the sample surface. Furthermore provided that measurements were performed on isolated crystallites, the residual lattice effects are not prohibitve.

- The main advantage of the caesium ion bombardment is easy to handle; pure Cs^+ ion beams are obtained even without any mass filter. The high value of the sputtering yield is favorable to rapid recording of depth profiles; but the equilibrium of the caesium concentration and the sputtering yield are only attained after erosion of about twice the mean range of Cs^+ ions (i.e. 150-180 Å), thus the initial profiles have no significant meaning. Furthermore, the caesium concentrations attained are generally too low to supply high ionization yields; and the lattice effects may hinder the use of Cs^+ for metallic samples. Nevertheless, for most applications to semiconductor materials, the targets are made amorphous by any ion bombardment so that lattice effect is remaining.

- For both procedures, the caesium coverage is not sufficient to hinder the charging up of insulating sample surface; electron feeding is needed.

- Some improvements can be made to the caesium ion source and to the caesium jet. From the discussion it appears that lighter alkaline ions Rb^+, K^+ or Na^+ could be used to reach higher superficial concentrations by direction implantation. In such a procedure, the gain in concentration probably overcomes the loss of the individual alkaline efficiency.

References

1. M.K. Abdulayeva, A.K. Ayukanov and Shamsiya, Rad. Effects 18, 167 (1973)
2. P. Vallerand and M. Baril, Int. J. Mass Spectrom. Ion Phys. 24, 241 (1977)
3. P. Williams, R.K. Lewis, C.A. Evans and P.R. Hanley, Anal. Chem. 49, 1999 (1977)
4. M. Bernheim and G. Slodzian, J. de Phys. (Lettres) 38L, 325 (1977)
5. M. Bernheim, J. Rebiere and G. Slodzian, in Secondary Ion Mass Spectromety SIMS-II, ed. by A. Benninghoven, C.A. Evans, Jr., R.A. Powell, R. Shimizu, H.A. Storms, Springer Series in Chemical Physics, Vol. 9 (Springer, Berlin, Heidelberg, New York 1979) p. 40
   J. Microsc. Spectrs. Electron. 5, 261 (1980)
6. G. Slodzian, Surface Sci. 47, 324 (1975)
7. M. Bernheim, Thesis Orsay (1973)
8. M. Bernheim and G. Slodzian, Nucl. Instr. Meth. 132, 615 (1976)
9. M. Bernheim and G. Slodzian, in Secondary Ion Mass Spectromety SIMS-II, ed. by A. Benninghoven, C.A. Evans, Jr., R.A. Powell, R. Shimizu, H.A. Storms, Springer Series in Chemical Physics, Vol. 9 (Springer, Berlin, Heidelberg, New York 1979) p. 29
10. S.P. Lindgren and L. Valden, Sol. St. Comm. 25, 13 (1978)
11. M. Bernheim and G. Slodzian, J. Microsc. Spectr. Electron. 6, 141 (1981)
12. M. Bernheim and G. Slodzian, Int. J. Mass Spectr. Ion Phys. 20, 295 (1976)

Secondary Ion Mass Spectrometry of Organic Compounds; A Comparison with Other Methods (EI, CI, FI, FD, FAB)

K.D. Klöppel

Department of Chemistry, University of Siegen,
D-5900 Siegen 21, Fed. Rep. of Germany

In recent years secondary ion mass spectrometry (SIMS) has attained an increasing importance for the detection of organic compounds deposited on solid surfaces. In the static SIMS-method very low primary ion current densities are used in order to minimize the extent of fragmentation of the organic compounds deposited on metallic surfaces [1,2]. However, relatively fragment-free mass spectra of organic compounds have been also obtained applying high primary ion current densities [3,4]. A definite advantage of these secondary ionization (SI) techniques over more conventional ionization methods, like electron impact (EI), chemical ionization (CI), and field ionization (FI) lies in the fact that evaporation of the organic sample prior to analysis is not necessary. In this respect SI resembles the field desorption technique (FD) and many thus be preferentially applicable to the structure elucidation of organic salts and other labile, non-volatile compounds. Fragmentation patterns and ion intensities obtained by SI are, however, more readily controlled by varying parameters, such as current density, energy, and kind of primary ions, as well as nature and pretreatment of the target material. It has been shown that intensive cationic molecular ions are obtained by adding alkali halides to the organic target material [2,5].

SI-spectra of very thick organic compound layers, i.e. bulk solids have been obtained replacing the primary ion bombardment by a fast atom bombardment (FAB-technique) [6]. Charging of the surface as in ion bombardment of nonconductive layers is said to be avoided hereby. The present work was, therefore, undertaken to probe the feasibility of combining SI with other established ionization methods in a commercial mass spectrometer making full use of its high level of technical sophistication.

A double focusing Varian 311 A instrument was adapted to SI-measurements in a way that a direct comparison of the mass spectra obtained by the different ionization techniques was guaranteed [4]. A schematic drawing of the apparatus is given in Fig. 1. Two types of ion sources are now available for mass spectrometric investigations: source 1 for EI, SI and FI/FD measurements; source 2 for EI and CI measurements. For carrying out SI-measurements of organic compounds drops of their solutions in suitable solvents were deposited on specially treated

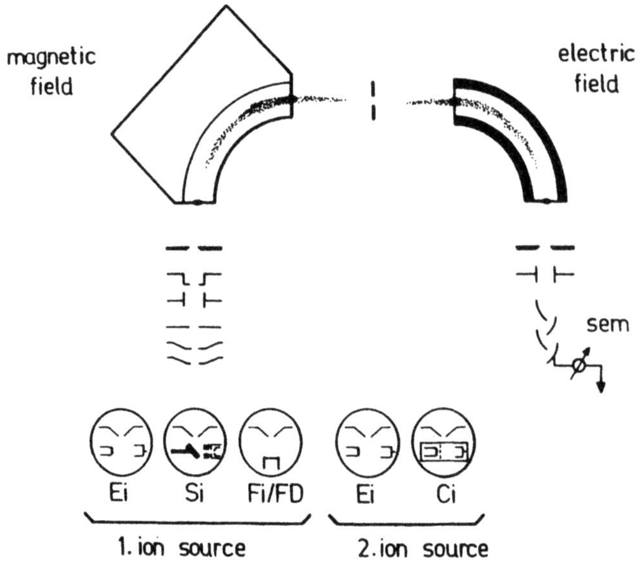

Fig. 1 Schematic representation of the Varian 311 A mass spectrometer equipped with combined ion sources EI: electron impact, SI: secondary ionization FI/FD: field ionization/desorption CI: chemical ionization

graphite and etched silver surfaces. SI-mass spectra were then studied as a function of kind and intensity of the primary ion beam. Ar^+- and Xe^+ ions of 2 to 5 keV energy were used with current densities of 0.1 to 10 $\mu A\ cm^{-2}$. Alkylhalides, ethers, amino acids, peptides, dyes, boron chelates, and sugars have been investigated. The results obtained from secondary ionization were compared with EI-, CI-, FI-, and FD-measurements and as much as possible with the published results of the FAB-technique.

SI-mass spectra of all investigated compounds exhibit surprisingly little molecular damage. Best results were obtained choosing a low primary ion current density and 2 keV energy. The recording speed of the SI-mass spectra was of the order of 2 seconds per 100 mass units corresponding to total primary ion dose densities of 2 to $3 \cdot 10^{-6}$ A s cm^{-2} during measurement. For some compounds the SI-emission was found to increase during primary ion bombardment. Keeping other variables constant, it was found that SI-intensities were higher when using the higher noble gas xenon, while the fragmentation pattern remained virtually unchanged. At low primary ion current densities M^+, $(M+1)^+$, and only few characteristic fragment-peaks are the most prominent features of the SI mass spectra, some similar to CI spectra using CH_4 as reactant gas. Others, especially the involatile compounds, show spectra similar to those obtained by the FD-technique. With increasing primary ion current densities the fragmentation pattern gradually approaches a peak distribution resembling EI-spectra.

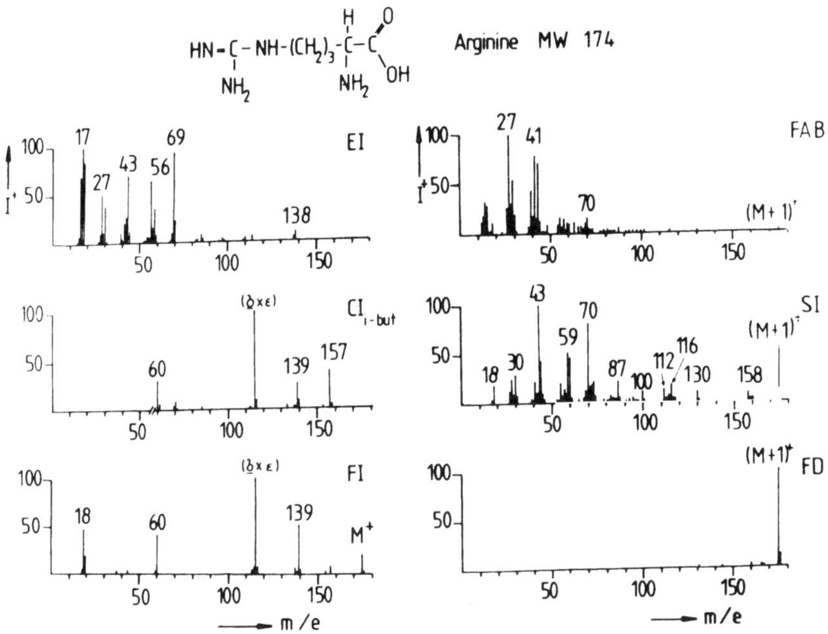

Fig. 2 Mass spectra of arginine obtained by applying different ionization techniques. EI: 70 eV, SI: Ar^+-ions at 2 keV and $5 \cdot 10^{-7}$ A cm^{-2}, CI: reactant gas i-butane, FAB: reproduced from SURMAN and VICKERMAN [6]

For illustrating the characteristic features of the different ionization methods, mass spectra of some substances of biological importance, such as amino acids and peptides, will be discussed here. The amino acid arginine constitutes a representative example of a compound whose thermal instability precludes volatilization without decomposition. In Fig. 2 EI-, CI-, FI-, SI-, FD-mass spectra of arginine obtained with the 311 A and the FAB-spectrum of SURMAN and VICKERMAN [6] (converted into a single intensity scale) are reproduced for comparison. It is seen that only the direct surface ionization methods lead to sufficient molecular intensities. The SI-spectrum of arginine has a well balanced ratio of high molecular, ion intensity vs. fragmentation, establishing the molecular weight, and providing sufficient data for structure analysis.

A similar result has been obtained for the tripeptide gly-gly-gly, cf. Fig. 3. By vapourisation of this tripeptide cyclic diketopiperazines are preferentially formed and characterize the peak distribution in EI and CI mass spectra. In contrast to the single $(M+H)^+$ molecular ion of the FD-spectrum, the SI-spectrum shows a number of characteristic fragment ion peaks while retaining an intensense $(M+H)^+$-peak. Further results demonstrate the applicability of SI to quantitative analysis. It was possible for example, to evaluate SI-mass spectra of multicomponent mixtures of amino acids. Fig. 4 shows

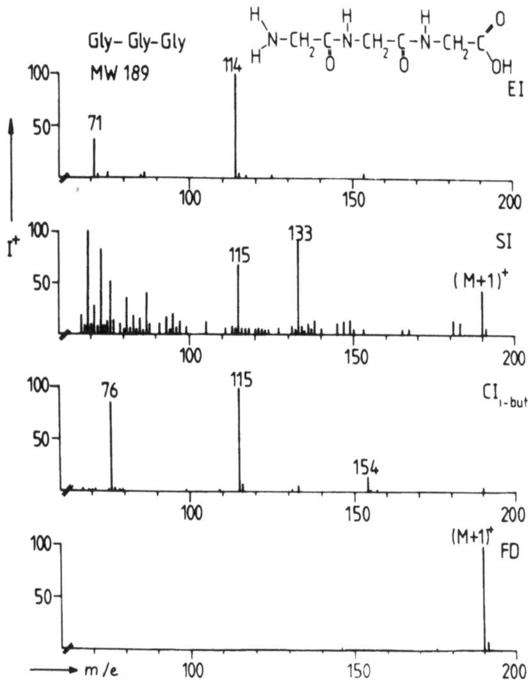

Fig. 3 Mass spectra of the tripeptide gly-gly-gly obtained by applying different ionization techniques. Conditions see Fig. 1

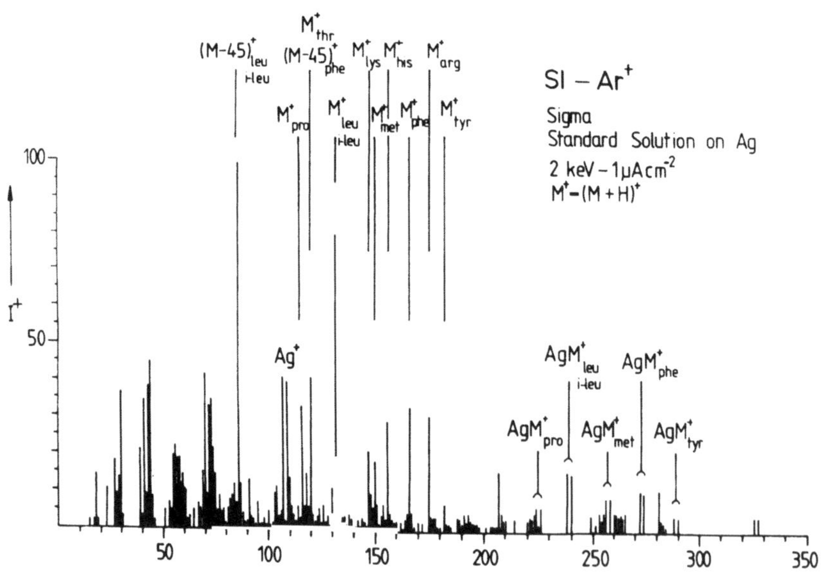

Fig. 4 Secondary ion mass spectrum of Sigma Standard Solution containing 17 amono acids in 0.1 n HCl-solution

the SI spectrum of 1 µl Sigma Standard Solution[1] containing 17 amino acids (0.1 n) and NH_4Cl in 0.1 n HCl solution deposited on an etched silver surface. In spite of the small concentration of each component and the relatively high primary ion current density of 2 µA cm^{-2} used here, intense protonated molecular ions and silver containing ions are observed. This spectrum is similar to that which has been recently obtained applying the static SIMS method [7]. The charging of thick organic layers in SI-investigations can be avoided by application of a suspension of the organic sample in a suitable conductive solvent having a low vapour pressure. Ion bombardment of such liquid target surfaces leads to results resembling those of the FAB-technique. The background pressure raises during these measurements up to $2 \cdot 10^{-3}$ Pa. Very high ion current densities can be maintained for long periods without causing an increased fragmentation of the organic sample. Fig. 5 shows a section of the SI-spectrum of the peptide Angiotensin II, which consists of eight amino acid entities. The peptide was dissolved in gylcerol and deposited on a silver surface. It is seen that the protonated molecular ion at mass number 1045 shows a high intensity despite the high current density (150 µA cm^{-2}) of the primary argon ions. Similar results have been obtained with other peptides at hand. It appears that solvents such as gylcerol may even interact with peptides forming protonated molecular ions prior to ion bombardment.

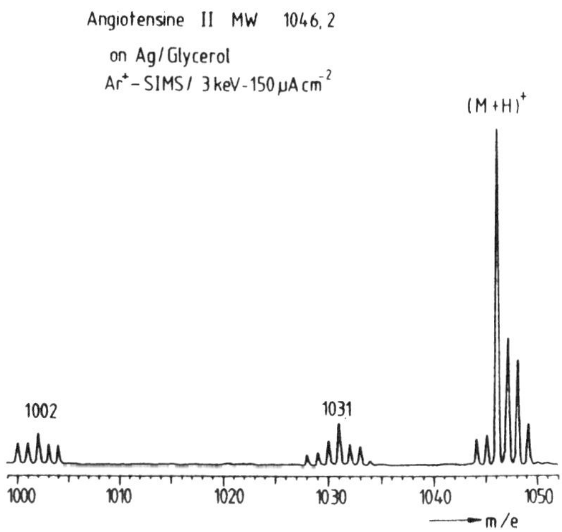

Fig. 5 Section of secondary ion mass spectrum of an Angiotensin II/glycerol-solution on silver surface

---

[1] The Sigma Standard Solution contains the following amino acids: ala, arg, asp, glu, gly, his, i-leu, leu, lys, meth, phe, pro, ser, thr, tyr, val, and cystine.

Further development of this sample preparation method is being investigated and expected to expand the range of applicability of secondary ionization to other biologically interesting materials. In conclusion, SIMS merits consideration as a powerful alternative tool in chemical analysis.

References

1. A. Benninghoven and W. Sichtermann, Org. Mass Spectrom. 12, 595 (1977)
2. H. Grade, R.G. Cooks, J. Am. Chem. Soc. 100, 5615 (1978)
3. K.D. Klöppel, W. Seidel, Int. J. Mass Spectrom. Ion Phys. 31, 151 (1979)
4. K.D. Klöppel and G. v. Bünau, Int. J. Mass Spectrom. Ion Phys. 39, 85 (1981)
5. H. Kambara and S. Hishida, Org. Mass Spectrom. 16, 167 (1981)
6. D.J. Surman and J.C. Vickerman, J. Chem. Soc., Chem. Commun. 324 (1981)
7. A. Benninghoven and W. Sichtermann, Int. J. Mass Spectrom. Ion Phys. 38, 351 (1981)

Secondary Ion Emission from NbV-Alloys

J. Schou[1], G. Flentje, W.O. Hofer, and U. Linke
I.G.V., Kernforschungsanlage Jülich, D-5170 Jülich, Fed. Rep. of Germany

Introduction

In an attempt to establish quantitative analyses of solids by secondary ion mass spectrometry (SIMS), a series of samples in the niobium-vanadium binary alloy system was studied. Ejection of cluster particles, both electrically charged and neutrals, was recently reported to confirm closely to the stoichiometry of the targets [1]. In an extension of this and related work [2,3,4] the behaviour of monoatomic particles as well as the influence of surface morphology was investigated here.

Experimental

Two rather different instrumental arrangements were used: a combined SIMS-AES instrument and a commercial r.f. quadrupole mass filter/electron impact ionizer equipped with both an ion-electron converter for particle detection and an energy filter. The latter system, described in detail elsewhere [1,5,6], was used in this study only when neutral particles were to be analysed. For the analysis of secondary ions and subsequent in-situ examination of the surface by Auber electron spectroscopy (AES), the instrument shown schematically in Fig.1 was used. In cases where the lateral resolution of the system (approx. 5 μm) was insufficient, scanning electron microscopy (SEM) investigations were carried out on a separate instrument.

Preparation of the NbV-samples proved to be very difficult when the same standard on surface finish was required over the entire Nb-concentration range (0.5 to 95 at. %). It turned out that these alloys became covered with a deformed surface layer after the initial polish. The crucial step is to etch this layer away before the final polishing procedure is performed. This was accomplished by vibratory polishing in a suspension of Al_2O_3 in an aqueous solution of $K_3(Fe(CN)_6)$.

Some samples showed laminar grain structure and individual grains even indicated slight segregation, both signs of incomplete thermal processing after production. These samples were annealed under UHV-conditions for 10 hours at 1900 °C. This procedure established a perfect globular structure, reduced the carbon content, and removed virtually all defects from the surface generated during the polishing process. It did not affect the cluster distributions, however.

[1]Risø National Laboratory, DK-4000 Roskilde, Denmark

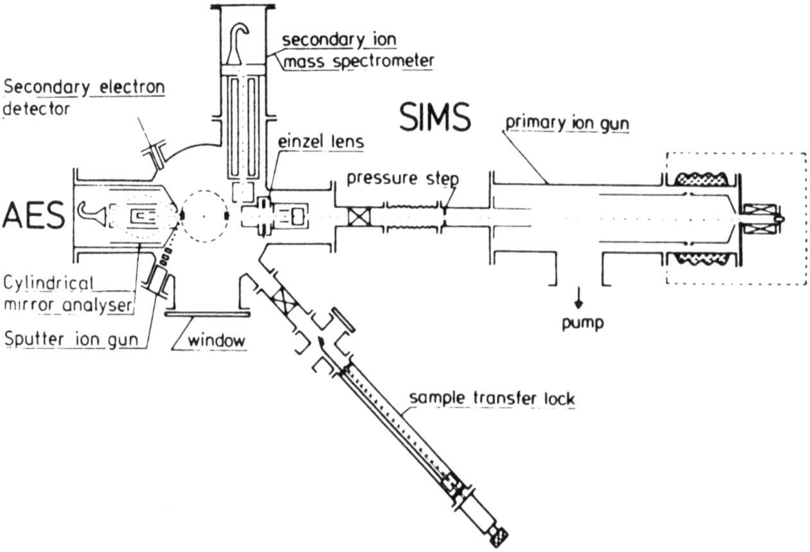

Fig. 1 The experimental set-up at KFA Jülich (Combined SIMS-Auger Analysis System)

Results and Discussion

A main result of Ref. [1] was the close correlation between cluster intensity and stoichiometry. In particular, this is valid for neutral clusters, hence proving that neutral particles are much less influenced by chemical enhancement effects. Fig.2 shows results for monatomic particles together with data for dimers [1]. Obviously the intensity of post-ionized Nb-atoms represents the composition of the samples better than particles ejected directly in the ionized state.

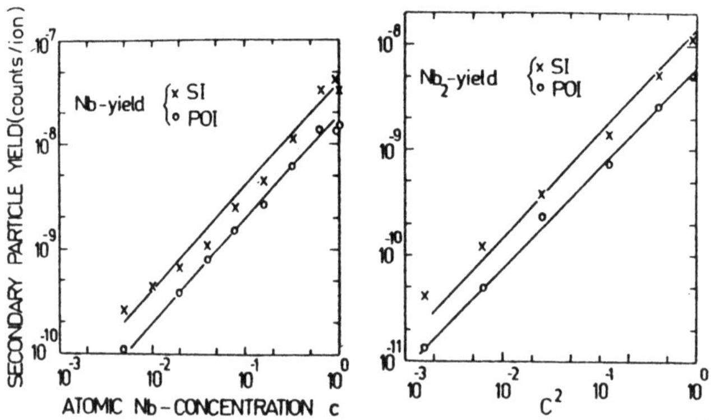

Fig. 2 Nb-monomer yields versus the atomic Nb-concentration c and Nb-dimer yields versus the Nb-concentration squared, c^2, in the binary NbV-alloy system. SI, secondary ions; POI, post-ionized neutrals. The samples were bombarded by 6 keV Ar^+-ions with a current density of 20 μAcm^{-2}

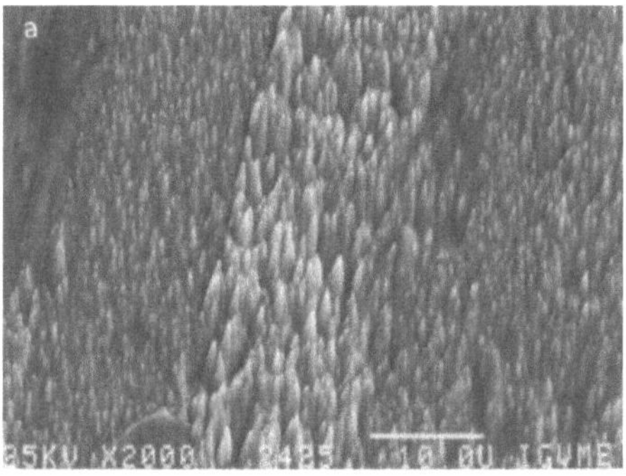

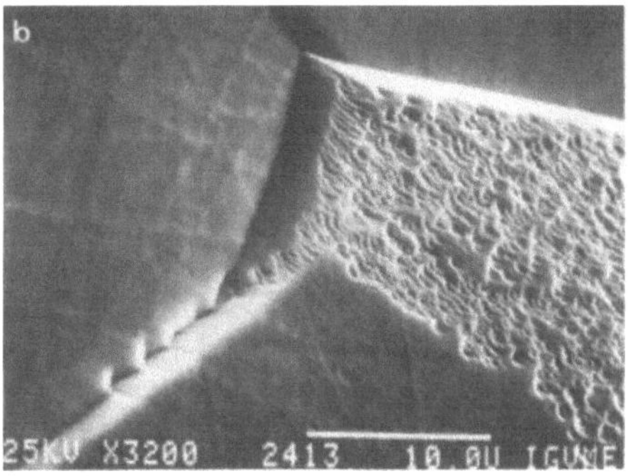

Fig.3 Scanning electron micrographs of (a) the alloy $Nb_{0.16}V_{0.84}$ after 6 keV Ar^+-ion bombardment. Fluence: $5\cdot10^{18}$ cm^{-2} without beam-scanning.
(b) The alloy $Nb_{0.32}V_{0.68}$ after 10 keV Xe^+-ion bombardment with beam-scanning. Fluence: 10^{18} cm^{-2}

In a view of the strongly varying surface structure among individual grains in the irradiated spot, this good correlation is surprising. How different the grains may be is illustrated by Fig. 3. A non-typical example with laminar cone-covered grains is shown in (a), whereas typical grain-structure is presented in (b). Extensive AES investigations were performed mainly for the sake of checking whether or not cone-covered grains deviate in composition from grains which remain flat during irradiation. The result is clearly negative: There is almost no difference in composition between grains of different surface morphology, either in the ratio of the main constituents (Nb or V) or in the impurity content. Carbon (and oxygen) is usually present. Graphitic carbon is effectively removed during ion bombardment, but carbidic carbon arises at the same time. The grain boundaries are surmised as a possible source.

Line scans through the sputtered crater vary markedly from grain to grain, but this is merely a topographic effect, influencing the electron yield but not the *ratio* of the Auger signals (Fig. 4). This effect is also evident in lateral element distribution images: Bright areas in the secondary electron Auger images evince stronger surface structures in SEM. It is therefore concluded that the presence of cone-covered grain surfaces neighbouring essentially flat ones after ion bombardment is a *lattice orientation* effect.

Since the clusters appear to be well suited to quantitative analysis, optimum conditions for cluster emission - both with respect to absolute intensity and abundance distribution - were sought. For this reason cluster distributions were measured for Ar^+-, and Xe^+-ions in the energy regime from 5 to 15 keV. There is a faint tendency for heavier clusters to increase in intensity with projectile mass (Fig. 5) but the effect is small. At different primary energies no clear tendency

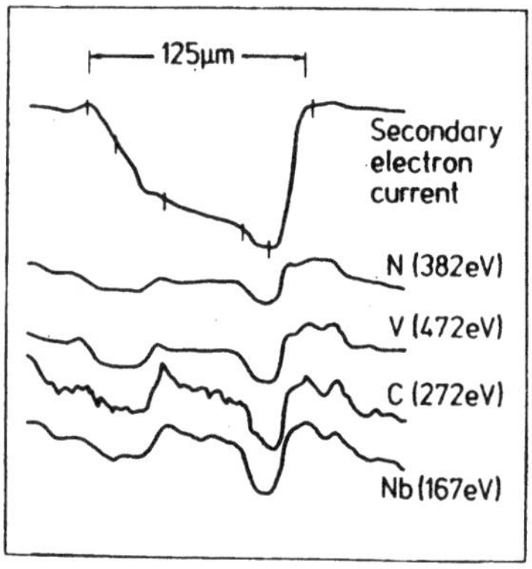

Fig. 4 Lateral scan over a crater on a $Nb_{0.16}V_{0.84}$ alloy generated by 10 keV nitrogen ions without beam-scanning. Fluence: $5 \cdot 10^{20}$ cm^{-2}

Fig. 5 Secondary ion yields for ion bombardment of a NbV-alloy

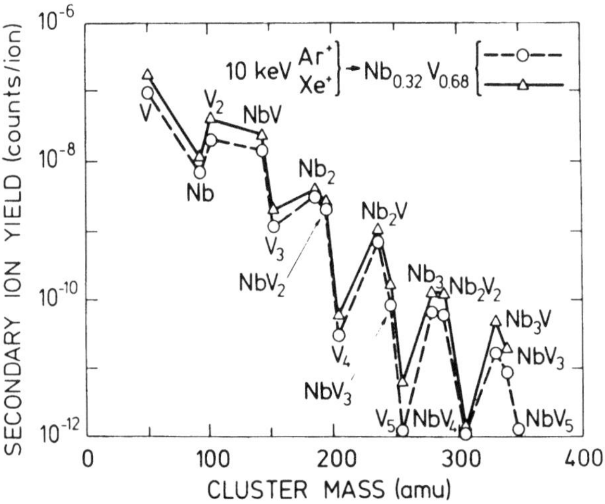

was observed. Thus it leaves these two parameters, mass and energy, free for optimisation with respect to other requirements, e.g. depth resolution or speed of erosion.

Bombardment by reactive gas ions such as O_2^+, on the other hand, strongly suppresses higher-order clusters [7] but adds complexity to the spectrum by producing a multitude of oxide lines. Hence little is gained by reactive projectiles except for enhancement of monatomic ions and suppression of pronounced surface topography.

Conclusion

The surface morphology has been studied by means of AES, SEM, and secondary electrons spectroscopy as well as by secondary particle emission. The concentration ratio of the constituents turned out to be almost constant from grain to grain in all cases, independent of the surface morphology. To a large extent the sputtered monatomic ion and neutral intensity is proportional to the Nb-concentration, the post-ionized neutrals showing considerably better agreement.

The Commission of the European Communities has supported a visit of J. Schou to KFA Jülich, where the work was initiated.

References

1. J. Schou and W.O. Hofer, to be published
2. G. Blaise, in "Material Characterization Using Ion Beams", eds. J.P. Thomas and A. Cachard (Plenum, New York, 1978) p. 143
3. Ming L. Yu and W. Reuter, J. Appl. Phys. 52, 1478 (1981)
4. M. Riedel, T. Nenadovic and B. Perovic, Acta Chim. Acad. Sci. Hung. 97, 197 (1979)
5. J. Giber and W.O. Hofer, in "Proc. of the Symposium on Sputtering", eds. P. Varga, G. Betz and F.P. Viehböck (Inst. für Allg. Physik, Technical University of Vienna, 1980) p. 697
6. W.O. Hofer and F. Thum, Nucl. Instr. Meth. 149, 535 (1978)
7. G. Modler and H.E. Beske, KFA-rep. 1651, Kernforschungsanlage Jülich (1980)

Part III

Fundamentals II.
Depth Profiling

Depth Profiling by SIMS: Depth Resolution, Dynamic Range and Sensitivity

C.W. Magee, R.E. Honig, and C.A. Evans, Jr.[1]
RCA Laboratories, Princeton NJ 08540, USA

1. Introduction

Secondary ion mass spectrometry (SIMS) utilizes a beam of energetic primary ions to sputter away a solid [1] surface producing ionized sputtered particles which can be mass spectrometrically detected. This technique provides in-depth information on atomic constituents by recording one or more mass peaks as the sputtering process erodes the sample, thus producing the detected signal from increasingly greater depths beneath the original sample surface. This depth profiling technique has become one of the most important uses of SIMS, but the accuracy with which the data reflect the original atomic profile within the sample depends on many parameters. Several critical reviews on depth profiling exist in the Ref. [2,3,36] which broadly cover most aspects of this mode of analysis. The present work is limited to three major aspects: 1) depth resolution, 2) dynamic range, and 3) sensitivity. The relationships between these areas will be discussed, as well as how they influence the accuracy of the final profile.

2. Depth Resolution

The quality of a depth profile, called "depth resolution", is best described in terms of the measured width of an interface between two layers. Fig. 1 compares the ideal "true" profile, rectangular in shape, with the measured profile which is conveniently represented by an integrated error function. The interface width Δt (in terms of sputtering time t) or Δz (in terms of depth z) is defined as the interval where the intensity drops from 84 % to 16 % of maximum signal, equivalent to two standard deviations (2σ) of the error curve [3]. Alternative definitions found in the literature are based on the 90-10 % or the 95-5 % interval corresponding respectively to widths of $2.564\,\sigma$ or $3.290\,\sigma$ [4].

There are two factors related to the sputtering process which influence the depth resolution of a profile. They are atomic mixing effects, and non-uniform sputtering within the analyzed area due to inhomogeneous primary ion current density and/or in homogeneous sputter yield across the bombarded area (e.g. in polycrystalline material). Atomic mixing results from

[1]Charles Evans and Associates, San Mateo CA 94402, USA

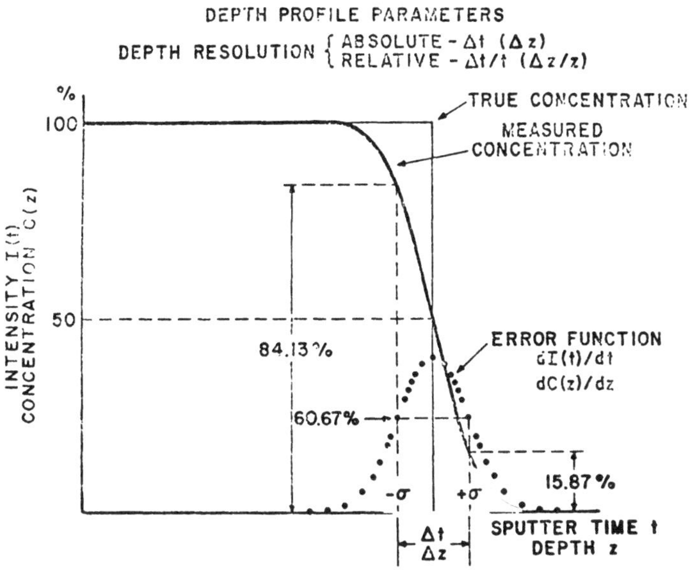

Fig. 1 Parameters for depth resolution

the energetic interaction of the primary ion with the sample lattice atoms. Two principal types of atomic mixing may occur. When the primary ion undergoes a small impact-parameter collision with a sample atom, the struck atom may recoil into the solid. This is usually referred to as "recoil implantation". When large impact-parameter collisions occur between the primary ion and the sample atoms, those atoms are displaced from their lattice positions and in turn strike neighbouring atoms which are also displaced. The net effect of this process is a general homogenization of all atoms in the near-surface region of the sample which are affected by the collision cascade. The thickness of this near-surface region is usually of the same order as the range of the primary particle in the sample. This homogenization process is referred to as "cascade mixing". Due to the probabilities of having large and small impact-parameter collisions, the cascade mixing effect is usually the dominant component of atomic mixing and will be discussed below.

Non-uniform erosion of the sample surface will also affect depth resolution by exposing various depths to the analyzing ion beam. While this contribution to depth resolution degradation can generally be neglected due to refinements in SIMS instrumentation, it will later be shown how this may still be the single most important factor in determining depth resolution.

A. Cascade Mixing

The depth resolution of any technique which employs sputtering by energetic particles is limited by the depth within which the primary ion collision cascade produces atomic displacements within the solid. This depth is dependent on the energy of the primary ion, the atomic numbers of bombarding ion and substrate, and the angle of incidence between primary beam and sample surface normal. Clearly, the use of low-energy, high-Z particles, and large incident angles with respect to the surface will minimize this cascade mixing depth. This atomic mixing process has been modelled theoretically [5,6] and detected experimentally [7,8], but previously never in an accurate manner free of artifacts. However, TSONG and co-workers [9] have recently published a simple, accurate, and widely applicable method of determining the depth of the cascade mixing layer. They sputter a series of SiO_2 films of known varying thickness, thermally grown on Si while monitoring the oxygen secondary ion signal as a function of time. Due to the presence of cascade mixing, the oxygen signal starts to decrease once the damaged layer of altered composition [10] contacts the SiO_2/Si interface. A plot of constant oxygen signal sputtering time vs. SiO_2 film thickness yields a straight line with an intercept equal to the full width of the mixed layer. The accuracy of the experiment depends only on the accuracy with which the SiO_2 films can be measured (usually ± 5 Å by ellipsometry) and is not influenced by an inhomogeneous current density distribution of the primary ion beam [9].

The results of such an experiment performed at RCA Laboratories are shown in Fig. 2. The experiment employed 5 keV $^{133}Cs^+$ ions incident on SiO_2 at 60° with the surface normal. The detected ion was $^{18}O^-$ and, utilizing a charge-compensation procedure [11], this signal remained constant from the start of sputtering

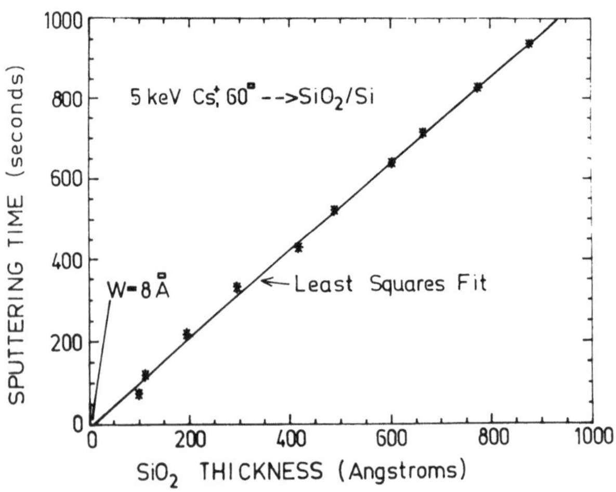

Fig. 2 Determination of cascade-mixing range. Samples were SiO_2 thermally grown on Si. Thicknesses were measured by ellipsometry (± 5 Å). Since the cascade mixing layer reaches the Si/SiO_2 interface before the SiO_2 layer is completely removed, the intercept of steady-state oxygen sputtering time vs. SiO_2 thickness yields the width of the cascade-mixing layer for SiO_2. (Experimental method from [9])

until the interface was reached (except for the time require to sputter away the outermost 50 Å during which charge equilibrium was established). The results indicate a cascade mixing range of ≤ 10 Å. The accuracy of the experiment can be assessed from the good least-squares fit of the straight line through the data. Clearly, other workers should repeat this experiment to measure atomic mixing ranges under different experimental conditions because the cascade-mixing range sets the limit on the optimum attainable depth resolution. As is shown below, however, there are other factors which may degrade depth resolution more seriously.

B. Non-uniform Sputtering

One of the factors which affects depth resolution is related to the total depth sputtered: with increasing depth, the resolution deteriorates. If the absolute depth resolution Δz is directly proportional to z, the total depth sputtered, then the relative depth resolution $\Delta z/z$ will be constant. If Δz is some fixed value, independent of z, then $\Delta z/z$ will vary as z^{-1}. Rather exhaustive compilations of this dependence exist [12-14] and most of the data fall somewhere between these two functions. This is not surprising considering that uneven sputter erosion (caused by either non-uniform primary ion current density within the analyzed area or by cone formation) will undoubtedly affect the measured depth resolution to varying degrees depending on experimental conditions. The different factors affecting depth resolution have previously been discussed [3] and measured [14], although in the latter case the inherent interface width of the Ni/Cu system studied was not sufficiently small to enable the investigators to evaluate the depth-independent portion of Δz attributed to atomic mixing caused by the 100 eV Ar sputtering ions.

An investigation was undertaken to study the depth-resolution characteristics of the RCA SIMS instrument [15]. When analyzing sharp interfaces by SIMS, one must choose appropriate samples to insure that the depth profiles will not be altered by artifacts. The sputtering rates and ion yields should not change from one layer to the next. Therefore, the ideal sample would involve only an isotopic change between layers. Such samples have been profiled by SIMS [15-17] even though they are difficult to make, especially when sharp interfaces are required as well as varying layer thicknesses.

For the present study, we chose InGaAs/GaAs structures grown by vapor phase epitaxy. Previous studies [18] had shown that abrupt III-V heterojunctions can be grown in controlled, widely varying thicknesses. In addition, they can be grown with only a small amount of indium. This is critically important to insure that the sputtering rate does not change between the InGaAs layer and the GaAs substrate (which otherwise would alter the depth scale). Also, since the amount of indium is kept small, there should not be an appreciable change in ion yield between the two layers.

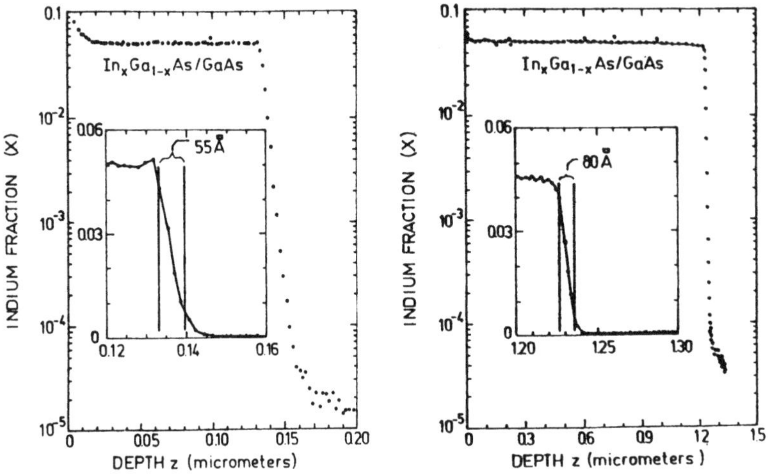

Fig.3 Indium depth profiles for layers of $In_{0.05}Ga_{0.95}As$ grown by vapor-phase epitaxy on GaAs substrates. The linear inserts show the measured widths of the interfaces. Layer thickness are (a) 0.13 μm and (b) 1.3μm. Sputtering conditions were: 5 keV Ar^+, incident at 60^0

The samples we decided to use were $In_{0.05}Ga_{0.95}As/GaAs$, with the InGaAs layers ranging in thickness from 0.015 to 1.3 μm. Indium profiles for the 0.13 and 1.3 μm samples are presented in Fig. 3, with interface widths shown in the linear inserts. They show that a tenfold increase in sputtered depth produces an interface broadening of only 25 Å. Interface widths obtained for all the samples in the series are shown in Fig. 4 plotted as a function of sputtered depth. The two-component nature of the Δz dependence upon z is readily seen. The data points for Ar^+ bombardment show how, up to a sputtered depth of approximately 0.13 μm, Δz shows no measurable dependence on z. Only at greater depths does Δz start to increase. We assume that this increase is due to an increasing unevenness of the sputtered crater bottom, and that at depths <0.13 μm, this contribution to Δz is too small to be detected in the presence of the cascade mixing effects.

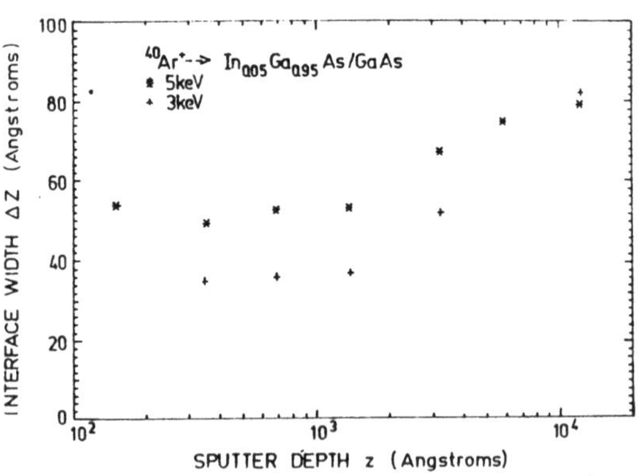

Fig. 4 Dependence of interface width Δz upon sputtered depth z for $In_{0.05}Ga_{0.95}$-As/GaAs samples. Samples varied in thickness from 150 to 13.000 Å. Ar^+ primary ion energies of 3 and 5 keV were used

These cascade-mixing effects can be separated more clearly from the non-uniform current density effects by reducing the energy of the primary ion and thus, its range in the solid. The data in Fig. 4 illustrate how the atomic mixing range is decreased from 55 Å to 35 Å by lowering the primary ion energy from 5 to 3 keV. However, as with the 5 keV bombardment data, when sputtering is carried to large depths, Δz is again dominated by the cumulative effect of uneven cratering of the sample in agreement with previous published data [4].

C. Depth Resolution for Non-ideal Sample Conditions

The above examples illustrate the limitations on depth resolution imposed by various parameters for ideal samples (i.e. smooth surface, uniform ion yield). The following example, (Fig. 5), shows how, in some cases, the measured profiles may bear little resemblance to the actual profile. This is a multi-element depth profile of a 1500 Å Pd/Si sample showing the detected signals for Pd and Si as a function of sputtering time. Analysis conditions were: 5 keV $^{40}Ar^+$ primary ion bombardment; positive secondary ion detection; and an oxygen jet directed towards the sample to increase positive ion sensitivity. As schematically shown in region (a) of Fig. 5 the Pd layer develops a rough surface as it is sputtered. It has long been known that sputter-etching a surface can dramatically change its topography, with cone formation being the feature most often observed when polycrystalline metal films are sputtered [19]. As the deepest areas between cones break trhough the Pd/Si interface, a sharp rise is detected in the Si signal. In region (b), both Pd and Si are exposed to the sputtering beam, resulting in a delayed rise in the Si signal until the cones have been etched almost completely into the Si substrate, corresponding to region (c) of Fig. 5. It is evident that the artifact of sputter-induced surface roughening has completely obliterated the true shape of the depth profiles.

A second artifact is apparent in Fig. 5: the raise in the Pd signal is region (b) of the depth profile. This can be

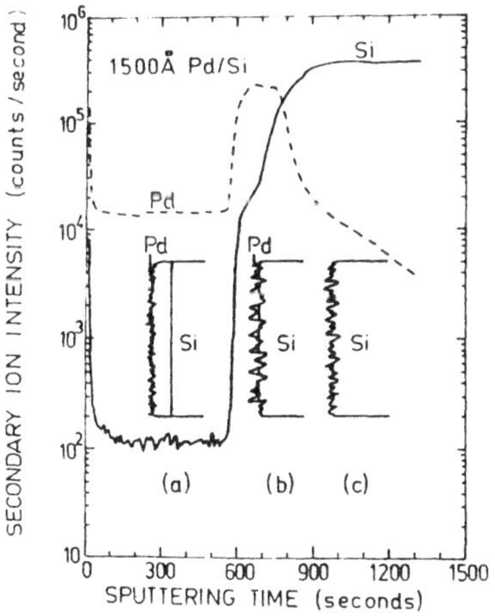

Fig. 5 SIMS multi-element depth profile 1500 Å Pd/Si. Region (a): surface totally within the Pd; region (b): surface intersecting both Pd and Si; region (c): sputtered down to the Si substrate. To be noted are: surface roughening, changes in secondary yield, and ion-induced atomic mixing. 5 keV Ar^+ with O_2 jet on sample, positive ion detection

attributed to the presence of Si which increases the amount of O sticking to the sample which in turn increases the Pd ion yield. This effect is less pronounced when using inert-gas bombardment without an O jet, but is still present to some degree due to the chemical yield-enhancement effect due to Si alone.

A third effect of non-ideal sputtering on depth resolution is shown in region (c) of the Pd depth profile of Fig. 5. The slow decrease in the Pd signal beyond 900 seconds of sputtering time can be attributed to atomic mixing of the Pd into the Si. Studies [8] have shown that certain atoms can be mixed with Si with great efficiency by ion bombardment, to a depth approaching the range of the bombarding ion in the solid. Models [20] have been put forward which propose that only those metal atoms on, or just beneath, the sample surface can be sputtered, and as they are, the remaining metal atoms become mixed to a slightly greater depth as the surface recedes into the bulk. This mechanism correctly predicts the exponential decay of the Pd profile seen in region (c) of Fig. 5, as well as profiles observed for other metals sputtered from Si [20].

3. Dynamic Range

One aspect of depth profiling by SIMS which is seldom discussed in the literature is the dynamic range in concentration over which the ion of interest can be monitored accurately. The factors which affect the dynamic range of an analysis are: A) crater-edge effects; B) neutral beam effects; C) absolute count rate; and D) background count rate.

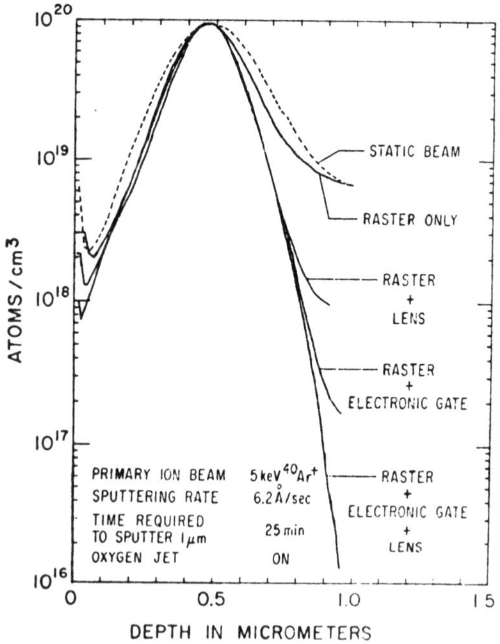

Fig. 6 150 keV boron implanted into silicon, profiled with increasingly sophisticated primary ion beam techniques and secondary ion optics

A. Crater-Edge Effects

When depth profiling is done by SIMS, the primary ion beam is usually focused and rastered several beam diameters in perpendicular directions to produce a flat-bottomed crater [21,22]. Ion optical and/or electronic apertures are then used to select data only from the central flat portion of the crater, thus rejecting ions which are emitted from the crater walls where various depths are exposed to the sputtering beam. The effectiveness of these means of crater edge rejection is shown in Fig. 6. These considerations are most important when profiling from regions of high concentration to regions of low concentration.

B. Neutral Beam Effects

One of the major limitations in obtaining depth profiles with a wide dynamic range is due to the neutral non-focused component of the primary beam which may strike the sample producing secondary ions from areas surrounding the sputtered crater [23]. A recent study [24] has shown how this component can be the limiting factor in dynamic range even if the neutral/ion ratio is 10^{-5}. This study ascertained that charge-exchange neutrals can be formed in the vicinity of the differential pumping aperture [24] as well as from the space surrounding the ion gun anode aperture [15]. To remove this non-focused component effectively, the bend in the beam-line, previously recognized as being necessary [15], should be placed on the *low-pressure* side of the pressure step. The impressive result of this instrumental improvement is shown in Fig. 7.

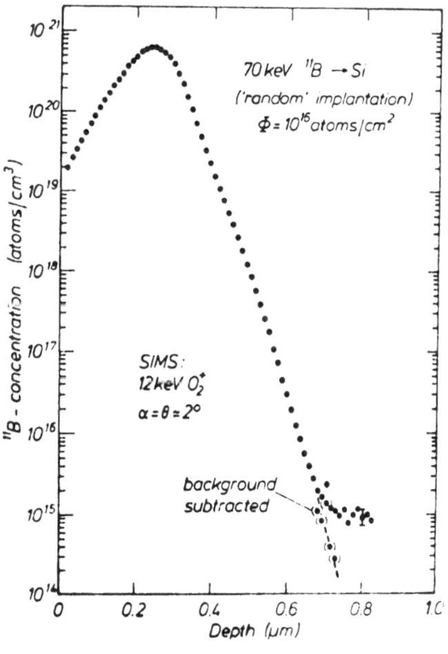

Fig. 7 Depth profile of a boron-implanted silicon sample after suppression of neutrals originating in the ion-gun section (beam current 1.5 μA, scan width 0.76 mm, linear gate 13%). Real time consumed per data point 60 s. The average background intensity corresponds to 15 counts/s

C. Instrumental Background

The dynamic range of a depth profile can also be limited by instrumental background which may be due to the following:

1) mass spectral overlap
2) residual gas contamination
3) re-deposition
4) detection of unfilterable particles (quadrupole MS only)
5) electron multiplier and detection circuits

The detection of background level is important because the dynamic range of a depth profile is always determined by the peak-to-background ratio, for example, for an ion-implanted sample. Since there are relatively few ways to increase peak count rates, reduction of background count rates becomes imperative.

1) Mass Spectral Overlap

High background at a particular mass of interest can be caused by a spectral overlap due to an element or combination of elements which make up the sample matrix. Such mass interferences due to matrix ions cannot be circumvented by monitoring multiply charged ions, dimer or trimer ions, or other molecular ions, and can only be eliminated by separating the ion of interest from the background ion. This requires the use of high-resolution, double-focussing mass spectrometers [25]. One technologically significant example of this kind of background is the analysis of Fe in Si. Iron always is a possible source for deep-level traps in Si and sensitive analyses for its presence are important. Unfortunately, ^{56}Fe is interfered with at that nominal mass by $^{28}Si_2$. In addition, ^{57}Fe is masked by $^{28}Si^{29}Si$, and ^{58}Fe by $^{29}Si_2$. Without high mass resolution, ^{54}Fe is the only isotope which is free from matrix-ion interference, and its isotopic abundance of 5.8 % results in unacceptably low sensitvity. Fortunately, new high resolution SIMS instrumentation is available which now makes the analysis of Te in Si possible [26].

2) Residual Gas Contamination

Another kind of spectral overlap problem can contribute to the background. It arises from mass spectral peaks which are due to residual gas contamination of the sample. Obviously, these interferences can be reduced by improving the vacuum in the sample chamber, thus decreasing the arrival rate of contaminating particles [27], or by increasing the sputtering rate. An example of such a background reduction is shown in Fig. 8. It illustrates how the analysis of H in Si   interfered with by adsorption of H_2O on the sample surface during profiling of a proton-implanted Si sample [28]. Detection limits similar to those shown for 10^{-10} Torr may be obtained at higher pressure (10^{-9} Torr) simply by increasing the sputtering rate from 5 to 300 Å/s [29].

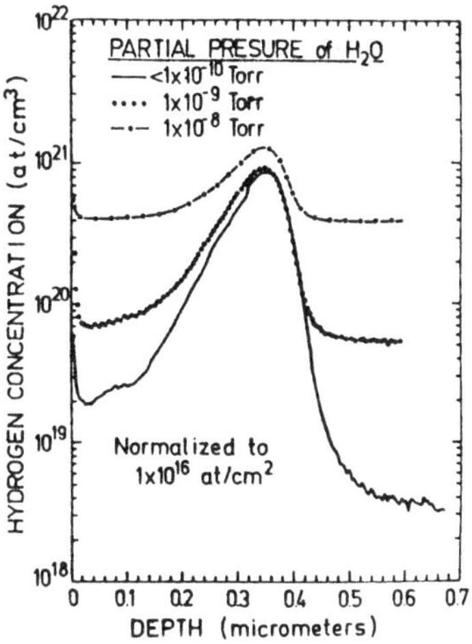

Fig. 8 Effect of H_2O partial pressure on H dynamic range. Sample is 35 keV H implanted into Si, dose: $1 \cdot 10^{16}$ at/cm^2. Hydrogen background levels can be estimated from concentrations down to which the profiles can be followed. Sputtering rate: 5 Å/s

3) Re-deposition and Memory

Another source of high background may be the redeposition onto the crater bottom of material from the crater walls [23] and from the extraction lens [30]. It is clear that these problems, like those of direct crater-edge rejection (Fig. 6) are more serious when the concentration of the element of interest is higher at the crater walls than in the analyzed center area.

The "memory" from extraction lens elements is a particularly difficult problem. For better ion extraction efficiency, the lens must be placed close to the sample, but this proximity also increases the amount of material deposited on the lens. This sputtered material can then be detected directly as secondary ions emitted from the lens [15], or after being re-sputtered back onto the sample surface [30]. Ions formed in the former case can be discriminated against by using stigmatic, or "imaging" secondary ion extraction optics [31]. Detection of material resputtered onto the sample can be minimized by increasing the sample-to-extraction lens distance [32], but with a sacrifice in instrument transmission.

4) Unfilterable Particles

Background levels may also be raised by detection of unfilterable particles when using quadrupole-type SIMS instruments. Photons and high-energy sputtered neutral species will be detected by electron multipliers, generating a mass-independent background. These neutrals and photons can be removed by the secondary ion energy analyzer [33] or by placing the electron multiplier off the quadrupole axis. Complete rejection of these particles is important if background from these sources is to be reduced to the level of intrinsic electron multiplier noise (typically 0.1 count/s for positive ion detection and 1 count/s for negative ion detection [15].

5) Electron Multiplier and Detection Circuits

When detecting negative ions from insulating samples which require an auxiliary electron beam for charge compensation [11], stray electrons may enter the electron multiplier. This signal level may limit the dynamic range of a depth profile. Such is the case in Fig. 9 which shows a depth profile of a 0.6 μm SiO_2 film implanted with ^{35}Cl at 180 keV to a fluence of $3 \cdot 10^{14}/cm^2$. Using negative-ion detection, the stray-electron background determines to what level the Cl concentration can be followed.

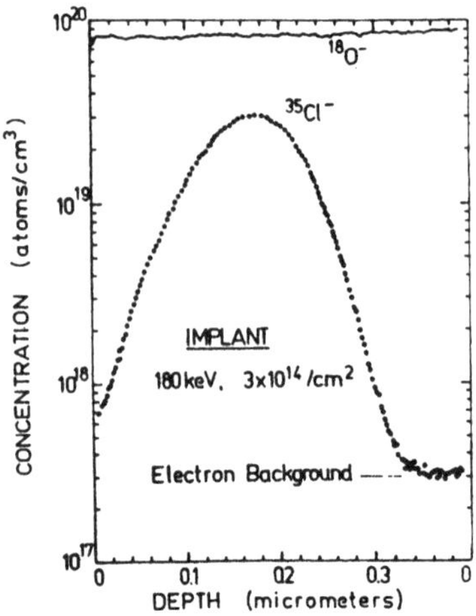

Fig. 9 Chlorine depth profile for SiO_2 sample implanted with ^{35}Cl at 180 keV to a dose of $3 \cdot 10^{14}$ atoms/cm^2. The primary ion bombardment was 5 keV Ar^+. Electron beam charge compensation was used. Detected stray electrons set the limit on dynamic range for this analysis

4. Sensitivity

The ultimate sensitivity of a depth-profiling analysis is determined by a signal-to-noise ratio. The factors which govern the noise in the data are the same as those discussed above with regard to dynamic range. The magnitude of the signal is, in large part, determined by the ion yield of the species monitored. This ion yield usually can be affected favourably by manipulation of the surface chemistry by using chemically active elements as the primary bombarding species [34,35]. In cases where high ion yields can be obtained, such as B in Si under O bombardment (Fig. 7), the ultimate sensitivity is usually determined by the background level as discussed above. There are instances, however, where the sensitivity of the profile is set entirely by the ion yield. Fig. 10 illustrates such an example. Shown is a $^{52}Cr^+$ depth profile of an undoped GaAs epitaxial layer grown on a Cr-doped substrate. The background is due to the electron multiplier, amounting to approximately 0.1 counts/s. Under these conditions, the ion yield for Cr is not high, and 10 s integration times were necessary even to obtain 2-3 counts, equivalent to approximately $2 \cdot 10^{15}$ Cr atoms/cm^3. Together with the 7 Å/s sputtering rate of this analysis, the profile is limited to 70 Å/data point. In order to obtain better ion statistics, even longer integration times would be necessary at an even greater sacrifice to depth resolution.

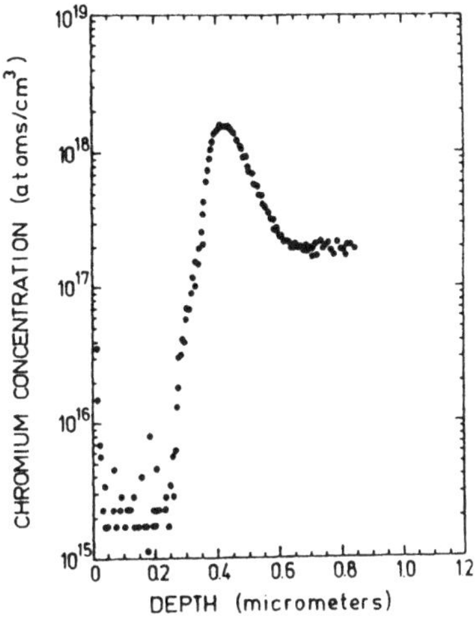

Fig. 10 $^{52}Cr^+$ depth profile for an undoped GaAs layer grown as a chromium-doped GaAs substrate. Analysis conditions were: 5 keV N_2, oxygen jet directed at sample; sputtering rate: 7 Å/s; integration time per data point: 10 s; 10^{15} at/cm^3: 1 count/integration period

The absolute count rate of a depth profiling analysis is also directly proportional to the primary ion current falling within the analyzed area. This has important implications for small-area analyses when small diameter primary beams must be used. Studies [21] have shown how detection limits vary roughly as the square of the primary ion beam diameter. Under extreme conditions (probe diameter = 2 μm, probe current = = $2 \cdot 10^{-10}$ A, 10 x 10 μm rastered area, 3 x 3 μm gated area), the detection limit for B in Si was found to be $5 \cdot 10^{17}$ at/cm^3, or two orders of magnitude greater than that obtainable under more normal depth profiling conditions (25 μm probe diameter, $3 \cdot 10^{-8}$ A).

5. Summary

The data presented here show that for SIMS depth profiling:

1) depth resolution is determined by the atomic-mixing range and the flatness of the sputtered crater within the analyzed area;

2) dynamic range is limited by crater-edge effects, neutral beam effects, spectral overlap, residual gas contamination, recontamination from previously sputtered material, and noise from detection of non-filterable particles; and

3) in the absence of measurable background, sensitivity is dependent on ion yield and analyzed area, and can be increased only at the expense of depth resolution and spatial resolution.

Acknowledgements

The authors wish to thank F. Kolandra for preparing the ion-implanted samples; and K. Wittmaack for providing data [36] prior to publication.

References

1. Recent studies (M. Barber, R.S. Bordoli, R.D. Sedgwick and A.N. Tyler, unpublished) have shown that liquids may also be sputtered to yield SIMS spectra.
2. E. Zinner, Scanning 3, 57 (1980)
3. S. Hofmann, Surf. Interface Anal. 2, 148 (1980)
4. H.W. Werner, Vacuum 20 (1974)
5. H.H. Andersen, Appl. Phys. 18, 139 (1979)
6. U. Littmark and W.O. Hofer, Nucl. Instrum. Meth. 168, 329 (1980)
7. J.A. McHugh, Rad. Eff. 21, 209 (1974)
8. Z.L. Liau, B.Y. Tsaur and J.W. Mayer, J. Vac. Sci. Tech. 16, 121 (1979)
9. I.S.T. Tsong, J.R. Monkowski and D.W. Hoffman, Nucl. Instrum. Meth. 182, 237 (1981)
10. H.F. Winters and J.W. Coburn, Appl. Phys. Lett. 28, 176 (1976)
11. C.W. Magee and W.L. Harrington, Appl. Phys. Lett. 33, 193 (1978)
12. R.E. Honig and W.L. Harrington, Thin Solid Films 19, 43 (1973)
13. R.E. Honig and C.W. Magee, "Proceedings of the 26th Annual Conference on Mass Spectrometry", p. 207, St. Louis (1978). Also included in Ref. 3.
14. J.W. Coburn, E.W. Eckstein and E. Kay, J. Appl. Phys. 46, 2828 (1975)
15. C.W. Magee, W.L. Harrington and R.E. Honig, Rev. Sci. Instrum. 49, 477 (1978)
16. D.E. Carlson and C.W. Magee, Appl. Phys. Lett. 33, 81 (1978)
17. Y. Limoge, R. Seguin and J.L. Séran, Proceedings of the 8th International Congress on "X-ray Optics and Microanalysis", D.R. Beaman et al., Eds., (Pendell, Midland, 1980) p. 347
18. R.D. Dupuis, P.D. Dapkus, C.M. Garner, C.Y. Su and W.E. Spicer, Appl. Phys. Lett. 34, 335 (1979)
19. W.O. Hofer and H. Liebl, Appl. Phys. 8, 359 (1975)
20. P. Williams, Appl. Phys. Lett. 36, 758 (1980)
21. T.A. Whatley, D.J. Comaford, J. Colby and P. Miller, in "Surface Analysis Techniques for Metallurgical Applications", R.S. Carbonara and J.R. Cuthill, Eds., ASTM STP 596, Am. Soc. for Testing Mat., Philadelphia, (1976) p. 114
22. K. Wittmaack, Appl. Phys. 12, 149 (1977)
23. J.A. McHugh in "Methods of Surface Analysis", S.P. Wolsky and A.W. Czanderna, Eds., (Elsevier, Amsterdam, 1975) p. 223
24. K. Wittmaack, Appl. Phys. Lett. 37, 285 (1980)
25. D.K. Bakale, B.N. Colgy and C.A. Evans, Jr., Anal. Chem. 47, 1532 (1975)
26. J.-M. Gourgout, in Secondary Ion Mass Spectrometry SIMS-II, ed. by A. Benninghoven, C.A. Evans, Jr., R.A. Powell, R. Shimizu, H.A. Storms, Springer Series in Chemical Physics, Vol. 9 (Springer, Berlin, Heidelberg, New York 1979) p. 286
27. C.W. Magee, J. Electrochem. Soc. 126, 660 (1979)
28. C.W. Magee and E.M. Botnick, J. Am. Vac. Soc. 19, 47 (1981)

29. C.A. Evans, Jr., unpublished
30. H.W. Werner, Acta Electron. 19, 53 (1976)
31. As in the instrument described in Ref. 15, and in direct imaging ion microscopes
32. K. Wittmaack, personal communication (1979)
33. R. Schubert and J.C. Tracy, Rev. Sci. Instrum. 44, 487 (1973)
34. V.R. Deline, C.A. Evans, Jr. and P. Williams, Appl. Phys. Lett. 33, 578 (1978)
35. V.R. Deline, W. Katz, C.A. Evans, Jr. and P. Williams, Appl. Phys. Lett. 33, 832 (1978)
36. K. Wittmaack, to be published

Disturbing Effects in Sputter Profiling

S. Hofmann

Max-Planck-Institut für Metallforschung, Inst. für Werkstoffwissenschaften, D-7000 Stuttgart, Fed. Rep. of Germany

1. Introduction

Sputtering by ion bombardment in combination with a suitable analysis technique is a universally applicable method to obtain the distribution of the elemental composition of a sample versus its distance perpendicular to the original surface. Although sputter depth profiling has found widespread application and numerous review articles on this subject have already appeared (e.g. [1-5,64,65]), we have not yet achieved a full understanding of the various distortions induced during profiling which impede a direct correlation between the measured profile and the original, true profile present in the sample prior to the analysis.

In the following, the basic considerations in the evaluation of measured profiles are summarized with respect to the detrimental effects in sputter profiling, their recognition, experimental minimization and correction by deconvolution procedures.

In its dynamic mode, SIMS appears to be the most direct way to get a correlation between the instantaneous surface concentration and the sputtered depth, since the sputtered matter itself is detected. The analytical information, however, depends strongly on the details of the secondary ion emission, i.e. on the state of the surface left behind. On the other hand, techniques such as AES, XPS, and ISS analyse the residual surface and are completely independent on the sputtering and ionization processes. It is mainly due to this reason, why - together with FIM and RBS - the electron spectroscopies have been employed very successfully to investigate detrimental effects of sputtering to profiles which will be discussed below.

2. Basic Considerations

The primary information obtained in a sputter profiling experiment consists of a measured signal intensity I_i of an element i as a function of the sputtering time ($I_i = f(t)$, Fig. 1). However, the expected quantitative result is the local concentration c_i of the detected element as a function of its distance perpendicular to the original surface, $c_i = f(z)$. Therefore, the main task in quantification of a measured sputtering profile is the correct conversion of the intensity/time relation into a concentration/depth relation.

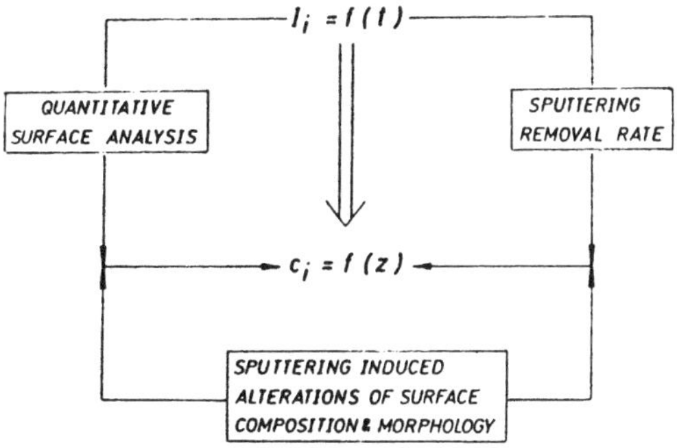

Fig. 1 Scheme of the basic considerations to convert a measured intensity/time profile, $I_i = f(t)$, to a "true" concentration/depth profile, $c_i = f(z)$

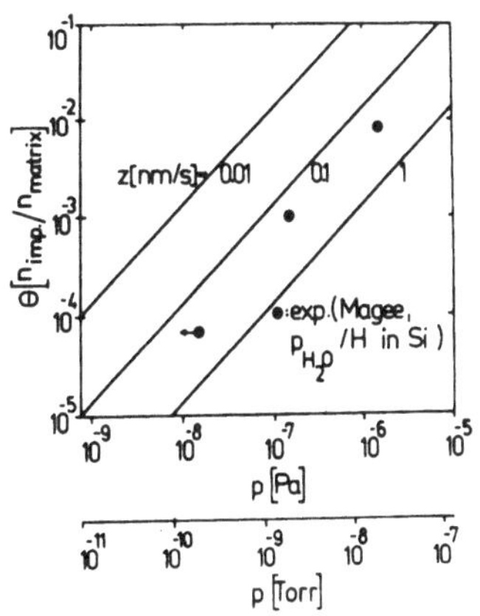

Fig. 2 Dependence of the relative impurity coverage Θ at the sputtered surface as a function of the partial pressure of the adsorbing gas, p. The decisive parameter is the sputtering rate ż according to Eq.(1). Measured values are taken from MAGEE and BOTNICK [14].

As schematically shown in Fig. 2 three steps of evaluation have to be performed: Calibration of the intensity scale and the time scale as well as the consideration of the sputtering induced effects on sample topography and composition.

Calibration of the intensity/concentration scale is the main task of quantitative surface analysis and the reader is refered to other reviews [6,7]. Again surface composition dependent matrix effects, which are much more pronounced in SIMS than in AES and XPS, together with the information depth of the respective methods are of primary importance here.

Calibration of the depth scale requires a knowledge of the sputtering rate ż as a function of t. Assuming a constant ż, z(t) = ż·t. Thus the simplest way is to sputter with the same primary ion beam parameters through a layer of similar composition and known thickness, such as an anodic oxide layer (Ta_2O_5/Ta) or a metallic evaporation layer and to compare the resulting sputtering times. A more accurate method is to measure the depth of the sputter crater of the analyzed sample by interferometric or mechanical styled techniques.

However, these approaches can only evaluate a mean value for the sputtering rate, and changes of $\dot{z}$ with sputter depth are neglected. Principally, those changes are expected at the beginning of sputter erosion in the build up regime of a damaged surface layer and in regions of strong composition changes, for example at interfaces. An example of this effect has been shown for the AES sputter profiling of Ni/Cr multilayers with N_2^+ ions [8]. Assuming a different sputtering rate for Ni and Cr given by the apparent thickness difference and a proportionality of the instantaneous sputtering rate with composition, a correction of this type of distortion was possible [8]. In general, however, an accurate correction of the time/depth scale can only be made if the instantaneous sputtering rate is measured in situ by interferometry [9] or - in the case of thin evaporation layers - by simultaneous X-ray emission analysis [10].

Having performed the calibration of the time/depth and the intensity/concentration relations, the obtained depth profile is still not a true image of the real, original profile in the sample before sputtering. The reason is that a variety of disturbing parameters will lead to changes of the structure, microtopography and composition of the surface layer due to ion bombardment induced effects and reactions with the ambience. All these effects lead to a deviation of the measured profile from the original "true" profile. They limit the accuracy and the precision of a depth distribution analysis and have to be considered in quantitative profile evaluation.

3. Parameters Disturbing Depth Profiles and the Minimization of Their Influence

Table 1 shows a summary of the effects distrubing sputtering profiles. They may be divided into instrumental factors, those

Table 1 Survey of the distrubing effects in sputter profiling

INSTRUMENTAL FACTORS	adsorption (residual gas atmosphere) redeposition of sputtered species crater wall effects impurities in ion beam neutrals from ion gun non-uniform ion beam intensity information depth
SAMPLE CHARACTERISTICS	original surface roughness crystalline and defect structure alloys, compounds, 2nd phases insulators
RADIATION INDUCED EFFECTS	primary ion implantation atomic mixing enhanced diffusion and segregation sputtering induced roughness preferential sputtering and decomposition of compounds

depending on the characteristics of the sample under study, and in effects depending on the fundamental interactions of the ion beam (and/or electron beam in AES) with the sample [11]. Whereas we have in general no choice to change the latter ones, the instrumental factors can be substancially optimized by a proper design of the instrument and the experimental set up.

With respect to the fundamental ion bombardment induced disturbances we are in a somehow intermediate position. Principally, they cannot be avoided. However, a better understanding of the sputtering process will enable us to choose the experimental conditions so as to minimize these effects with respect to a given sample.

In the following, the effects listed in Table 1 are briefly discussed.

3.1 Instrumental Factors

A prerequisite in sputter depth profiling is the avoidance of chemical alterations of the sample surface during sputtering [12].

Considering adsorption of reactive gases from the ambience the sample surface must be under UHV ($<10^{-9}$ Torr or 10^{-7} Pa) conditions with respect to their partial pressure. Assuming a sticking probability of one, a sputtering probability equal to that of the matrix, a simple calculation shows that the mean surface coverage of the impurity, Θ is given (for $\Theta<<1$) by the ratio of adsorption rate and sputter desorption rate [13,14]:

$$\Theta \approx 10^6 \cdot p[\mathrm{Torr}]/\dot{z}[\mathrm{\AA/s}] \quad (1)$$

where $\dot{z}$ is the sputtering rate, p is the partial pressure and Θ = ratio of adsorbed atoms/matrix atoms. In Fig. 2, this ratio is plotted against the partial pressure p for different rates (for a silicon matrix). It shows that even for rather high sputtering rates (1 nm/s) and good UHV (10^{-8} Pa) we can hardly reach the ppm region. This was demonstrated recently by MAGEE [14,64] for profiling hydrogen in silicon with SIMS as a function of the partial pressure of H_2O. With respect to the simplifications in Eq.(1) the results correspond fairly well to the predictions in Fig. 2 for the sputtering rate of about 0.5 nm/s and assuming dissociative adsorption (2 H atoms per H_2O molecule) at the silicon surface. For lower sputtering rates, (e.g. 10^{-2} nm/s) as more frequently used in AES, considerable higher impurity concentrations are expected. However, since the detection limit in AES is between 10^{-2} and 10^{-3} the conditions are less stringed and can be more easily fulfilled. MATHIEU et al. have demonstrated [15] that CH_4 partial pressures in the HV range (10^{-4}-10^{-5} Pa) can severely alter AES depth profiles of Ta_2O_5/Ta. Another source of impurities at the analyzed surface is the redeposition of sputtered species (due to resputtering at construction parts close to the sample surface) which may lead to memory effects. They are less pronounced in defined ion beam sputtering as compared to glow discharge mass spectrometry. Impurities in the primary ion beam are implanted into the sample surface, which may come from contaminations in the ion formation chamber and from an impure sputter gas. To avoid the latter, differentially pumped ion guns are used in modern profiling equipment even in AES profiling. The more severe conditions of high sensitivity SIMS have led to the design of ion guns with mass separation in the beam (e.g. a Wien-filter for velocity separation) [16].

A non-uniform ion beam intensity is recognized in the shape of the crater formed during sputtering. Any deviation from a uniform spatial distribution means that the sample is analyzed at different depths at the same time resulting in a loss of depth resolution [11,17,18]. Therefore, rastering of the primary ion beam is frequently used to obtain a flat crater bottom. However, neutrals formed before the electrostatic deflection may still give rise to a distortion. This effect is avoided by a small bend in the ion beam after the deflection plates [16]. Crater wall effects are commonly reduced by raster gating of the output signal in SIMS, so that only the inner part of the crater is analyzed. An extraction lens system may further improve the situation, as shown by MAGEE, who has given an optimized instrumental design for SIMS analysis [16,64].

The mean information depth is an inherent property of the analysis method. It is at minimum (first monolayer) in ISS, of the order of 0.5 nm or less in SIMS [19] and between 0.4 and 3 nm depending on the detected electron energy in AES and XPS [20].

3.2 Sample Characteristics

Composition and structure of the sample cause certain limitations of the quality of a sputtering profile. First of all there is the influence of the original surface roughness on depth resolution. There are only a few experiments considering this influence in detail [21,22], which show a decrease in the depth resolution due to surface roughness. Recently, SEAH and LEA [23] have proposed a model based on the angle dependence of the sputtering yield [24] which predicts an influence proportional to the sputtered depth increasing with the deviation from normal incidence of the primary ions. An illustration of the detrimental effect of roughness is shown in Fig. 3 for the sputtering profile of a NiCr multilayer sample with an original roughness of 10 nm and 1 μm respectively [22]. The conclusion is that we may expect optimum profiles only for flat polished samples.

The dependence of the sputtering yield on surface orientation [25] is responsible for the development of a surface with height steps at the grain boundaries of a polycrystalline material [26]. A depth resolution degradation increasing proportional to the sputter depth is due to this effect, which was even found for sputtering of copper with oxygen ions [27]. In the case of more reactive materials, however, it is generally observed that oxygen sputtering (or flooding the surface with oxygen during argon sputtering) leads to a smoother sputtered surface. This is probably due to the build-up of a quasi-amorphous oxide layer which is sputtered more homogeneously [28].

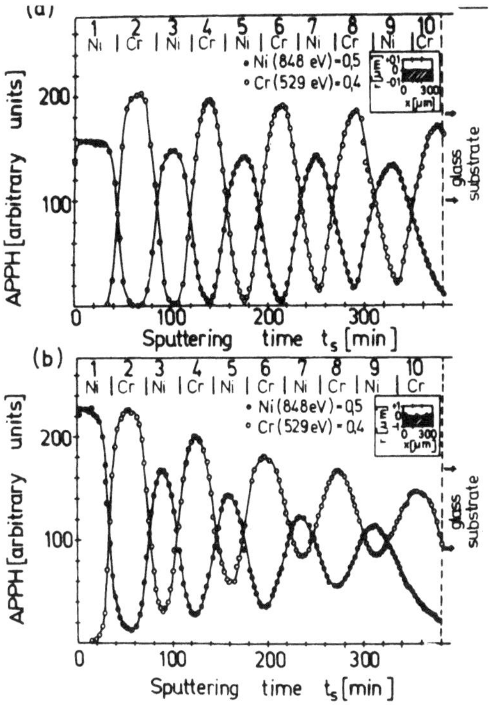

Fig. 3 AES sputtering profiles on Ni/Cr sandwich multilayers (37 nm single layer thickness) demonstrating the degradation of depth resolution caused by the original surface roughness effect [22]. (a) smooth surface (roughness <10 nm), (b) roughness ≈1 μm

Defect structures such as dislocations or stacking faults may have a similar effect as polycrystalline structures on the development of sputtered surface roughness [27]. Even in the case of perfect single crystals, cone formation may develop under bombardment [29,30]. This again can be explained by the orientation dependence of the sputtering yield and is reduced by the use of sample rotation or of sputtering with two guns at different incidence angle [21].

Alloys and multiphase materials are prone to variations in sputtering yield with local composition leading to increased roughness and to profile distortions [65].

Contamination of the surface can also result in cone formation due to reduced sputtering yields of impurity covered surfaces [32,33].

The general observation, that sputtering of non-crystalline single phase materials with smooth surfaces, which tend to amorphize during sputtering (such as semiconductors or oxides) and gives much better profiling results than metallic materials, is qualitatively consistent with the statements made above. Therefore it is not surprising that profiles with optimum depth resolution have been obtained for dopants (small concentration) in silicon, and for semiconductors and oxides in general.

3.3 Ion Bombardment Induced Effects

In sputter profiling, we only use a by-effect of the more general aspect of radiation damage by ion (or neutral) bombardment, as well described by the Sigmund theory of sputtering [34]. The incoming ion gradually looses its energy by elastic and inelastic scattering at the target atoms and the electron shells, respectively. During the collisional process, target atoms are displaced from their original sites, and recoiling atoms displace others so that a complex collision cascade is formed. Only a small part of the atoms of the collisional cascade is sputtered. The major part contributes to a mixing of atomic layers together with a formation of structural defects (vacancy and interstitial clusters). Primary ions and recoil atoms are implanted into deeper layers depending on their angle of incidence, masses and binding energies. Thus together with enhanced diffusion due to the induced structural defects a rather complicated "altered layer" is formed. The recently developed theoretical models have led to at least a semi-quantitative understanding of the effect of ion bombardment induced compositional changes on depth profiling [35-39].

The primary implantation is gradually build-up until equilibrium between sputtering and implantation is reached. This zone, which is of the order of the projected range, corresponds to the formation of the zone of mixing. It is characterized by a change in composition and structure which causes a change in the sputtering rate until the steady state radiation damage is achieved [10,40,41].

Implantation of primary ions can be analyzed by EAS [10,42]. Whereas noble gas ions give some problems with insufficient retention, reactive ions such as N_2^+ stick to the matrix. Their concentration profile has been recently found to correspond to the predictions of the WITTMAACK-SCHULZ model for the steady state [42]. Fig. 4 shows the projected range as a function of energy for N_2^+ bombardment of cobalt.

A fundamental contribution to the broadening of depth profiles is caused by the build-up of a collisional cascade. Primary recoils are displaced into deeper layers ("knock-on"

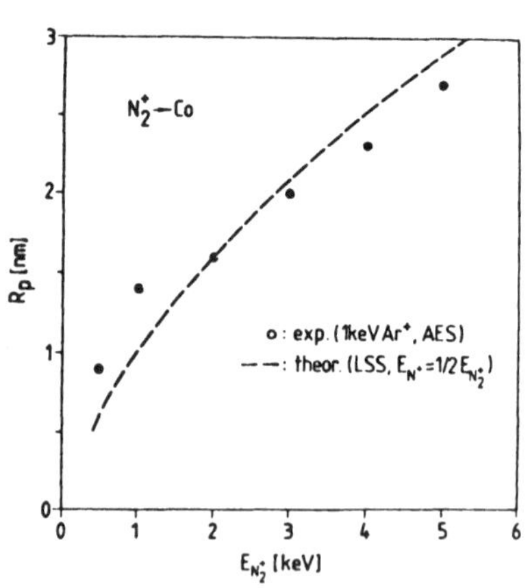

Fig. 4 Mean projected range of nitrogen during sputtering of cobalt with N_2^+ ions between 0.5 keV and 5 keV as compared with LSS theory [42]

effect) and secondary recoils give rise to a more random distribution of relocated target atoms ("atomic or cascade mixing") [35-41]. This effect directly depends on the energy of the primary ions and the displacement energy of the target atoms. Random walk cascade mixing was treated by ANDERSON [35], whereas calculations TAIKANG et al. [36] and LITTMARK and HOFER [37] have shown that a broadening of an interface occurs which is skewed towards the interior of the sample (even for isotropic cascade mixing). This was experimentally confirmed by ETZKORN and KIRSCHNER [10,39]. The development of the broadening increases with increasing ion doses (i.e. the sputtered depth) until a steady state is achieved [37]. For larger depths, the atomic mixing contribution is constant (for constant sample composition). It is approximately proportional to the square root of the primary ion energy [35] and is less for higher sputtering yield [36]. An example of the energy dependence of the interface broadening is shown in Fig. 5 for the anodic oxides Nb_2O_5/Nb and Ta_2O_5/Ta with 30 nm thickness. The influence of atomic mixing on depth resolution is of the order of the projected range. Consequently the disturbing effect of knock-on and atomic mixing can be minimized by using heavy ions (e.g. Xe^+) at low energy and oblique incidence.

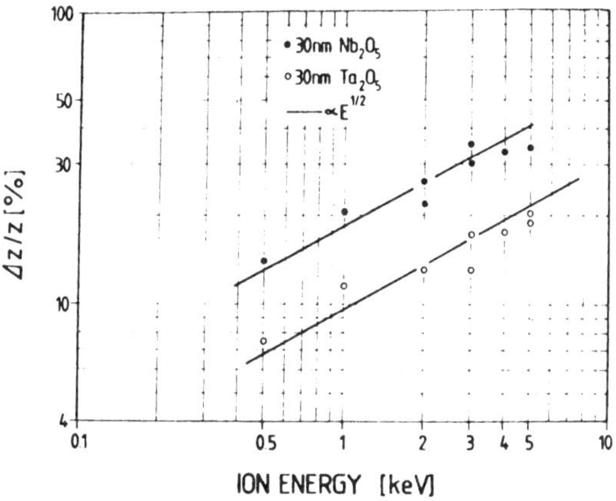

Fig. 5 Dependence of the relative depth resolution Δ/z on the primary ion energy for the sputtering of anodic oxide films Nb_2O_5/Nb and Ta_2O_5/Ta with Ar^+ ions. In both cases, the increase is approximately proportional to $E^{1/2}$

Atomic mixing within the collisional cascade is inevitably linked with the generation of structural defects which give rise to enhanced diffusion [43]. The driving forces can be due to originally existing or induced concentration gradients, (bulk and surface diffusion) segregation [44,45], electric fields generated by ion (or electron) induced charging of insulators (electrotransport) [46]. Whereas an induced surface diffusion may improve the sputtering induced surface roughness and consequently depth resolution [47], enhanced bulk diffusion broadens sputtering profiles [48,66]. More systematic studies of temperature dependent profiling are needed to clearly differentiate the contributions of thermally activated transport from the atomic mixing as well as their interactions.

Sputtering induced surface roughening may develop in profiling due to the various inhomogeneous sample characteristics (crystalline orientation, defects, 2nd phases) as discussed above. Even at homogeneous, single crystalline surfaces it can be expected due to the incidence angle and lattice orientation dependence of the sputtering yield [24,29,30]. Semiconductors which amorphize under bombardment are less prone to sputter induced roughening as compared to metals.

A first attempt to describe the microtopography obtained due to sputtering statistics is the sequential layer sputtering model [12], which predicts an increasing surface roughness with depth. However, experiments with reactive ion sputtering and particularly in semiconductors and oxides have shown that a constant depth resolution can be attained. Recent model calculations predict that a site dependent sputtering probability [49] as well as a certain surface mobility [47] will cause a smoothing of the sputtered surface at larger sputter depths. Therefore it is concluded, that the profile broadening due to sputtering statistics is limited to a few monolayers, which is obtained already after a sputter depth of about 10 monolayers. The often found $z^{1/2}$ dependence of depth resolution on depth may be explained by a superposition of constant depth resolution parameters and those with linear depth dependence [50].

In any case, quasi-amorphous evaporation layers, semiconductors and oxides are less sensitive to induced roughening. For metals, reactive primary ions (O_2^+, N_2^+) are frequently used to supress surface roughening [51].

Due to the different atomic sputtering yields, the composition of the surface layer of multicomponent samples is generally changed during sputtering [52-56]. If the sputtering yield of the components is independent of their bulk concentration, the surface composition is changed inversely proportional to the sputtering yields [52,55]. Therefore, quantitative analysis with residual surface detecting methods (AES, XPS) is affected whereas SIMS can still detect the correct bulk composition. The build-up of the altered layer [56] during sputtering causes a transient in surface composition, which superposes the time profile. Therefore the depth resolution at an interface will be worse but there are no quantitative experiments concerning the magnitude of this effect. Model calculations by LIAU et al. [41] predict (together with atomic mixing) a profile broadening of the order of R_p/r with R_p = projected range and r = ratio of the sputtering yields of the two respected components. Of course the sputtering time/depth relation gets asymmetric with changing composition if preferential sputtering occurs. The most detrimental effects with respect to depth profiling is that an apparent profile may show up when there is no concentration gradient (see Fig. 6b) and vice versa. The influence of preferential sputtering is reduced if the thickness of the altered layer is small (low primary ion energy,

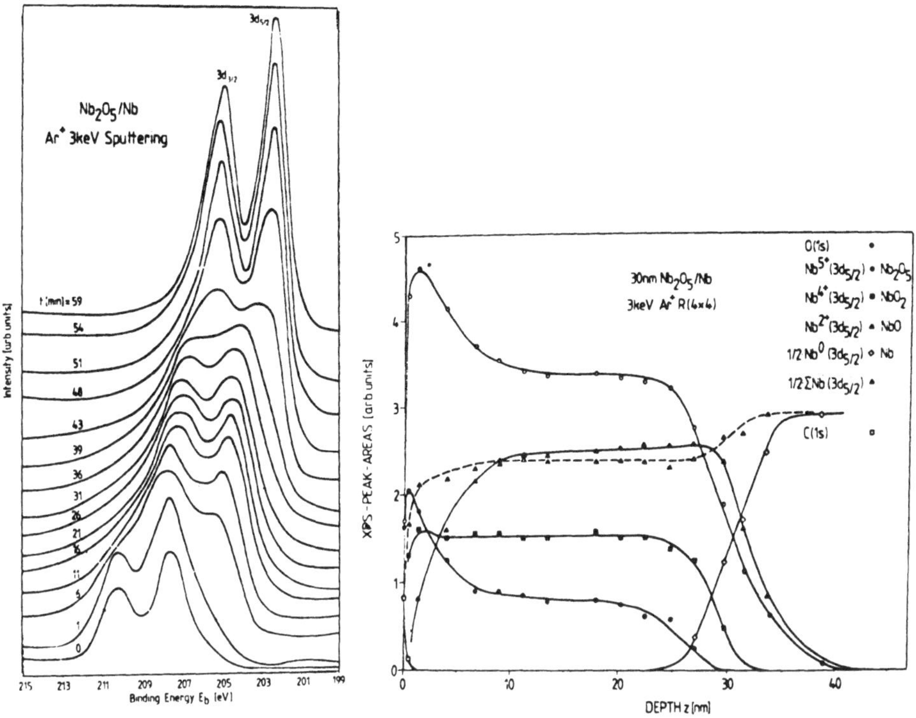

Fig. 6 Decomposition of Nb_2O_5 during sputtering with Ar^+ ions (3 keV) studied by XPX. (a) shift of the Nb 3 d peaks as a function of sputtering time; (b) XPS sputtering profile showing preferential sputtering of oxygen corresponding to a decrease of Nb_2O_5 and increase of NbO in the surface layer until a steady stage is reached. (The hump at 1 nm is an artefact caused by carbon contamination as seen in the C 1s peak)

heavy ions). It can be neglected in the case of small absolute composition changes, as in profiling of dopants in semiconductors.

Another fundamental process induced by sputtering is the change of the chemical state of the sample constituents. In the case of compounds, in general, a decomposition occurs due to the atomic displacements in the collision cascade. Many oxides are reduced to lower oxidation states [57,58]. As an example, Fig. 6 shows the influence of Ar^+ bombardment on the sputtering of a Nb_2O_5/Nb layer (30 nm thickness). Fig. 6a depicts the Nb 3d XPS peaks as a function of sputtering time. The shift of the peaks can be ascribed to lower oxidation states until the pure metal substrate is reached. Fig. 6b shows the corresponding evaluation of the deconvolved peak areas (proportional to concentration) for Nb_2O_5, NbO, NbO_2 and Nb. This effect causes difficulties in the interpretation of XPS sputtering profiles. Oxides with higher binding energies of the lighter metals (SiO_2, MgO, Al_2O_3) show negligible reduction by ion bombardment [57].

4. Depth Resolution and Principles of Deconvolution

The disturbing effects discussed above are the cause of an observed broadening of a measured profile with respect to the true profile which can be described by an appropriate resolution function [59,60,63]. Because of the complexity of the sputtering process an exact determination of the resolution function is difficult to obtain [13]. However, many profiling experiments on thin sandwich layers and interfaces have shown that the resolution function can be approximated by a Gaussian function. In case of a rectangular true profile, this is broadened during sputtering into an error function shape. This function is defined by only one parameter, the variance σ, and the depth resolution is generally defined as $\Delta z = 2\sigma$ [3]. According to these definitions, Δz is obtained from the difference in z for 84 % and 16 % of the maximum intensity. In case of constant sputtering rate the relative depth resolution $\Delta z/z$ (where z is at 50 % intensity) corresponds to $\Delta t/t$ on the sputtering time scale and can be directly taken from the experimental results. Relative depth resolution values obtained with different methods at different systems were complied by HONIG [61] and are discussed in [4].

The depth resolution achieved in a specific sputter profiling experiment limits the accuracy by which a measured, evaluated profile matches the true profile. Therefore it is of importance to know the different parameters which determine Δz so that their detrimental influence can be minimized.

If the single contributions Δz_j are independent of each other the total depth resolution Δz measured experimentally is described by a quadrate addition [62]:

$$\Delta z = [\Sigma(\Delta z_j^{\,2})]^{1/2} . \qquad (2)$$

Eq.(2) is only exact for Gaussian resolution functions. It can be shown, however, that even for exponential functions (e.g. Δz_λ) Eq.(2) is a good approximation [60].

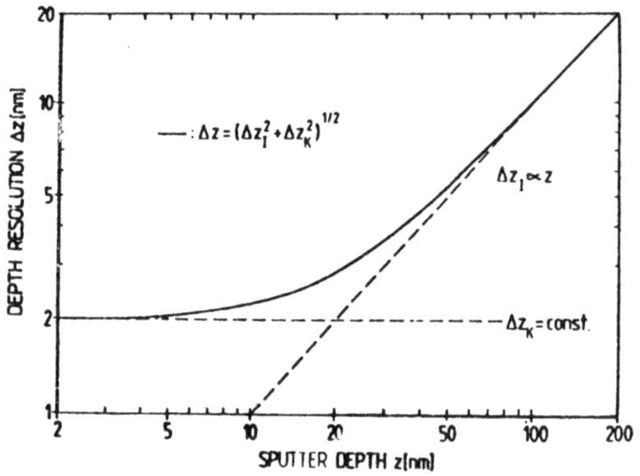

Fig. 7 Generalized dependence of the total depth resolution Δz on depth z as predicted by Eq.(2) for a superposition of constant and depth dependent contributions (see Table 2). The shown example is for an instrumental effect $\Delta z_I = 0.1 \cdot z$ and an atomic mixing effect of $\Delta z_k = 2$ nm

The different contributions can be characterized by their dependence on the total dose or the sputtered depth. Table 2 shows how the basic contributions can be divided in constant, depth dependent and sputter rate or time dependent ones. For larger sputter depths, those Δz_j which increase with depth will be prevalent, whereas for small sputter depths the constant contributions will limit the observed sputter broadening [62]. Fig. 7 shows the generally predictable shape of the depth dependence for a superposition of an instrumental effect ($\Delta z_i \sim z$, e.g. $\Delta z_I/z = 10$ %) and a constant contribution (e.g. atomic mixing, Δz_k = const = 2 nm). According to an argument by WERNER [50], due to this type of superposition a dependence $\Delta z \sim z^{1/2}$ ($\Delta z/z \propto z^{-1/2}$), which is often found experimentally [62], is approximated in the region of intermediate depth.

Table 2 Dependence of different contributions to depth resolution Δz on sputtered depth z (total ion dose)

information depth	Δz_λ = const
atomic mixing	$\Delta z_k \rightarrow$ const ($z \gg R_p$)
preferential sputtering	$\Delta z_p \rightarrow$ const ($z \gg R_p$)
non-uniform ion beam intensity	$\Delta z_I \sim z$
sputter-induced topography	$\Delta z_r \sim z^{1/2} \ldots z$
radiation enhanced diffusion	$\Delta z_T = f(D_b, D_s, \dot{z}, t \ldots)$

With the knowledge of the total experimental depth resolution (or, more precisely, the depth resolution function) a deconvolution procedure can be applied [59,60] to get an improved approximation of the true profile. This means the solution of the convolution integral:

$$I(z) = \int_{-\infty}^{+\infty} c(z')g(z-z')dz' \tag{3}$$

where g(z-z') is the resolution function. The solution is found either by Fourier transfromation or by the method of successive iterations after VAN CITTERT [59]. Convergence proofs of this method (for a Gaussian resolution function) [60] have shown that it is limited to profiles, for which the maximum slope $(dI/dI_0/dz)$ is less than about 56 % of the Δz_q of the resolution function (Fig. 8, region II). For higher slopes it has been found useful to assume a real profile and solve Eq.(3) directly with an appropriate resolution function (Fig. 8, region III). This procedure of trial and error has to be repeated until the measured profile is reproduced with sufficient precision. It should be noted, however, that any approximation in region III (Fig. 8) imposes an increasing ambiguity for increasing $(dI/I_0/dz) \cdot \Delta z_q$. Therefore it is essential to optimize the disturbing effects in profiling to get at least into region II (Fig. 8) or preferably to region I, where the relative deviation of the measured profile from the true profile is less than 1 %.

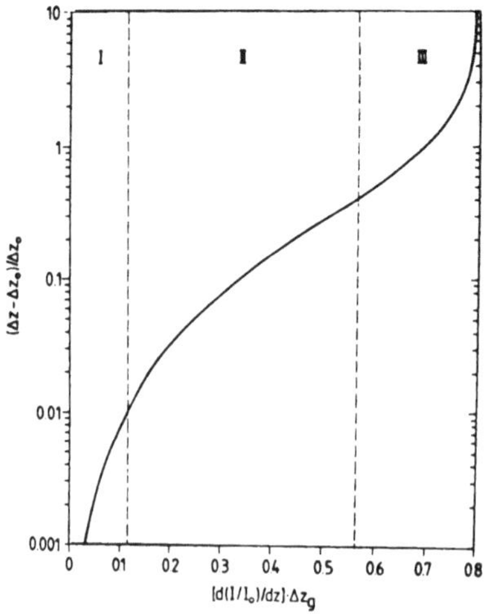

Fig. 8 The relative deviation $(\Delta z-\Delta z_o)/\Delta z_o$ between the depth resolutions of the measured (Δz) and the true profile (Δz_o) as a function of the product of maximum slope $d(I/I_o)/dz$ of the measured profile and the depth resolution Δz_g of the resolution function for error function profiles [60]. The three regions indicate the cases where: I: no deconvolution is necessary; II: the VAN CITTERT deconvolution can be applied; III: deconvolution is ambiguous (true profile assumption and comparison with measured profile)

5. Summary and Conclusion

Many of the discussed effects which disturb the accuracy of sputtering profiles cannot be fully eliminated. A better understanding of their origin, however, allows their minimization. The effects of surface contaminations are reduced by the use of a high purity primary ion beam and a base pressure of contaminating reactive gases in the test chamber below 10^{-8} Pa. On the other hand, sputtering with reactive ions (N_2^+, O_2^+) reduces microtopographical changes due to crystalline structure and impurities and favours quantification in SIMS. Two ion guns at different angles and rotation of the sample will supress incident angle dependent texturing and cone formation.

Recoil implantation and atomic mixing influence on depth profiling is reduced using heavy primary ions with low energy (< 1 keV) and at oblique incidence to the sample surface. Ion beam rastering should be used to provide a uniform sputtering within the analyzed area. Raster gating is necessary to avoid edge effects in SIMS and ISS.

With respect to the sample, optimal results are expected for flat, smooth surfaces, an amorphous structure without second phases, and components with a similar sputtering yield. Sufficient electrical and heat conductivity prevents charging and temperature increase during analysis. With respect to the experimental set up, a test of the profiling system with a standard sample (e.g. Ta_2O_5/Ta) with a sharp interface should be performed to obtain a comprehensive knowledge of one's instrument with respect to depth profiling.

References

1. R.E. Honig, Thin Solid Films 31, 89 (1976)
2. A. Benninghoven, Thin Solid Films 39, 3 (1976)
3. H.W. Werner, Acta Electronica 19, 53 (1976)
4. S. Hofmann, Surf. Interface Anal. 2, 148 (1980)

5. E. Zinner, Scanning 3, 57 (1980)
6. H.W. Werner, Surf. Interface Anal. 2, 56 (1980)
7. K. Wittmaack, Nuclear Instr. a. Meth. 168, 343 (1980)
8. S. Hofmann and A. Zalar, Thin Solid Films 60, 201 (1979)
9. J. Kempf, in Secondary Ion Mass Spectrometry SIMS II, ed. by A. Benninghoven, C.A. Evans, R.A. Powell, R. Shimizu, H.A. Storms, Springer Series in Chemical Physics, Vol. 9 (Springer, Berlin, Heidelberg, New York 1979) p. 97
10. J. Kirschner and H.W. Etzkorn, Appl. Surf. Sci. 3, 251 (1979)
11. S. Hofmann, Analysis 9, 181 (1981)
12. S. Hofmann, Appl. Phys. 9, 59 (1976)
13. P.H. Holloway and H.J. Stein, J. Electrochem. Soc. 123, 723 (1979)
14. C.W. Magee and E.M. Botnick, Vac. Sci. Technol. 19, 47 (1981)
15. H.J. Mathieu and D. Landolt, J. Microsc. Spectr. Electron 3, 113 (1978)
16. C.W. Magee, W.L. Harrington and R.E. Honig, Rev. Sci. Instr. 49, 477 (1978)
17. H.W. Werner, Vacuum 24, 493 (1974)
18. J.B. Malherbe, J.M. Sanz and S. Hofmann, to be published in Surf. Interface Anal.
19. T. Ishitani and R. Shimizu, Appl. Phys. 6, 241 (1975)
20. M.P. Seah and A.W. Dench, Surf. Interface Anal. 1, 2 (1979)
21. H.J. Mathieu, D.E. McClure and D. Landolt, Thin Solid Films 38, 281 (1976)
22. S. Hofmann, J. Erlewein and A. Zalar, Thin Solid Films 43, 275 (1977)
23. M.P. Seah and C. Lea, Thin Solid Films 81, 257 (1981)
24. H. Oechsner, Appl. Phys. 8, 185 (1975)
25. M. Bernheim and G. Slodzian, Int. J. Mass Spectr. Ion Phys. 12, 93 (1973)
26. W. Hauffe, this Conference
27. V. Naundorf and M.P. Macht, Nuclear Instr. a. Meth. 168, 405 (1980)
28. K. Tsunoyama, T. Suzuki, Y. Okashi and H. Kishidaka, Surf. Interface Anal. 2, 212 (1980)
29. L.T. Chadderton, Rad. Eff. 33, 129 (1977)
30. W. Szymak and K. Wittmaack, "Proc. Symp. on Sputtering" (eds.: P. Varga et al., Perchtolsdorf 1980, p. 236)
31. J.M. Walls, D.E. Sykes and D.D. Hall, Proc. 8th Int. Vac. Congr. Vol. I, p. 287, Cannes (1980)
32. G.K. Wehner and D.J. Haijicek, J. Appl. Phys. 42, 1145 (1971)
33. J.L. Vossen, J. Appl. Phys. 47, 544 (1976)
34. P. Sigmund, in Sputtering by Particle Bombardment I, ed. by R. Behrisch, Topics in Applied Physics, Vol. 47 (Springer, Berlin, Heidelberg, New York 1981) p. 9
35. H.H. Andersen, Appl. Phys. 18, 131 (1979)
36. S. Taikang, R. Shimizu and T. Okutani, Jap. J. Appl. Phys. 18, 1987 (1979)
37. U. Littmark and W.O. Hofer, Nuclear Instr. a. Meth. 168, 329 (1980)
38. P. Sigmund and A. Gras-Marti, Nucl. Sci. a. Meth. 182, 25 (1981)

39. H.W. Etzkorn, U. Littmark and J. Kirschner,"Proc. Symp. on Sputtering", eds.: P. Varga, G. Betz and F.P. Viehböck, Perchtolsdorf 1980, p. 542
40. P. Williams, Nucl. Instr. a. Meth. 182, 15 (1981)
41. Z.L. Liau, B.Y. Tsaur and J.W. Mayer, J. Vac. Sci. Technol. 16, 121 (1979)
42. I.B. Malherbe and S. Hofmann, Surf. Interface Anal. 2, 187 (1980)
43. R.R. Hart, H.L. Dunlap and O.J. Marsh, J. Appl. Phys. 46, 1947 (1975)
44. S. Hofmann, Mater. Sci. Eng. 42, 55 (1980)
45. R. Kelly, in Proc. Symp. on Sputtering, eds.: P. Varga, G. Betz and F.P. Viehböck, Perchtolsdorf 1980, p. 390
46. F. Ohuchi, M. Ogino, P.H. Holloway and C.G. Pantano, Surf. Interface Anal. 2, 85 (1980)
47. J. Erlewein and S. Hofmann, Thin Solid Films 69, L39 (1980)
48. S. Myers, Nucl. Instr. a. Meth. 168, 265 (1980)
49. M.P. Seah, J.M. Sanz and S. Hofmann, Thin Solid Films 81, 239 (1981)
50. H.W. Werner, to be published
51. W.O. Hofer and H. Liebl, Appl. Phys. 8, 359 (1975)
52. H. Shimizu, M. Ono and K. Nakayama, Surf. Sci. 36, 817 (1973)
53. G. Betz, Surf. Sci. 92, 283 (1980)
54. H.J. Mathieu and D. Landolt, Surf. Sci. 53, 228 (1975)
55. H.W. Werner and N. Warmoltz, Surf. Sci. 57, 706 (1976)
56. P.S. Ho, J.E. Lewis, H.S. Wildmann and J.K. Howard, Surf. Sci. 85, 19 (1979)
57. R. Kelly, Nucl. Instr. a. Meth. 149, 553 (1978)
58. H. Schoof and H. Oechsner, Proc. 4th Int. Conf. Solid Surf. a. 3rd Europ. Conf. Surf. Sci., Cannes (1980) Vol. II, p. 1291
59. P.S. Ho and J.E. Lewis, Surf. Sci. 55, 335 (1976)
60. S. Hofmann and J.M. Sanz, Proc. 8th Int. Vac. Congr., Cannes 1980, Vol. I (Thin Films), eds.: F. Abeles and M. Croset, Paris 1980, p. 90
61. R.E. Honig and C.W. Magee, Proc. 26th Ann. Conf. on Mass Spectrometry, St. Louis 1978, p. 207 (for references see [4])
62. S. Hofmann, Appl. Phys. 13, 205 (1977)
63. P. Laty, D. Seethanen and F. Degreve, Surf. Sci. 85, 353 (1979)
64. C.W. Magee, R.E. Honig and Ch. A. Evans, this Conference
65. F. Degreve, this Conference
66. H.M. Nitz, O. Ganschow, U. Kaiser, L. Wiedmann and A. Benninghoven, Surf. Sci. 104, 365 (1981)

The Theory of Concentration Depth Profiling by Sputter Etching

W.O. Hofer and U. Littmark
Institut für Grenzflächenforschung und Vakuumphysik,
Kernforschungsanlage Jülich*
D-5170 Jülich, Fed. Rep. of Germany

Concentration depth profiling by controlled erosion of surface layers by sputtering is generally hampered by distorting effects due to recoil mixing and radiation enhanced migration. Emphasis is placed here on *recoil mixing* where we distinguish between two different collisional relocation processes: primary recoil mixing, on the one hand, and cascade recoil mixing on the other. In spite of the higher average energy/range of primary recoils, *cascade recoils* are the dominating cause for recoil mixing; this is by virtue of their much higher number [1].

The alteration of internal concentration profiles $P_k(\Phi,x)$ with fluence Φ is determined by numerical solution of the general balance equation

$$\frac{dP_k(x,\Phi)}{d\Phi} = -\frac{P_k(x,\Phi)}{N}\int_{-\infty}^{\infty} W_k(x,x')dx' + \int_{z(\Phi)}^{\infty} \frac{P_k(x',\Phi)}{N} W_k(x',x)dx' \qquad (1)$$

where the transport function $W_k(x,x')$ describes the mass transfer from a depth element at x to an element at x' per incident ion; $x = z(\Phi)$ denotes the position of the surface, and N the target density. $W_k(x,x')$ can be determined from transport theory [1] or by simulation procedures [2]. Both approaches are in good agreement; this is remarkable since not only the way of finding the transfer function is different, but also the interatomic potentials used in [2] differ from the power potentials used here (projectile-target interaction: $m = 1/3$, target-target collisions: $m = 1/4$). The recoil velocity spectrum is described by

* Euratom Association

$$F_{vk}(E_k,\vec{e}_k) = C_1 \frac{F_E(x)}{E_k^2} + C_2 \frac{\vec{F}_P(x)}{E_k^{3/2}} \cdot \vec{e}_k \qquad (2)$$

where E_k is the energy of recoils of type k and $\vec{e}_k$ the direction of their velocity; F_E and $\vec{F}_P$ are the depth distribution of deposited kinetic energy and momentum, respectively; C_1 and C_2 are numerical constants. The second term describes the deviation from isotropic collision cascades; it turned out to be of importance only for near-surface processes [1].

The integro-differential equation (1) in general constitutes a system of coupled, non-linear equations. This is because the density distribution of tracer atoms $P_k(x,\Phi)$ influences the transfer function. Strictly speaking, already the factorization used in (1) is an approximation towards separation of mutual influences of the tracer atom distributions. Linearization and decoupling of the system can be achieved, however, in two important special cases:

- the collision dynamics of all target components are so similar that an equal-mass case can be considered; this is generally true if the mass ratios do not deviate by more than a factor of 2-3.
- the tracer concentrations are so small that the collision cascade is essentially determined by the matrix [3].

The target density N and its alteration during irradiation is of concern for an assessment of a geometrical depth scale, i.e. the conversion from an area density (atoms/cm^2) to a length unit (nm), but this additional complication will not be discussed here (see [1]).

It is noted, furthermore, that the loss term in (1) includes the sputtering yield ($x' < 0$); the speed of erosion is therefore not an external input quantity, but part of the solution of the fundamental balance equation. This is a natural consequence of the fact that sputtered particles are just the fraction of relocations which intersect the surface.

Both internal concentration distributions and external "sputter" profiles, i.e. the surface concentration or the partial sputtering yield of tracer atoms as a function of fluence or eroded depth, are obtained in a stepwise numerical solution of (1). Apart from the equal-mass approach there are no restrictions regarding the tracer, neither in profile shape nor in concentration. In addition, also implanted projectiles are taken into account, a feature which is particularly important when in-depth profiling is performed with chemically active projectiles such as N^+, O^+, Cs^+, J^-. These ions are often preferred in SIMS because of their enhanced secondary ion yield. They are, in addition, often used in general surface erosion techniques owing to their ability of suppressing the development of pronounced surface structures, such as facets and cones [4,5]. The formation of chemical bounds, however, prevents thermal release of these implanted projectiles; they are accomodated in the lattice and must be sputtered in order to be removed from the solid. This increases the fluence (ions per unit area) required for erosion of a given depth element Δz

$$N \cdot \Delta z = \begin{cases} Y \cdot \Delta\Phi & \text{without implantation} \\ (Y - I_I) \cdot \Delta\Phi & \text{with implantation} \end{cases} , \qquad (3)$$

Y is the sputtering yield and I_I the implantation probability. It is obvious from (3) that the required fluences are the larger the more the sputtering yield approaches I_I, a situation not too far from sputtering with O^+ or N^+ ions. As a consequence, mixing is significantly enhanced compared to bombardment with particles which are released by non-collisional mechanisms. Hence, if for reasons of surface morphology, matrix effects or ion yield enhancement, reactive projectiles are to be used in depth profiling, *heavy* ions are clearly of advantage owing to their larger sputtering yield ($Y \gg I_I$).

Heavy ions are to be preferred also for reasons of depth resolution: since it is essentially the dimension of the collison cascade which determines profile broadening, shallow cascades are generally aimed at in high resolution depth profiling. This is achieved by low-energy, heavy-ion bombardment, especially under oblique angle incidence.

Calculated depth profiles of thin (δ-function-like) tracer layers generally are of a distinct *asymmetry* [1]. This holds true even for perfectly *symmetric* internal relocation mechanisms such as isotropic collision cascades. Two processes are mainly responsible for this asymmetric broadening:

- the gradient of the recoil density (in collisional mixing represented by the $F_E(x)$-function), and
- the gradual transfer of the internal profile through the surface.

The effect of gradients in the recoil density has been discussed in a previous communication [1].

The transformation of any (internal) concentration distribution into an (external) sputter depth profile will never be shape conserving. When surface-near parts of the profile have already "passed through" the surface, deeper-lying regions will still be subject to mixing. This is illustrated in case of a highly simplified – but often discussed – situation, namely that of diffusional mixing during sputter erosion. The broadening of internal Gaussian profiles

$$F(z,x) = \frac{1}{\sqrt{2\pi\sigma^2(z)}} \exp\left[-\frac{(x-m)^2}{2\sigma^2(z)}\right] \qquad (4)$$

is simulated by letting the standard deviation σ develop with eroded depth z according to

$$\sigma(z) = \text{const. } z^p, \; p = 1/2,\ 1,2 \qquad (5)$$

simultaneously with the recession of the surface to $x = z$.

Obviously, the case $p = 1/2$ corresponds to ordinary diffusion, while $p = 1$ or 2 is more representative for collisional mixing. In Figure 1 the surface concentration $F(z,z)$ is plotted in these three cases. The constant in (5) was adjusted so that $\sigma = 60$ Å at $z = m$, i.e. when the eroded depth equals the location of original δ-function tracer ($\sigma = 0$). The resulting depth profiles clearly show asymmetry although the internal profiles are steadily symmetric. The skewness is the more pronounced the larger the ratio: mixing/erosion.

This effect is small when the final σ is small, but this is generally *not* the case in collisional mixing where σ is of the order of the cascade dimensions. [The $\sigma(z = m)$-value used in Fig.1 corresponds to erosion of Si by 2 keV Ne ions]. Whatever the magnitude of the distortion, this example of asymmetric sputter profiles from perfectly symmetric concentration distri-

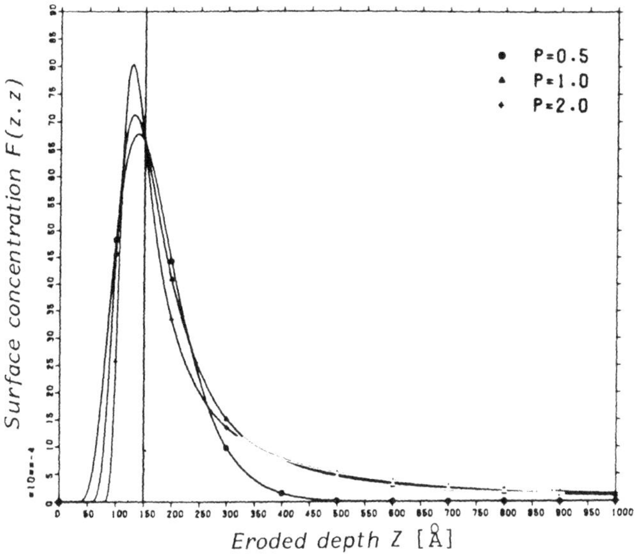

Fig.1. Diffusional mixing of a δ-function like tracer positioned at a depth of 150 Å. Internal profiles are assumed to be given by (4) and (5) with m = 150 Å and $\sigma(z) = (z/m)^P$ 60 Å. The curves show the surface concentration F(z,x = z) as a function of eroded depth z

butions demonstrate the necessity of calculating the entire shape of the external profile [7]. Too much information is lost when characterizing sputter profiles with a few parameters (shift and broadening) [3]. This sinple model calculation finally urges caution to simply adopt the shape of internal concentration distributions for sputter-erosion profiles [3,6,8,9]

References

1 U. Littmark, W.O. Hofer: Nucl. Instrum. Methods *168* (1980) 329
2 T. Ishitani, R. Shimizu: Appl. Phys. *6* (1975) 241
3 P. Sigmund, A. Gras-Marti: Nucl. Instrum. Methods *168* (1980) 389
4 W.O. Hofer, H. Liebl: Appl. Phys. *8* (1975) 359
5 J.L. Whitton, G. Carter: Proc. Sympos. Sputtering, Perchtoldsdorf, Vienna, 1980, ed. by P. Varga et al., p.552
6 H.H. Andersen: Appl. Phys. *18* (1979) 131
7 H.W. Etzkorn, U. Littmark, J. Kirschner: Proc. Sympos. Sputtering, Perchtoldsdorf, Vienna, 1980, ed. by P. Varga et al., p.542
8 S. Matteson: Appl. Phys. Lett. (to be published); see also the publications of this author in ref.9
9 Numerous recent publications on Recoil Mixing can be found in the Proceedings 2nd Int. Conf. Ion Beam Modification of Materials, Albany, N.Y., 1980, ed. by R.E. Benenson et al. (North-Holland, Amsterdam 1981)

Surface Topography Development During *SIMS* Investigations and Using it to Get Additional Information on Polycrystalline and Heterogeneous Solids

W. Hauffe

Section of Physics, Dresden Technical University,
GDR-8010 Dresden

1. Introduction

SIMS investigations are often confronted with problems of surface topography development resulting from the selective sputter-etch process in the bombarded region. Mainly the dynamical SIMS with emission or scanning ion microscopy of the sample surface has to consider contrast effects in the pictures, which are not only caused by concentration or orientation differences but also by topographic effects. However, also in macro-SIMS systems the sputtering process is accompanied by relief formation influencing the analysis, but, on the other hand, being also useful for subsequent scanning electron microscopical (SEM) investigations. The extended group of heterogeneous polycrystalline samples which are initially smooth and cleaned show characteristic ion-etch structures at phase and grain boundaries (Figs 1 and 2). Its knowledge is necessary to separate the topographic contrast from other influences and is favourable for getting information on the grain structure of the sample. Micro-scale features of the relief as facets are neglected here but can also influence the ion emission characteristics [4].

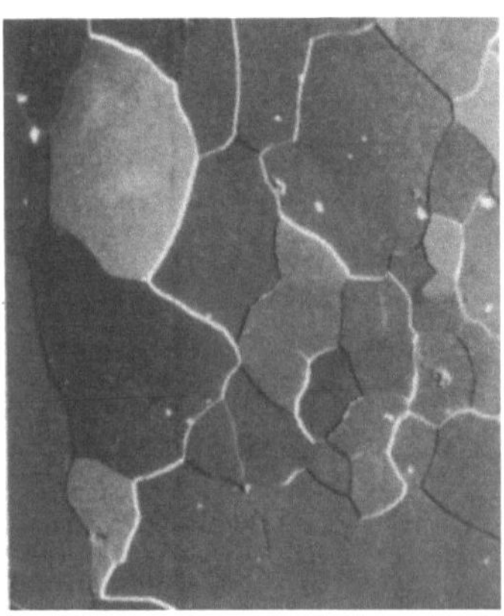

Fig. 1 Polycrystalline Fe surface after bombardment with 10 keV Kr^+ ions at 50°

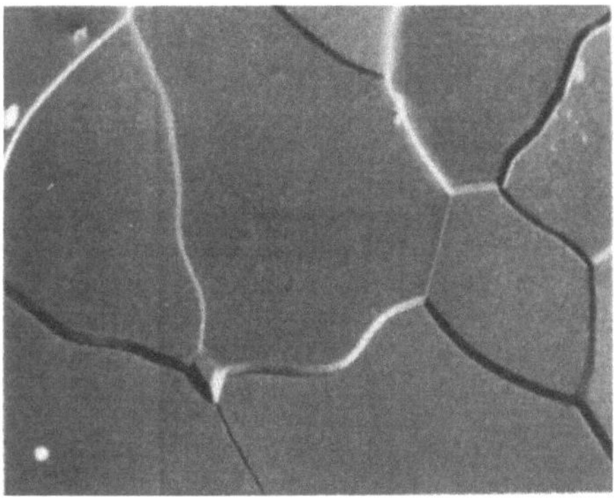

Fig. 2 Detail from Fig. 1

2. Topography on Heterogeneous and Polycrystalline Samples due to Ion Bombardment

2.1 Basic Effects

The bombardment of a smooth surface of a heterogeneous or polycrystalline material by a parallel beam of energetic inert gas ions leads to a surface relief due to the variation of the erosion rates, i.e. the displacement velocities of the grains in consequence of the following effects [1]: different materials show different sputtering yields depending on their atomic mass and binding energy [2]. Additionally, grains of the same phase differently oriented with respect to the ion beam give also rise to various sputtering yields. This selective etching leads to a marking of the grain structure, which is schematically shown in Fig. 3.

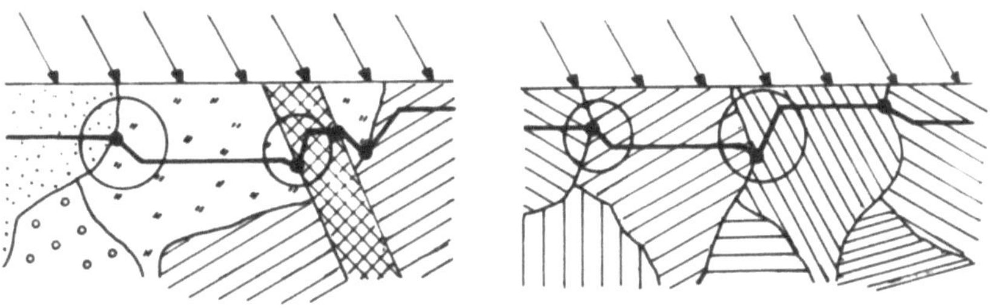

(a) Heterogeneous material (b) Polycrystalline material

Fig. 3 Selective etching during ion bombardment, schematically shown on heterogeneous (a) and polycrystalline material (b). The circles mark typical etch structures at phase and grain boundaries (see 2.2)

2.2 Development of Relief Structures at Phase and Grain Boundaries

At phase and grain boundaries according to a model given earlier by the author [1,3], steps (single slope) or grooves (two slopes) are developed depending on the ion beam direction, the displacement velocities of the grain surfaces, and the special grain boundary arrangement. Standard cases of the development of single slopes A_1 and A_2 are schematically shown in Fig. 4, depending on the lower (case (1)) or the higher displacement velocity (case (2)) of grain 2 compared with grain 1. Fig. 5 shows both types of structures on Fe for oblique ion incidence. According to this model, the intersection line of the grain boundary with the surface B is situated in the upper or lower edge of the slope, and therefore the exact position of the grain boundary is well known. The slope angle can be determined from the sputtering yield data and the three-dimensional grain boundary position [1,3].

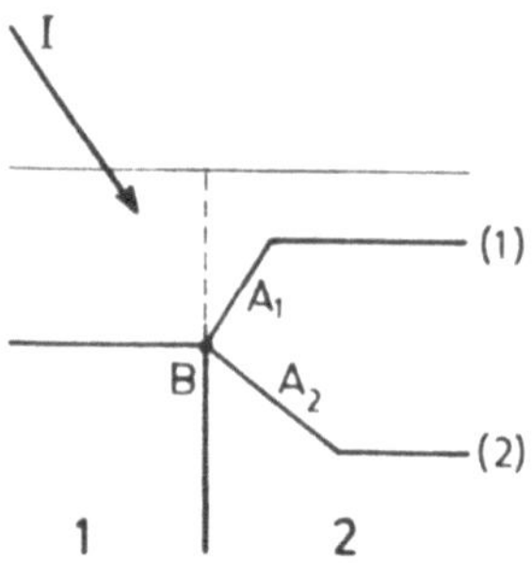

Fig. 4 Development of slopes A_1 or A_2 on grain 2 depending on a lower (case (1)) or a higher displacement velocity (case (2)) of grain 2 compared to grain 2 B: intersection line of the grain boundary; I: Ion beam direction

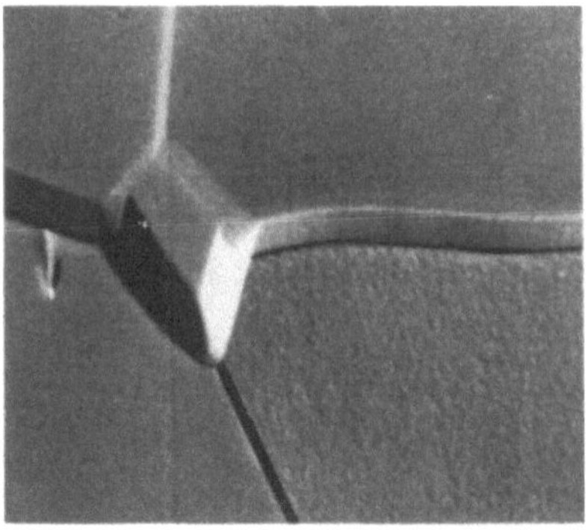

Fig. 5 Surface topography (detail of Fig. 2) at grain boundaries on Fe of slope types described in Fig. 4. Bombardment conditions: see Fig. 1

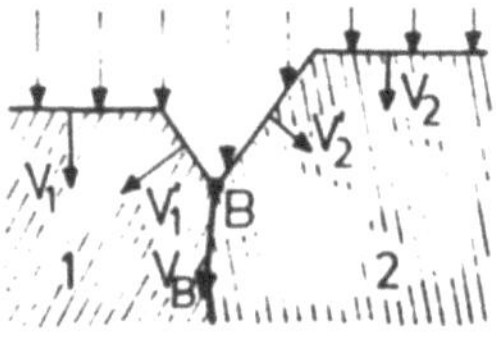

Fig. 6 Groove formation at a grain boundary (intersection line B) at normal incident ions

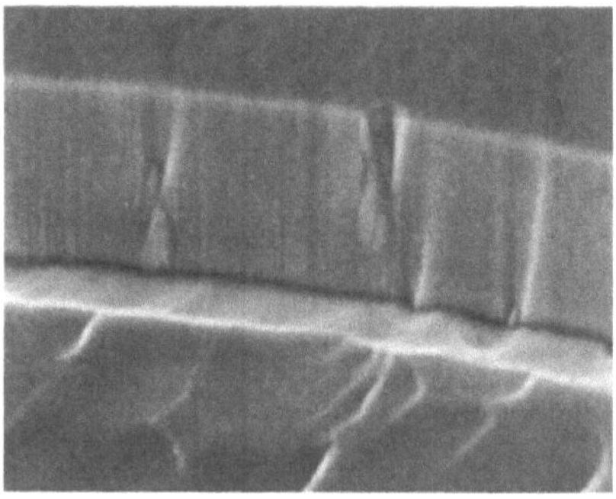

Fig. 7 Groove formation at a grain boundary on Ag, bombarded with 10 keV Ar^+ ions normally to the initial surface

The conditions for groove formation (two slopes) are fulfilled at nearly normal ion beam incidence on grain boundaries directed into the material almost perpendicular to the sample surface as schematically shown in Fig. 6 and on a real surface in Fig. 7.

The fact that atoms immediately belonging to the grain boundary are sputtered preferentially is a necessary condition, but it is not sufficient for developing a visible groove [1]. Additionally, the velocities v_1' and v_2' must reach sufficiently high values.

2.3 Special Cases

In the model used here a lot of special cases can be discussed, especially with respect to different grain and phase boundary arrangements and bombardment conditions. In Fig. 8 we give a schematic example of the possibility developing only a single slope at normal ion incidence at a grain boundary slightly inclined to the initial surface because the second slope cannot reach the velocity which is sufficiently high. A further consequence of the model of importance for SIMS investigations consists in the etching-behaviour at grain boundaries if the primary ion beam direction is changed. Then, according to the change of all sputtering yields, new slopes occur. Fig. 9a and b illustrate some examples of the development of new slopes starting from the grain boundary, Fig. 10 gives the experimental evidence for the situation of Fig. 9b. Bombardment intervals with more than two different ion beam directions produce very complex relief structures. A continuous changing of the ion beam direction (e.g. only azimuthal) diminishes the anisotropy and leads to a more uniform removing of material.

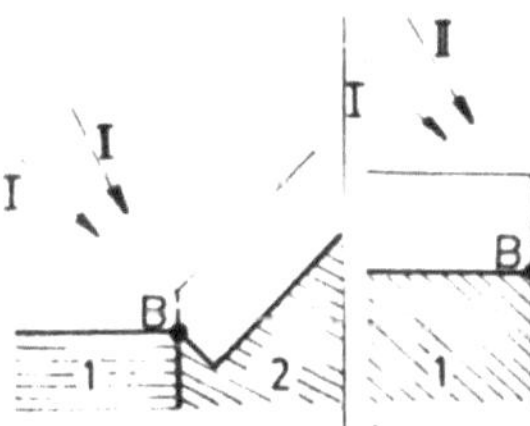

Fig.8 Formation of a single slope at the grain boundary (B) in the case of normally incident ions

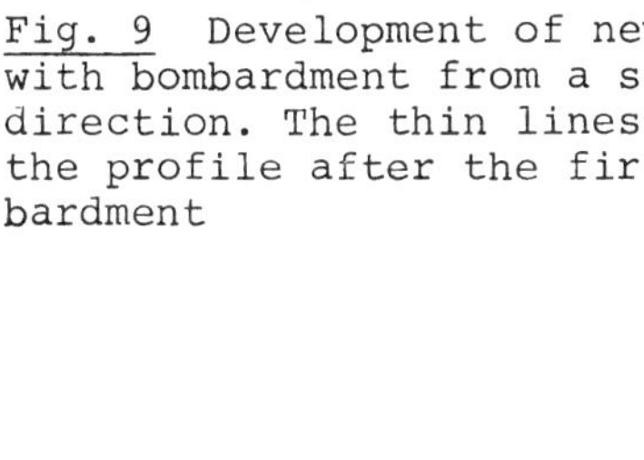

Fig. 9 Development of new slopes with bombardment from a second direction. The thin lines show the profile after the first bombardment

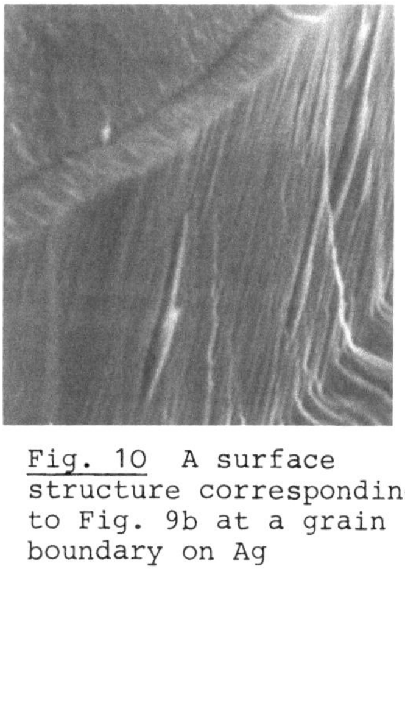

Fig. 10 A surface structure corresponding to Fig. 9b at a grain boundary on Ag

3. Contrast Phenomena

The ion-bombardment-induced surface relief depending on the special SIMS method influences the secondary ion signal and gives rise to a topography contrast in secondary ion pictures. The two main standard cases of secondary ion imaging are seen in Fig. 11 and 12, both combined with the typical relief structure for oblique (Fig. 11) and normal ion incidence (Fig. 12). However, also the scanning electron microscopy is well suited to investigate the ion-bombarded samples, and we have to consider the topography contrast in the SEM. In this case one has the advantage of the higher lateral resolution of the electron microscope. For discussing the main effects of relief contrast in the SI mode we can use the analogy between ion and electron pictures. The slopes near grain boundaries

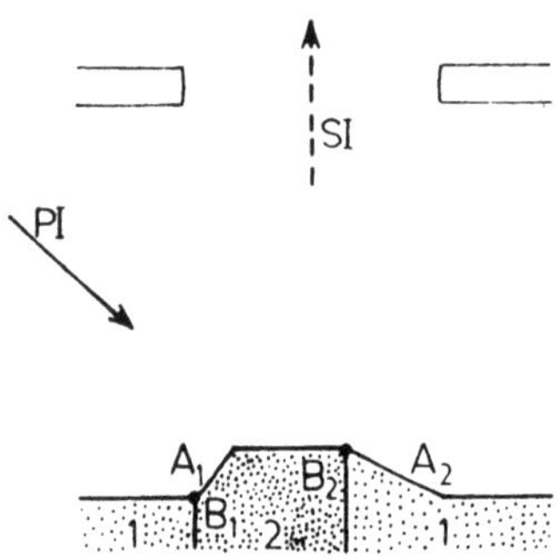

Fig.11 SIMS arrangement with oblique ion incidence. (PI) primary ions,(SI) secondary ions. B_1, B_2 intersection lines of phase of grain boundaries. A_1, A_2 slopes

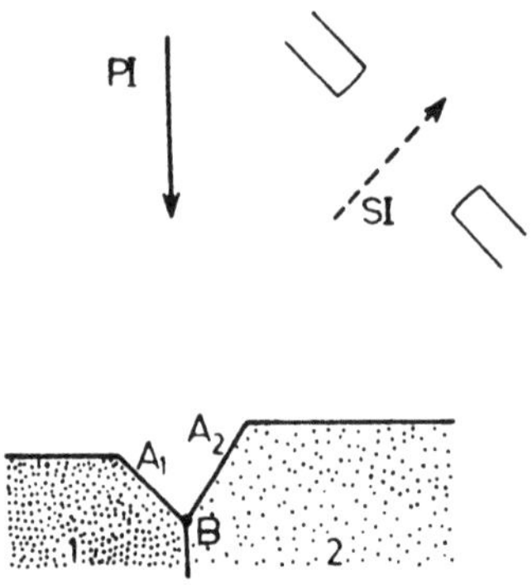

Fig.12 SIMS arrangement with normally incident primary ions (signs see in Fig.11)

give rise to changed values of emitted and collected ions compared to those of the horizontal areas. Corresponding to the SEM pictures (see Figs 1, 2 and 5), the slopes will appear bright or dark depending on their inclination with respect to the primary ion beam direction and the detector position (e.g. slopes A_1 and A_2 in Figs 11 and 12). These contrasts should not be misinterpreted as concentration effects. But with the knowledge of the development of the sputter-etch-induced relief near grain and phase boundaries discussed above, not only the topography contrast can be separated, but from the well-known real positions of intersection lines of grain boundaries the grain shapes and distributions can be determined. These consequences from topography effects are the more important the more the SIMS microprobe instruments go to a higher spatial resolution, as is expected for the future.

1. W. Hauffe, Thesis A, Technical University of Dresden, 1971
2. P. Sigmund, Phys. Rev. 184, 383 (1969)
3. W. Hauffe, phys. stat. sol. (a) 4, 111 (1971)
4. W. Hauffe, Thesis B, Technical University of Dresden, 1978

Sputtering of Metals with 20 keV O_2^+; Characteristic Etch Patterns, Sputtered Atom Yields and Secondary Ion Mass Spectra

K. Tsunoyama, T. Suzuki, Y. Ohashi, and M. Konishi

Research Laboratories, Kawasaki Steel Corp.,
Chiba, Japan

In secondary ion mass spectrometry an oxygen primary beam is frequently used because it enhances and stabilizes the yield of secondary sample ions [1,2]. But its role in the process of secondary ion emission is not yet fully understood [3-6]. The purpose of the present paper is to investigate the surface structure, sputtered atom yields and secondary ion mass spectra of metals bombarded with O_2^+. Accumulation of these fundamental data will help to elucidate the effect of oxygen ion on sputtering and to establish the method of quantitative interpretation of mass spectra.

The specimens used were polycrystalline pure metals provided by Johnson-Matthey Corp.. They were ground on metallographic papers and polished with fine alumina powder [7]. The ion bombardment was accomplished with an ion microporbe mass analyzer made by ARL. 20 keV O_2^+ impacted the target surface at normal incidence and were raster scanned over the area of 75x60 μm or 100x80 μm. The beam diameter was about 10 μm and the intensity was 50 or 90 nA. Total ion dose was in the range of $4\cdot10^{17}$-$4\cdot10^{19}$ ions/cm^{-2}. In the case of secondary ion detection the incident ion beam was fixed and the spot size was enlarged to 100 μm. The beam intensity was 10 nA and the sample voltage was 1.5 kV. The sputtered ions were mass analyzed in a double-focusing mass spectrometer and detected with a Daly-type ion counting system.

A strong chemical effect of the implanted oxygen atoms was found for the surface structure of the ion bombarded area. The etch patterns observed by a SEM (Figs. 1-5) were classified according to the binding energy of mono-oxide B_{MO} of target atom M and oxygen atom. In the case of Si and Ge, for example, the values of B_{MO} are relatively high (8.3, 6.8 eV, respectively) and quite uniform for the ion bombarded area. Meanwhile for Ag and Au the values of B_{BO} are less than 2.0 eV and cones were seen to develop. For Ti, Cr, Mn, Fe, and Mo the values of B_{MO} are in the range of 4.3 - 6.8 eV and their surfaces were relatively smooth or a few facets were produced. And for Co, Ni, Cu, and Zn, the values of B_{MO} are 2.7 - 3.8 eV and lots of hillocks or pits developed.

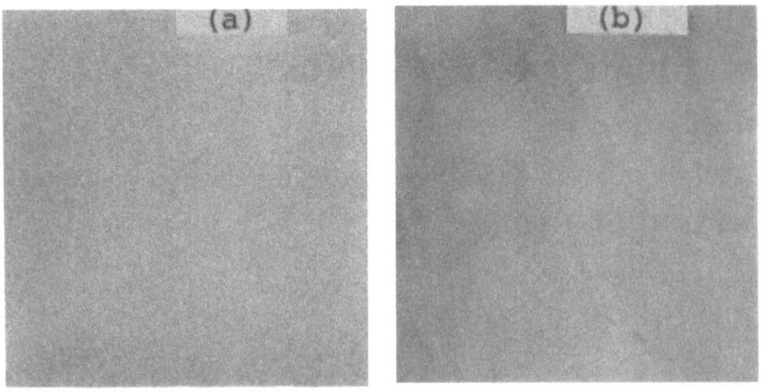

Fig. 1

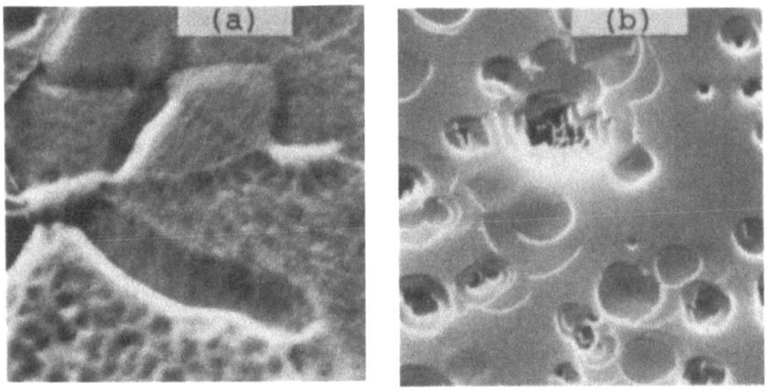

Fig. 2

Fig. 1 Scanning electron micrographs (magnification 3000x) of samples bombarded with 20 keV O_2^+: (a) silicon, dose $3.2 \cdot 10^{19}$ ions/cm^{-2}: (b) germanium, dose $1.4 \cdot 10^{19}$ ions/cm^{-2}

Fig. 2 Scanning electron micrographs (magnification 3000x) of samples bombarded with 20 keV O_2^+: (a) tin, dose $5.0 \cdot 10^{18}$ ions/cm^{-2}: (b) aluminium, dose $1.1 \cdot 10^{19}$ ions/cm^{-2}

Target	Sputtering Yields	Target	Sputtering Yields
Al	0.52 ± 0.13	Cu	3.0 ± 0.4
Si	0.54 ± 0.03	Ge	2.7 ± 0.20
Ti	0.27 ± 0.07	Nb	0.34 ± 0.03
V	0.71 ± 0.06	Mo	0.82 ± 0.09
Cr	0.74 ± 0.09	Ag	9.7 ± 0.2
Mn	0.55 ± 0.13	Sn	2.5 ± 0.4
Fe	0.57 ± 0.05	W	0.97 ± 0.23
Co	0.85 ± 0.05	Au	7.4 ± 0.1
Ni	1.0 ± 0.1		

Table 1 Sputtered atom yields for metals under 20 keV O_2^+ bombardment

The sputtered atom yields were obtained by measuring the depth of the crater with an interferoscope (Table 1). Except Ge, Sn, Cu, Ag and Au, the yields are less than 1.0 atoms/ion. The sputtered atom yields showed similar Z-dependence to those obtained by noble gas ion bombardment (Fig. 6). Several types of molecular ions were observed in the secondary ion mass spectra and the most pronounced were MO^+, M_2^+ as well as M^+.

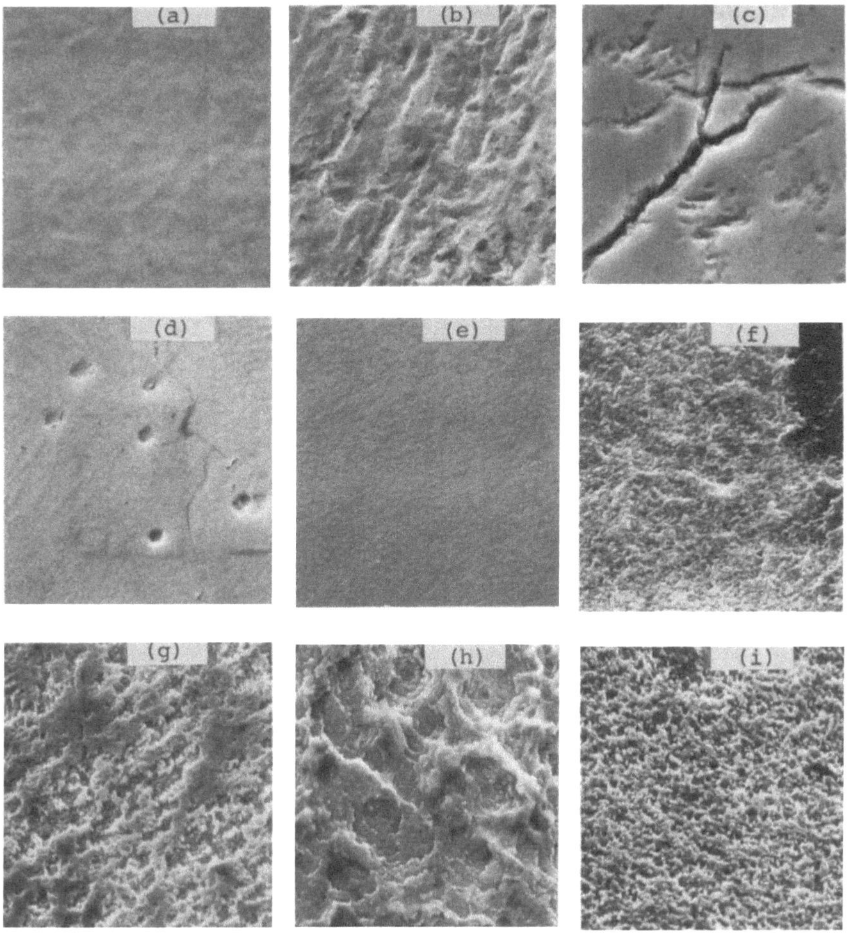

Fig. 3 Scanning electron micrographs (magnification 3000x) of metals of the first transition period bombarded with 20 keV O_2^+: (a) titanium, dose $1.7 \cdot 10^{19}$ ions/cm^{-2}; (b) vanadium, dose $8.4 \cdot 10^{18}$ ions/cm^{-2}; (c) chromium, dose $8.6 \cdot 10^{18}$ ions/cm^{-2}; (d) manganese, dose $1.2 \cdot 10^{19}$ ions/cm^{-2}; (e) iron, dose $2.0 \cdot 10^{19}$ ions/cm^{-2}; (f) cobalt, dose $4.2 \cdot 10^{18}$ ions/cm^{-2}; (g) nickel, dose $4.2 \cdot 10^{18}$ ions/cm^{-2}; (h) copper, dose $8.4 \cdot 10^{18}$ ions/cm^{-2}; (i) zinc, dose $8.4 \cdot 10^{18}$ ions/cm^{-2}

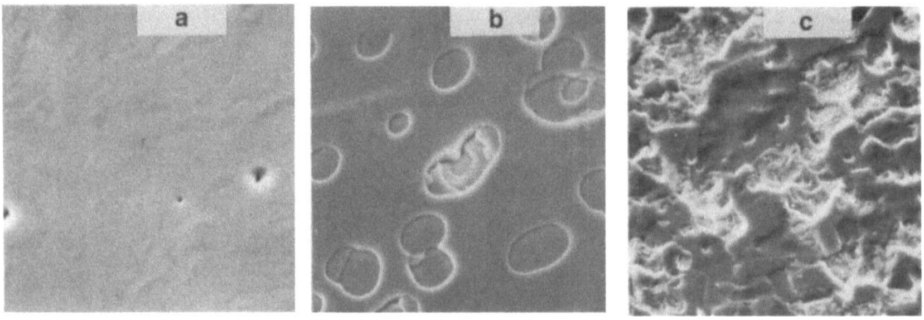

Fig. 4

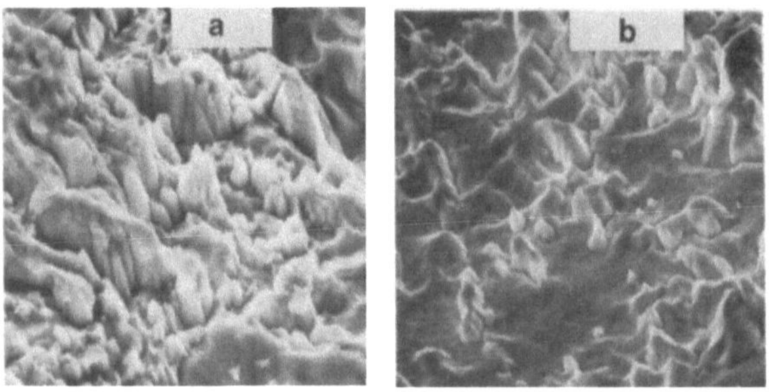

Fig. 5

Fig.4 Scanning electron micrographs (magnification 3000x) of refractory metals bombarded with 20 keV O_2^+: (a) molybdenum, dose $5.6 \cdot 10^{18}$ ions/cm^{-2}; (b) niobium, dose $2.5 \cdot 10^{-19}$ ions/cm^{-2}; (c) tungsten, dose $4.4 \cdot 10^{18}$ ions/cm^{-2}

Fig.5 Scanning electron micrographs (magnification 3000x) of noble metals bombarded with 20 keV O_2^+: (a) silver, dose $8.4 \cdot 10^{18}$ ions/cm^{-2}; (b) gold, dose $4.2 \cdot 10^{18}$ ions/cm^{-2}

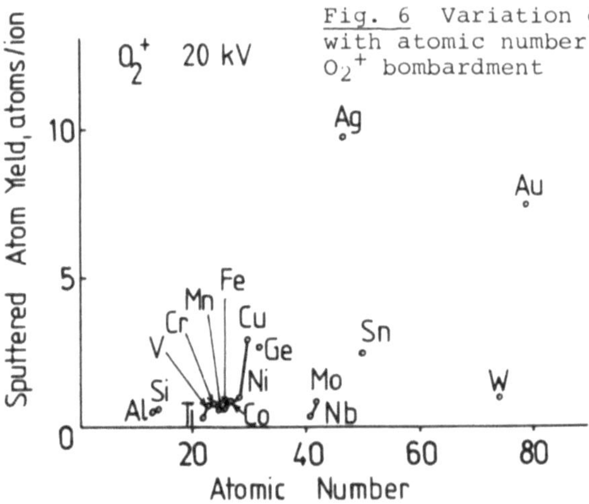

Fig. 6 Variation of sputtered atom yield with atomic number of the metal for 20 keV O_2^+ bombardment

The intensity ratio of MO^+ to M^+ showed similar Z-dependence as those of B_{MO}, but no correlation was found between the intensity ratio of M_2^+ to M^+ and B_{MM}. More detailed inspection of the observed correlations is, however, required.

References

1. C.A. Andersen, Int. J. Mass Spectrom. Ion Phys. 2, 61 (1969)
2. K. Tsunoyama, Y. Ohashi and T. Suzuki, Anal. Chem. 48, 832 (1976)
3. C.A. Andersen and J.R. Hinthorne, Anal. Chem. 45, 1421 (1973)
4. G. Blaise and G. Slodzian, Surf. Sci. 40, 708 (1973)
5. V.R. Deline, W. Katz, C.A. Evans, Jr. and P. Williams, Appl. Phys. Lett. 33, 832 (1978)
6. R. Shimizu, T. Okutani, T. Ishitani and H. Tamura, Surf. Sci. 69, 349 (1977)
7. K. Tsuruoka, K. Tsunoyama, Y. Ohashi and T. Suzuki, "Proc. of the 6th International Vacuum Congress", Kyoto (March 1974); Japan J. Appl. Phys. Suppl. 2 Pt. 1, 387 (1974)

Depth Resolution of Ion Bombardment Technique Applied to NiPd, NiPt, PtPd, Thin Layer Systems

J. Giber, D. Marton, J. Lászlô, and J. Mizsei[1]

Institute of Physics, Technical University,
H-1111 Budapest, Hungary

Introduction

A lot of studies have been published already about the important questions in depth profiling: how the ion bombardment process itself can influence the accuracy of the measured depth profile, what depth resolution can be reached in an optical case, which parameters play a significant role in the profile measurement and how to get the real concentration profile from the measured one [1-4]. As the accuracy and reliability of SIMS measurements have to be subjects of both theoretical and experimental studies, one must examine depth profiles with respect to:

a.) different layer producing techniques
b.) material, and
c.) sputtering rate dependence.

Based on these results one can deal with the analysis of the distortion of complete depth profiles to find a transmission or a weight function (depending on function parameters, of course) to the linear box, whose input is the real depth profile of the sample and whose output is the measured one. It seems also to be useful to examine SIMS and Auger depth profiles of similar material systems, because they are complementary techniques. The present paper reports on the first steps of a research program outlined above.

Experimental

We used Balzer's-made SIMS equipment [5] and Ar^+ primary ions of 3 keV primary energy and a current of 10^{-6} A. The diameter of the primary beam is 2.5 mm (FWHM), and to avoid the crater wall effect we applied a diaphragm of 1.4 mm diameter. The distribution of the current density of the primary beam is quasi-Gaussian. (The maximal decrease of the current density within a diameter of 2.4 mm is 3 % of its maximal value). The angle of incidence was 30° to the surface. The ion gun is a modified Finkelstein-type. The Ar pressure during the measurement was about 10^{-4} Pa, the background pressure was about 10^{-7} Pa. In the case of reactive sputtering experiments the oxygen was blown into the chamber through a jet. The oxygen pressure was $4 \cdot 10^{-4}$ Pa. The depth profiles of $^{195}Pt^+$, $^{106}Pd^+$, $^{58}Ni^+$, $^{28}Si^+$ ions have been detected.

---

[1]Chair of Electronic Devices, Technical University, Budapest, Hungary

The samples we have chosen were from a triplet of elements from the 8th column of the periodic system of the elements. Our sample series consisted of some unrepeated variations of the second order of this triplet. This choice was made because the sputtering rates of the elements of the same column of the periodical system are quite similar [6]. The principle of the measurement is as follows: we deposited one material of the triplet on a substrate in a thin layer (in about 10 nm thickness), and on top of this film we deposited the other material of the pair in different thicknesses (from 0 to about 400 nm) sample by sample in 5-6 steps in one series. Thus we could follow the altering of the SIMS profile of different buried layers as the function of the thicknesses of different outer layers. We have examined these series by reactive Ar^+-sputtering in oxygen environment and non-reactive Ar^+-sputtering measurements.

The sample were prepared by the following procedure: Layers under investigation were deposited in a conventional RF-diode sputtering equipment of Alcatel-type (APC 1.120) in Ar gas at 0.13 Pa pressure. The targets were discs of 100 mm diameter and the target-substrate distance was 50 mm. Chemically cleaned silicon monocrystal wafers were used as substrates. The contaminated or damaged layer of the substrate surface was removed by thermal oxidation and etching of the SiO_2 film. The vacuum chamber was pumped down to a pressure of about $5 \cdot 10^{-4}$ Pa. The sputtering process was performed in 7 steps. The first step was the buried layer deposition, then the target was changed and the second layer was sputtered in 5-6 different thicknesses. Every layer thickness and the sum of the inner and outer layer thicknesses could be measured with Talystep independently by means of a suitable photoresist pattern. The choice of the substrate material is justifiable as it provides an extremely good quality of surface flatness.

Results

In every series of measurements we were interested in the relative depth resolution depth functions. The subjects of our measurements were of two kinds. Partly we have analysed the slope of the ion yield coming from the buried layer by the convention of 84-16 % (supposing a real profile of a unit pulse function), partly the slope of the outer layer, similarly. In the case of the investigation of the buried layer ΔZ is the abscissa interval between the ordinate values $(0.84 \cdot c/c_{max})$ and $(0.16 \cdot c/c_{max})$, Z is obtained from the Talystep measurement. Probably this definition of Z involves slightly different results compared with the case where Z is also determined from the measured profile. We supposed that the measured CPS values are proportional to the concentration values. The transformation of the time scale to a depth scale was determined by a constant multiplicator (sputtering rate s') in the case of the examination of the buried layer, while there was a mean sputtering rate value <s'> defined in the case of the outer layer in the following way: by dividing the Z value of the Talystep measurement by the proper time interval (this is the abscissa value of the ordinate of c/c_{max} = 0.5 on the slope) we obtained a sputtering rate, and the mean value became the mathematical average of these rates for one series. In Fig. 1 the particular points of the analysis are shown in an example. Our results can be seen on Figs 2, 3, 4 and 5. In these figures the lines connecting the measured points serve only for a better synopsis.

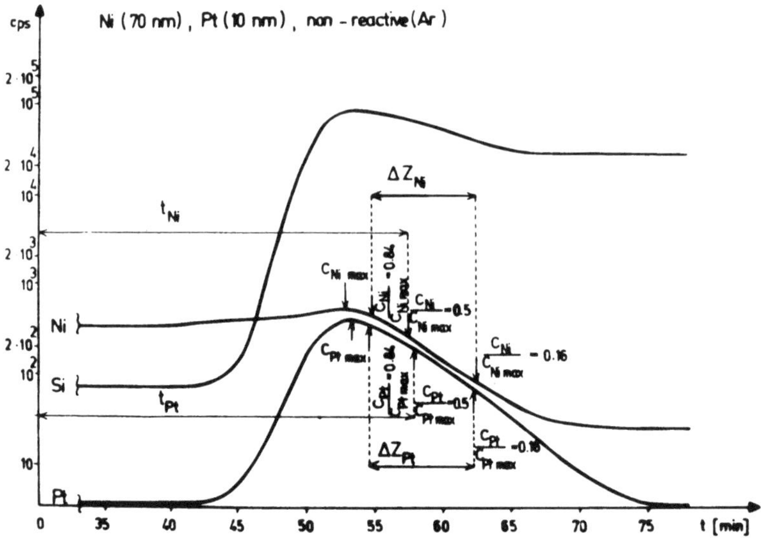

Fig. 1 The particular points of analysis of a measured SIMS depth profile. Ni is the outer layer (70 nm), Pt is the buried one (10 nm). Non-reactive measurement (Ar^+-sputtering)

Discussion

Figures 2 and 4 prove above all that the deviation of the features of the different pairs of materials is not significant considering the reproductive experiments (denoted by (1) and (2) on the figures) of the Pd-Pt series. The data of Figs 3 and 5 are derived from reactive measurements. It is seen that the deviations between different series are bigger and the $\Delta Z/Z = f(Z)$ functions are more complicated. No conclusion can be drawn whether the reactive sputtering is really preferable for the examination of depth resolution or not, as in Figs 4 and 5 the reactive measurement seems to give better results while the comparison of the data of Figs 2 and 3 is contradictory to this statement also found in the literature [7]. Namely, Fig. 3 shows a detrimental influence of the O_2 on the resolution especially in the case of the Pt-Pd system and on the other hand in a surprising way in the case of thick outer layers. This might be due to the fact that Pd and Pt have no stable oxides and Ni produces only a weak bond with oxygen. Therefore no homogeneous oxides are formed. Returning to the non-reactive examinations it is worthwile evaluating the measured data mathematically by fitting linear functions to the

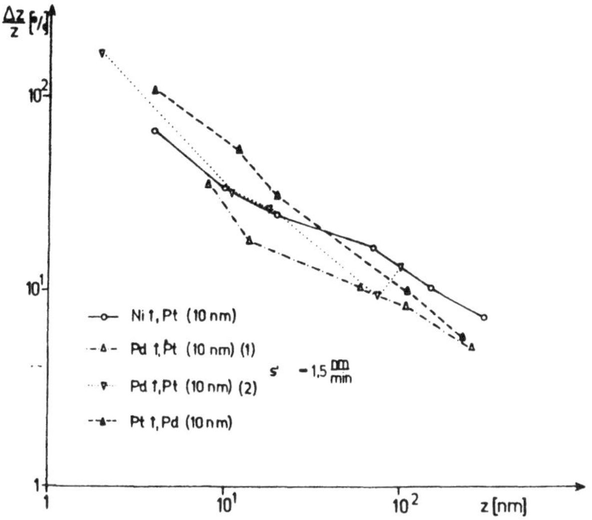

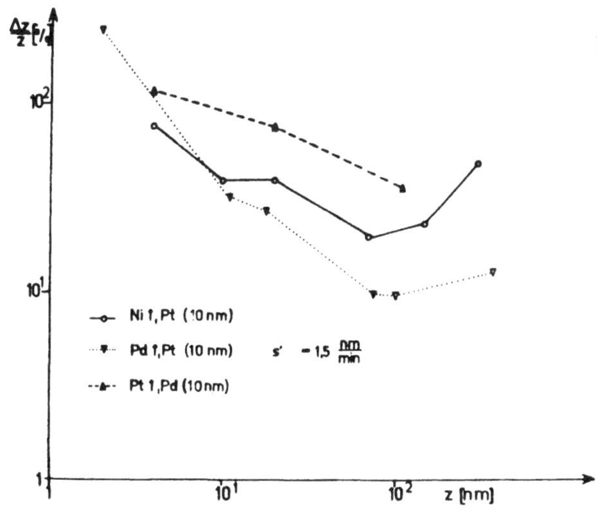

Fig. 2 The relative depth resolution as the function of depth. The subject of investigation is the slope of the ion yield from the buried layer. (1) and (2) mean measured results on two sample series of the same kind. s' is a constant sputtering rate. Non-reactive measurement (Ar^+-sputtering). Mark ↑ denotes the element whose thickness is changing in the given series

Fig. 3 The same examination as on Fig. 2. Reactive measurement (Ar^+-sputtering in oxygen environment)

points in a logarithmical coordinate system. Table 1 contains the regression parameters a and b defined by the equation

$$\lg \frac{\Delta Z}{Z} (\%) = a + b \cdot \lg Z,$$

and r^2 which is the so called coefficient of determination, being closer to 1 if the fitting is better. It can be seen that with the exception of one case, the fitting is convenient and the parameters a and b are often approaching the data of S. HOFMANN, where b = -0.5 [8]. The fact that there are no pronounced deviations between the results of different pairs of materials can be explained partly by the similar sputtering rates of these elements (according to our choice) and partly by the fact that the depth resolution is essentially dependant on instrumental factors [1].

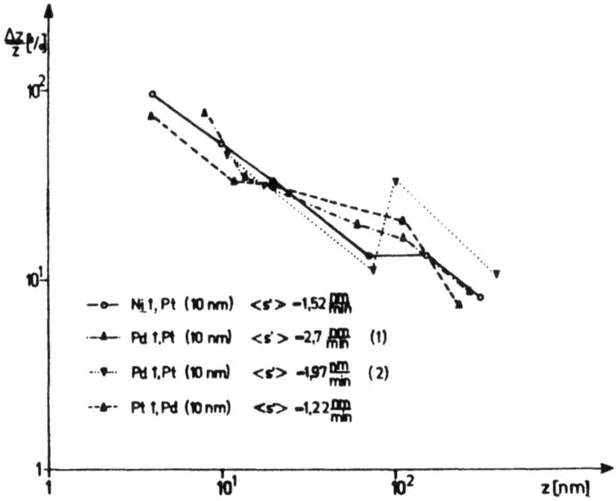

Fig. 4 The relative depth resolution as the function of depth. The subject of investigation is the slope of the ion yield from the outer layer. Notes (1) and (2) see in Fig. 2. <s'> is the mean sputtering rate. Non-reactive measurement (Ar^+-sputtering). Mark ↑ denotes the element whose thickness is changing in the given series

Table 1 Regression parameters

	a	b	r^2
Slope of the buried layer			
Ni-Pt	2.061	-0.482	0.983
Pd-Pt (1)	1.933	-0.507	0.965
Pd-Pt (2)	2.324	-0.680	0.935
Pt-Pd	2.459	-0.718	0.996
Slope of the outer layer			
Ni-Pt	2.269	-0.561	0.974
Pd-Pt (1)	2.290	-0.552	0.951
Pd-Pt (2)	1.983	-0.366	0.599
Pt-Pd	2.121	-0.483	0.908

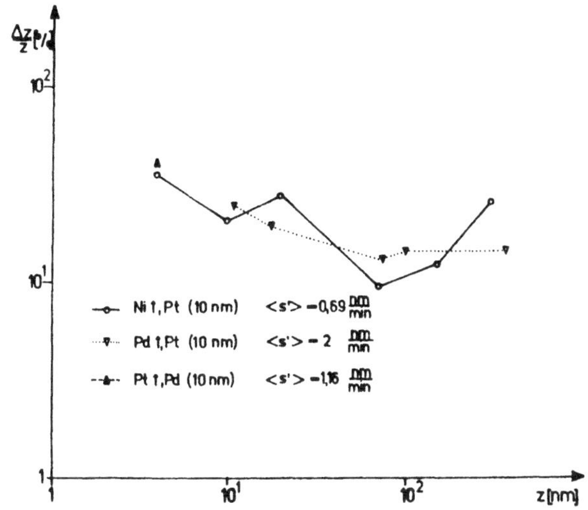

Fig. 5 The same examination as on Fig. 4. Reactive measurement (Ar^+-sputtering in oxygen environment). Mark ↑ denotes the element whose thickness is changing in the given series

The depth resolution of the sputter profiling technique has been investigated mostly by means of AES measurements so far. Since the ion sources and detection techniques are both different in SIMS instruments from the ones in AES, authors regard it important to investigate the problem of SIMS depth resolution by SIMS.

References

1. S. Hofmann, Surface Interface Anal. 2, 148 (1980)
2. P.H. Holloway and R.S. Battacharya, Surface Interface Anal. 3, 118 (1981)
3. W.O. Hofer and U. Littmark, Phys. Lett. 71A, 457 (1979)
4. H.H. Andersen, Appl. Phys. 18, 131 (1979)
5. F. Pavlyák, L. Bori, J. Giber and R. Buhl, Jap. J. Appl. Phys. 16, 335 (1977)
6. M. Kaminsky, "Atomic and Ionic Impact Phenomena on Metal Surfaces", Academic Press, New York (1965)
7. M. Bernheim and G. Slodzian, Int. J. Mass Spectrom. Ion Phys. 12, 93 (1973)
8. S. Hofmann, Appl. Phys. 13, 205 (1977)

The Influence of Ion Bombardment on the Results of AES-Depth Profiling on Silicides

Th. Wirth

Zentralinstitut für Elektronenphysik, Akademie der Wissenschaften der DDR, GDR - 1199 Berlin

Introduction

The determination of quantitative concentration ratios of silicides with AES-depth profiling leads to results which are influenced by a strong sputter effect. We have investigated aspects of this problem by experiments on silicide-silicon systems. During sputtering two essential effects appear simultaneously, namely preferential sputtering and the knock-on effect [1,2]. In the literature preferential sputtering is regarded as the main effect [1,3-7]. However the knock-on effect can also influence concentration ratios [1,8,9]. I am reporting about investigations on the knock-on effect.

According to the theory of SIGMUND [10-12] the sputter yield depends on ion energy, the mass ratio of ion and target atom, and on binding energy. As sputtering of material from the surface and the knock-on effect appear simultaneously, the latter effect supposedly depends on the same parameters.

Three types of experiments are carried out, namely the independent variations of ion energy, current density and bombardment time.

The aim of this paper is to show that the knock-on effect can essentially influence the results of Auger depth profiles. We shall give a more extensive report elsewhere

Experimental

The experiments are carried out on NiSi and PtSi layers on silicon substrates. The film thickness was some hundred nanometers and the stoichiometric composition of Ni/Si as 48 at% to 52 at% and Pt/Si as 50 at% to 50 at% was controlled with RBS and in case of NiSi also with X-ray diffraction.

The Auger parameters were: electron energy: 5 keV, rastered area: 100 μm^2. The concentration ratios are determined using low energetic peaks with similar escape depths as well as high energetic and low energetic peaks with different escape depths [13]. For peak interpretation the relative sensitivity factors were used. These can be influenced essentially additional to sputtering by the escape depth ratio and backscattering effects. However, in case of ADP using peaks with equal escape depths these effects are negligible [14,15]. This assumption is supported by ADP on NiSi using the low energetic peaks where the concentration ratio nearly agrees with the results of RBS and X-ray diffraction measurements.

In case of ADP using peaks with different escape depths these effects can influence the relative sensitivity factors and so the concentration ratios additional to sputtering. In order to explain the degree of influence on the relative sensitivity factors we will investigate both effects at unsputtered and sputtered in-situ-prepared silicides.

In the following report for better representation of the measurements for converting peak height ratios into atomic percent plots in all cases the standard sensitivity factors were used [16].

Results

The surfaces of NiSi and PtSi were bombarded by Ar^+ ions with energies from 2 keV to 5 keV for about 200 s. After that a 1 keV Ar^+ depth profiling was carried out in the same bombardment crater. The current density was 45 $\mu A/cm^2$ in both cases.

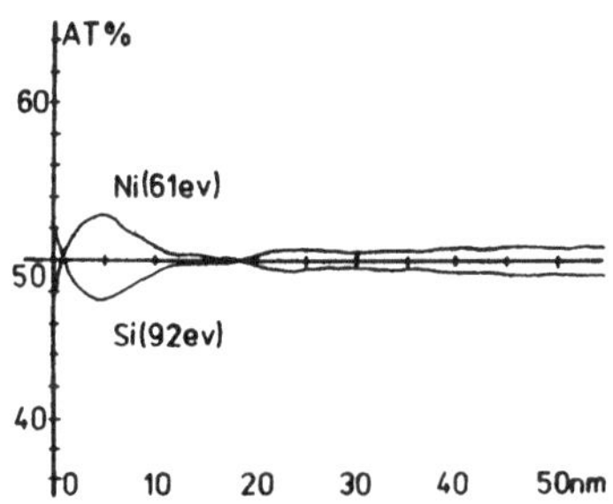

Fig. 1 1 keV Ar^+ depth profile after 3 keV Ar^+ ion bombardment

The concentration versus sputter depth for a 1 keV Ar^+ depth profile after 3 keV Ar^+ ion bombardment is shown in Fig. 1. The Ni concentration at first increases, then attains a maximum and decreases with greater depths to a nearly steady-state value characteristic of 1 keV ion energy. The depths corresponding to the maxima and the differences in metal concentrations between maximum and plateau are evaluated from these AES-knock-on profiles and are shown in the following figures. In Fig. 2 shows the difference in Ni concentration (at%) versus the ion energy.

The concentration difference increases_ higher ion energy. The Ni maximum is weakly shifted from 2.5 nm to 4.5 nm with increasing ion energy. Similar results are obtained from experiments with 3 keV and 5 keV ion energy on PtSi. The concentration difference rises from 6.2 at% at 3 keV to 6.8

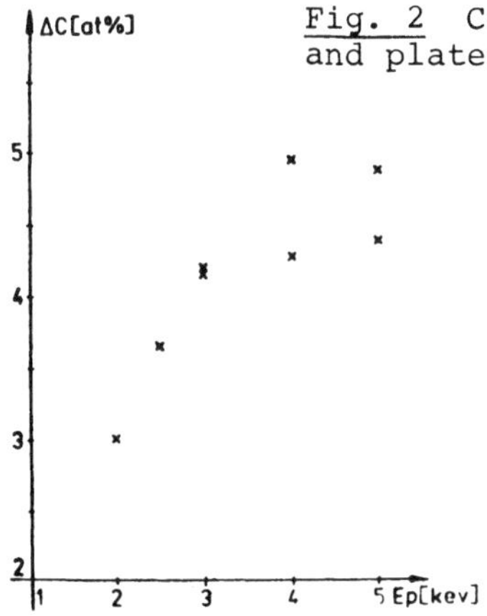

Fig. 2 Concentration differences between maximum and plateau

at% at 5 keV ion energy. The maximum is shifted from 0.8 nm for 3 keV to 2.4 nm for 5 keV ion energy. However, the concentration ratio of Pt/Si is 2:1 in the plateau which differs from the RBS results.

The following conclusions can be drawn from these results.

1. The number of ion-target atom knock-on strokes increases with increasing beam energy. Therefore the metal concentration increases the region of the concentration maximum.

2. The maximum is shifted to larger depths with increasing ion energy due to higher energy transfer by strokes.

3. The deviating Pt/Si value in the plateau region is supposed to be due to a mass effect. Because Pt is heavier than Ni we observe smaller penetration depths in PtSi. Therefore the implantation maximum lies closer to the surface and thus influences the low energetic peaks.

In order to demonstrate the influence of the knock-on effect on the concentration ratios determined from depth profiles we have measured ADPs on NiSi and PtSi with beam energies from 1 keV to 5 keV. The current density was 45 $\mu A/cm^2$.

Differences were found in the plateau concentrations as a function of the ion energy (Fig. 3). For NiSi we obtained only a small increase in concentration deviations from the volume ratio with rising ion energy using low energetic peaks with equal escape depths. The Si concentration in this case comes out to be greater than the Ni concentration.

Strong deviations from the bulk ratio are observed using peaks with different escape depths. In this case the Ni concentration is higher than the Si concentration.

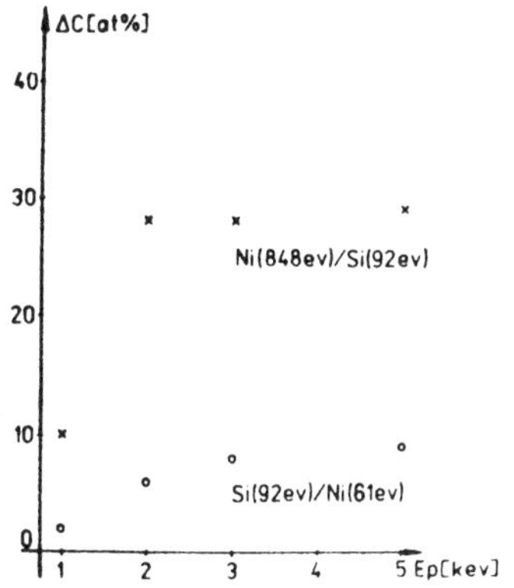

Fig.3 Concentration differences between Ni and Si plateau of ADPs

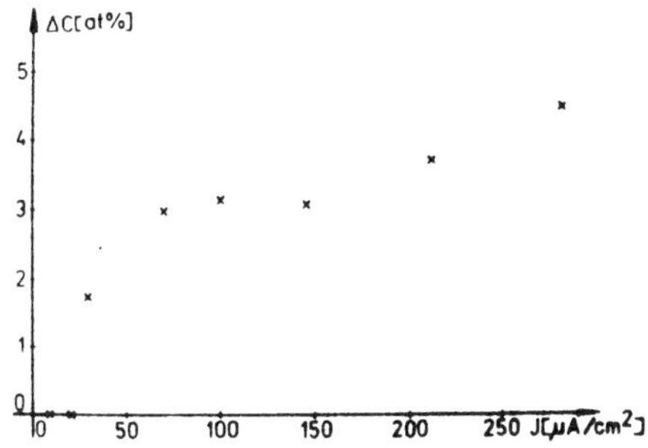

Fig.4 Concentration differences between maximum and plateau versus current density

From ADP on PtSi the concentration ratios of Pt/Si are 2:1 for all ion beam energies used in case of equal escape depths and 3:1 in case of different escape depths. Unlike NiSi, the Pt concentration is always greater than the Si concentration. The following conclusions can be drawn from these results. In case of NiSi the ADPs are essentially influenced by knock-on effects. Due to knock-on the Ni concentration is depleted at the surface and enriched in the depth. Therefore using Auger peaks with different escape depths (Si: 0.4 nm, Ni: 1.6 nm) the intensity of the Ni peak is determined by implanted metal atoms. This leads to a difference in plateau values up to 30 at%. With peaks of equal escape depth the intensity of the Ni peak is determined by the depletion layer. In case of PtSi a strong superposition of knock-on and preferential sputtering takes place because of the essential mass differences. This implies that concentration ratios are determined by both effects which cannot be clearly distinguished.

In order to investigate the influence of current density NiSi was bombarded by 2 keV Ar^+ ions with current densities from 10 $\mu A/cm^2$ to 300 $\mu A/cm^2$ about 200 s with subsequent 1 keV Ar^+ depth profiling.

Figure 4 shows the concentration differences (at%) versus the current densities ($\mu A/cm^2$). A concentration maximum cannot be detected with current densities from 10 to 20 $\mu A/cm^2$. From 20 to 70 $\mu A/cm^2$ a maximum evolves that increases weakly up to 300 $\mu A/cm^2$. The position of the maximum corresponding to the given ion energy is nearly at a constant depth of 2.5 nm. Finally, in order to investigate the influence of knock-on effect in context of the dynamical equilibrium the surface of NiSi was bombarded by 5 keV Ar^+ ion beam and current density of 45 $\mu A/cm^2$ and subsequent 1 keV Ar^+ depth profiling.

From these results it can be concluded that the dynamical equilibrium is obtained after 200 s bombardment time. This is confirmed by the beginning of the plateau of a 5 keV Ar^+ ADP on NiSi. After 50 nm the dynamical equilibrium is obtained, which nearly agrees with the value of 48 nm obtained from the 200 s bombardment time.

Conclusion

From these results it can be concluded that AES depth profiling of silicides is essentially influenced by the knock-on effect. An important result of the present studies is the necessity of using Auger peaks with equal escape depth in order to make data analysis meaningful. The aim of the subsequent work is to minimize the sputter effects on silicides either by optimization and/or introducing appropriate correction factors.

Acknowledgements

The author wishes to thank Dr. Möhling for X-ray diffraction measurements on NiSi and Mr. Linke and his colleagues for RBS measurements as well as Dr. Lange and Mr. Atzrodt for helpful discussions.

References

1. G.K. Wehner: Methods of Surface Analysis, Methods and Phenomena 1, ed. by A.W. Czanderna (Elsevier, Amsterdam 1975)
2. J.W. Coburn: Thin Solid Films 64 (1979) 371
3. J.M. Poate, W.L. Brown, R. Homer, W.M. Augustiniak, J.W. Mayer, K.N. Tu, W.F. van der Weg: Nucl. Instr. and Methods 132 (1976) 345
4. Z.L. Liau, J.W. Mayer, W.L. Brown, J.M. Poate: J. Appl. Phys. 49 (1978) 5295
5. G. Betz: Surface Sci. 92 (1980) 283
6. G. Betz, J. Marton and P. Braun: Nucl. Instr. and Methods 168 (1980) 541
7. G. Betz, M. Arias and P. Braun: Nucl. Instr. and Methods 170 (1980) 347
8. M. Opitz, G. Betz and P. Braun: Acta Phys. Acad. Sci. Hung. 49 (1980) 119
9. P. Sigmund: J. Appl. Phys. 50 (1979) 7261
10. P. Sigmund: Phys. Rev. 184 (1969) 383
11. P. Sigmund: J. Vac. Sci. Technol. 17 (1980) 396
12. P. Sigmund, A. Gras-Marti: Nucl. Instr. and Methods 168 (1980) 389
13. G. Ertl and J. Küppers: Low Energy Electrons and Surface Chemistry (Verlag Chemie, Weinheim, F.R.G. 1974)
14. P.M. Hall and J.M. Morabito: CRC Critical Review in Solid State and Material Sci. 8 (1978) 53
15. P.M. Hall and J.M. Morabito: Surf. Sci. 83 (1979) 391
16. L.E. Davis, N.C. MacDonald, P.W. Palmberg, G.E. Riad, R.E. Weber: Handbook of Auger Electron Spectroscopy (Physical Electronics Industries, Inc., Eden Prarie, MN. 1977)

A Study of Secondary Ion Energy Distributions During Sputtering of MIS Layer Structures

G.Ph. Romanova, P.I. Didenko, A.A. Yefremov, V.G. Litovchenko, and R.I. Marchenko

Institute of Semiconductors of the UkrSSR Academy of Sciences, SU-252650 Kiev, USSR

For several years we investigated the secondary ion mass spectra during the sputtering of MIS structures. It is known that these systems contain thin layers of metals, dielectrics and semiconductors, their properties (especially the structure) being strongly dependent on the growth conditions and subsequent treatment. There may be polycrystalline substances with the grain of various dimensions and monocrystalline ones. Such materials as SiO_2 have a large number of modifications. In this work along with the results on the energy distribution of the secondary ions from the semiconductors and dielectrics also a qualitative model is presented of "the ensemble of cascades", which makes it possible to explain non-contradictorily all the results obtained.

Experimental

The experimental conditions were described earlier [1,2]. Incident Ar^+ ion beam of $E = 2$ keV, $j = 20\ \mu A/cm^2$ was used, providing the sputtering rate of 60 Å/min. Energy and mass resolution made up 1 eV and 200 correspondingly.

Results

Figure 1 shows the energy distributions (ED) obtained during the sputtering of the films, which form an MIS structure. ED for Al film on a conducting substrate (SiO_2 with large leakages) coincides with that for polycrystalline Al. A broader ED is observed for an insulating substrate. A splitting of the Al^+ peak takes place at Al-SiO_2 interface which is caused by a superposition of the emission from the metal and alumosilicate grains. SiO_2 layers give different ED depending on the manufacture technique. SiO_2 layer is known to be completely amorphous if grown in the presence of HCl vapours, the electrical strength of both thermal and "chloride" SiO_2 being the same. ED becomes substantially narrower for SiO_x films, if $x < 2$. The heterogeneity at SiO_2-Si interface is felt in ED of Si^+, if the conditions (slow cooling) necessary for a formation of microcrystalline inclusions in the amorphous film are met. And, finally, the shape of ED for the semiconductor layers of Si depends on the degree of the target crystallinity.

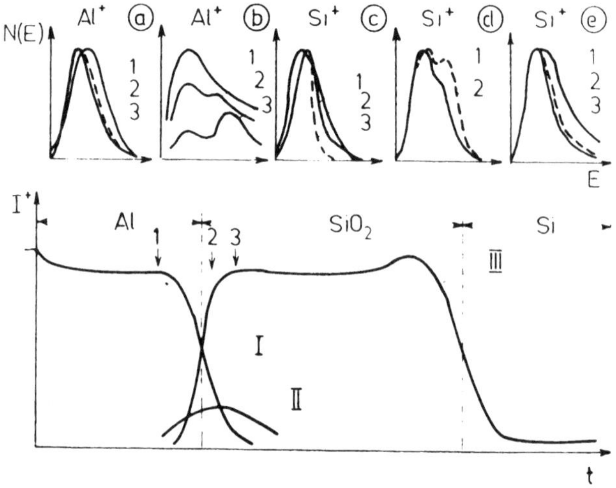

Fig. 1 Secondary ion yield profiles and their ED during the sputtering of the layers forming Al-SiO_2-Si structure: I - Al^+ (emission from the metal film), II - Al^+ (emission from the alumosilicate grains at Al-SiO_2 interface), III - Si^+ (emission from SiO_2 layers and from Si substrate). (a) ED for Al^+: 1 - Al, polycrystalline, 2 - thin Al film on a conducting SiO_x substrate, 3 - thin Al film on an insulating SiO_2 substrate.
(b) A splitting of ED for Al^+ on inhomogeneous interface between the phases. The arrows 1,2,3 on Al^+ profile indicate the moments of ED measurements. (c) ED for Si^+: 1 - thermal "dry" SiO_2, 2 - "chloride" SiO_2, 3 - SiO_x film with $x \simeq 1$.
(d) A splitting of ED for Si^+ at the inhomogeneous SiO_2-Si interface: 1 - fast cooling of SiO_2-Si system after the oxidation, 2 - slow cooling. (e) ED for Si^+: 1 - single crystal Si, 2 - epitaxial Si layer, 3 - polycrystalline Si layer

The Model

The difficulties in the explanation of the observed dependences of ED shape upon the structure arise if secondary ion emission is considered as a pure surface phenomenon, because the incident beam disturbs considerably both the surface itself and the adjacent layer of about tens of Å. The consequences of this disturbance result in the direct structural perturbation due to radiation defects and the changes in the stoichiometry [3] up to a complete reconstruction of the electron band structure [4]. These effects would have to smooth down to a considerable extent all the structural differences, however, data mentioned indicate that such a situation is not realized. Therefore one can assume, that deeper layers of the matrix take part directly or indirectly in the creation of the secondary ions and the formation of their ED [5]. The model presented below considers the creation of the secondary particles taking into account the influence of the bulk properties of the matrix. The data on the computer simulation of the atomic collisions [6] and on the non-equilibrium plasma investigations [7] are also used.

The model includes the following basic assumptions:

1.) The process of sputtering at any moment is a result of a superposition of a great number of separate interaction acts between the incident ions and the matrix. The observed yield of the secondary particles is the result of such microinteractions (named below as "the cascades"), which are averaged over the whole ensembles and the observation time.

2.) Each cascade is born, formed and relaxes within some definite macro-abient, namely the matrix of the target. The properties of this macro-environment as a whole determine the most probable dynamics and the geometry of the cascade.

3.) The experimentally observed ED is closely connected with cascade energetics and, hence, reflects all the stages of the cascade evolution, whose form and dynamics are close to the most probably ones.

4.) The charge of the secondary particles is determined by a) their initial effective charges, b) the electron exchange at the interface cascade-matrix. The electronic properties of such a many-particle system as the cascade is determined also by the macro-ambient, i.e. by the electronic properties of the matrix.

A solution of the problem of ED expressed in terms of the ensemle of the cascades consists of the following three steps: 1. The establishment of the connection between the cascade geometry and the matrix properties (taking into account the presence of the amorphous layer at the target surface). 2. The establishment of the connection between the cascade geometry (it is the most probable form), i.e. its energetics and the ED of the secondary particles regardless their charge state. 3. The establishment of the connection between the electronic characteristics of the matrix and those of the particles involved in the cascade and leaving the target.

Let us consider briefly each of these steps.

1.) The computer simulation of the cascade collisions had shown that the cascade geometry is sensitive to the matrix peculiarities. Heating, the impurities, and denser stacking of the atoms cause a localization of the cascade and hinder the energy drain. The existence of the long-range order opens additional channels for the energy drain from the region of the cascade in the form of high-energy branches (focusons and crowdions).

2.) The existence of the channel for the energy drain or, vice versa the factors localizing the cascade must affect the energy spectra of the secondary particles. The particles in more localized cascades have higher mean energy and smaller energy scatter. The existence of high-energy brances leads to an asymmetrical increase in ED and to an appearence of a high-energy tail. Table 1 summarizes the effect of separate matrix parameters on the shape of cascades and the influence of their shape on ED. It is to be noted that as was shown in [8], the

Table 1

	Cascade geometry	Energy spectrum	Matrix peculiarities
1	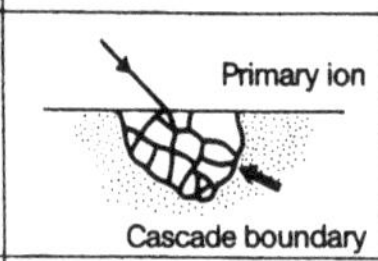	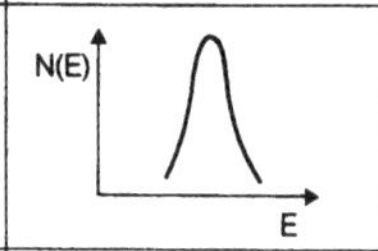	Cascade localization: 1. Temperature increasing 2. Dense amorphous matrix
2	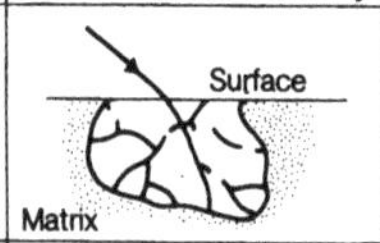	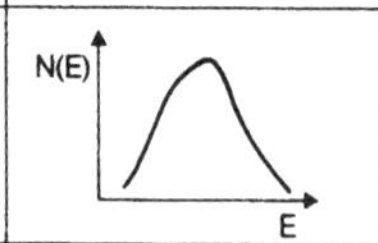	Cascade delocalization: Weak amorphous matrix
3	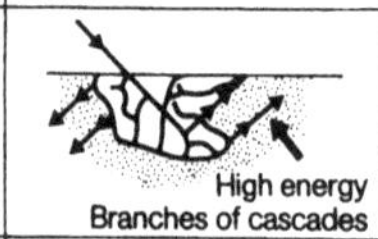	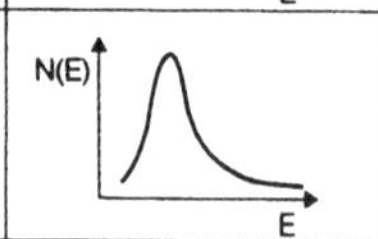	Cascade in a long-range matrix high energy branches of cascade are shown
4	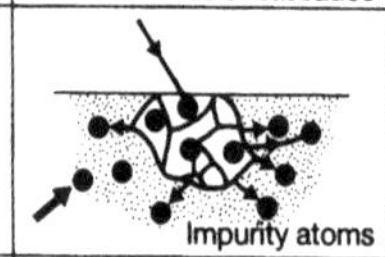	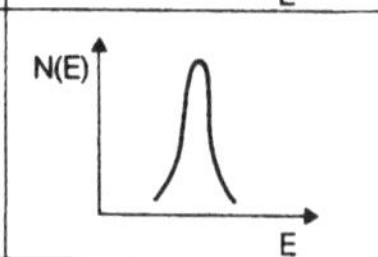	Influence of impurity: The cutoff of high energy branches
5	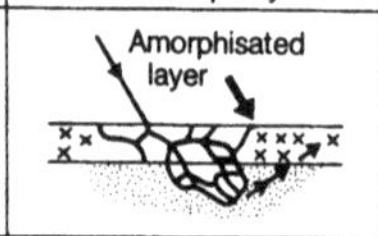	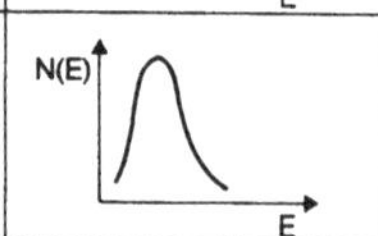	Surface amorphization cascade is divided into two parts
6	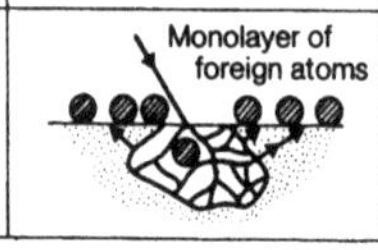	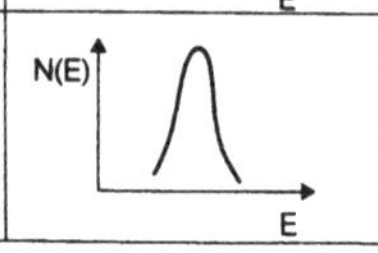	Surface monolayer of foreign atoms is screened high energy branches of cascade

most probably energy of the secondary ions is close to that of defect formation in the matrix.

3.) The charge state of the particles leaving a cascade is affected by the following factors. The type of the bond (ionic or covalent), which determines the effective charges at the matrix atoms. The same effective charges (if considering the collisions in solids as an instant disturbance) will have shifted particles just after collision. Thus the type of the bond determines the initial charge state of every particle involved in a cascade of the collisions. The initial charge state is non-steady-state under these conditions and begins to change. This may occur due to a neutralization and is determined (for positive ions) by the electron concentration in the matrix and in the cascade itself. The final charge state of a particle is determined also by its electron affinity and duration of its stay in the cascade: the longer staying particles lose more information about their initial state. The reconstruction of the electron shells of the particles in a cascade is accompanied by the excitation energy dissipation. The more developed long-range order the matrix possesses, the sooner the excitation disappears. The degree of long-range order determines the shape of the spectrum of the matrix elementary excitation, which are able to stimulate by taking off the excitation of the particles in a cascade. Among the particles which can be involved in

Table 2

	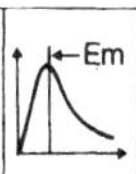 E_m	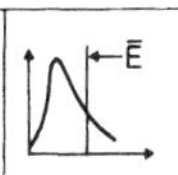 $\bar{E}$	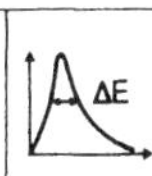 ΔE	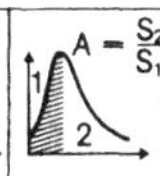 $A = \frac{S_2}{S_1}$	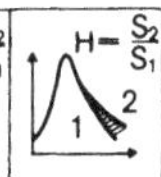 $H = \frac{S_2}{S_1}$	Ionation probability β	
	The most probable energy	Average energy	FWHM	Assymetry of ED	Part of HE particles	β+	β−
I_0 Current density	−	−	+	−	−	+	+
E_0 Primary ionenergy		−	+	−	−		
Impurity atoms		−	−	−	−	Electroneg. + Electropos. −	Electroneg. − Electropos. +
Ed or Ea	+	+	+			+	+
Long-range degree	+	+		+	+	−	−
n_0 Electron concentration						−	+
f Polarity of bond						+	+

+ Correlation − Ant.correlation

this process along with the phonons, also the exitons [9] and the plasmons [10] can be named. These considerations allow us to predict the changes in ED and the ionization probabilities depending on the matrix properties and the incident beam parameters (Table 2).

Discussion

A comparison of the experimentally obtained ED with the data of Tables 1 and 2 reveals their correlation with the structural properties of the matrix. In particular, more disordered matrices show broader ED (compare: poly—monocrystalline Si, "chloride" — "dry" thermal SiO_2). For thermal SiO_2 possessing a definite degree of long range order a characteristic high-energy tail appears in ED. An introduction of the impurity into the matrix (e.g. O into Si) leads to a localization of the cascade and to a narrowing of ED. The charge accumulation and its inhomogeneity due to the corresponding microscopic inhomogeneity in the target matrix [11] cause the deformation of ED up to its splitting into several separate maxima. This situation is often realized near the boundaries between the phases.

References

1. V.G. Litovchenko, R.I. Marchenko and G.Ph. Romanova, Proc. of the 23rd Int. Coll. TH Ilmenau (GDR), Heft 6, p.71 (1978)
2. V.G. Litovchenko, G.Ph. Romanova, "Surface Diagnostique with Ion Beams". Donetsk State University, Donetsk, p.139 (1980)
3. V.Ye. Yurasova and Yu.V. Linnik, ibid. p. 67
4. J.W. Corbet, Surf. Sci. 90, 205 (1979)
5. D.E. Harrison et al., Surf. Sci. 76, 311 (1978)
6. V.V. Kirsanov, "Defects of Crystalls and Their Computer Simulation", Nauka, Leningrad, 1980
7. D.I. Slovetsky, "The Mechanisms of Chemical Reactions in Non-equilibrium Plasma", Nauka, Moscow, 1980
8. V.G. Litovchenko, G.Ph. Romanova, R.I. Marchenko and P.I. Didenko, Proc. of the VII Intern. Conference on Atomic Collision in Solids. Moscow, 1980, Vol. 2, p. 162
9. R. Kelly, "Ion Implantation in Semiconductors and Other Materials", Mir, Moscow, 1980, p. 194
10. G.M. Philippov, Fiz. Tverdogo Tela 19, 1920 (1977)
11. V.G. Litovchenko, "The Physical Fundamentals of the Semiconductor Layer Systems", Naukova dumka, Kiev, 1980

Structural Effects in SIMS at the Depth Profiling of Boron Implanted Silicon

V.T. Cherepin, A.K. Kosyachkov, A.D. Krasyuk, and M.A. Vasiliev

Institute of Metal Physics, Academy Sciences UkrSSR,
SU-Kiev 142, USSR

Introduction

Secondary ion mass spectrometry is currently widely used to study the in-depth distribution of alloying additions in thin films and, in particular, for the in-depth profiling of implanted atoms in semiconducting materials.

The ion implantation and subsequent physico-thermal treatment cause modifications in the physico-chemical state of Si substrate. But the mechanism of secondary ion formation, affecting the results of the analysis by the SIMS method, depends strongly on just this state. Most practical in-depth analyses of implantation profiles ignore this factor. It is assumed that the use of oxygen ions as bombarding particles serves to suppress completely the possible structural or matrix effects of the target under study.

The objective of the present work was to establish a correlation between the physico-chemical state of a semiconducting surface and the secondary ion emission (SIE) intensity under atomically pure conditions and also in the case when the implanted layer is sputtered by oxygen ions under conditions typical for the in-depth analysis.

Experimental

Single crystalline Si plates KEF-10 doped with $^{11}B^+$ (50 keV, the dose 10^{15} cm^{-2}) at room temperature were used as targets. In order to suppress the channeling effect which complicates the character of implanted atoms distribution the doping with B ions was performed in the direction [111] with the use of mechanical scanning of Si samples with respect to the direction of the bombardment.

The physico-chemical state of doped layers was varied by annealing at different temperatures and by additional irradiation by $^4He^+$ ions with the dose 10^{15} cm^{-2}. The data on regimes and sample preparation are given in Table 1. The target heating (the 5-th column in the Table) up to 600°C during the surface irradiation treatment served to avoid formation of vacancy complexes hampering B atoms migration in silicon crystal lattice. The boron SIE was studied in the ion microanalyzer.

Table 1 Treatment of the samples

No. sample	Annealing temperature [°C]	Treatment $^4He^+$			Depth profiling with:
		E [keV]	Φ [cm^{-3}]	$T_{impl.}$ [°C]	
1	300	-	-	-	$^{40}Ar^+$
2	400	-	-	-	
3	500	-	-	-	
4	600	-	-	-	
5	700	-	-	-	
6	no annealing	-	-	-	O_2^+
7	800	-	-	-	
8	800	40	10^{15}	600	
9	800	30	10^{15}	600	

Ar^+ primary ions were used to analyze the samples 1-5, but oxygen ions where chosen for the samples 6-9. Their energy was 5 keV at the current density 10^{-4} A cm^{-2}. It should be stressed that the use of oxygen helps to suppress the effect of the surface physico-chemical state, while the use of Ar ions is most favourable for the study of this effect.

Results

Figures 1 and 2 show the boron range profiles obtained at the sputtering with Ar^+ and O^+ ions. Two conclusions that may be drawn from these data are important for the in-depth analysis of doped semiconductors. The first one implies that the B range profiles in all the samples subjected to the homogenizing annealing (Fig. 1) are close to Gaussian for both types of primary ions. This result agrees well with the LSS theory and also with the data of other authors. Hence the kinetic curves for the emission intensity of boron secondary ions depict well the distribution character of implanted boron atoms.

The second result shows that the integral currents of B^+ secondary ions proportional to the areas limited by the concentration curves differ strongly in cases of sputtering with Ar^+ ions (Fig. 1) and O^+ ions (Fig. 2) despite the fact that the concentrations of implanted boron ions were kept the same in all the samples.

Thus, for instance, when Ar ions were used for the analysis, the B integral fluxes were changed within one order of magnitude with the increase of the annealing temperature from 300 °C to 700 °C (samples 1-5, Fig. 1). The use of oxygen does not help in a complete suppression of the lattice effects (Fig. 2). Moreover, a strict correlation is observed between the growth of integral ion flux and the increase of the physico-chemical influence of the doped layer structure: the

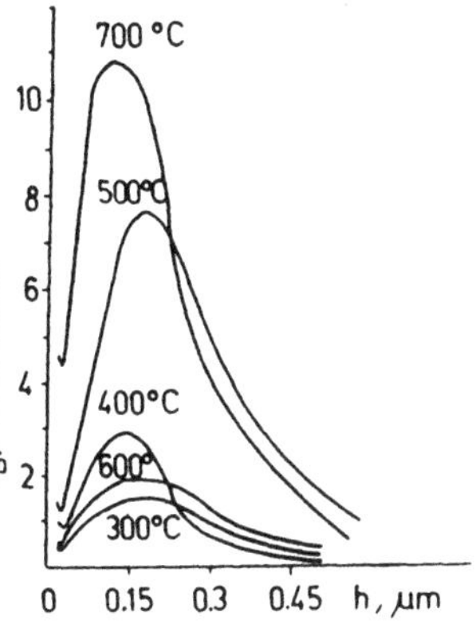

Fig.1 Depth profiles of samples 1-5 obtained at Ar^+bombardment

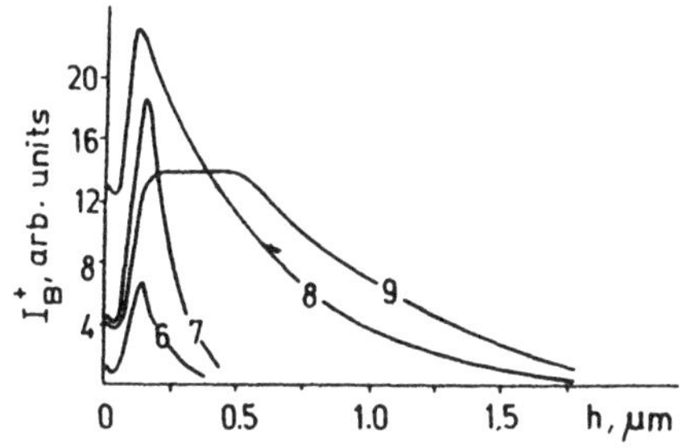

Fig.2 Depth profiles of samples 6-9 obtained at O^+ bombardment

initial state (6), annealing at 800 ^oC (7), similar annealing with subsequent irradiation by helium ions (8,9).

Thus our second important result from the analysis of the data in Figs. 1 and 2 involves the fact that the emission intensity of boron secondary ions depends not only on concentration of B atoms, but also on the physico-chemical properties of the layer under study which are responsible for the mechanism of ion formation.

Discussion

Now we consider some of the obtained results from the point of view of SIMS application to the depth profiling. As it was mentioned above, for the annealed samples (1-7) the distribution of implanted B atoms is close to the Gaussian. Profiles obtained on the samples (8) and (9) (Fig. 2) subjected to additional treatment with $^4He^+$ ions are distinguished by the presence of "tails" in B distribution due to the radiation-induced diffusion. We estimated the diffusion coefficient as $D = (8\text{-}9)\cdot 10^{-14}\ cm^2\cdot s^{-1}$ this being in a good agreement with the data obtained from the works where the diffusion parameters were determined for the case of surface irradiation by protons. Hence it may be concluded that, similarly to the case of proton irradiation, the bombardment of Si with He ions causes formation of point defects enhancing the diffusion process. It should also be mentioned that the presence of "plato" in B distribution for the sample (9) (Fig. 2) comes from the fact that the spacing between the maximum of the defected region, created by He ions injected with the energy 30 keV, and the maximum in the distribution of implanted boron ions is not as large as for the sample (8) where He ions had the energy 40 keV.

Thus the character of obtained profiles may be explained well enough in terms of generally accepted notion about the mechanisms of implantation and diffusion in doped semiconducting layers.

However, it may be seen from the figures that the boron secondary ion current does not correspond adequately to the concentration of this element in the doped layer. This implies a connection between the physico-chemical state of the layer and the secondary ion emission. It is known that for pure metals, metal alloys and compounds the SIE depends strongly on the crystallographic and electronic structure of the object under study. Let us compare in this connection some characteristics responsible for electrical properties of implanted semiconducting layers and the SIE characteristics.

Figure 3 shows the integral yield of boron secondary ions as a function of the annealing temperature for the samples (1-5). Here also the temperature dependence of the surface concentration of active centers, n_s, is given, that is, the dependence of concentration of B atoms transferred from interstitials into the modes of Si crystal lattice. It may be seen that the temperature dependence of active center concentration correlates with that of the B integral flux within the accuracy of the experiment. Thus the drop in active center concentration in the range 500-600 ^{o}C known as the inverse or negative annealing is accompanied by the corresponding drop in B ion integral flux.

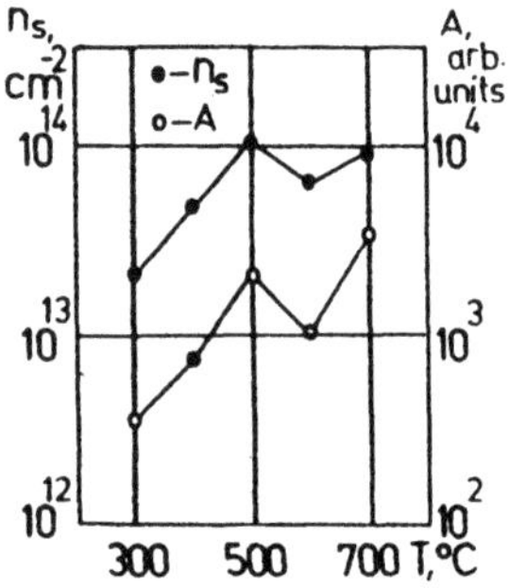

Fig. 3 Active atoms concentration and integral current of B^+ as a function of annealing temperature of samples 1-5

Correlation between these dependences is also observed for the samples subjected to the irradiation treatment. The areas under the B profiles for the samples (8) and (9) implanted with He ions (Fig. 2) are equal to each other with the accuracy of experimental error. We have used the 4-probe method for additional measurements of active center concentrations and have found that they are rather similar for these samples and amount to $8.3 \cdot 10^{19}$ cm^{-3} and $8.6 \cdot 10^{19}$ cm^{-3}, respectively.

Similar interconnection between the growing number of active centers and the rise in the integral yield of B secondary ions may be found from the analysis of the data published in the literature. There are works on SIMS depth profiling of implanted B atoms carried out with oxygen ions, while the 4-probe method was used to study relevant distribution profiles of active centers, that is, the concentration of boron atoms in substitution sites. The initial Si samples were doped with B ions in the direction [111] at room temperature and were then subjected to the homogenizing annealing, as in our experiments.

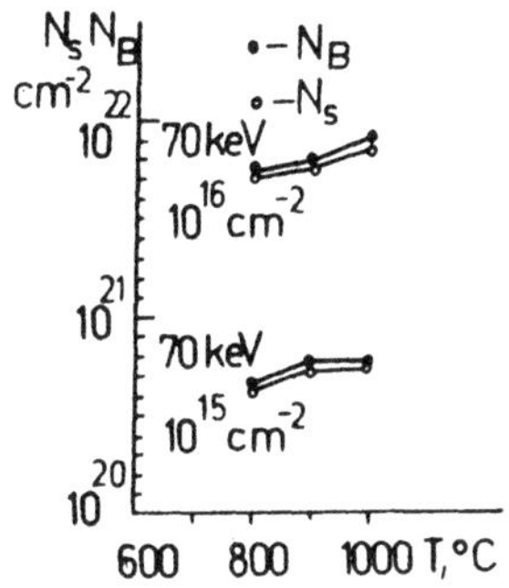

Fig. 4 Integral concentration of active atoms and integral current of B^+ as a function of treatment of the samples [1]

While analyzing these results we estimated the integral number of active centers and the integral yield of B secondary ions under different regimes of thermal treatment. The obtained data are shown in Fig. 4. The correlated increase may be seen on the total yield of B ions (N_B) and in the total number of active centers (N_S) with the rise of the annealing temperature.

Hence the analysis of published data supports the conclusion of the present work about the interconnection between the structure and the electronic and emissive properties of semiconducting materials. The transfer of B atoms from interstitial positions into the crystal lattice nodes and the corresponding change in their charged state (the active centers formation) is accompanied with the increased ionization probability of B atoms at the sputtering of doped layer.

The surface physico-chemical state of semiconducting materials affects strongly the emission character of B secondary ions. This may be of importance when using primary ions of both inert and active gases. Hence the treatment can significantly influence the emission intensity of secondary ions and should account for the depth profiling of semiconducting materials using SIMS method.

References

1. W.K. Hofker, H.W. Werner, D.P. Oosthoek and N.L. Koeman, Appl. Phys. 4, 125 (1974)

Comparison of Compositional Thin Film Depth Profiling Obtained by SIMS, IIR and AES

E. Hauser, G. Hobi, K.H. Guenther, and E. Brandstaetter[1]

BALZERS AG., FL-9496 Balzers, Fürstentum Liechtenstein

Introduction

Ion etching used in connection with surface-sensitive analytical techniques is a common method of determining the chemical composition of thin films and substrates. The method of surface analysis employed depends on the aims of the investigations and the information required. In this paper the advantages and limitations of AES, SIMS and IIR for the analysis of dielectric materials will be discussed. Each of the three analytical methods has been employed for parallel compositional investigations in a broad-band antireflection coating as well as in a dichroic beamsplitter.

Ion Etching Properties

The results of any surface-sensitive analytical technique used with ion etching are very dependent on the distribution of the beam current density. Irregular distribution of the ions may cause a non-uniform removal of the layers, thus leading to alterations of the elemental signal as compared to the true elemental composition. In a similar way, elemental information from the crater edge can influence the depth profiles, if the edge shows a slope rather than a step-like shape. A slope will develop particularly when an oblique angle of the incidence for the ion beam is chosen. The influence of particles emitted from the slope must be taken into consideration especially for depth profiling by SIMS or IIR. Besides these instrumental effects there are other artifacts which are attributed to the sputtering process. Material can be scattered from the rim and the walls to the center of the pit. Non-uniform removal can further be caused by preferential sputtering because of different phases or compounds contained in the sample. In addition, the original material distribution may be altered in the specimen due to impact and diffusion processes known as atomic mixing, preferential sputtering and recoil implantation. Although all these effects cannot be avoided completely, their influence can be reduced by an appropriate choice of the primary ion current density distribution. However, the requirements are sometimes contradictory. As an example, for reducing the effect of preferential sputtering, the primary ion energy should not be too low, whereas low primary ion energy is necessary for reducing the effects of atomic mixing.

---

[1] on leave from Institut für Werkstoffprüfung und Metallkunde, Montanuniversität, A-8700 Leoben, Austria

Instrumental Precautions

In order to obtain relevant depth profiles from dielectric coatings and glass substrates, some precautions should be taken. First, it is highly desirable to have an ion source available which allows the controlled adjustment of the ion beam profile as well as its energy within a wide range. These requirements are met by the newly developed Penning-type ion source of the Balzers IEU 100, where the ion beam profile can be varied continuously from a peak-like shape to a sharp-edged rectangular profile [1]. The energy of the primary ions ranges from 3 keV to 8 keV, and both inert gases (argon, nitrogen) and oxygen, as well as mixtures of the two can be used.

The use of oxygen ions as the sputtering species is most valuable in the investigations of oxides, as it prevents them from being reduced to suboxides or even to metallic states under the influence of the impinging ion beam. Moreover, it is also known that the excitation probability for optical radiation and the emission yield for secondary ions are much higher for oxides than for their metals. Thus, in the case of weak signals stemming from metals or suboxides it is highly advisable to add some oxygen to the sputtering gas in order to enhance the signal to noise ratio.

In the case of depth profiling by detecting IIR, the use of an interference filter or a manually operated monochromator allows the continuous tracing of the intensity of only one spectral line at the time. In order to simultaneously obtain the depth profiles of several elements, one must use a programmable monochromator which is able to jump from one to the next spectral line within a rather short time interval. Of course, a multichannel or a point plotter which records the intensities of the respective spectral lines, yielding depth profiles comparable to those familiar multiplexed profiles obtained with Auger spectrometers, is also necessary.

Depth profiling of dielectric materials is certainly more problematic than of metals or semiconductors, because of charging effects caused by emitted ions and electrons. Whereas these disturbing effects can be reduced by special arrangements of the sample and shielding stops at a certain level tolerable for AES and IIR depth profiling, the use of an auxiliary electron gun is unavoidable for charge compensation in the case of SIMS. Such an electron gun must allow the continuous control of the electron energy up to at least 500 eV as well as of the electron current which might be as high as the incident primary ion beam current.

Experimental Results and Discussion

A broad-band antireflection coating and a dichroic beam-splitter were investigated by AES, SIMS and IIR. The AES depth profiling was performed with a PHI instrument which employs a cylindrical mirror analyzer (CMA). The SIMS and IIR depth profiles were obtained on a Balzers IEU 100 [2]. For all depth profiles, Ar^+

ions were used as the sputtering species. The energy E_p as well as the angle of incidence Ø of the primary ions were different for the three techniques: AES (E_p = 3 keV, Ø = 30°), SIMS (E_p = 6 keV, Ø = 50°), IIR (E_p = 8 keV, Ø = 70°). During the ion bombardment, the sample surface was flooded with electrons of appropriate energy, eliminating the charging of the dielectric samples.

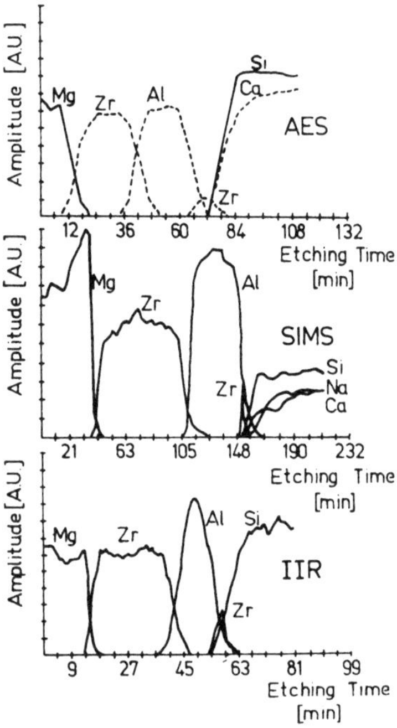

Fig. 1 Comparison of AES-, SIMS- and IIR-depth profiles (broad-band antireflection coating $MgF_2/ZrO_2/Al_2O_3/ZrO_2$ on glass)

Figure 1 shows the depth profiles of a broad-band antireflection coating consisting of $MgF_2/ZrO_2/Al_2O_3/ZrO_2$-layers on a glass substrate. At first glance, the three profiles obtained by the different methods look very similar, but the sharp transitions within the elemental profiles indicating the layer interfaces are most striking in the SIMS profile. A closer look reveals an increase of the etching time ratios for the ZrO_2 and the Al_2O_3 layers as taken from the respective elemental profiles in the AES, SIMS and IIR analyses. This increase goes parallely with the increasing primary energies of the incident beams. In Fig. 2, similar results were found with a three layer dielectric optical coating ($TiO_2/SiO_2/TiO_2$), also vacuum deposited onto a glass substrate. The aforementioned increase in etching time ratios with increasing primary energy is not as pronounced for the TiO_2/SiO_2-layers as in the other case with the ZrO_2/Al_2O_3-layers. However, in both cases the ratio increase in the etching times for the two respective layer pairs can be qualitatively explained by their sputtering yields. It is known [3] that ZrO_2 and SiO_2 have higher sputtering yields than Al_2O_3 and TiO_2, respectively. As this finding cannot be explained

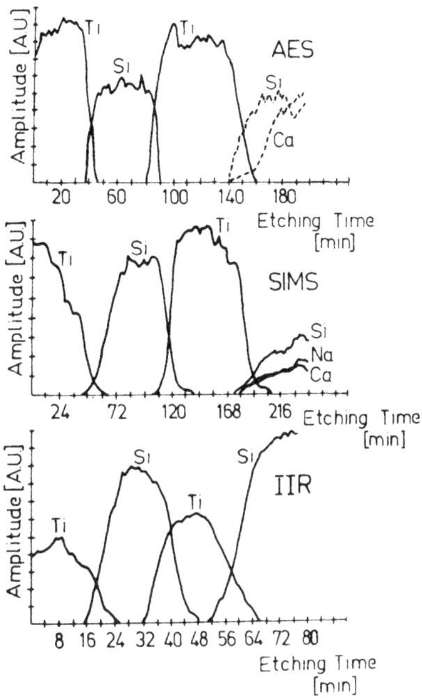

Fig. 2 Comparison of AES-, SIMS- and IIR-depth profiles (dichroic beamsplitter $TiO_2/SiO_2/TiO_2$ on glass)

properly by the molecular masses and the specific weight of the compounds involved, it is more likely that chemical binding plays a dominant role which can be deduced from their heat of formation. In this respect, a higher primary ion energy (as used in our case with IIR) may eliminate the differences in sputtering yields of the oxides, thus leading to a more realistic thickness ratio. However, higher ion energies will cause atomic mixing at the layer interface regions, yielding broader transitions from one elemental depth profile to the other.

Conclusion

Depth profiling with AES, SIMS and IIR, each used for parallel compositional investigations in optical coatings, gave similar results. The preferential use of a particular method over the others depends mainly on the material under investigation, the information required, and last but not least also on the experience gained in the use of the analytical equipment.

We would like to thank Dr. H. Bach from Schott & Gen., Mainz for practical hints and valuable discussions.

References

1. K.H. Guenther, E. Hauser, G. Hobi, P.G. Wierer and E. Brandstaetter, these Proceedings
2. A detailed description is given by H. Bach, Beitr. Elektronenmikroskop. Direktabb. Oberfl. 12/1, 277 (1979)
3. R. Kelly and N.Q. Lam, Radiat. Eff. 19, 39 (1973)

Part IV

Quantification

Quantifications of SIMS

G.H. Morrison

Department of Chemistry, Cornell University,
Ithaca, N.Y. 14853, USA

Introduction

The title of this paper refers to the quantification of SIMS or secondary ion mass spectrometry. More specifically we will be discussing the ion microprobe, a form of SIMS. The ion microprobe is presently one of the most sensitive methods for the *in-situ* characterization of elemental distributions in samples of limited dimensionality, often with detection limits in the parts per million range. The capabilities of lateral elemental imaging for defining microfeatures of interest(1 μm spatial resolution), and of elemental depth profiling with resolution of less than 100 Å make the ion microprobe unsurpassed in the field of three-dimensional trace element localization.

The expansion of the general utility of the ion microprobe, however, has been hampered by the difficulty in obtaining quantitative measurement of concentration in samples not extremely well characterized. This difficulty is due to several considerations: 1) secondary ion yields vary over four orders of magnitude from element to element, 2) each element's ion yield is affected by the matrix in which it is contained, and 3) instrumental parameters and ion collection efficiencies vary from case to case. These factors have, in general, limited the ion microprobe analysis of unknown samples to qualitative elemental characterization.

The methods of quantification in ion microprobe analysis will be briefly reviewed followed by a description of studies in our laboratory to evaluate the various methods. Finally, recent efforts will be described for the quantification of ion images using computerized digital image processing of microfeatures and fields of view.

Two basic approaches have been developed for the quantification of secondary ion intensities. The empirical calibration approach utilizes external standards to derive secondary ion ratios in terms of sensitivity factors which calibrate the unknown sample elemental secondary ion intensities [1]. There are a number of modifications of this approach, including matrix correction methods [2,3].

The other approach to quantification is the use of semi-theoretical models. Various researchers have tried to develop theoretical models that simulate ion-surface interaction that predict concentrations through the use of internal standards. The most widely used semi-theoretical method to date has been the local thermal equilibrium (LTE) model as exemplified by ANDERSON and HINTHORNE's computer program CARISMA [4]. It uses the Saha-Eggert equation describing a plasma at thermal equilibrium. Elemental concentrations are calculated by using two or more internal standards to computer fit the two unknown parameters, temperature and electron density. After these parameters are established, the ionization probabilities of unknown elements are calculated and their respective ion intensities converted into elemental concentrations. A number of simplified LTE modifications have been proposed by SHIMIZU et al. [5], MORGAN and WERNER [6], RUDENAUER, STEIGER and WERNER [7], and SIMONS, BAKER and EVANS [8]. Other physical models have also been proposed by a number of investigators including a nonthermal equilibrium model [9], a molecular orbital model [10], a quantum mechanical model [11], and one based on auto-ionization [12].

Table 1 Problems in Quantification of SIMS

A. Physical model approach
   1. Secondary ions formed by several processes.
   2. Must compensate for instrumental discrimination.
B. Empirical calibration approach
   1. Only works when sample and standard match in matrix composition.
   2. Must reproduce experimental conditions.
   3. Need homogeneous standards.

Table 1 summarizes the basic problems that have hindered quantitative results to date. The physical model approach has yielded results only good to a factor of 2 or 3. One explanation is that the secondary ions are formed by several processes, so that no one model defines the situation. A second limitation to the accuracy of these computational methods has been the failure to include the effects of instrumental discrimination on the relative ion intensities.

The empirical approach has been used in most physical analytical methods with considerable success; however, there are a number of considerations of particular importance in ion microprobe analysis. Firstly, the approach only works when the sample and standard match in matrix composition. Secondly, one must reproduce the experimental conditions in order to relate the standard measurements to the unknown samples, and finally, one needs homogeneous standards at the micrometer level.

Semi-Theoretical Approach

A number of studies have been performed in our laboratory at Cornell University to address some of these problems. First, an investigation was performed to evaluate the semi-theoretical model approach [13]. A shortcoming of previous examinations of the accuracy of these computational methods has been the failure to include the effects of instrumental discrimination on the relative ion intensities. Furthermore, no direct comparison of these methods by using a single data set obtained from one instrument has previously been made, making comparison of claims of accuracy difficult to perform.

The evaluation study by RUDAT and MORRISON [13] was designed to serve several functions. Most importantly, the effects of instrumental discrimination were virtually eliminated so that unencumbered accuracy of the investigated models could be clearly seen. If the overall accuray is not improved by removing these effects, then the analytical usefulness of the methods must be brought into question. Secondly the results of different semi-theoretical models were directly compared using the same data sets, so that no confusion regarding different analysis conditions in different laboratories can exist.

The models examined included the simplified LTE [8], and those of WERNER et al. [6,14], SCHROEER [11,15], GRIES and RUDENAUER [16], JURELA [17], and a normalization procedure. A CAMECA IMS-300 instrument was used to study a number of standard metal matrices including low alloy steels, copper, and aluminium. The abundance-corrected integrated intensities were corrected for instrumental discrimination. The details of the computer calculations and other experimental details can be found in the original papers [13,18].

Figure 1 presents the distribution of error factor for some of the models evaluated. Also included are results using an oxygen backfill in the source chamber. The results have been adjusted so that the area within each bar accurately represents the population for that error factor range. Negative error factors are underestimations, positive factors are overestimations. Crosshatched areas are the contributions from the standard elements. The best results were obtained with LTE-3, which normalized the sum of the concentrations.

To summarize this study, it demonstrates that even with instrumental effects removed from the data, semi-theoretical methods of quantification in SIMS are inadequate. This is particularly true for approaches that do not remove additional degrees of freedom, namely, the methods not requiring "oxide corrections" or normalization to 100 %. These analytical approaches provide the true test of the applicability of the models and the methods to the analytical situation, and it is clear that none of the presently useable models are adequate.

Empirical Approach

As mentioned earlier, limitations to this approach to quantification result from the non-availability of homogeneous standards that match the matrix composition of the unknown samples.

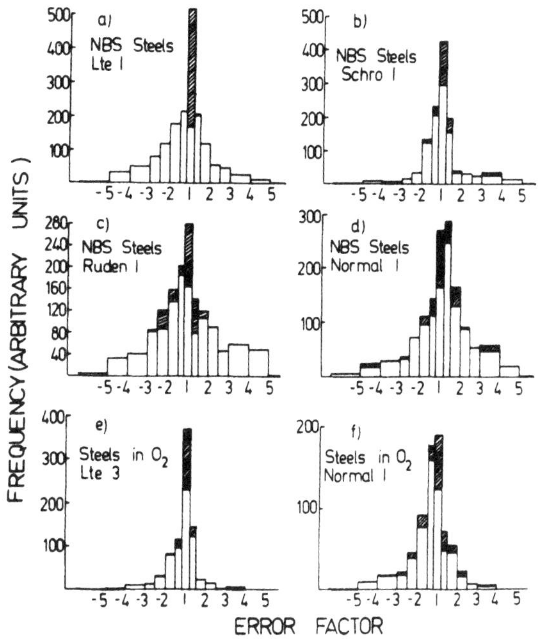

Fig. 1 Error factor distribution in calculated concentrations of elements in NBS steels by (a) LTE I program (residual vacuum); (b) by SCHROEER I program (residual vacuum); (c) by RÜDENAUER I program (residual vacuum); (d) by Normalization I program (residual vacuum); (e) by LTE 3 program (oxygen backfill); (f) by Normalization I program (oxygen backfill). Crosshatched areas are the contributions from the standard elements. Negative error factors are underestimations [13]

First, let us examine the problem of homogeneity. Most available standards have been certified for bulk concentrations by methods that are insensitive to concentration gradients on a microscale. Since microprobe analysis does not allow homogenization of large amounts of material to eliminate sampling problems, as is the common practice in bulk techniques, sampling error may constitute a large fraction of the total analytical error if these standards are indiscrimately applied to microprobe measurements.

In our laboratory we have used the imaging capability of the instrument to evaluate the degree of heterogeneity of samples. Fig. 2 shows the distribution of various elements in NBS SRM-664 alloy steel. The field of view is 115 μm diameter $^{56}Fe^+$, $^{27}Al^+$, $^{48}Ti^+$, and $^{51}V^+$. Aluminium is the most heterogeneous element in the sample, located primarily in small, widely spaced inclusions. The distribution of Ti is also characteristic of Nb and Zr. The inclusion density is much greater than for Al, although on a microscale the element is still quite heterogeneously dispersed. The V is considerably more homogeneous than the others; however, inclusions are still easily visible.

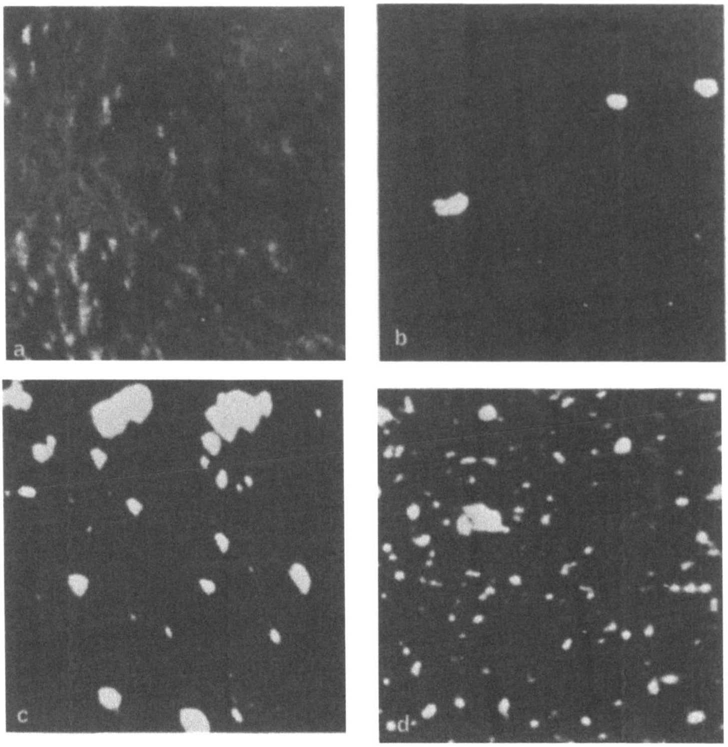

Fig. 2 Ion images from NBS SRM-664 steel: (a) $^{56}Fe^{+}$; (b) $^{27}Al^{+}$; (c) $^{48}Ti^{+}$; (d) $^{51}V^{+}$. Fields of view: 115 μm [20]

These images provide qualitative information but a more rigorous approach is needed. Our laboratory has developed a sampling constant concept to represent the level of heterogeneity in microanalysis. The basic assumption is that the element of interest is located in the sample as randomly dispersed inclusions. An equation was developed to determine the sampling constant for a given element. It utilizes size, distribution, and intensity information from the ion image to calculate the sampling constant (K_S) [19].

Digital image processing techniques were used to make the measurements [20]. To a first approximation, a greater number of inclusions results in a smaller sampling constant, which indicates a more homogeneous material. Of somewhat less importance are size and intensity of the inclusions, and the randomness of their dispersion throughout the sample. The resultant sampling constant can be used to estimate the precision of a measurement obtained from a particular sampling area, or it can be used to determine the required sampled area to achieve a desired precision. It can also serve as a basis for comparison of the sampling situation of one element to another in a particular sample, or to compare the same element in different sample materials.

Table 2 Calculated sampling constants (K_S, μm) for alloying elements in NBS SRM-664 low alloy steel [20]

Element	Al	Ti	V	Cr	Mn	Zr	Nb
K_S	77.5	25.5	19.4	14.0	7.2	35.1	34.9

Table 2 lists the sampling constants for a number of elements in NBS SRM-664 steel, thus providing a more objective approach to microanalysis. As an example, the error expected for 10 analyses of 4000 μm^2 areas of Ti would be 24.8 %. Thus, if one wants quantitative analysis of heterogeneous species in a sample, the sampling constant approach provides the necessary guidance. Using this sampling approach with the Matrix Ion Species Ratio Method (MISR) developed in our laboratory we were able to achieve an accuracy of 10 % relative error for a variety of elements in steels, aluminium alloys, and copper alloys [21].

In addition to the availability of homogeneous standards, we mentioned earlier the need for a good sample/standard matrix match in the empirical approach to quantification. To date the only well characterized standards that are available for ion probe analysis are the NBS SRM-600 series of steels and the RM-30 glass series. Fortunately, we now have at our disposal a means to remedy the lack of standards through the use of the technique of quantitative ion implantation [22]. This technique is used in the preparation of electronic semiconductor devices.

In the ion implantation process an ion beam of almost any element can be created, mass filtered and accelerated to substantial energies, normally between 10 and 600 keV, and then "implanted" into the near surface region of a sample material. Additionally, the ion beam is rastered to ensure uniformity of doping in the lateral directions, and more critically, the current delivered to a fixed area of the sample stage is monitored and integrated to provide an accurate and controllable dopant concentration in the implanted sample. Although this concentration is originally known only as the amount of implanted-ion fluence (cm^{-2}), the use of the SIMS depth profiling capabilities to determine the depth distribution and the integrated ion signal, combined with an accurate measurement of the depth of analysis allows the necessary conversion to concentration in atoms/cm^3 [23,24,24a].

Thus is is potentially possible to implant any element into any matrix to provide a known concentration of the dopant element. Both the implant energy and the fluence are the important parameters to be established. The implant energy must be sufficient to assure that the near-Gaussian depth distribution of the implant is essentially subsurface, i.e., the following tail of the distribution does not extend to the sample surface to any large degree. Additionally, it should not be so deep that the SIMS depth profile takes an excessive amount of time. Energies in the 100 to 300 keV range satisfy both of these requirements. To decide the fluence to be implanted, as a general rule-of-thumb it may be assumed that the maximum concentration of the implanted species will be approximately 10^5 times the implant fluence.

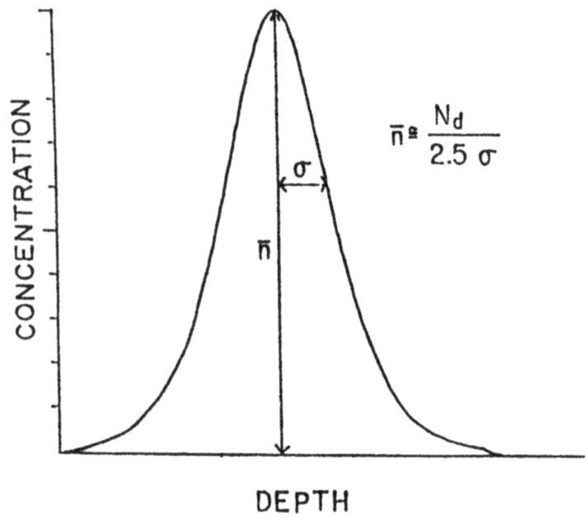

Fig. 3 Parameters for the use of the Gaussian approximation for the calculation of implanted concentrations [23]

As Fig. 3 shows, the distribution of the implant is Gaussian (in small polycrystalline or amorphous materials). The actual maximum concentration of the implanted element may be calculated from the fluence by several methods. To match the concentration obtained to that desired, it is necessary to know in advance the standard deviation of the Gaussian concentration distribution and make use of the relationship:

$$\bar{n} = N_d/(2.5\sigma) \qquad (1)$$

where $\bar{n}$ = maximum concentration (atoms/cm^3), N_d = implant fluence (ions/cm^2), σ = standard deviation of the Gaussian distribution (cm). It is equal to one-half the peak width at 0.606 maximum. The standard deviation of the implant distribution may be calculated from the LSS theory [25]. Tables exist for certain implant/substrate combinations as a function of implant energy. Such calculations are generally accurate to within about 30 %.

After the samples have been ion implanted, there are several methods, with varying degrees of accuracy, with which to determine the actual maximum concentration. The method used in our study involves integrating the secondary ion signal derived from the implanted species and combining it with the measurement of the analysis depth to determine the concentration of the implant. It is assumed here that the SIMS ion intensities are linear with concentration and that an accurate measurement of the analysis crater is possible. It is also assumed that the sputtering rate is constant throughout the analysis.

The pertinent parameters for this method are shown in Fig. 4. With the aim of calculating the peak concentration of

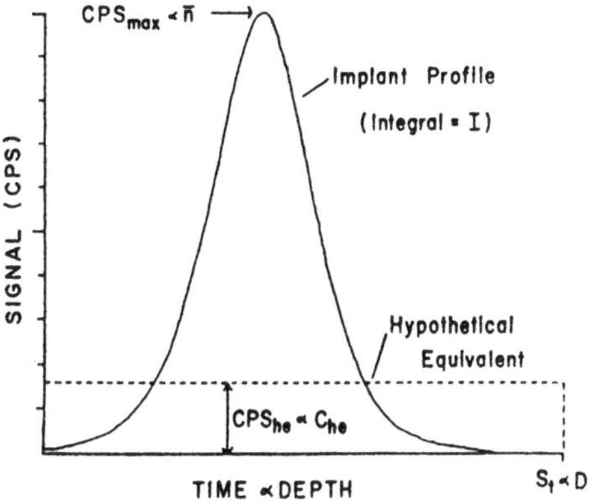

Fig. 4 Simulated SIMS depth profile of an ion implanted standard material showing the pertinent parameters: CPS_{he} and C_{he} are the hypothetical equivalents of the integrated secondary ion signal and the average concentration of the implanted species in the profiled volume, respectively [23]

the implanted standard in mind, it is inconvenient to perform calculations which depend on the exact shape of the depth distribution of the implant. This distribution is normally very close to a Gaussian function but may be somewhat distorted due to the effects of channeling which can occur during implantation. Therefore, with the assumption of linearity of the SIMS signal with concentration, the implanted quantity may be considered as hypothetically equivalent with an integrated ion signal, as reported earlier [24a], which is equal to that of the actual implant distribution. The CPS_{max} is proportional to $\bar{n}$ and the CPS_{he} is proportional to C_{he}.

If the hypothetical equivalent is considered to have a concentration which is homogeneous with depth and which ends abruptly at a depth D, which is defined as the measured depth of the analysis crater (cm), it may be seen that its concentration (atoms/cm^3) is equal to the ion implanted fluence, N_d, divided by D. Therefore the implanted concentration at the maximum $\bar{n}$ is defined by:

$$\bar{n} = C_{he} \cdot \frac{CPS_{max}}{CPS_{he}} = \frac{N_d}{D} \cdot CPS_{max} \cdot \frac{S_t}{I} \qquad (2)$$

where C_{he} = concentration of the hypothetical equivalent, CPS_{max} = maximum signal from implant profile, CPS_{he} = signal from the hypothetical equivalent, N_d = fluence, S_t = analysis time, and I = integrated implant signal (CPS^2).

Since the accuracy of the implant fluence is better than 5 %, the accuracy of such a calculation is dependent primarily on the depth measurement. It is therefore our belief that the concentrations found by such a method are accurate to better than 15 % as a conservative estimate. Additionally, such a method of calculation is widely applicable to many types of implanted materials because of its independence from the shape of the implanted distribution.

The use of ion implanted standards for quantification of SIMS can be taken a step further. Owing to the characteristic shape of the depth concentration distribution of an ion implanted species, it is possible to distinguish the secondary ion signal arising from the implant from that due to the same element originally present in the sample. In the absence of mass spectral interferences, it is therefore possible to use ion implantation to perform solid state standard addition analysis to obtain quantitative elemental concentrations in microsamples such as epitaxial layers, or in a particular microfeature of a surface [23].

Image Quantification

Until recently, the major emphasis in quantification of SIMS analysis has been the use of bulk-type correction schemes for local analyses and depth profiling. However, the imaging mode of SIMS offers much to the study of elemental distributions in the surface of a sample. Here we shall restrict ourselves to an ion microscope analysis producing a large amount of "multi-dimensional" information, both spatial and elemental, which can be assimilated only qualitatively by visual interpretation. Recent efforts in our laboratory have shown that it is feasible to extract quantitative information from ion images using the method of computerized digital image processing [26-28].

An ion image normally consists of a 250 μm diameter field of view with a 1 μm point-to-point resolution. Thus, an ion image potentially delivers simultaneously over 50,000 information bits where each bit is related to the number of secondary ions of a particular mass sputtered from a particular point on the sample. Using computer techniques it is possible to perform multielement micro-feature and field analysis of heterogeneous solids.

Quantitative evaluation of ion images requires the ability to evaluate each image at the picture element (pixel) level and then to recombine pixels to yield meaningful information about microstructures within the image field. Previously, images have been recorded photographically, digitized off-line with a microphotodensitometer, and then processed by digital image processing techniques [28]. We have recently developed a new system for direct acquisition and digitization of ion images using a low light level silicon intensified target camera coupled with an image intensifier (ISIT). The ion image produced on the fluorescent screen of the ion microscope is viewed by the camera for real-time digital image processing. The system, microscopic image digital acquisition system (MIDAS), facilitates high-speed acquisition of ion images at low light levels, real-time digitization and image processing, and storage in a format compatible with off-line digital image processing routines [29]. The process of image analysis is summarized in Table 3, including steps of acquisition, amendment, analysis, and display. In the amendment step one can convert image intensity to ion intensity, and with the availability of standards, to concentration.

Two approaches to ion image quantification have been evaluated in our laboratory [30]. The *first* is a theoretical model correction of the one-parameter LTE type requiring one internal standard, similar to that used by STEIGER and

Table 3 Procedure for Image Analysis

Acquisition:

- Direct digitization
- View in real-time
- Archive

Amendment:

- Correct for instrumental artifacts
- Convert image intensity to ion intensity to concentration

Display:

- Contour plot
- x-y intensity plot
- 3-D contour
- 3-D full image

RÜDENAUER [31], but incorporating several modifications. The *second* is an application of the matrix ion species ratio method (MISR), a sensitivity factor approach by GANGEI et al. [21], again, with some modifications to facilitate its use with ion images.

A series of ion images for various constituents of NBS-664 steel was recorded on electron film and digitized via a scanning microphotodensitometer mentioned earlier. Results for ^{48}Ti, ^{51}V, and ^{52}Cr are summarized here but the original study should be consulted for details [30].

Table 4 LTE and MISR Corrected Results for NBS SRM-664 [30]

element	certified conc. [%]	relative error[a] [%]			expected sampling error [%]
		LTE(Fe)[b]	LTE(F)[b]	MISR	
Ti	0.24	-109	-156	-30	48
V	0.10	+77	+40	+46	34
Cr	0.06	-50	-22	0	24

[a] Integrated image area compared to certified bulk concentration
[b] Internal standard element

Results for the LTE calculations shown in Table 4 are generally within the range of accuracy expected from this type of algorithm. As mentioned earlier, LTE corrected results are quite often in error by a factor of 2 or more, but the ability to apply a semi-theoretical approach of this type to a sample with little *a priori* knowledge other than one internal standard element makes it useful for providing semiquantitative concentration information.

The MISR corrected results shown in Table 4 are overall a much closer fit to the certified values, as was anticipated due to the success of this method when applied to bulk-type analyses [21]. These results are also very close to, or within, the ranges of error expected from sampling uncertainties. The apparent success of the method is its ability to compensate for changes in sampling environment by using a varying set of relative sensitivity factors.

Table 4 also contains the results for the LTE correction using implanted fluorine as an internal standard element. It is interesting to compare the results from different choices of internal elements.

Figure 5 shows a three-dimensional representation of the ^{48}Ti, $^{51}V^+$, and $^{52}Cr^+$ images before and after correction to concentration by an LTE calculation using iron as the internal standard. Although absolute numerical values are less accessible in this type of output, the combination of qualitative and quantitative information present makes possible an evaluation of relative sample heterogeneity and concentrations which is often difficult to obtain from an intensity-modulated display, in addition to a picture of the spatial relationships and the microfeatures.

This is just one example which should suggest to the reader the multitude of possibilities for the application of digital

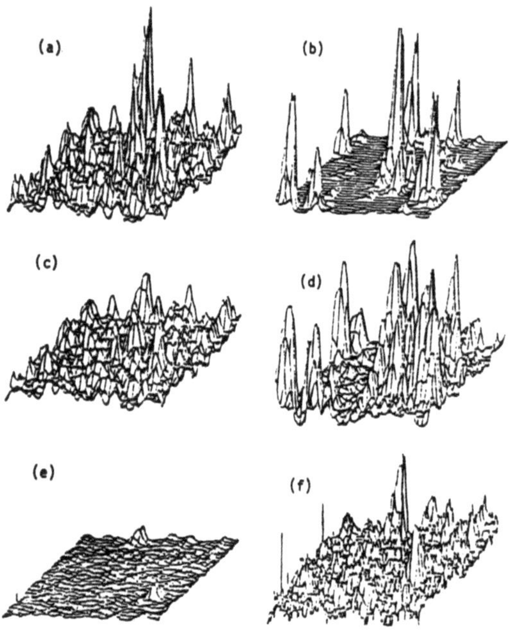

Fig. 5 (a), (c), (e): Uncorrected 3-D plots of $^{48}Ti^+$, $^{51}V^+$, and $^{52}Cr^+$ in NBS SRM-664 steel. (b), (d), (f): Corrected plots of $^{48}Ti^+$, $^{51}V^+$, and $^{52}Cr^+$. Field of view is 100 μm [30]

ion image analysis to materials characterization problems. Digital image processing is a dynamic, fast growing area which will have a dramatic effect on the future of ion microanalysis.

In summary, this paper attempts to show the present status of quantification of SIMS. While the semi-theoretical approach is, in principle, easier to apply, the evaluation shows that more work needs to be done to make it an analytically useful method. In contrast, the empirical approach can be used for quantitative results on the order of 10 to 15 % accuracy provided attention is paid to heterogeneity and standards of comparable matrix composition to the unknowns are available. In implantation appears to be a convenient and practical solution to this problem. Finally, digital image processing techniques now make it possible to extract useful information from ion images for microfeature analysis. Thus, with further improvements in quantifications SIMS can be equally competitive with other techniques.

Acknowledgements

This work was supported by the National Science Foundation, the Office of Naval Research, and the National Institutes of Health.

References

1. J.A. McHugh, in "Methods of Surface Analysis", S.P. Wolsky and A.W. Czanderna, eds. Elsevier Publ. Co., N.Y. 1976
2. J.J. Larsson, A. Lodding, H. Odelius and L.G. Petersson, "Calibration Methods for Ion Probe Application to Fluorine in Biological Hard Tissue", U.S.-Japan Joint Seminar on SIMS, Honolulu, Hawai, 1975
3. H. Tamura, T. Ishitani and I. Kanamota, "Correction for Ionization Yield", U.S.-Japan Joint Seminar on SIMS, Honolulu, Hawai, 1975
4. C.A. Andersen and J.R. Hinthorne, Anal. Chem. 45, 1421 (1973)
5. R. Shimizu, T. Ishitani and Y. Ueshima, Japan J. Appl. Phys. 13, 249 (1974)
6. A.E. Morgan and H.W. Werner, Anal. Chem. 48, 699 (1976)
7. F.G. Rüdenauer, W. Steiger and H.W. Werner, Surf. Sci. 54, 553 (1976)
8. D.S. Simons, J.E. Baker and C.A. Evans, Anal. Chem. 48, 1341 (1976)
9. N.J. Coles, Surf. Sci. 55, 721 (1976)
10. Z. Sroubek, Surf. Sci. 44, 47 (1974)
11. J.M. Schroeer, T.N. Rhodin and R.C. Bradley, Surf. Sci. 34, 571 (1973)
12. P. Joyes, J. Physique 29, 774 (1968); 30, 243 (1969); 30, 365 (1969)
13. M.A. Rudat and G.H. Morrison, Anal. Chem. 51, 1179 (1979)
14. A.E. Morgan and W.H. Werner, Surf. Sci. 65, 687 (1977)
15. J.M. Schroeer, Vacuum 22, 603 (1972)
16. W.H. Gries and F.G. Rüdenauer, Int. J. Mass. Spectr. Ion Physics 18, 11 (1975)
17. Z. Jurela, Int. J. Mass. Spectr. Ion Physics 12, 33 (1973)

18. M.A. Rudat and G.H. Morrison, Int. J. Mass Spectr. Ion Physics 27, 249 (1978); 29, 1 (1979); 32, 53 (1979)
19. G.J. Scilla and G.H. Morrison, Anal. Chem. 49, 1529 (1977)
20. D.M. Drummer, J.D. Fassett and G.H. Morrison, Anal. Chim. Acta 100, 15 (1978)
21. J.D. Ganjei, D.P. Leta and G.H. Morrison, Anal. Chem. 50, 285 (1978)
22. W.H. Gries, Int. J. Mass. Spectr. Ion Physics 49, 2322 (1979)
23. D.P. Leta and G.H. Morrison, Anal. Chem. 52, 277 (1980)
24. D.P. Leta, G.H. Morrison, G.L. Harris and C.A. Lee, Int. J. Mass Spectr. Ion Physics 34, 147 (1980)
24a. H.W. Werner, Acta Electronica 19, 53 (1976)
25. J. Lindhard, M. Scharff and H.E. Schiott, Mat.-Fys. Medd. Dan Vid. Selsk 33, 14 (1963)
26. J.D. Fassett, J.R. Roth and G.H. Morrison, Anal. Chem. 49, 2322 (1977)
27. J.D. Fassett and G.H. Morrison, Anal. Chem. 50, 1861 (1978)
28. J.D. Fassett, D.M. Drummer and G.H. Morrison, Anal. Chim. Acta 112, 165 (1979)
29. B.K. Furman and G.H. Morrison, Anal. Chem. 52, 2305 (1980)
30. D.M. Drummer and G.H. Morrison, Anal. Chem. 52, 2147 (1980)
31. W. Steiger and F.G. Rüdenauer, Anal. Chem. 51, 2107 (1979)

Quantitative Chemical Analysis of Standard Iron Alloys by SIMS

Z. Jurela

Boris Kidric Institute of Nuclear Sciences, Laboratory for Atomic Physics, Beograd, Yugoslavia

Introduction

SIMS technique is often used for the quantitative chemical analysis of alloys. It is considered to be a semiquantitative method. The models of plasma in the local thermal equilibrium at the surface (LTE) for 50 % of the results have the factor of discrepancy smaller than 2, and only for 5 % of the results it is higher than factor 5 [1-8]. The application of the non-equilibrium surface ionization model for the quantitative chemical analysis is of recent date [9,10]. Good agreement of theoretical and experimental results for ALCAN aluminium, ferro-niobium and basalt rock has been obtained [10]. For the majority of constituents the factor of discrepancy is smaller than 1.5, and only for niobium it is approximately 4.

Non-equilibrium Surface Ionization Model

Essential characteristics of the non-equilibrium surface ionization model are:

- Solid surface has a dominant influence on the charge of sputtered particles by means of its work function, Φ, and intense charge exchange with sputtered particles at distances smaller than the critical distance for charge exchange.
- The correction of the ionization potential for hyperthermal rates $\Delta I = -2/r_i$ for conductive surfaces and $\Delta I = -0.265(2/r_i)$ for non-conductive surfaces depends on the normal components of a sputtered particle: the increase of $v_\perp$ increases $|\Delta I|$ [11], but usually the low energy yields of some components are used, and thus the quoted expressions for ΔI are mostly sufficient.
- Every sputtered particle leaves the surface passing through a ring of neighbouring particles which make a center of sputtering and their energy of oscillation is determined by the temperature of sputtering centers, T_{sc}. Sputtered particles regularly require their own energy in collision processes, and therefore have a very wide energy distribution: from 0-3000 eV, and even wider [12].
- The temperature of sputtering centers determines the degree of ionization of sputtered particles, not their energy, with the exception of that part in the low energy spectrum which appeared as a result of the thermal evaporation [13-15].

- Mathematical interpretation is very simple, what is not the case with other models of secondary ion emission.

Let us start with the known equations for the degree of ionization of the non-equilibrium surface ionization model for pure elements [16,17]

$$\alpha_+ = (Q_p/Q_a)\exp\{5040(\Phi+2/r-I)/T_{sc}\},$$

$$\alpha_- = (Q_n/Q_a)\exp\{5040(\Phi-A)/T_{sc}\},$$

where $\Delta I = -2/r$ is the correction of the ionization potential at the critical distance for charge exchange, $x_c = 1.8\ r$ and r-radius of atoms. The equations for alloys, compounds and minerals are obtained from these equations. The equations for the degree of ionization of the i-constituent read

$$\alpha_i^+ = (Q_i^n/Q_i^a)\exp\{5040(\Phi_{exp}+2/r_i-I_i)/T_{sc}\},$$

$$\alpha_i^- = (Q_i^n/Q_i^a)\exp\{5040(A_i-\Phi_{exp})/T_{sc}\}.$$

For an alloy which consists of n elements, a system of n-pairs of such equations is obtained.

There are two approaches for the comparison of experimental results with theoretical predictions. The first and the most common one is to measure the currents of constituents and correct them due to their isotopic composition, J_i^+, and then to draw a diagram of dependences $J_i^+Q_i^o/c_iQ_i^+ = f(I_i+\Delta I_i)$, where c_i is the concentration of the i-constituent in atoms %, and the right hand side presents (the corrected) ionization potential of the i-constituent. The temperature is determined from the slope of a line. In dependence of the applied model, the temperature has different names such as: "apparent" temperature, "plasma temperature", "ionization temperature" etc. In the case of the non-equilibrium surface ionization model it is the temperature of sputtering centers, T_{sc}. Then the discrepancies of some constituents from the drawn line in the diagram are determined and the possible reasons and explanations for it are sought. Furthermore, the relative sensitivity factors of the matrix for every constituent as well as the concentration of some constituents from the following equation are calculated

$$c_i/c_{i+1} = (J_i^+Q_{i+1}^pQ_i^a/J_{i+1}^+Q_i^pQ_{i+1}^a)\ \times$$
$$\times\ \exp\{5040(I_i+\Delta I_i-I_{i+1}-\Delta I_{i+1})/T_{sc}\}$$

The second approach, in this paper, is carried out with the aim to find out what experimental results should be expected when the NBS No.461-468 low-alloy steels are bombarded with the 15 keV Xe^+ ion beam intensity of 1 μA. For such a case the sputtering coefficient is S = 6 atoms/ion and T_{sc} = 2500 K [16,17]. Let us start from Table 1 which presents the composition of NBS low-alloy steels in atoms %. As the concentration of Fe in the quoted steels is higher than 95 % it may be

Table 1 Composition of NBS No.461-468 low-alloy standard steels in atomic percent as quoted by NBS

Element	Atomic number	Low-alloy steels NBS No. [atom %]							
		461	462	463	464	465	466	467	468
B	5	0.0013	0.00258	0.00623	0.0258	0.00052	0.00103	0.00103	0.00103
C	6	0.699	1.860	2.280	2.507	0.1719	0.302	0.509	1.212
N	7	0.0239	0.0319	0.0240	0.0278	0.0199	0.0239	0.0159	0.0240
O	8	0.0700	0.0209	0.0245	0.0209	0.0104	0.0174	0.0139	0.0140
Al	13	0.0104	0.0476	0.0561	0.0103	0.393	0.0310	0.330	0.0871
Si	14	0.0937	0.556	0.819	0.952	0.0576	0.0497	0.515	0.149
P	15	0.0958	0.0811	0.0561	0.0306	0.0144	0.0216	0.0592	0.0415
S	26	0.0348	0.0348	0.0350	0.0348	0.0174	0.0174	0.0173	0.0349
Ti	22	0.0117	0.0431	0.0117	0.00465	0.233	0.0664	0.302	0.0128
V	23	0.0264	0.0636	0.110	0.317	0.0219	0.00767	0.0447	0.187
Cr	24	0.140	0.794	0.280	0.0836	0.00429	0.0118	0.0385	0.581
Mn	25	0.367	0.955	1.174	1.339	0.0325	0.115	0.278	0.479
Fe	26	96.639	95.944	96.460	96.670	99.239	99.431	97.066	96.791
Co	27	0.2470	0.104	0.0124	0.0265	0.00757	0.0436	0.0443	0.152
Ni	28	1.649	0.666	0.373	0.128	0.0247	0.0485	0.0833	0.982
Cu	29	0.299	0.176	0.415	0.0824	0.0167	0.0290	0.0586	0.229
Ge	32	0.00116	0.00231	0.00193	0.00115	0.00269	0.00231	0.00230	0.00077
As	33	0.0209	0.343	0.0749	0.0134	0.00744	0.0104	0.1039	0.0597
Zr	40	0.00307	0.0385	0.123	0.00611	0.00122	0.00306	0.0573	0.00307
Nb	41	0.00663	0.0577	0.1177	0.0222	0.00060	0.00300	0.1735	0.00361
Mo	42	0.175	0.0465	0.0702	0.0168	0.00291	0.00640	0.1217	0.1166
Ag	47	0.00078	0.00010	0.00010	0.00155	0.00013	0.00023	0.00206	0.00010
Sn	50	0.0104	0.0310	0.00614	0.0202	0.00047	0.00235	0.0468	0.00424
Ta	73	0.00062	0.0111	0.0465	0.0212	0.00031	0.00062	0.0707	0.00155
W	74	0.00365	0.0161	0.0320	0.00667	0.00030	0.00182	0.0605	0.0234
Pb	82	0.00081	0.001	0.00325	0.00538	0.00013	0.00035	0.00016	0.00013

concluded that they have the same work function. Let us take the tabular value which is recommended for the pure surface $\Phi = 4.31$ eV [18]. In this calculation the following sources have been used: ionization potentials from WEAST's CRS Handbook [19], the correction of ionization potential $\Delta I = -2/r_i$ was calculated according to SLATER's radii [20], partition functions Q^p, Q^n and Q^a according to CORLIESS [21] and electron affinity according to LINEBERGER [22] and ZOLLWEG [23]. For the current of a primary ion beam of 1 μA and the sputtering coefficient S = 6 atoms/ion it comes out that $3.74 \cdot 10^{13}$ atoms/s is being sputtered from the spot. In Table 2 the calculated values of the following magnitudes for all constituents are presented: $J_i^+ Q_i^a / c_i Q_i^p$ (column 3) the degree of ionization α_i^+ (column 4), theoretical matrix sensitivity factor, TMSF, defined as $\alpha_i^+/\alpha_{ME}^+ = \alpha_i^+/\alpha_{Fe}^+$ (column 5) and total ion currents of the constituents of NBS low-alloy standard steels (column 6-13).

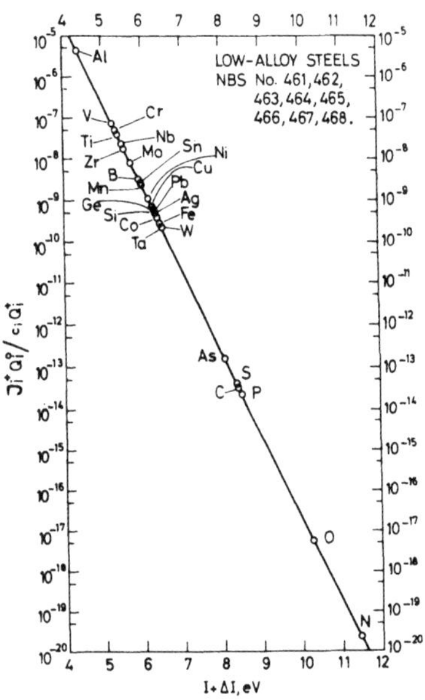

Fig. 1 $J_i^+ Q_i^o / c_i Q_i^+$ dependence on the corrected ionization potential for NBS No. 461-468 low-alloy steels bombarded by 15 keV Xe^+ ion beam

The diagram of dependences $J_i^+ Q_i^a / c_i Q_i^p = f(I_i + \Delta I_i)$ is shown in Fig. 1. It is characteristic that aluminium is at the top of the diagram and other amphoteric elements are placed in the upper part of the diagram. Metallic elements are grouped in three orders of magnitude at the upper half of the diagram, and non-metallic elements are spread at the lower half of the diagram. It means that, in principle, non-metallic elements due to their high ionization potential could hardly be detected as positive ions, especially oxygen and nitrogen. The degrees of

Table 2 Values of the calculated characteristic magnitudes $J_i^+ Q_i^a / c_i Q_i^p$, degree of ionization, theoretical matrix sensitivity factor and total ion currents of different constituents for NBS low-alloy steels bombarded by 15 keV Xe^+ ion beam and the current intensity of 1 μA

Element	Atomic number	$I_i^+ Q_a / c_i Q_p$	α_+ ions/atom	Matrix sensitivity factors	Ion current of different constituents in low-alloy steels NBS No: in A							
					461	462	463	464	465	466	467	468
B	5	3.06E-9	8.51E-5	1.10E0	5.28E-13	1.32E-12	3.18E-12	1.31E-11	2.63E-13	5.27E-13	5.25E-13	2.38E-11
C	6	3.43E-14	3.68E-9	4.78E-5	1.53E-14	4.11E-14	1.96E-14	5.54E-14	3.80E-15	6.68E-15	1.13E-14	2.70E-14
N	7	2.39E-20	8.95E-15	1.16E-10	1.29E-21	1.71E-21	1.29E-21	1.50E-21	1.07E-21	1.28E-21	8.53E-22	1.29E-21
O	8	5.50E-18	4.08E-13	5.29E-9	1.71E-19	5.12E-20	6.00E-20	5.11E-20	2.56E-20	4.27E-20	3.40E-20	3.42E-20
Al	13	4.26E-6	1.20E-1	1.56E3	7.48E-9	3.43E-8	4.05E-8	7.45E-9	2.83E-7	2.24E-8	2.38E-7	6.28E-8
Si	14	5.06E-10	5.06E-5	6.57E-1	2.84E-11	1.69E-10	2.48E-10	2.89E-9	1.75E-11	1.51E-11	1.56E-11	4.53E-11
P	15	2.31E-14	7.01E-9	9.10E-5	4.03E-15	3.41E-15	2.36E-15	1.29E-15	6.06E-16	9.10E-16	2.49E-15	1.75E-15
S	16	4.16E-14	3.08E-9	4.00E-5	6.44E-16	6.44E-16	6.47E-16	6.43E-16	3.22E-16	3.22E-16	3.21E-16	6.46E-16
Ti	22	3.95E-8	1.21E-2	1.57E2	8.49E-10	3.13E-9	8.51E-10	3.38E-10	1.69E-8	4.83E-9	2.19E-8	9.34E-10
V	23	7.33E-8	1.08E-2	1.40E2	1.70E-9	4.10E-9	7.11E-9	2.05E-8	1.41E-10	4.95E-10	2.89E-9	1.21E-8
Cr	24	5.12E-8	5.88E-3	7.64E1	4.94E-9	2.80E-8	9.90E-9	2.95E-9	1.51E-10	4.17E-10	1.36E-9	2.05E-8
Mn	25	2.31E-9	4.61E-4	5.99E0	1.01E-9	2.64E-9	3.25E-9	3.71E-9	8.99E-11	3.18E-10	7.70E-10	1.32E-10
Fe	26	3.02E-10	7.7E-5	1.00	4.47E-8	4.43E-8	4.46E-8	4.47E-8	4.58E-8	4.59E-8	4.48E-8	4.47E-8
Co	27	4.05E-10	6.11E-5	7.49E-1	9.06E-11	3.82E-11	4.54E-12	9.71E-12	2.77E-12	1.60E-11	1.64E-11	5.57E-11
Ni	28	1.16E-9	6.95E-5	9.03E-1	6.88E-10	2.78E-10	1.55E-10	5.35E-11	1.03E-11	2.02E-11	3.48E-11	4.09E-10
Cu	29	7.61E-10	5.28E-5	6.88E-1	9.49E-11	5.57E-11	1.31E-10	2.61E-11	5.28E-12	9.19E-12	1.86E-11	7.26E-11
Ge	32	6.41E-10	6.10E-5	7.92E-1	4.23E-13	8.45E-13	7.07E-13	4.22E-13	9.84E-13	8.44E-13	8.41E-13	2.82E-13
As	33	1.57E-13	3.41E-8	4.43E-4	4.29E-15	7.02E-15	1.53E-14	2.74E-15	1.52E-15	2.14E-15	2.13E-14	1.22E-14
Zr	40	1.90E-8	4.18E-3	5.43E1	7.69E-11	9.66E-10	3.08E-9	1.53E-10	3.06-E11	7.66E-11	1.43E-9	7.68E-11
Nb	41	2.38E-8	3.63E-3	4.71E1	1.44E-10	1.25E-9	2.56E-9	4.83E-10	1.31E-11	6.54E-11	3.78E-9	7.86E-11
Mo	42	8.58E-9	1.23E-3	1.60E1	1.29E-9	3.44E-10	5.19E-10	1.25E-10	2.15E-11	4.73E-11	9.00E-10	8.63E-10
Ag	47	5.22E-10	4.35E-5	5.61E-1	2.03E-13	2.70E-14	2.72E-14	4.05E-13	3.37E-14	6.08E-14	5.38E-13	2.71E-14
Sn	50	2.79E-9	2.81E-4	3.65E0	1.75E-11	5.23E-11	1.03E-11	3.40E-11	7.91E-13	3.96E-12	7.89E-11	7.15E-12
Ta	73	2.30E-10	5.16E-5	6.69E-1	1.91E-13	3.44E-12	1.44E-11	6.58E-12	9.54E-14	1.91E-13	2.19E-11	4.78E-13
W	74	2.32E-10	4.25E-5	5.39E-1	9.11E-13	4.01E-12	7.99E-12	1.66E-12	7.56E-14	4.54E-13	1.51E-11	5.84E-12
Pb	82	5.73E-10	1.25E-4	1.62E0	6.10E-13	1.22E-12	2.44E-12	4.05E-12	1.01E-13	2.63E-13	1.21E-13	1.01E-13

Matrix sensitivity factors based on Fe = 1.00

ionization are in a wide range from 0.12 for aluminium to $8.95 \cdot 10^{-15}$ for nitrogen. Aluminium is followed by: Ti, V, Cr, Zr, Nb, Mo, Mn, Sn, Pb etc. Nitrogen is preceeded by: O, S, C, P, As etc. If a diagram were made it would be similar to that in Fig. 1, but the ordinate would be increased by approximately 5 orders of magnitude.

Theoretical matrix sensitivity factors show how the atoms of different constituents in alloys could be more easily or with more difficulty ionized than the matrix atoms, in this case iron atoms. As may be seen in Table 2 aluminium has a rather high value TMSF = 1560. For one order of magnitude the following values are lower Ti(157) and V(140). Then follow Cr(76.4), Zr(54.1), Nb(47.1), Mo(16.0), Mn(5.99) and Sn(3.65). Two elements have TMSF somewhat higher than 1: Pb(1.62) and B(1.10). A large group of constituents of low-alloy steels No.461-468 has TMSF slightly smaller than 1. They are: Ni(0.903), Co(0.793), Ge(0.72), Cu(0.686), Ta(0.669), Si(0.657) and W(0.539). Non-metallic elements have very low values for TMSF: from 10^{-5} for carbon to 10^{-10} for nitrogen. The lower limit of detection for all constituents in different matrixes can be easily determined in ppm from the TMSF values and the sensitivity of the apparatus. Total ion currents for different elements for the cited NBS standard steels calculated according to $J_i^+ = J_p \cdot S \cdot c_i \cdot \alpha_i^+$ in A are presented in the columns 6-13.

In the above procedure of the application of the non-equilibrium surface ionization model for the quantitative chemical analysis of alloys, compounds and minerals by determining the currents of positive ions of different constituents, J_i^+, are presented. The same procedure can be carried out with the currents of negative ions, J_i^-. It is convenient for the determination of the concentration of non-metallic alloys, especially carbon and nitrogen. Due to a limited space these results are not quoted in the paper. Though the analysis is usually carried out by means of positive ion currents, J_i^+, it is very useful to measure at least the current of negative ions of the matrix element, J_{ME}^-, as it enables the work function of the spot, Φ_{exp}, and the temperature of the sputtering center, T_{sc}, at the given experimental conditions, to be determined exactly.

The observed change of the sensitivity factor of some elements in dependence on the concentration in low-alloy steels bombarded by ions of reactive gases [4,5,8] shows that, due to that bombardment, the carbides, nitrides or oxides are formed which behave completely in a different way during sputtering and the secondary ion emission. This effect is particularly important for the ingredients: Ti, Zr, Nb, Mo, Ta and some others, almost in all matrixes. All these difficulties in the quantitative chemical analysis of alloys by SIMS can be avoided by bombarding the targets with inert-gas ions. Consequently, smaller positive ion yields are obtained, but it can be easily compensated by using a more sensitive detection system.

References

1. C.A. Andersen and J.R. Hinthorne, Anal. Chem. 45, 1421 (1973)
2. A.E. Morgan and H.W. Werner, Anal. Chem. 48, 699 (1976)
3. D.S. Simons, J.E. Beker and C.A. Evans, Jr., Anal. Chem. 48, 1341 (976)
4. J.M. Schroeer, J. Vac. Sci. Technol. 14, 343 (1977)
5. D.H. Smith and W.H. Christie, Int. J. Mass Spectrom. Ion Phys. 26, 61 (1978)
6. H.W. Werner, Surf. Interface Anal. 2, 56 (1980)
7. L. Illgen, H. Mai, U. Seidenkranz and R. Voigtmann, Surf. Interface Anal. 2, 77 (1980)
8. A.E. Morgan, Surf. Interface Anal. 2, 123 (1980)
9. A.G. Lototskii and F.A. Gimel'farb, Zh. Anal. Chem. 31, 433 (1976)
10. Z. Jurela, Int. J. Mass Spectrom. Ion Phys. 37, 67 (1981)
11. Z. Jurela, Int. J. Mass Spectrom. Ion Phys. 12, 33 (1973)
12. Z. Jurela, Radiat. Eff. 19, 175 (1973)
13. J.B. Sanders, Radiat. Eff. 51, 43 (1980)
14. R.E. Johnson and R. Evatt, Radiat Eff. 52, 187 (1980)
15. M. Szymonski, 9th ICACS, Lyon'81, to be published in Nucl. Instrum. and Meth.
16. Z. Jurela, 9th ICACS, Lyon'81, to be published in Nucl. Instrum. and Meth.
17. J. Jurela, to be published
18. B.S. Fomenko, I.A. Podcernyaeva, "Emissionnye i adsorbcionnye svojstva vescestv i materialov, Spravocnik", Atomizdat, Moskva, 1975
19. "CRC Handbook on Chemistry and Physics", 57th edition, 1976-1977, ed. R.C. Weast, CRC Press, Cleveland
20. J.C. Slater, J. Chem. Phys. 41, 3199 (1964)
21. C.H. Corliss, J. Rez. Phys. Chem. 66A, 169 (1962)
22. W.C. Lineberger, IEEE Trans. Nucl. Sci. NS-23, 934 (1976)
23. R.J. Zollweg, J. Chem. Phys. 50, 4251 (1969)

Application of the LTE Model to Quantifying the Secondary Ion Spectra of Steels

J. Suba and A. Stopka

SKODA Central Research Institute,
316 00 Plzen, Czechoslovakia

Introduction

The quantification of SIMS spectra obtained from steels was performed using the model based on the LTE concept, as proposed by ANDERSEN [1,2]. Since the calculation performed by this model requires the values of a large number of constants for each element and since all the constants are available with a sufficient accuracy, the calculations are done assuming some simplifications. SIMONS [3] described a model neglecting a correction for negative ions and introduced an empirical correction for molecular ions. The results obtained by such an analysis have approximately the same accuracy as those obtained on working with the more complex Anderson model. MORGAN and WERNER [4-8] introduced a further simplification, by assuming that the positive singly charged secondary ions only are representative for the sample composition. This simplification allowed to reduce the number of fitting parameters for the calculation of comparative concentrations. The quantity T described by ANDERSEN as the absolute temperature of the plasma generated by an impinging was denoted by the authors as a fitting parameter without a precise physical meaning. The method gives good results only when the bombarded surface is oxygen saturated. It fails in the determinations of elements showing high yields of monoxide ions (Cr, elements belonging to the IIIb to VIb groups of the periodic system, including lanthanides and actinides). In spite of numerous discussions resulting in critical comments [9-11] the LTE model useful for practical application gives results in practice.

Experimental

In the present work a simplified model as proposed by SIMONS [3] was used. The calculations were done by a modified Saha-Eggert equation

$$c_i = CI_i^+(1 + n_e T^{-3/2} Y_i^{(a_i T^{-1})} \exp(-16.100 n_e^{1/2} T^{3/2}))$$

where c_i is the concentration of i-th element in the matrix, C is a proportionality constant, I_i^+ is the isotope corrected ion current of i-th element, n_e is $2.07 \cdot 10^{16}$ concentration of electrons in the plasma (cm^{-3}), T is plasma temperature (K), Y_i is the partition - function ratio for the i-th element ions and atoms, and a_i is $11.6 \cdot 10^3 \cdot E_i$ where E_i is the first ionization potential of the i-th element (eV). In the calculations of element concentrations, the experimental intensities of secondary ions were first reduced into corrected intensities. In case when the sum of $(MO^+ + MO_2^+)$ molecular ion intensities obtained from element M as observed in the spectrum was lower that 10 % of M^+ the moelcular ions could be neglected. When the intensity of $(MO^+ + MO_2^+)$ molecular ions was higher that 10 % of M^+ a sum $(M^+ + MO^+ + MO_2^+)$ was considered in the calculations. For monoisotopic elements the corrected value of intensity is equal to the measured intensity. In case of the elements with 2 or more isotopes the corrected intensity was calculated as a ratio of the experimental isotope intensity to the percent isotope concentration in the element. For the calculations of normalizing constant C, electron concentration in the plasma, and plasma temperature, known data on the concentrations of two internal standards in the matrix determined by an independent chemical analysis were used. The optimum values of parameters for calculating the element concentration in the sample matrix were identified as extremes of the residual function.

In the work a SIMS spectrometer [12] as designed in the Instrumentation Institute, Czechoslovak Academy of Sciences, Brno, CSSR was used. Xe^+ and O_2^+ ions with an energy of 3 keV and current intensities of several mA/cm^2 were used. The chemical compositions of the steel samples were determined by XRF spectrometry, atomic absorption spectroscopy, and wet analysis procedure. The 15x15x2 mm samples were metallographically polished and prior to their introduction into a vacuum chamber they were subjected to an ultrasonic cleaning procedure. The steel spectra were measured in a mass range of 1-100 and recorded only after a steady state of the secondary ion emission was reached which was the case after ≈30 min. In Fig. 1 the SIMS spectrum of high alloyed steel is given.

Results

The results of analyses are given in Table 1. In Fig. 2 the effect of the standards used - on final results - is represented. The final results are heavily affected in the case when middle - concentration elements, e.g. Cr and V or Cr and Mn, are selected as the internal standards. The calculated concentrations of the main steel element, i.e. iron, are then extremely low (16.75 %) where the real content is 95.4 %. A considerable improvement of the results could be achieved in using more than two standards which are, however, of low value for a majority of applications to practice. In addition to the results of analyses also the main parameters of the correction program, T and n_e, together with their mutual relationship were investigated. In Fig. 3 the dependence of T as a function of n_e is given. Our results given in Fig. 3 show that the quantitative evalution of steel spectra gives T and n_e values scattered in the respective ranges of 3500 to 6000 K and 10^{15} to 10^{17} cm^{-3}. The form of the dependence (T and n_e) obtained by us is similar to that reported by SIMONS [3] and RÜDENAUER [10].

Table 1 Results of calculations

Steel		30228		22059			89322/3		
Ions		Xe^+	O^+		Xe^+	O^+		Xe^+	O^+
standard		Fe,Cr			Fe,Cr			Fe,Cr	
T (K)		4700	4500		5000	4040		4650	5000
n_e-(cm^{-3})		$4.21 \cdot 10^{15}$	$3.01 \cdot 10^{15}$		$2.7 \cdot 10^{16}$	$1.78 \cdot 10^{15}$		$2.08 \cdot 10^{16}$	$2.32 \cdot 10^{15}$
Wt %	ACTUAL	CALCULATED		ACTUAL	CALCULATED		ACTUAL	CALCULATED	
Fe	98.58	98.40	98.90	95.40	95.49	95.35	70.65	70.67	70.46
Cr	0.10	0.10	0.10	2.80	2.80	2.80	17.40	17.40	17.45
Ni	-	-	-	0.07	-	6.22	9.50	9.90	7.98
Mn	0.77	1.22	1.47	0.49	0.79	0.92	1.25	2.30	1.25
Si	0.26	0.65	-	0.20	0.81	1.22	0.42	-	-
Cu	0.20	0.21	0.14	0.09	0.11	0.32	0.09	0.77	0.18
Al	0.0017	0.09	0.14	0.003	0.01	0.00	0.19	1.09	0.62
V	-	-	-	0.31	0.22	0.26	-	-	-
Ti	-	-	-	-	-	-	0.50	0.47	0.22
Na	0.0001	0.01	0.01	$2.3 \cdot 10^{-4}$	0.01	0.00	$2.3 \cdot 10^{-4}$	0.00	-

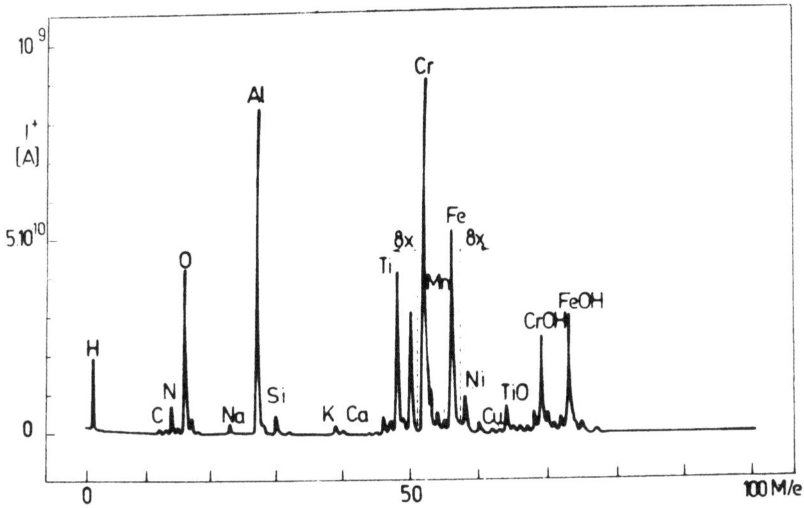

Fig. 1 SIMS spectrum of 89322/3 steel

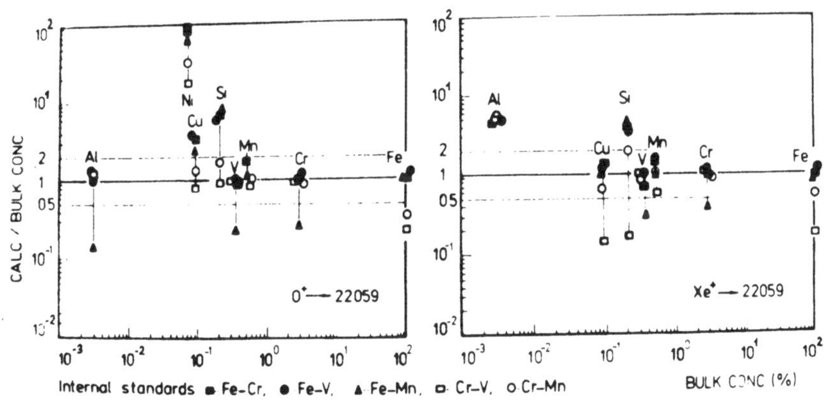

Fig. 2 Effect of internal standards choice 22059 steel, on ratio (calculated/specimen bulk concentration) plotted as ratio of bulk concentration (%)

A comparison of the absolute T values obtained on using various correction programs and also considerations involving the appropriate literature data shows that the T and n_e values are of interest for the quantitative evaluation of SIMS spectra only as adjustable parameters of the correction program. The results obtained for Xe^+ show that the LTE concept-based model can also be applied to cases where the sample surface is bombarded with noble gas ions. The same level of accuracy as in the case of application of O_2^+ ions was achieved.

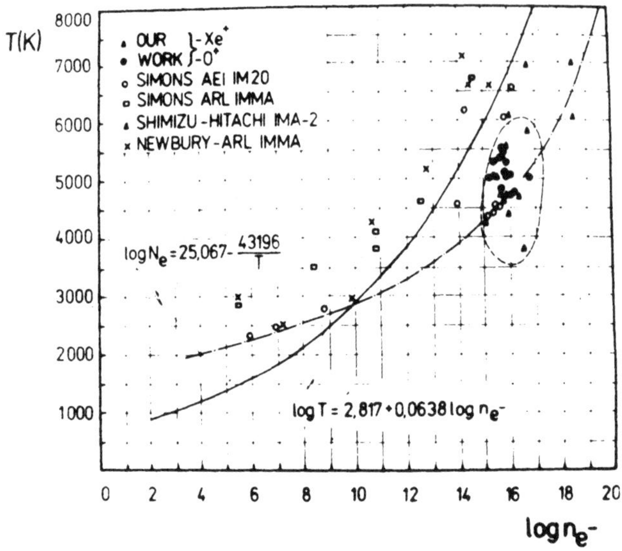

Fig. 3 Calculated values of T and log n_e (curves drawn) compared with experimental data [3]

References

1. C.A. Andersen and J.R. Hinthorne, Anal. Chem. 45, 1421 (1973)
2. C.A. Andersen, Int. J. Mass Spectrom. Ion Phys. 2, 61 (1969)
3. D.S. Simons, J.E. Baker and C.A. Evans, Anal. Chem. 48, 1341 (1976)
4. A.E. Morgan and H.W. Werner, Anal. Chem. 48, 699 (1976)
5. A.E. Morgan and H.W. Werner, Anal. Chem. 49, 927 (1977)
6. A.E. Morgan and H.W. Werner, Surf. Sci. 65, 687 (1977)
7. A.E. Morgan and H.W. Werner, Microchim. Acta (Wien) 31 (1978)
8. A.E. Morgan and H.W. Werner, J. Micr. Spectr. Electron. 2, 285 (1977)
9. F.G. Rüdenauer, Microchim. Acta (Wien) Suppl. 7, p. 85 (1977)
10. F.G. Rüdenauer, W. Steiger and H.W. Werner, Surface Sci. 54, 553 (1976)
11. J.N. Coles, Surf. Sci. 55, 721 (1976)
12. K. Hladil, P. Vasina, R. Salek, J. Fiser and D. Stulik, Czechoslovak J. Phys. A27, 490 (1977)

Modification of the MISR Method with the Use of Implantation of Standard Elements

J. Giber, A. Sólyom, L. Bori, and J. Gyulai[1]

Institute of Physics, Technical University,
H-1111 Budapest, Hungary

Introduction

The most accurate calibrational method for SIMS to date is that of the matrix ion species ratio indexed sensitivity factors (MISR) [1,2]. In this method the RSF's are recorded against the relative secondary yields of two clusters or multiple charged atomic ions of the matrix species which characterize better the reactive gas surface concentration than the partial pressure of the reactive gas in the vacuum chamber [3]. The greatest problem in the practice of the calibrational methods is the selection of the most suitable standard sample with a close match to the one to be measured [4]. It has been demonstrated that ion implantation techniques may be used to incorporate a known amount of almost any element into the subsurface region of the sample [5], therefore LETA and MORRISON have proposed the application of ion implantation for the fabrication of internal standards in the sample itself [6].

In this work we propose a method that combines the advantages of the MISR method and the implantation of standard elements and demonstrates its applicability for boron implanted silicon single crystals.

Principle of the Quantification

In their "standard addition analysis" [6], LETA and MORRISON determined the concentration of Al and Si in homogeneous NBS low alloy steels and of Te in GaAs. They integrated the depth profile of the implant and computed the concentration - cps ratio R:

$$R = \frac{F \cdot t}{D_o \int^{\tau} I^+(t)\,dt} \qquad (1)$$

where D is the depth of analysis (in cm) determined by Talystep measurement, $I^+(t)$ is the secondary ion current of the Al^+, Si^+, Te^+ respectively (in cps), F is the implant fluence (in $atom/cm^2$) and τ is the time of the analysis. The original Al, Si or Te concentration in the sample is then:

$$C_o = R \cdot I_o^+ \qquad (2)$$

[1] Central Research Institute for Physics, Budapest, Hungary

where I_o^+ is the secondary ion yield before the implantation (in cps).

For the analysis of nonhomogeneous samples this method is not the most suitable one, because the variation in the original concentration with the depth makes the integration difficult and inaccurate. Using a high enough implant dose to overcome this difficulty may lead to radiation damage causing distortion of the original concentration profile.

The method we propose needs two identical unknown samples; one of them is implanted with the element of interest. Then both of them are depth profiled at least to the depth of the maximal concentration of the implant. The implantation energy has to be chosen so that the position of the implant peak would be in the depth region of interest. To obtain depth profiles that are easy to evaluate, the implant dose must be high enough to produce an implant concentration peak about one to ten times as much as the one to be measured at that depth. After reaching the implant maximum, spectra have to be taken on both both samples at the same indexing matrix ion ratio. From the two spectra the unknown C_o concentration at that depth can easily be computed from the equation:

$$(C_o + C_i)/C_o = I_i^+/I_o^+ \qquad (3)$$

where I_i^+ is the secondary yield (in cps) of the element of interest in the implanted sample and C_i is the implant maximum concentration. Since the implanted profile with good approximation is Gaussian,

$$C_i = \frac{F}{\sqrt{2\pi}\,\Delta R_p} \qquad (4)$$

where ΔR_p is the standard deviation of the Gaussian concentration distribution, computed by the theory of LINDHARD, SCHARFF and SCHIOTT (LSS) [9] and are available for the most commonly used species/substrate combinations.

Supposing that the secondary ion yields are proportional to the concentration to be measured, which in the case of low concentrations is a quite reasonable assumption, from the knowledge of C_o at a certain place the whole profile can be quantified.

Another possible method would be the derivation of the RSF calibration curve according to the MISR method from the ion yields of some higher concentration matrix constituent as a reference and the implanted maximum concentration and secondary ion yield. This can be especially useful in the case where many similar samples are to be analysed.

Experimental

To prove the applicability of these methods it was natural to choose previously implanted single crystal semiconductor samples as "unknown samples", because the peak concentration of the implanted atoms is known with good enough accuracy. We selected an easily available n-type silicon (100) single crystal (2 Ωcm) pair as substrates implanted with ^{11}B and ^{31}P.* The internal standards were produced by secondary implantation of one of the pairs with same element. The measurements were performed on Balzers made SIMS equipment [7]. The data of the samples and of the experimental conditions can be seen in Table 1. The measured depth profiles are presented in Fig. 1.

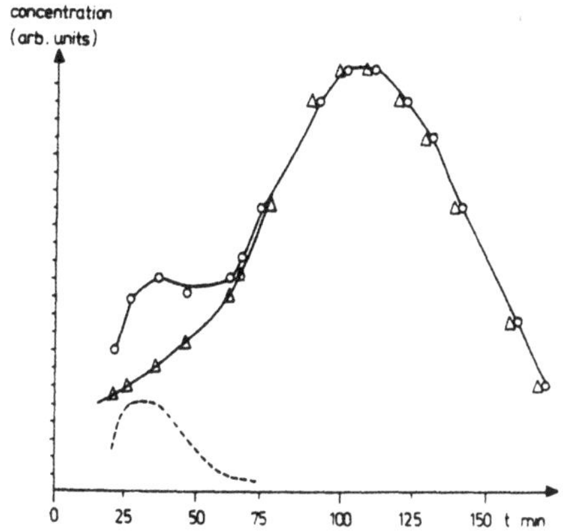

Fig. 1 Measured concentration profiles of the once and twice implanted sample

Table 1 Experimental data

Sample I. Si(100) implanted with	^{11}B (80 keV, $1\cdot10^{16}$ atom/cm^2)
Sample II. Si(100) implanted with	^{11}B (80 keV, $1\cdot10^{16}$ atom/cm^2 + ^{11}B (25 keV, $1\cdot10^{15}$ atom/cm^2)
Experimental parameters:	
primary ion:	Ar^+
primary energy:	3 keV
primary current:	1 μA
beam diameter:	1.8 mm
O_2 partial pressure:	$3\text{-}3.6\cdot10^{-4}$ Pa
background pressure:	$1\cdot10^{-6}$ Pa

In our case C_i in Eq.(3) is the maximum concentration of the second implant. Its position is determined by the substraction of the depth profile of the boron in the once implanted sample (sample I) from that of the twice implanted one (sample II).

*The implantations were performed in the Central Research Institute for Physics, Budapest, Hungary

For this deconvolution the usually applied logarithmic scale is not at all suitable, therefore our equipment was used in the linear mode. (For the every-day practice of the proposed method a built-in computer facility can be extremely useful.) A spectrum was taken on sample I after the peak, then the deconvolution was accomplished by graphical subtraction. After the deconvolution, sample II was again depth profiled at a different spot, and spectrum was taken at the place of the second implant peak. The oxygen pressure was adjusted to get the same SiO^+/Si^+ ratio as in the measurement of sample I. It was carried out with 40 % accuracy.

Results

The measured and calculated concentrations can be seen in Table 2 showing the feasibility of the proposed method. The mean projected range (Rp) and standard devitaion (ΔRp) values were taken from [8]. We computed these parameters for both implants using the other one as a standard. We applied a calculation method similar to the one WERNER had used in this studies on implanted layers [10].

Table 2 Results and errors of the analysis

^{11}B in Si	Rp tab.	Rp calc.	Rp tab.	Rp calc.	calc.con.[atom/cm^2] Gaussian	calc.con.[atom/cm^2] calc.	c/c rel. err.
25 keV	830	904(a)	390	375(b)	$1.02 \cdot 10^{20}$	$1.26 \cdot 10^{20}$(a)	+23.5%
80 keV	2740	2520(b)	730	1170(a)	$5.46 \cdot 10^{20}$	$4.43 \cdot 10^{20}$(b)	-18.9%

(a) determined from the tabulated data of the 80 keV implant
(b) determined from the tabulated data of the 25 keV implant

For further examination of the applicability of our proposed method we have planned the independent measurement of the second implant depth profile on a sample implanted only with 25 keV boron.

In their paper [6] LETA and MORRISON have discussed the advantages of the solid state standard addition analysis over the other empirical calibrational methods with homogeneous materials. These advantages are in keywords: the elimination of the effects of changing instrumental parameters (residual gas pressure, ion transmission, etc.) and of the matrix effects by the fact that both the standard quantity and the one to be measured are determined exactly in the same matrix. In our case the two matrices are somewhat different, but if the element(s) of interest are contained in solid solution in the sample - as it is closely true for the implanted quantity - this small difference cannot lead to the changing of the secondary ionization probability.

For future use of this method it would be very useful to incorporate an additional ion gun for ion implantation in the measuring system itself. In that case the proposed method can be modified in the following way: only one sample is needed. After depth profiling the unknown sample it is implanted by the auxiliary ion gun, then again depth profiled on a different spot and Eq.(3) (or the MISR method) can be applied for the spectrum taken on the pre- and post-implanted samples.

References

1. J.D. Ganjei, O.P. Leta and G.H. Morrison, Anal. Chem. 50, 285 (1978)
2. H. Tamura, US-Japan Joint Seminar, Takarazuka, Japan, 1978
3. V.R. Deline, US-Japan Joint Seminar, Takarazuka, Japan, 1978
4. W.H. Christie, International Lab. 8, 9 (1978)
5. W.H. Gries, Int. J. Mass Spectrom. Ion Phys. 30, 113 (1979)
6. D.P. Leta and G.H. Morrison, Anal. Chem. 52, 277 (1980)
7. F. Pavlyák, L. Bori, J. Giber and R. Buhl, Jap. J. Appl. Phys. 10, 355 (1977)
8. "Ion Implantation Range Date for Silicon and Germanium Device Technologies" Ed. B.Smith, Harwell, Learned Information (Europe) Ltd. Oxford, 1977
9. J. Lindahrd, M. Scharff and H.E. Schiott, Mat.-Fys. Medd. Dan. Vid. Selsk. 33, No 14 (1963)
10. H.W. Werner, Acta Electronica 19, 53 (1976)

Use of Ionic Implantation for Quantification of SIMS Analysis in Metals and Oxides - Application to Corrision Studies

J.C. Pivin, D. Loison, C. Roques-Carmes[1], J. Chaumont[2], A.M. Huber, and G. Morillot[3]

Laboratoire de Métallurgie Physique, B. 413, Université Paris-Sud
F-91405 Orsay Cedex, France

Ionic implantation is a simple and reliable method for the preparation of standards containing well defined concentrations and in-depth distributions C(z) of dopants. This method, which is commonly used for quantification of secondary ion emissions of dopants in semiconductors [1,2] may be also of great interest in the field of corrosion studies, since transport of atoms and electric charges in oxide films is generally controlled by impurities low in concentration.

Before calibrating ionic emission yields, deviations of the dopant distributions from Gaussian curves were determined by statistical analysis of SIMS profiles. Profile distortions occuring respectively during implantation itself and SIMS profiling were compared on basis of RBS data. The influence of the implanted dose and the surface roughness were also considered.

The implanted matrices were Si, Al, Ni, NiO and Cr_2O_3 and the implanted dopants ^{13}C, ^{34}S, ^{29}Si, ^{89}Y. Si was chosen as a reference material and Al as a metal of mass comparable to that of Si. The influence of C, S, Si, Y impurities on properties of NiO or Cr_2O_3 films is of fundamental interest in corrosion, because such dopants determine the endurance of Ni electrodes for electrolysis of water or that of superalloys which develop Cr_2O_3 films at high temperatures.

SIMS Profiles in Implanted Standards

Procedure

The materials were monocrystalline Si and polycrystalline Ni, Al, of high purity (99.999 %). The oxides were polycrystalline NiO and Cr_2O_3 films about 1 μm thick, formed respectively on Ni oxidized 5 h at 600°C and on a pure 66 Ni/34 Cr alloy oxidized 5 h at 800°C. The purity of these films was determined quantitatively by SIMS (using calibrations from [3]): the chromia film contained less than 50 ppm Ni and less than 10 ppm

---

[1]Laboratoire de Microanalyse des Surfaces, E.N.S.M.M., F-25030 Besançon Cedex, France

[2]Laboratoire René Bernas, C.S.N.S.M., F-91406 Orsay, France

[3]L.C.R. Thomson C.S.F., F-91401 Orsay, France

of other metals. Metallic materials were polished to 1/4 μm with diamond paste before implantation or oxidation. The implantation parameters are given in Table 1. The projected range Rp, range straggling ΔRp, full width at half maximum 2ΔR and the maximum of concentration C (z = Rp) were calculated using the L.S.S. theory and tables extracted from [4]. The distribution was assumed to be Gaussian, and the range parameters in compounds NiO, Cr_2O_3, were approximated on the basis of a mean value of atomic number $Z = \sum_i C_i Z_i$.

Table 1 Implantation parameters

ions	matrix (Z)	Si (14)	Al (13)	Ni (28)	NiO (18)	Cr_2O_3 (14.4)
^{13}C	E(keV)			80	80	80
	Rp(Å)			1087	1101	1349
	ΔRp(Å)			456	362	383
	* C(Rp) (atom%)			0.90	1.00	1.04
^{34}S	E(keV)	100	100	100	100	100
	Rp(Å)	1336	1121	487	507	603
	ΔRp(Å)	460	378	218	187	209
	* C(Rp)(atoms%)	1.72	1.75	1.87	1.94	1.84
^{29}Si	E(keV)		100	100	100	100
	Rp(Å)		941	580	603	698
	ΔRp(Å)		336	245	210	234
	* C(Rp) (atom%)		1.92	1.67	1.74	1.64
^{89}Y	E(keV)	60	60			
	Rp(Å)	369	308			
	ΔRp(Å)	130	106			
	* C(Rp) (atoms%)	6.2	6.2			

*C(Rp) = dose/ $\sqrt{2\pi}$ x ΔRp x d cible (atoms/cm^{-3}) are indicated only for a dose of $1 \cdot 10^{16}$ ions/cm^{-3}

**implanted doses were $2.5 \cdot 10^{14}$, $2.5 \cdot 10^{15}$, $5 \cdot 10^{15}$ and $2.5 \cdot 10^{16}$ for ^{89}Y, ^{29}Si, ^{34}S and $5 \cdot 10^{14}$, $5 \cdot 10^{15}$, $1 \cdot 10^{16}$, $5 \cdot 10^{16}$ for ^{13}C

***the F.W.H.M. may be deduced from ΔR = 2 x 1.17 x ΔRp

All SIMS experiments were carried out in a CAMECA microanalyser. Bombardment conditions were the following ones:

- for $^{89}Y^+$, $^{29}Si^+$ and $^{34}S^-$ profiles, oxides were sputtered by 5.5 keV Ar^+ ions under residual pressures ranging from $5 \cdot 10^{-8}$ to $1 \cdot 10^{-7}$ Torr.

- for $^{89}Y^+$, ^{29}Si and $^{34}S^-$ in metals, the surfaces were sputtered by 5.5 keV Ar^+ ions under controlled oxygen pressure, such that ionic emissions were maximum. The large variations of ionic emissions in metallic targets with the oxygen coverage (i.e. the residual oxygen pressure, but also the content of oxidable dopants and the bombardment density) precludes the establishment of calibration curves under low but uncontrolled residual pressures.

- for $^{13}C^+$ profiles, the surfaces were sputtered by 5.5 keV Cs^+ ions, because of the very low emission yield of carbon ions in metals and oxides under Ar^+ or O_2^+ bombardment.

Results and Discussion

Figure 1 shows for example ^{34}S experimental profiles and calculated distributions (dashed curves) obtained for a same dose ($5\cdot 10^{15}$ ions/cm^{-2}) implanted in various matrices; the coordinates Rp, C(Rp) of the maximum of the theoretical Gaussian were normalized to the maximum of the experimental distribution. Except for the Si target, the experimental profiles are not fitted by Gaussian functions. Similar results were obtained for other implanted species. The experimental projected ranges (deduced from calibrated sputtering yields) correspond to the calculated ones (see Table 2), but the experimental F.W.H.M. are larger than the theoretical ones.

The shapes of the experimental distributions were more precisely characterized by PEARSON distributions [5,9] with four moments [($\mu_n = \sum_{i=1}^{n} C(z_i)\cdot(z_i-\bar{z})^n$ is the centered moment of order n]. Firscher's or Pearson's coefficients of asymmetry ($\gamma_1 = \mu_3/\mu_2^{3/2}$) and oblateness ($\gamma_2 = \mu_4/\mu_2^2 - 3$) cannot be calculated exactly for distributions with a finite limit (distributions are cut at the surface) and values of $n \leq 100$. They are estimated by coefficients g_1 and g_2 (defined in Table 2) [9]. The disymmetry of profiles is characterized by large and positive values of g_1: the implantation profiles are steeper towards the surface than towards the bulk. The small values of g_2 indicate that the profiles are wider than Gaussian distributions. Note that their values are very comparable for Si and S implanted in all targets.

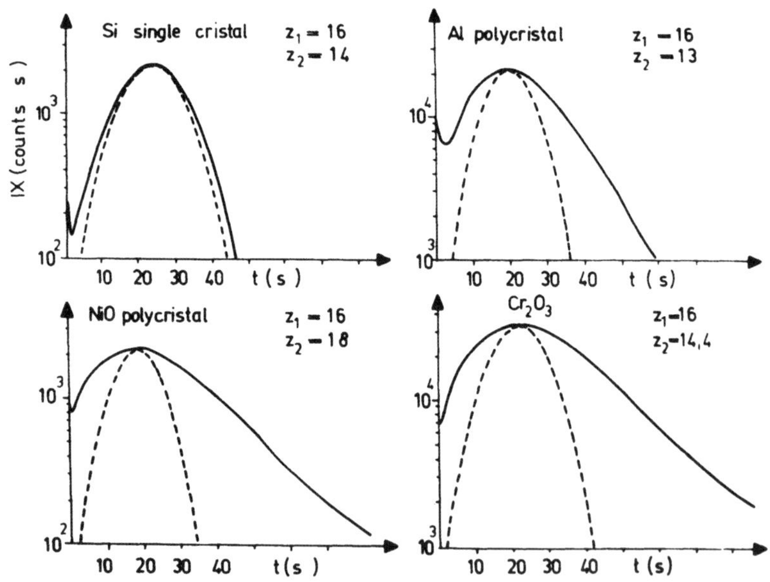

Fig. 1

Table 2 Moments of the in-depth distribution

Moments		80 keV Y		100 keV S				100 keV Si	
		Si	Al	Si	Al	NiO	Cr_2O_3	NiO	Cr_2O_3
Rp(A)	L.S.S.	369	308	1336	1121	507	603	603	698
	measured	400+30	310+20	-	-	-	-	650+40	680+50
ΔRp(A)	L.S.S.	130	106	460	378	187	209	210	234
	measured	155+10	175+10	-	-	-	-	390+24	475+50
parameter of asymmetry g_1		12.8	12.7	4.4	4.0	4.8	4.9	2.9	7.0
parameter of oblateness g_2		173	169	19	15	14	24	7.4	51

g_1 and g_2 were deduced from n = 100 values of $(z_i, C(z_i))$ with: $g_1 = k_3/k_2^{3}$; $g_2 = k_4/k_2^{2}$

$k_2 = \Sigma(z_i-\bar{z})^2/n-1$; $k_3 = [n/(n-1)(n-2)]\,\Sigma(z_i-\bar{z})^3$

$k_4 = [n(n+1)\,\Sigma(z_i-\bar{z})^4 - 3(n-1)[\Sigma(z_i-\bar{z})^2]^2]/[(n-1)(n-2)(n-3)]$

Fig. 2

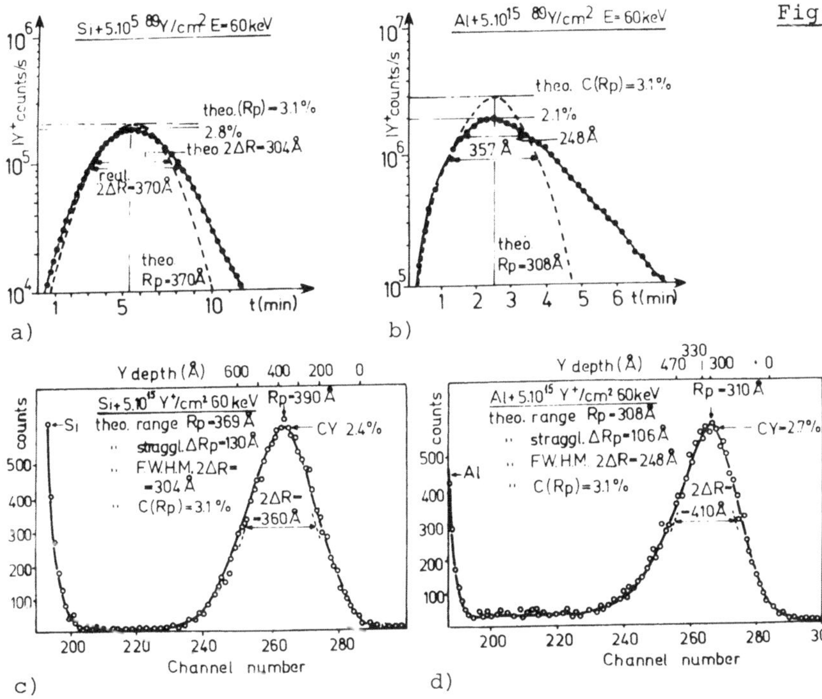

It was verified on aluminium targets with various surface polishing that differential sputtering effects due to the surface roughness do not account for the profile distortion. Moreover it appears that the main part of this distortion does not occur during SIMS profiling but during the implantation itself, on account of the comparison of SIMS and RBS data for Y implanted in silicon and aluminium (Fig. 2). The difference between the measured and calculated F.W.H.M. of the Y distribution in Si may be due to the lack of precision in range straggling calculations. The widening of the distribution is probably due to a diffusion of the dopant induced by radiation damages; such an effect was already noticed for N, P or Yb implanted in Ni, Fe or steels [6].

Calibration Curves

However these standards may be used for SIMS calibration:

- the form of the profile does not depend on the implanted dose (Fig. 3), so that the implanted distribution depends only on the concentration at the first order.

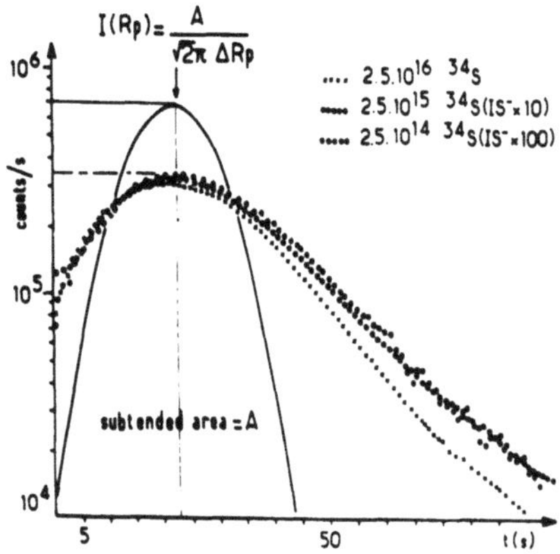

Fig. 3

- the secondary ion intensity at the maximum of the distribution is almost a linear function of the implanted dose, so that the SIMS profile may be considered as the real distribution. Therefore the concentration at the maximum of the distribution may be determined by measuring the area subtended by the experimental profile and drawing a Gaussian curve with a corresponding area and moments Rp, ΔRp corresponding to the L.S.S. theory (Fig. 3). The useful yield I (number of collected ions per sputtered atom) may also be calculated directly by dividing the total number of collected ions (subtended area) by the implanted dose. Emissions of the implanted species $A^{\pm}$ ($^{13}C^-$, $^{34}S^-$, $^{29}Si^+$) were normalized to an ionic emission $B^{\pm}$ (M^+ or O^-) of the matrix M. I(B) was measured on the same area but after sputtering the implanted layer, in order to obtain a ratio IA /IB which varies only with the concentration C_M^A and the ionisation probability $P_M^{A\pm}$ of atoms A in matrix M.

$$\frac{IA^{\pm}}{IB^{\pm}} = \frac{P_M^{A\pm} \cdot C_M^A}{P_M^{B\pm} \cdot C_M^B} = \frac{P_M^{A\pm} \cdot C_M^A}{\text{Constant}} \qquad \text{(see Ref. 3)}$$

since the ionisation probability $P_M^{B\pm}$ of species B and the concentration C_M^B in pure metal M or oxide M_O are constants. Note that strong dependences of the Ni^+, Cr^+ emission with the silicon content have been observed in Ni, NiO and Cr_2O implanted with $^{29}Si^+$, and the calibration curves would have been hyperbolic functions of C^{Si} if measuring both intensities Si^+, Ni^+ (or Cr^+) simultaneously.

The value of $IA^{\pm}/IB^{\pm}$ (pure M) at the maximum of the in-depth distribution was plotted as a function of the corrected concentration C(z = Rp) (Fig. 4, 5, 6). It appears that $P_M^{A\pm}$ is quite a constant for this range of concentrations C_M^A ranging from 10 ppm to 2 %.

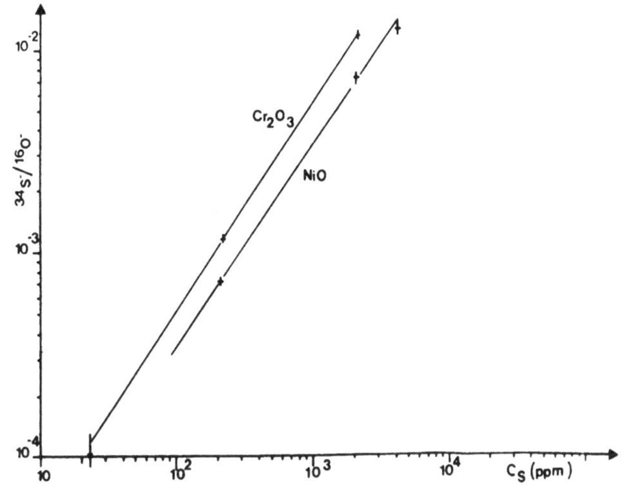

Fig. 4

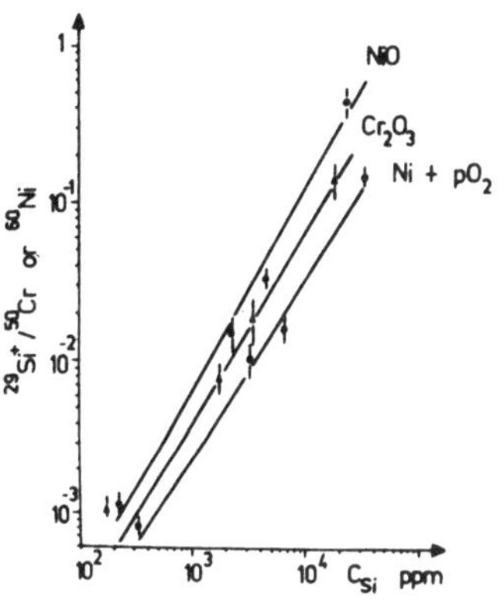

Fig. 5

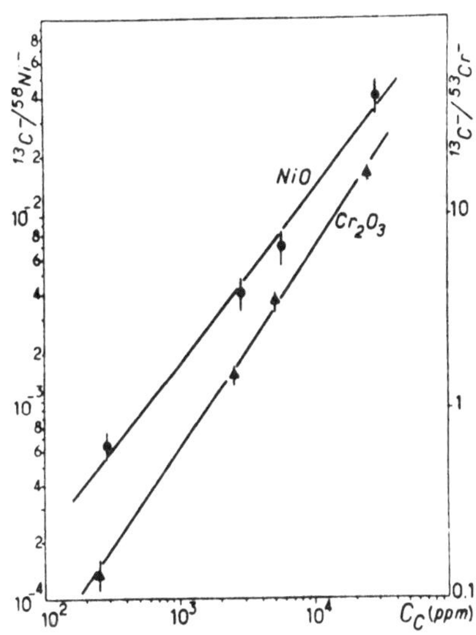

Fig. 6

Application to Corrosion Studies

These calibrations were used for quantitative analysis of dopents in oxide films formed on implanted Ni and NiCr alloys. These analyses, combined with thermogravimetry data and SIMS profiles on samples oxidized alternately in $^{16}O/^{18}O$ provided interesting results on the influence of S, Si, Y impurities on the nucleation and growth of oxide films [7,8]. Fig. 7 shows for example the analysis of a 66 Ni/34 Cr alloy doped with sulfur then oxidized successively 24 h in natural oxygen,

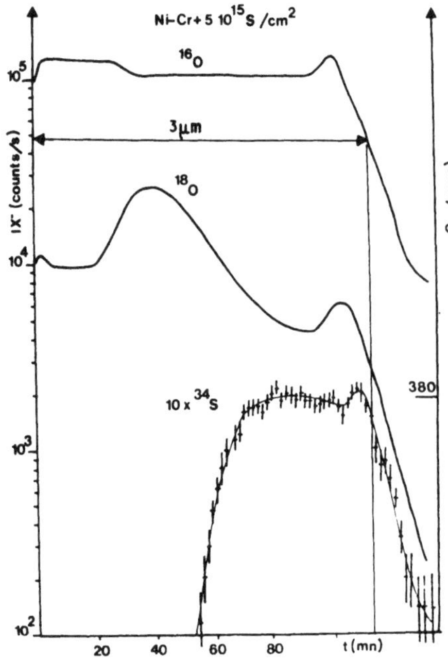

Fig. 7

24 h in a 33 % ^{18}O enriched atmosphere, and 60 h in natural oxygen. The dopant implanted in a thin layer near the surface (Rp = 600 Å, ΔRp = 200 Å) dissolves in the Cr_2O_3 film (on a thickness of 1 µm) during its formation and accelerates the diffusion flux of chromium towards the surface: the oxidation kinetics increases but the ^{18}O profile is unchanged [8]. Similar experiments with a silicon implanted alloy show that silicon segregates at the metal-oxide interface and does not affect the oxidation kinetics. On the contrary, yttrium dissolves in the Cr_2O_3 film formed on an yttrium doped alloy and modifies its growth mechanism: the film grows by inward diffusion of oxygen (as deduced from ^{18}O profiles) and the oxidation kinetics is reduced.

References

1. D.P. Leta and G.H. Morrison, in Secondary Ion Mass Spectrometry SIMS II, ed. by A. Benninghoven, C.A. Evans, Jr., R.A. Powell, R. Shimizu, H.A. Storms, Springer Series in Chemical Physics, Vol. 9 (Springer, Berlin, Heidelberg, New York 1979) p. 61; Anal. Chem. 52, 277 (1980)
2. T.W. Sigmon, ibid, p. 80
3. J.C. Pivin, C. Roques-Carmes and G. Slodzian, ibid p.160 and J. Appl. Phys. 51, 4158 (1980)
4. G. Dearneley, J.M. Freeman, R.S. Nelson and J. Stephen, "Ion Implantation", North-Holland Publ. Co., Amsterdam (1973)
5. F. Jahnel, H. Ryssel, G. Prinke, K. Hoffmann, K. Müller, J. Biersack and R. Henkelmann, Nucl. Instr. and Meth. 182-183, 233 (1981)
6. L. Thome, J. Chaumont, H. Bernas, to be published

7. J.C. Pivin, C. Roques-Carmes, J. Chaumont and H. Bernas, Corr. Sci. 20, 947 (1980)
8. D. Loison, J.C. Pivin, J. Chaumont, Proc. of the Conf. Ion Implantation into Metals, held at UMIST, UK, 23-26 June 1981, Pergamon Press Ltd, publication date Spring 1982, paper no. 46
9. Statistique appliquiée à l'exploitation des mesures, C.E.A. Editions, Masson Publications Paris 1, 53 (1978)

Secondary Ion Emission from Binary and Ternary Amorphous Alloys

H. Gnaser[1], M. Riedel[2], J. Marton[1], and F.G. Rüdenauer [3]

[1]Inst. of General Physics, Techn. Univ. Vienna
A-1090 Wien, Austria

[2]Department of Physical Chemistry and Radiology, Eötvös University,
H-1111Budapest, Hungary

[3]Austrian Research Center, Seibersdorf,
A-1082 Wien, Austria

Introduction

It is well known that quantitation of the SIMS method by the use of theoretically founded models results in only semi-quantitative analytical data. The evaluation of SIMS spectra yields better analytical results when the method of relative sensitivity factors (RSF) is used [1-4]. Difficulties have however been reported concerning the commonly used standard samples [1,2,4,5]. We consider the metallic glasses to be suitable standards for quantitative SIMS for several reasons:

1. The standard alloys are often inhomogeneous on a micro-scale owing to the segregation of several elements and to the presence of insoluble phases. The metallic glasses can serve as better standards because they are single phase systems and are homogeneous also at the μm scale. The structure of metallic glasses is very close to that of liquids, i.e. without any crystalline order, but with internal friction of the same order of magnitude as that of solids [6].

2. The concentration range of the alloying components is generally limited in crystalline standards. Metallic glasses can be prepared in a broader concentration range, and even insoluble elements can form an amorphous phase under suitable conditions [6].

3. The ion yield of elements from crystalline samples depends also on the orientation of the bombarded surface [7]. In the isotropic amorphous alloys such effects which may cause errors in the analysis, cannot occur.

4. Amorphous alloys are of metallic character. For similar reasons, NEWBURY [2] considers the silicon-based "conventional" glasses as suitable standards.

It was the aim of this work
- to provide relative sensitivity factor data to be used for quantitative analysis of Fe-B-type metallic glasses using well characterized ternary metallic glasses as calibration standards,
- to check, whether these data can be used for quantitation of other Fe-based alloys,

- to investigate the matrix effect of the secondary ion emission from binary and ternary metallic glasses.

Experimental

Experiments were performed on the Seibersdorf Quadrupole Scanning Ion Microprobe [8,9]. A mass filtered primary O_2^+-beam of 10 keV energy rastered the sample. A surface layer of about 100 nm thickness was sputtered during the precleaning procedure. The secondary ion yield measurements were performed on two areas of each sample. Secondary ions were analysed by a commercially available energy analyzer/mass filter arrangement (RIBER SQ 156) and detected in the pulse counting mode. The isotope peak heights were automatically extracted from the mass spectra and appropriately averaged for each set of measurements. Instrument control and data acquisition was fully computerized using a stand-alone PDP 11/34 computer and a CAMAC I/O system [9]. The primary beam and sample operating parameters can be seen in Table 1.

Table 1 Standard operating conditions (SOC)
(Primary beam and sample related parameters)

primary ions	10 keV $^{16}O_2^+$, mass filtered
primary ion current	ca. 20 nA
beam diameter	ca. 50 μm
angle of incidence	0° with respect to target normal
secondary ion take off	45°
presputter raster	200 x 200 μm, 10 min
analysed area	100 x 100 μm
residual gas pressure in sample chamber	$<10^{-6}$ Pa
oxygen flooding	no

Two series of metallic glass alloys were studied:

1. Fe-B alloys with 11.7, 16.6 and 21.6 atomic per cent boron content.

2. $Fe_{80}B_{17}X_3$-type alloys, where X stands for a transition metal of the 4., 5. and 6. periods.

The samples have been prepared in the Central Research Institute for Physics of the Hungarian Academy of Sciences using the method of rapid cooling (10^6 K/s) of the molten metal alloys. The purity of the samples was checked by spectrochemical methods, and our SIMS measurements proved that the concentration of impurities was below the detection limit. The amorphous state of the samples has been verified by X-ray diffraction [10].

Data Evaluation

In the context of this work absolute and relative practical sensitivities of a large number of elements, differing in mass number up to 20 amu have to be compared. The data were taken during more than two weeks, so that special care had to be taken to avoid the errors connected with instrument drift and reproducibility of measurements. Therefore the following phenomenologic standardization procedure ("PCS method") has been used:

1. A "primary calibration standard" (PCS) is chosen. This is a well characterized homogeneous sample containing elements with mass numbers distributed over as large mass range as possible. The metallic glass sample $Fe_{80}B_{17}W_3$ fulfills these requirements.

2. "Standard operating conditions" (SOC) are defined. These include primary beam parameters, target properties and mass analyzer parameters. A part of the parameters of the SOC is given in Table 1. The SOC are chosen to approximate the mass-independent transmission [11].

3. "Original Apparent Yields" (OAY(S)) (and "original calibration ratios" (OCR(S)) (are determined under SOC for all elements (S = Fe, B, W) in the PCS:

$$OAY(S) = I^{+\prime}(S)/I_p$$

$$OCR(S) = I^{+\prime}(S)/I^{+\prime}(Fe)$$

where $I^{+\prime}$ is the secondary ion current and I_p the primary ion current. These are determined once with great accuracy and will be a reference for any future quantitative measurements. Now, if at any later time quantitative measurements on the Fe-B-W system are to be compared to another set of measurements, the analyzer has to be readjusted so that data taken on the CPS can always be reproduced as accurately as possible. For this purpose:

4. The SOC are again restored.

5. The PCS is reanalyzed and "actual apparent yields" (AA(S)) and "actual calibration ratios" (ACR(S)) are determined which now are characteristic for the actual set of measurements only:

$$AAY(S) = I^{+\prime}(S)/I_p$$

$$ACR(S) = I^{+\prime}(S)/I^{+\prime}(Fe) \quad .$$

Due to unavoidable inaccuracies in adjustments and due to instrumental drift with time, the original values of the OAY(S) and OCR(S) are reproduced within ±9 % after restoring SOC. The statistical accuracy of the determination of an AAY is typically 7 % and of an ACR 5 %. The instrument drift within one day (2-4 %) has been neglected in the data evaluation procedure. The "yield correction factor" (YCF) has been determined:

$$YCF(S) = AAY(S)/OAY(S)$$

which generally deviates from the value 1.

6. On a particular day the "actual yield" (AYU) of the element X in the unknown sample has been determined:

$$AYU(X) = I^{+\prime}(X)/I_p \ .$$

7. The YCF of the element is determined by interpolation using th YCF of the calibration elements (Fe, B, W) valid for that particular day, and finally the "corrected yield" (CYU(X)) of the element is calculated:

$$CYU(X) = AYU(X)/YCF(X) \ .$$

8. The practical sensitivity of element X is obtained from the CYU(X):

$$S_p(X) = CYU(X)/c(X)$$

where c(X) is the fractional elemental concentration of element X in the sample.

9. The"relative sensitivity factor" (RSF) of element X is calculated, referred to Fe (R = Fe):

$$RFS(X) = S_p(X)/S_p(R) = \frac{CYU(X)/c(X)}{CYU(R)/c(R)} \ .$$

A detailed discussion of the above procedures can be found in [11].

Results

Figure 1 shows the dependence of practical sensitivity for Fe and B and the relative sensitivity factor in the *binary metallic glasses* on the fractional atomic concentration of boron. It can be seen, that these values depend on composition of the alloy, which means the mutual influence of the constituents on the secondary ion emission.

Figure 2 and Table 2 show the RSF of various elements (X) in the *ternary Fe-B-X metallic* glass samples. In the three-component alloys the element X is emitted in each case from similar matrix: the neighbourhood of one X-atom contains about 27 Fe- and 6 B-atoms because of their random distribution in the amorphous state. Therefore the ion yields of the X-elements can be directly compared. Generally the ion yield decreases with increasing atomic number and in the same column the ion yields are lower in higher periods. The atomic number dependence of the RSF measured in this work agree qualitatively and quantitatively reasonably well with data determined by other authors

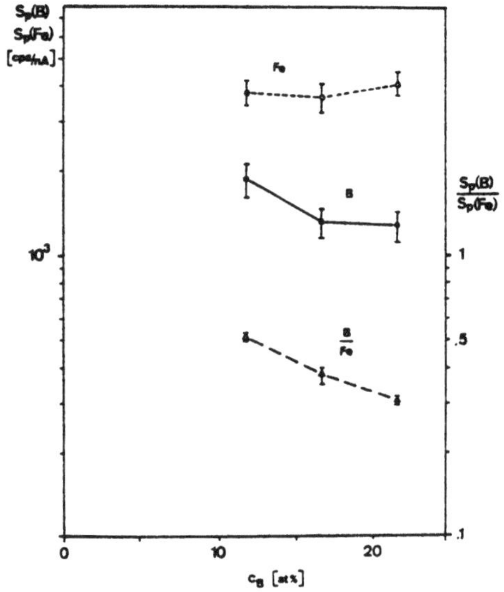

Fig.1 Absolute practical sensitivities and relative sensitivity factors of Fe and B in binary Fe-B metallic glasses

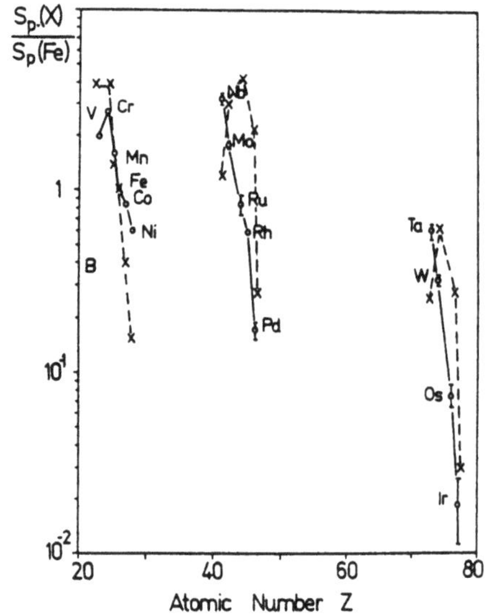

Fig.2 Relative sensitivity factors (referred to Fe) of transition metals in $Fe_{80}B_{17}X_3$ metallic glasses. (x) Pure metals, taken from [12]

using oxygen primary ions [1,5,12,13]. These comparative data were obtained from polycrystalline, multicomponent low-concentration iron-based alloys and from pure elements. On Fig. 2 also the RSF values measured by STORMS et al. [12] on pure metals are presented. The absolute values of [12] are about 2 orders of magnitude higher than our values. The dominant reason for this reduced sensitivity appears to be the lower transmission of the quadrupole when compared to a magnetic sector-field instrument.

Table 2 Relative sensitivity factors of transition metals in $Fe_{80}B_{17}X_3$ metallic glasses

electronic structure / period	d^3s^2		d^4s^2		d^5s^2		d^6s^2		d^7s^2		d^8s^2	
4	V	2.0	Cr	2.7	Mn	1.75	Fe	1.0	Co	0.84	Ni	0.60
5	Nb	3.2	Mo	1.8	Tc	-	Ru	0.83	Rh	0.59	Pd	0.17
6	Ta	0.61	W	0.32	Re	-	Os	0.076	Ir	0.018	Pt	-

According to our further investigations, at least for the elements studied, the RSF values determined on metallic glasses could also be used to quantify trace elements in stainless steel [16].

In the ternary systems the admixture element causes a small but statistically significant *matrix effect* in the sensitivity of iron and boron. Both $S_p(B)$ and $S_p(Fe)$ are far from being independent of the admixture element: in the group Nb-Pd there seems to be a similar monotonous decrease of S_p of B and Fe with atomic number of the element X as it was found in the S_p of the admixture elements themselves. Some recent publications [3,14,15] about highly concentrated alloys render it probable, that in the homogeneous alloys the element of higher degree of ionization increases the practical sensitivity of the other component, and the latter one have a reverse influence on the previous one. Also according to our measurements generally the sensitivity of B and Fe is increased by the admixture of elements with higher sensitivity and decreased by admixture of elements with lower S_p. Note, however, that the number of exceptions is big, and that the practical sensitivities of the elements X vary within a factor of about 200, while the practical sensitivities of B and Fe in dependence on the admixture element vary only within a factor of about 6.

References

1. D.H. Smith and W.H. Christie, Int. J. Mass Spectrom. Ion Phys. 26, 61 (1978)
2. D.E. Newbury, Pittsburgh Meeting on Anal. Chem., Cleveland (1977) no. 234
3. M. Riedel, PhD Thesis, Hung. Acad. Sci., Budapest (1978)
4. H.W. Werner, Surf. Interf. Anal. 2, 56 (1980)
5. J.M. Schroeer, J. Vac. Sci. Technol 14, 343 (1977)
6. R.W. Cahn, Contemp. Phys. 21, 43 (1980)
7. M. Bernheim and G. Slodzian, Int. J. Mass Spectrom. Ion Phys. 20, 295 (1976)
8. R.G. Rüdenauer, W. Steiger and U. Kraus, Microchim. Acta (Wien) Suppl. 8, 51 (1979)
9. F.G. Rüdenauer and W. Steiger, Microchim. Acta (Wien) in print
10. A. Lovas, private communication. *The authors wish to express their thanks to Dr. A. Lovas for supplying the samples.*
11. M. Riedel, H. Gnaser and F.G. Rüdenauer, ÖFZS-Report, to be published, PFZS, Vienna
12. H.A. Storms, K.F. Brown and J.D. Stein, Anal. Chem. 49, 2023 (1977)
13. C.A. Andersen, Int. J. Mass. Spectrom. Ion Phys. 3, 413 (1970)
14. M.L. Yu and W. Reuter, J. Appl. Phys. 52, 1478 (1981)
15. G. Modler and H.E. Beske, Berichte der Kernforsch. Jülich Nr. 1651 (1980)
16. M. Riedel, H. Gnaser and F.G. Rüdenauer, submitted to Anal. Chem.

Experimental Procedures for Quantitative Analysis of Silicate Minerals

A. Havette and G. Slodzian

Laboratoire de Physique des Solides, Université Paris-Sud
F-91405 Orsay, France

Elemental analysis of silicon bearing minerals by secondary ion mass spectrometry is particularly useful when dealing with small mineral phases and low concentration elements. Many problems have to be solved before getting reliable results. In this paper, we describe some of the experimental procedures we have been using. The most important problems are the electrical charging up effects on insulating materials, the presence of polyatomic ions and matrix effect.

1. Electrical Charging up

It is well known that the bombardment of insulating material produces electrical charging up which disturbs the measurements and induces migration of the most mobile ions (sodium, potassium...). In our experiments the sample is usually covered with a vacuum deposited metallic grid (150 μm mesh) which sets the potential of the object at a given voltage and collects the electrical charges. Then the primary beam density is altered until a nearly uniform potential is obtained over the free surface of the bombarded sample. The progress of the adjustment is followed by observing the ion images; the electrostatic mirror of the CAMECA IMS 300 allows to check that the potential variations are smaller than 1 volt inside a square of the grid. When the right conditions are obtained, the secondary ion currents are stable, the ion drift is considerably reduced and reliable measurements can be made. Many standard measurements have been repeated at time intervals of several months and yielded the same values.

When the sample is bombarded with positive ions (mainly O_2^+) and positive ions are collected, the charge balance depends upon the primary beam density (Fig. 1). One can observe that at high and low primary density the electrical charging up is respectively positive and negative [1]. Thus, it is possible to find an intermediate density at which the charging up in cancelled. This behaviour is due to a subtle balance between positive primary charges arriving on the sample, positive secondary charges leaving it and secondary electrons produced on the neighbouring extraction electrode which are attracted by the sample. It is very likely that those electrons not only help to discharge (or charge) the sample but also produce point defects which induce some conductivity in the superficial layer.

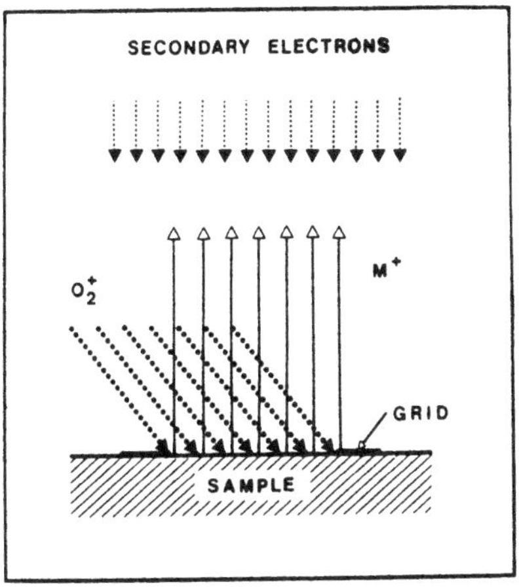

Fig. 1 Charge balance of the sample

It should be noted that in our experimental conditions if O^- ions are used as primary species the charging up is always negative and the sample potential is depressed by an amount of several tens of volts.

2. Interfaces Due to Polyatomic Ions

It is well known that mass interferences disturb the measurements of monoatomic species of trace elements. Such a disturbance may also affect major elements in some specific cases (for instance iron in a pyroxene mineral highly concentrated in calcium). High mass resolution (5000) is often sufficient for removing the most conspicuous interfering masses but in several instances much higher mass resolution would be desirable (isobares such as ^{48}Ti, ^{48}Ca, ^{58}Fe, ^{58}Ni, hydride ions from titatnium on, polyatomic species in the REE region, ...). Using high mass resolution reduces the secondary ion currents which produces a loss in sensitivity. When measurements must be performed in small areas (diameter smaller than 50 μm) the loss in sensitivity is even more drastic. However, in the latter the loss can be avoided by a proper matching between the collecting optical system and the mass spectrometer (transfer optics).

Another experimental procedure has been used. The energy window into which secondary ions are collected is hifted by about 150 volts [2]. The intensities of polyatomic species drop more rapidly than those of simple ions. Here again there is a loss of sensitivity (which cannot be overcome on small mineral phases by the use of a transfer optics). Moreover in some specific cases it does not work at all; for instance, hydride ions and multicharged ions have an energy distribution similar to that of simple ions. On the other hand the intensities of polyatomic species and the shape of the energy distribution curves of simple ions may vary strongly with the composition of the sample [3] therefore a careful preliminary study is necessary if one wishes to determine the amplitude of the energy shift which has to be performed in order to reduce the polyatomic peak height to the desired value.

Table 1 Cationic composition (%) of the standard minerals. K_X^{Al}: relative ionization yields of aluminium in X = Si, Al, Na...; measured yield: $R_{Si}^{Al} = I(Al^+)/I(Si^+) \cdot [Si]/[Al]$; calculated yield: $R_{Si}^{Al}[Si] + K_{Al}^{Al}[Al] + \ldots$

minerals \ X[%]	Si	Al	Na	K	Cs	Fe^{2+}	Mg	Ti	Fe^{3+}	measured yield	calculated yield	relative error[%]
glass	69.2	15.1	6.2	4.8	1.3	0.5	0.8	0.2	0	4.09	4.15	1.4
glass	68.4	14.5	9.3	5.3	0.4	2.1	0	0	0	4.31	4.39	1.8
albite	60.1	20.2	19.7	0	0	0	0	0	0	4.42	4.37	-1.2
leucite	51.0	24.8	0	24.2	0	0	0	0	0	4.45	4.57	2.7
oligoclase	55.0	25.1	13.5	1.1	5.0	0.3	0	0	0	4.46	4.62	3.5
sanidine	59.8	20.4	5.3	14.5	0	0	0	0	0	4.51	4.35	-3.7
oligoclase	56.3	24.0	15.3	1.2	3	0	0	0	0	4.61	4.50	-2.3
labradorite	45.7	33.5	6.6	0	13.5	0	0	0	0	4.78	4.92	3
glass	70.1	10.6	9.0	5.6	0	4.6	0	0	0	4.86	4.64	-4.6
bytownite	42.7	37.0	2.9	0	16.7	0.5	0.2	0	0	5.09	5.16	1.4
glass	68.4	10.5	10.6	5.3	0.3	3.0	0	0	1.6	5.11	5.02	-1.7
anorthite	41.3	38.7	1.3	0	18.7	0	0	0	0	5.15	5.16	0.3
cyanite	33.3	66.7	0	0	0	0	0	0	0	5.24	5.06	-3.5
glaucophane	48.0	10.8	11.5	0	0.3	8.3	11.1	0	0	5.61	5.91	5.4
cordierite	43.7	35.2	0.1	0.2	0	9.0	9.2	0	0	5.96	5.96	0.1
staurotide	25.5	57.4	0	0	0	10.9	1.9	0.2	0	6.46	6.59	2.0
glass	46.6	15.4	5.7	0.7	11.3	8.2	9.7	1.8	0	6.62	6.66	0.6
chlorite	21.4	10.9	0	0	0	1.6	30.0	0	1.4	6.69	6.56	-2.0
salite	45.3	4.3	0	0	22.0	6.9	19.9	1.7	0	6.96	6.98	0.3
melilite	36.9	12.9	7.3	0	32.5	0	7.5	0.1	2.9	7.12	6.74	-5.4
garnet	38.0	24.3	0	0	10.4	25.8	1.6	0	0	8.60	8.32	-3.2
glass	43.3	19.1	0	0	25	0	0	0	12.5	9.11	9.42	3.4
fassaite	34.5	14.8	0.4	0	23.8	0	11.3	2.2	10.8	10.03	9.89	-1.4
sphene	31.7	1.3	0.7	0	31.3	1.5	0	27.3	0	13.59	13.57	-0.2
K_X^{Al}	3.26	5.96	6.13	5.91	7.67	18.2	8.54	35.7	39.7			

3. Matrix Effects

For a given primary beam density the secondary ion intensities depend on the sputtering yield of the mineral being bombarded. Thus it is necessary to calculate normalized intensities; in the case of solicates the emission of Si ions is used as a reference; a relative ion yield can be defined as [4]:

$$R_{Si}^{X} = \frac{I(X^+)}{I(Si^+)} \cdot \frac{[Si]}{[X]}$$

where $I(X^+)$ and $I(Si^+)$ are respectively the intensities of the secondary currents of X^+ and Si^+ ions, [X] and [Si] cationic concentrations of X and Si in the mineral.

The secondary current depends upon the composition of the mineral but, to a first approximation follows a proportionality law:

$$R_{Si}^{X} \simeq K_{Si}^{X}[Si] + K_{Al}^{X}[Al] + \ldots$$

where K_{Si}^{X} = relative ionization yield of X element in a silica matrix...

A calibration procedure with twenty five known standards has been used for determining the relative ionization coefficients involved in this linear relation (Table 1). In this table it can be seen that the values of R_{Si}^{Al} varies from 4 to 13 and that silicon rich and iron poor minerals have smaller values of R_{Si}^{Al} than iron (or titanium) rich minerals. Table 1 shows that there is a fair agreement between the measured values of R and the values calculated from the empirical linear law. It is also worthwhile nothing that the degree of ionization of the cations in the sample may influence the yield of Al^+ ions. In particular K_X^{Al} value for $X = Fe^{3+}$ is twice that for $X = Fe^{2+}$. Once the K_Y^X coefficients are known, it is possible to calculate the practical ionization yields R of the elements present in the sample provided its composition is known. For unknown samples one has to use an iterative procedure as it has been explained elsewhere [5]. It is then possible to determine the cationic composition of mineral phases with an accuracy which is comparable to that given by the electron microprobe. It should be emphasized that the ion images help to localize and analyze mineral phases as small as 2 μm whereas the electron penetration very often prevents the compositional determination of such small phases because their thickness may be much smaller than their lateral extension.

References

1. G. Slodzian, R. Dennabouy and A. Havette, XVIII Colloquium Spectroscopicum Internationale, II, 590 (1975)
2. N. Shimizu, Earth and Planet. Sc. Let. 49, 398 (1978)
3. J.C. Lorin, A. Havette and G. Slodzian, these Proceedings
4. G. Slodzian, these Proceedings
5. D. Massare and A. Havette, J. Microsc. Spectrosc. Electron., to be published

SIMS Isotopic Measurements at High Mass Resolution

E. Zinner[1] and M. Grasserbauer

Institute for Analytical Chemistry, Technical University of Vienna
A-1090 Wien, Austria

One of the most promising capabilities of the ion microprobe mass analyzer which has not been fully realized so far is the measurement of isotopic ratios in small samples. In situ measurement of mineral assemblages in the ion probe avoids the time consuming desaggreagation of the sample into its mineral components [1]. However, the ubiquitous presence of interfering species (oxides, hydrides and other polyatomic ions as well as multiply charged ions) in the secondary ion spectrum has in the past restricted isotopic measurements by SIMS to elements like Li, Be, B [2] or Mg [3-8] where interferences are either virtually absent or their contributions can be corrected for. In all these studies interferences were small and/or the measured isotopic effects large. In the case of Pb in zircons [9] the number of interfering species is large and their contributions considerable. As a consequence the validity of correction procedures is questionable. Energy filtering offers the possibility of eliminating molecular interferences [10] but the method suffers from a large loss in the secondary ion signal and cannot be applied in the case of interfering hydrides and multiply charged ions. Another approach is to separate off interfering ions by using a secondary ion mass spectrometer capable of high mass resolution. The recently commercially available CAMECA IMS 3F achieves a mass resolving power which allows the elimination of most of the main molecular interferences in the atomic mass range 1-60. This opens the possibility to extend isotopic measurements to other elements and to cases where contributions from interfering ions are dominant [11].

We are presently engaged in the development of a computer software system for the measurement of isotopic ratios at high mass resolution. With the CAMECA IMS 3F it is possible to obtain a mass resolving power $R = m/\Delta m$ of more than 10^4. Fig. 1 shows a Si spectrum measured at $R > 5000$ which easily allows the separation of ^{28}SiH from ^{29}Si ($m/\Delta m = 3508$). The isotope measuring routine takes the following problems into consideration: a) fluctuation of the secondary ion signal; b) dead time losses of the ion counting system (electron multiplier); c) hysteresis effects of the magnetic mass analyzer; d) instability of the magnetic field.

[1]Permanent address: McDonnel Center for the Space Sciences, Washington University, St. Louis, MO 63130, USA

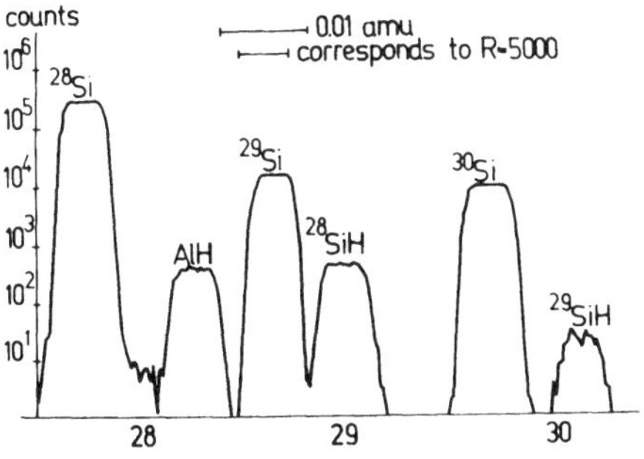

Fig. 1 High mass resolution silicon spectrum measured in Lake County plagioclase

Isotopic ratios are measured by peak jumping from one mass to the next. The peak jumping sequence is repeated many times (typically 50-100) in order to average over the statistical fluctuations of the secondary ion signal.

The dead time loss of the pulse counting system is corrected for each measurement. For this purpose the dead time of the whole counting system is determined in an operational way by measuring the isotopic ratios in a pure titanium sample. Measured are 4 isotopic ratios with ^{48}Ti as reference isotope at low mass resolution. Fig. 2 shows the result of such a measurement. Plotted are the deviations (Δ-values) from the normal terrestrial ratios [12]. The deviations are attributed to dead

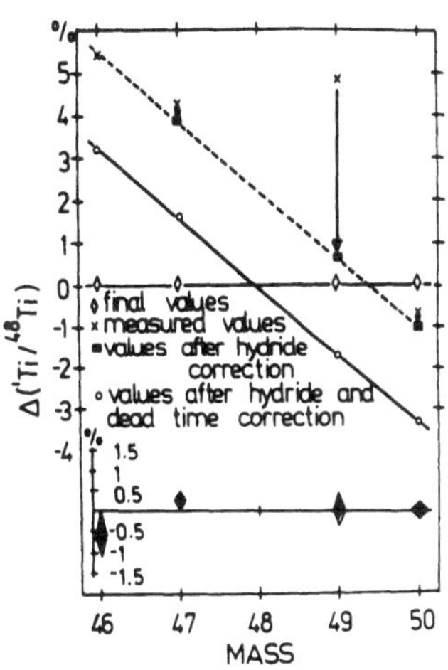

Fig. 2 Dead time determination from the measurement of Ti isotopic ratios. The three parameters hydride contribution, counting system dead time and isotopic mass fractionation are obtained from a least squares fit to 4 measured ratios. The fitted Δ-values are shown in the lower part of the figure with the statistical errors of the measurement. The values of the parameters obtained from this measurement are: dead time = 21.4 ns; hydride contribution at mass number n+1 = $3.1 \cdot 10^{-3}$ times signal at mass number n; mass fractional coefficient = -1.66 % per unit mass number. The count rate of $^{48}Ti^{+}$ was $1.13 \cdot 10^{6}$ cts/s

time losses, fractionation effects (instrumental fractionation during analysis and possible chemical fractionation during fabrication of the sample) and contributions from TiH. Careful inspection of the mass spectrum in the mass range 1-150 and measurements at high mass resolution did not reveal any other possible contributions. A least squares fit is performed under the assumption that the mass fractionation effect is linear which yields the parameters dead time, mass fractionation coefficient and hydride contribution. The deviations of the fitted values compare with the statistical errors of the measurement which are typically of the order of a few hundredths of a percent for a measurement of 100 cycles.

Hysteresis effects of the magnet can play a large role during measurements at high mass resolution when magnetic field values have to be set with a precision corresponding to less than 0.001 amu shift in mass. The magnetic field experienced by the secondary ions depends on the history of the magnet even if the value measured by the Hall probe is the same. Fig. 3 shows the change of the position Δm of the Al^+ peak during cycling through the mass sequence 25, 26 and 27 for 4 cases distinguished by different magnetic histories. In order to eliminate these hysteresis effects the field is cycled many times through the isotopic measurement sequence before final calibration of the peak position for the measurement.

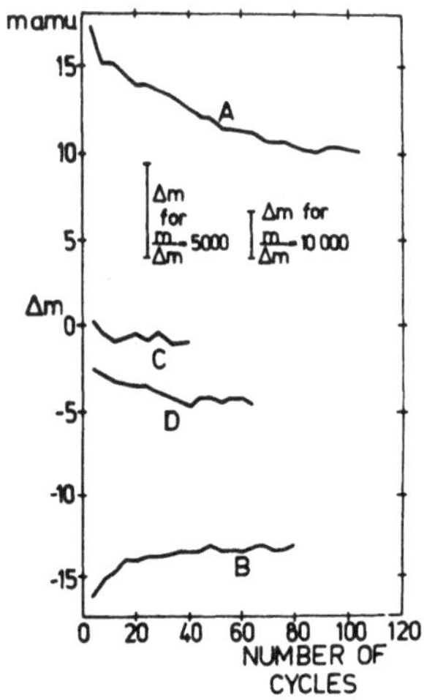

Fig. 3 Hysteresis effect of the magnetic field as seen by the secondary ions. Plotted is the shift of peak position of Al^+ ions in 10^{-3} mass units while cycling the magnetic field through a sequence of values corresponding to the mass numbers 25, 26 and 27. The peak position as defined by the rising edge of the high resolution (R = 5000) mass peak has been measured every 4 cycles relative to an arbitrary origin. In case A the field has been set to a value corresponding to mass 1 before starting with the 25, 26, 27 cycle, in B to mass 150, in case C the sequence of masses before the cycle was 1 and 50 and D it was 1, 150, 10, 70, 20, 50, 23 and 30

Instabilities of the magnetic field, the electrostatic analyzer voltages and possibly charging effects lead to a drift of the peak positions besides the initial change due to hysteresis effects. This drift is automatically compensated by repeated centering of the mass peaks during the measurement sequence. The program can either jump from peak center to peak center and count at the peak tops of scan across peaks, add scans from different cycles and determine the total counts from the flat top portions of the peaks. In either mode it is possible to measure isotopic ratios at a mass resolving power R of more than 8000 (separation of ^{56}FeH from ^{57}Fe).

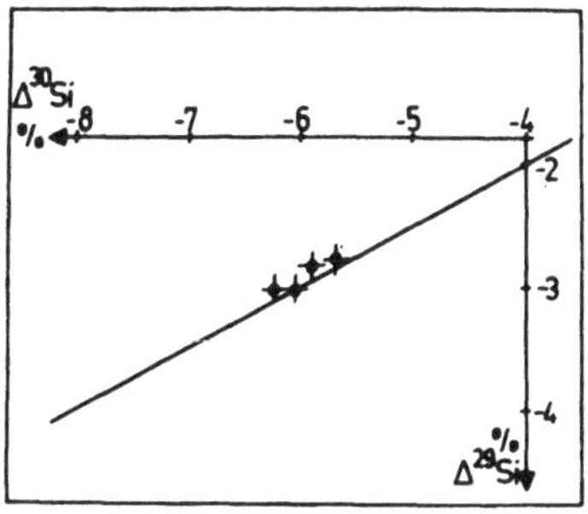

Fig. 4 Silicon isotopic ratios measured at a mass resolving power R = 5000 in lunar plagioclase. Plotted are the Δ-values in % obtained from 100 cycles with the statistical errors. The plotted line is the linear mass fractionation line through the origin corresponding to normal terrestrial composition [13].

A large number of tests have been performed in terrestrial and lunar minerals. The precision which can be achieved depends on concentration and ionization yield of the element in question, the size of the analyzed volume, the relative intensities of interfering species and the mass resolving power necessary for their elimination. A fundamental problem for isotopic measurements is the large mass fractionation effect observed in the measured ratios. The magnitude of this effect is different for different elements but also varies with the matrix and depends on instrumental conditions like the energy and angular selection of the secondary ions. Fig. 4 shows Si isotopic ratios measured in a lunar feldspar. The individual data points lie within less than 0.1 % of the mass fractionation line but scatter along this line by sometimes more than 1 %. The result is similar to Mg which shows a much smaller magnitude of the mass fractionation but the same scatter along the fractionation line and also for Ti where the mass fractionation coefficient varies from -1.3 % to -1.7 % per amu. SLODZIAN et al. [14] have presented a qualitative model which explains the mass fractionation effect as originating in the differential ionization process during sputtering. Additional work has to be performed to separate the sputter ionization contribution to the mass fractionation from possible instrumental effects. As long as deviations from the mass fractionation line (excesses of one isotope) are measured for an element with 3 or more isotopes (3-6) the relatively large error in the magnitude of the mass fractionation is of lesser consequence. In cases where the measurement of absolute ratios is sought this error has to be reduced by a better understanding of the fractionation effect and a better control of instrumental conditions.

References

1. T. Lee, D.A. Papanastassiou and G.J. Wasserburg, Geochim. Cosmochim. Acta 41, 1473 (1977)
2. D. Phinney, B. Whitehead and D. Anderson, Proc. Lunar Planet. Sci. Conf. 10th, (1979) p. 885
3. I.D. Hutcheon, I.M. Steele, J.V. Smith and R.N. Clayton, Proc. Lunar Planet. Sci. Conf. 9th (1978) p. 1345
4. I.D. Hutcheon and I.M. Steele, Lunar Planet. Science XI, 496 (1980)
5. I.D. Hutcheon and R.C. Newton, Lunar Planet. Science XII, 491 (1981)

6. J.G. Bradley, J.C. Huneke and G.J. Wasserburg, J. Geophys. Res. 83, 244 (1978)
7. J.D. Macdougall and D. Phinney, Geophys. Res. Lett. 6, 215 (1978)
8. J.C. Lorin and M.C. Michel-Levy, Lunar Planet. Science IX, 660 (1978)
9. J.R. Hinthorne, C.A. Andersen, R.L. Conrad and J.F. Lovering, Chem. Geology 25, 271 (1979)
10. N. Shimizu, M.P. Semet and C.J. Allegre, Geochim. Cosmochim. Acta 42, 1321 (1978)
11. J.C. Huneke, J.T. Armstrong and G.J. Wasserburg, Lunar Planet. Science XII, 482 (1981)
12. F.R. Niederer, D.A. Papanastassiou and G.J. Wasserburg, Lunar Planet Science XI, 809 (1980)
13. I.L. Barnes, L.J. Moore, L.A. Machlan, T.J. Murphy and W.R. Shields, J. Res. Nat. Bur. Stand. 79A, 727 (1975)
14. G. Slodzian, J.C. Lorin and A. Havette, J. de Phys. Lett. 41, L555 (1981)

Computer Peak Identification and Evaluation of SIMS Spectra

J. Antal, S. Kugler, and M. Riedel[1]
Physical Institute of the Technical University
H-Budapest, Hungary

The most important problems in computer peak identification and evaluation of SIMS spectra of multicomponent systems are:

(i) to find the possible ion species: ("variables") atomic, cluster and molecule ions,
(ii) the determination of the total isotopic currents of these species,
(iii) the reliability of the identification and evaluation process.

A few publications appeared recently concerning computer programs for evaluation of SIMS spectra [1-4]. We also developed a program package containing 3 modules which perform the above main tasks. The modules are called accordingly "LINE ANALYSIS", "SOLVE" and "SIMULATE".

In the following the most important aspects of these modules will be summarized and experiments using the package will be reported.

ad i): Most of the SIMS devices in operation have too low mass resolution to separate interfering isotopes of different ions having the same mass number. It was assumed that variables can be atomic ions, metallic clusters and molecule ions. The modul "LINE ANALYSIS" has to be initialized prescribing a noise level (back-ground level B.G.L.) and a cluster producing level (cluster limit, C.L.). Ion peaks below the B.G.L. will not be processed, while only those metallic ions are assumed to produce possible clusters and compound molecules for which the peak heights of the most abundant isotope (in CPS) are higher than C.L. Variables will be declared as *possible* after estimating their total isotopic currents using the CPS at the isotopic mass of highest abundance, *if* the estimated currents are higher that B.G.L. and if there is CPS input on all isotopes of that variable.

For isotope-currents lower than B.G.L. no CPS input is required (except of course for the main isotope). The module uses a basic library of the natural isotope masses and abundances for elements Z = 1 to 92 and the same data for some molecules generally present in the background: H_2, O_2, C_2, N_2. CH, CH_2, CH_3, CH_4, OH, OH_2, CO, CO_2. For all other clusters and

[1]Department of Physical Chemistry and Radiology, Eötvös University, H-Budapest, Hungary

molecules the program individually calculates these data starting from the atomic masses and abundances. Possible clusters and molecules are presented in the form M_n; M_nX where M is a metallic element above the cluster limit and X is any element previously identified in the spectrum with the exception of the noble gases. Singly and doubly charged ions are also taken into account.

In finding possible variables difficulties may occur especially in the case of very low input counting rates. This is because some of the isotopes of lower abundances cannot be taken into account to distinguish between overlapping ion-species.

ad ii): The variable list determines an abundance matrix $\underline{G}$, which will be decomposed if possible for disjunct not overlapping groups. In this way the evaluation of total ionic currents requires only the solution of smaller linear systems. In general, the least-square fitting method was used in the overdetermined (more equation than variables) as well as in the underdetermined case though here a linear programming should be more appropriate in principle. Standard matrix manipulation and inversion routines were used. When negative total currents are present in the results the negative variable of highest absolute value will be left out and the matrix solution will be iterated until all remaining variables are positive definite. In undetermined case the variables are previously ordered according to decreasing CPS values at their main isotopic masses and in deleting one of them a new variable is automatically introduced.

We have had rather good experiences with the algorithm as described. Nevertheless, we are planning to build in an optional linear programming routine.

In Table 1 the results of identification and evaluation of SIMS measurements performed by one of the authors is shown.

ad iii): It is rather difficult to test the method for reliability. We used for this purpose the "SIMULATE" module, which reconstructs the CPS spectra for a given ionic species current input. After evaluation of an input spectrum a simulated spectrum is reconstructed from the solution. Original and reconstructed spectrum are compared and CPS-differences for each line as well as the meansquare error are computed.

Comparison of the original and reconstructed CPS any way is a very direct test for the reliability of the method. One has to mention, on the other hand that different species inputs may in some cases result in the same CPS list as shown e.g. in Table 2. The authors hope however that this is a rather extreme example, not very common in normal practice.

The program was coded on a HP-9825 desktop computer. A complete program run requires some minutes depending on the number of variables in the decomposed groups. The implemented

Table 1a Measured and reconstructed mass spectra and lists of identified variables

	Measured		Reconstruction	
Mass	CPS 1.	CPS 2.	1.	2.
10	165	203	157	185
11	640	756	642	760
12	4	-	-	4
16	21	33	21	33
23	334	7	334	7
24	11	-	11	-
27	175	-	175	-
28	11	-	11	-
32	6	10	6	10
39	115	7	115	7
40	82	-	82	-
41	10	-	9	1
46	-	-	1	-
47	-	-	1	-
48	10	-	9	-
49	-	-	1	-
50	74	-	77	-
52	1487	-	1488	-
53	176	-	170	-
54	1164	1737	1134	1559
56	17143	24477	17143	24477
57	413	586	406	579
58	55	76	58	83
Error			33	179

Table 1b

Element	CPS 1.	CPS 2.
Fe	18701	26711
Cr	1777	-
B	799	946
Na	334	7
Al	175	-
K	124	7
Ar	80	-
Ca	2	-
O	21	33
N_2	11	-
C_2	11	-
Fe^{++}	1	-
Ti	13	-
O_2	6	10
C	-	4

Table 2 Different species inputs may result in same CPS list

Element	%	Element	%	Mass	CPS
H	4.17	H	4.17	1	658437
Al	20.83	Al	20.83	2	99
Si	41.67	Si	42.30	27	3292680
N_2	8.33	N_2	24.42	28	10000000
SiH	8.33	SiH	8.28	29	2548092
AlH	8.33	P	0.0018	30	265345
CO	8.33			31	40204
				32	6

program is limited to handle a maximum of 100 input CPS values in the mass range of 1 to 3000 a.m.u. and a variable list of 110 items maximum.

We are indebted to Prof. F. Rüdenauer for valuable advice and criticism and also for the opportunity for one of the authors [M.R.] to perform SIMS measurements at Seibersdorf. A more detailed discussion of these topics and other comparative spectrum identification development made in cooperation with Prof. Rüdenauer and his staff will appear in a joint publication in the near future.

References

1. R.Dobrozemsky, J. Vac. Sci. Technol. 9, 220 (1971)
2. W. Steiger, F.G. Rüdenauer, Vacuum 25, 409 (1975)
3. B.R. Kowalski, Anal. Chem. 47, 1152A (1975)
4. V.T. Cherepin, A.A. Kosyachov, G.I. Gudzenko, Int. J. Mass Spectrom. Ion Phys. 36, 225 (1980)

Part V

Application I. Depth Profiling

Depth Profiling of Copper Atoms Gettered in Ion-Damaged GaP

M. Griepentrog, H. Kerkow, H. Klose, and U. Müller-Jahreis
Humboldt-Universität zu Berlin
GDR-1040 Berlin

Introduction

The gettering process was introduced in the field of semiconductor technology by SHOCKLEY and GOETZBERGER [1]. The designation "gettering" comes from the valve technology. Gettering processes in semiconductors include

(i) the binding of unwished-for electrically active impurities and/or crystal defects into electrically inactive complexes,
(ii) the outdiffusion of metal atoms from space charge regions into surface layers and into interfaces,
(iii) the outdiffusion of dopants or the addition of atoms to bind unwished-for electrically active atoms.

Today, gettering processes include thermal annealing as well as gettering by phosphorous or boron silicate glass in silicon technology. Recently damage gettering processes were developed.

This backside gettering takes place during a thermal treatment in an inert gas atmosphere as a consequence of a preceding damage process which can be induced by mechanical treatment, ultrasonic, laser radiation or ion implantation. Sometimes the results are improved by a defined oxidation of the sample. In these cases oxygen is added to the intert gas. The application of gettering processes leads to marked higher yields in the production of integrated bipolar circuits as it was shown by KEVASAN et al. [2], to higher efficiency of polycrystalline solar cells [3], and to higher quantum efficiency of green LED's in GaP [4].

Conditions for gettering

In all backside gettering processes the following conditions must be performed for a good gettering:

(1) A region of higher solubility exists in comparison with bulk solubility, where in the final gettering state the free enthalpy reaches a minimum.
(2) The diffusion coefficient for the migration mechanism of atoms in bulk material D to the damage region must be characterized by

$$D(T_a) > L^2/t_a,$$

(L - sample thickness, t_a - annealing time).

The first condition explains the gettering process in dependence on pressure, temperature and time in connection with special material parameters: $G = U + pV - TS$.[1]

From the condition (2) the annealing parameters follow from a given material and sample geometry, for example of L = 200 μm and $t_a \equiv 1$ h we obtain a minimal diffusion coefficient of $D = 10^{-7}$ cm^2s^{-1}. The example shows that a getter effect only can be reached in the cases of fast diffusing atoms. Such elements are Cu, Fe and Au in Si, Cu and Cr in GaAs and Cu in GaP.

Ion beam gettering

From the first publication in the field of ion beam gettering in 1972 [6] and its first application in device technology 1973 [7] this new gettering method developed to an effective technological procedure. The advantages of the ion beam gettering compared to other methods are

(i) the high reproducibility of the damage conditions by the precise adjustment of dose, energy, temperature, and ion species,
(ii) the application of ions which become electrically inactive atoms inside the crystals,
(iii) the high vacuum conditions of the process,
(4i) the damage production through planar layers,
(5i) the gettering by implanted atoms themselves, and
(6i) the lower annealing temperatures in comparison with the glass gettering.

The possibilities of ion beam gettering were mainly studied for Cu in GaP and in first measurements for Ni and Cr in GaP.

Experimental

In the investigations of gettering processes the secondary ion mass spectroscopy was used for the determination of gettered atom concentrations. The high detection sensitivity for a great number of elements, the mass resolution and the possibility of depth profiling are advantages, which recommend SIMS for the investigation of gettering processes.

In the experiments the secondary-ion analyzer IMS 300 (CAMECA) was applied. The primary oxygen ions had an energy of

1) By AUTH [5] the terms were estimated for Si: $pV \approx 100$ Jkg^{-1}, $U = (10^6-10^7)$ Jkg^{-1}, and the entropy term about some 10^5 Jkg^{-1}.

5.5 keV. For the detection of Cr, Ni and Cu in GaP the $^{52}Cr^+$, $^{58}Ni^+$, and $(^{69}Ga^{63}Cu)^+$ ions were analysed. Implanted profiles in GaP served as standards for calibration. The depth calibrations were carried out by measuring the sputtering crater depths (Talystep). All ion implantations were performed by the special 350 keV accelerator of the Humboldt University.

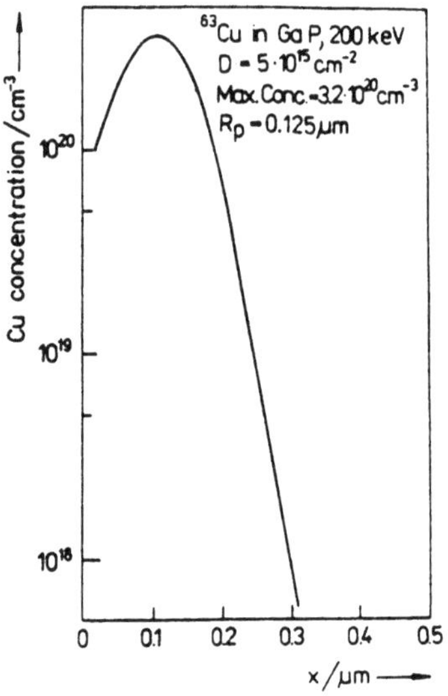

Fig. 1 SIMS depth profile of Cu implanted in GaP

In Fig. 1 the profile of Cu implanted in GaP is shown. The detection limit of the apparatus reaches $(3\text{-}5)\cdot 10^{17}$ cm^{-3}. This exceeds the solubility at the annealing temperatures applied here. The solubility at 650 °C is about $2\cdot 10^{16}$ cm^{-3} [8] and at 950 °C about $1\cdot 10^{17}$ cm^{-3} [9]. Ion damaged regions were produced by Ar implantation in (100)-GaP (VPE material). All annealing processes have take place in a nitrogen atmosphere. On the backside of the samples metal layers of 3000 Å thickness were vacuum deposited. For the control of the annealing and amorphization states measurements of RBS and PIXE were used.

Results

In Fig. 2 the copper gettering is shown in comparison to an undamaged and to a "copper-free" sample, respectively. The gettered copper decorates the ion-damaged region and its concentration is higher than the solubility at 500 °C. An analogous gettering effect is observed after an Ar implantation followed by an annealing process in homogeneously copper diffused GaP (Fig. 3).

Using the same gettering procedure, Ni and Cr have not been detected in the ion-damaged region. The dependence of the copper gettering on annealing conditions is represented in Figs. 4 and 5. In the case of T_a = 600 °C the expected time dependence of gettering concentration in the form of

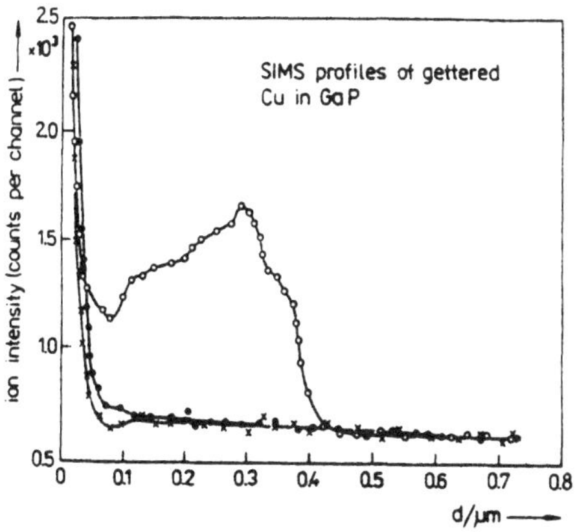

Fig.2

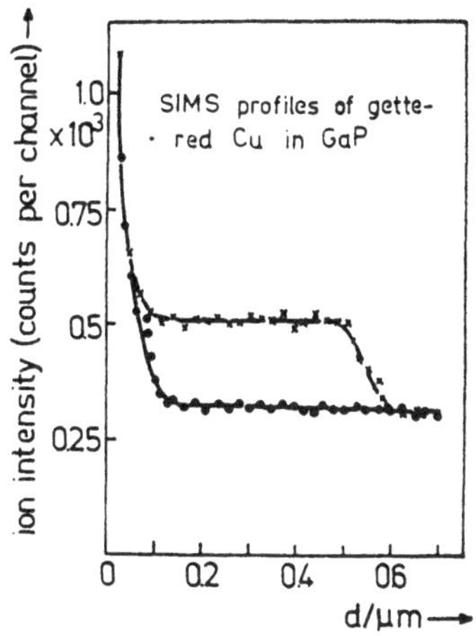

Fig.3

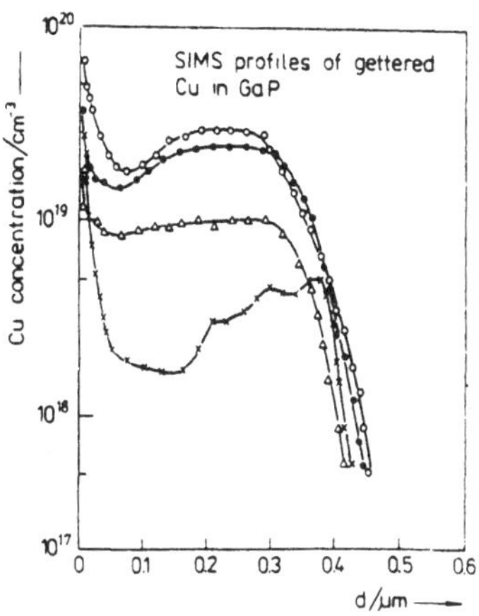

Fig.4

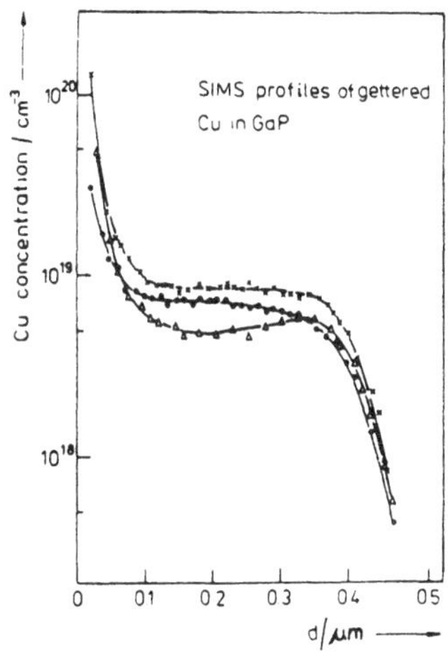

Fig.5

Fig. 2 SIMS depth profiles of Cu gettering in GaP after annealing for 5h at 500 °C. ooo - sample copper deposited and argon implanted (E = 300 keV, D = $2\cdot10^{15}$ cm^{-2}); ... - sample copper deposited and undamaged; xxx - sample without copper deposition and undamaged

Fig.3 SIMS Cu depth profiles of getting in Cu diffused GaP (3000 Å Cu deposited on the backside and for 2 h at 600 °C annealed). (xxx) sample Cu diffused and argon implaned (E = 300 keV, D = $5\cdot10^{15}$ cm^{-2}) and 1 h at 600 °C annealed, (ooo) sample undiffused and undamaged, 1 h at 600 °C annealed

Fig. 4 SIMS depth profiles of copper in GaP showing the temperature dependence of gettering by Ar implantation (E = 300 keV, D = $2\cdot10^{15}$ cm^{-2}) and annealing for 1 h (xxx - 600 °C, ΔΔΔ - 650 °C, ... 700 °C, ooo 750 °C)

Fig.5 SIMS depth profiles of copper in GaP showing the dependence the gettering by Ar implantation (E = 300 keV, D = $5\cdot10^{15}$ cm^{-2}) from the annealing time (annealed at 600 °C for ΔΔΔ 30 s, ooo 1h, xxx 2h)

$$N_{Cu} = N_O \exp (D \cdot t_a/L^2)^{1/2}$$

has been observed. RBS- and PIXE-measurements show a very high degree of amorphization (95-100 % in the sense of RBS-experiments) for an annealing interval of 500 °C ≤ T_a ≤ 750 °C in a nitrogen atmosphere. ($D_{Cu}(600\,^{o}C)_{meas.} \approx 10^{-7} cm^2 s^{-1}$.)

Discussion

As is shown in Fig. 6, the gettered copper atoms decorate distinctly the ion-damaged region and their profile corresponds to the Ar-profile determined by means of SIMS measured with the $^{40}Ar^{69}Ga^{+}$ cluster) and with the damage profile determined by RBS. Under these annealing conditions, the argon profiles remain unchanged. With increasing Ar dose the copper decoration of ion-damaged regions is more homogeneous (see Figs. 4 and 5).

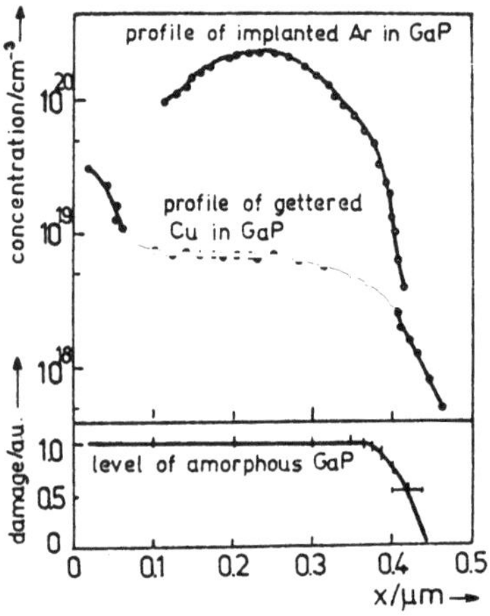

Fig. 6 SIMS depth profiles. ooo - Ar implanted in GaP (E = 300 keV, D = $5 \cdot 10^{15}$ cm^{-2}) annealed 1 h 600 °C; ... - Cu gettered in GaP by this Ar implantation after annealing for 1 h at 600 °C; +++ damage profile in GaP after Ar implantation and annealing for 1 h at 600 °C calculated from RBS spectra

For the migration process of copper atoms through the GaP bulk, an activation energy can be derived by means of the equation

$$N_{Cu} = N_o \exp(- E_a/kT_a)\exp(a\ t_a)^{1/2}.$$

From the measurements presented here a value of E_a = (1.57±0.15) eV is received. Im comparison to values of SAH and WANG for midgap centers (E_a = 2.15 eV) in silicon and MAGEE et al. for Cr in GaAs (E_a = 0.88 eV) this value represents a typical energy for fast diffusion processes [10,11]. If the annealing temperature exceeds 700 °C, the saturation of copper in damaged layers begins which can be explained by the limited solubility and the beginning of recrystallization (Fig. 4).

From the experimental results of copper gettering in GaP, one may conclude that

(1) copper is gettered by ion-damaged regions in GaP under conditions of T_a = 500 °C and an annealing time of one hour more,

(2) damaged layers can be decorated by copper in GaP similarly to crystal defects in silicon, and
(3) the gettering conditions are likely to be compatible with technological processes of LED's production.

Further investigations have to prove the utility of copper gettering by ion-damaged layers in electronic devices as in the case of the Al gettering in GaP [12,13].

The authors thank the VEB Werk für Fernsehelektronik Berlin for supporting of these investigations, and also to Dr. W. Frentrup.

References

1. A. Goetzberger and W. Shockley, Journ. Appl. Phys. 31, 1821 (1960);
   W. Shockely, Solid State El. 2, 35 (1961)
2. R. Kevasan et al., J. Electrochem. Soc. 126, 642 (1979)
3. T. Saitoh et al., IEEE Trans. El. Dev. ED-27, 671 (1980)
4. B. Wessels, El. Letters 15, 748 (1979)
5. J. Auth, Berichte Humboldt Universität zu Berlin (to be published)
6. T.M. Buck et al., Appl. Phys. Letters 21, 485 (1972)
7. C.M. Hsieh et al., Appl. Phys. Letters 22, 238 (1973)
8. M. Schneider and E. Nebauer, Phys. Stat. Sol. (a) 46, K71 (1978)
9. H. Klose et al., Phys. Stat. Sol. (a) 16, K39 (1973)
10. C.T. Sah and C.T. Wang, Journ. Appl. Phys. 46, 1767 (1975)
11. T.J. Magee et al., Appl. Phys. Letters 37, 53 (1980)
12. P.C. Mürau and R.N. Bhargava, J. Electrochem. Soc. 123, 728 (1976)
13. H. Klose and M. Rieth, Phys. Stat. Sol. (a) 66, (1981)

The Optimisation of SIMS for the Analysis of Semiconductor Materials

J.B. Clegg

Philips Research Laboratories
Redhill, England

The electronic properties of semiconductor materials such as conductivity, carrier mobility and carrier lifetime depend on the interactions between deliberately added impurity elements and contaminants, introduced during materials preparation and processing, at concentrations in the range 10^{13} to 10^{20} atoms cm^{-3}. The correlation of physical properties with chemical concentrations in the thin films typically used for industrial electronic devices requires the determination of the distribution of trace impurity elements as a function of depth within layers that may be only 0.1 μm thick. Secondary ion mass spectrometry (SIMS) can in principle measure the in-depth distribution of a broad range of elements but in practice the absolute accuracy of the concentration measurements, the spacial accuracy of the profile, the dynamic range of the measurement and the limits of detection can be distorted by apparently minor secondary processes occurring in the system or at the sample. These include contamination by impurity ions in the primary ion beam or sputtered from the system or surrounding sample, memory effects and atomic mixing of the sample matrix atoms during the sputtering process.

In a previous publication [1] we reported the analytical performance of a dedicated UHV quadrupole SIMS system (the a-DIDA ion microprobe manufactured by Atomika) in the analysis of GaAs using oxygen and argon primary ion bombardment. The detection limits of 10^{14} to 10^{17} atoms cm^{-3} are sufficient for many semiconductor characterisation studies but significant improvements were predicted with minor instrument modifications. This paper describes a number of instrument modifications and measurement techniques which improve the elemental detection capability of the system. Experimental evidence, based on the analysis of standard samples calibrated by independent techniques, is given to show the increased sensitivity and precision.

System Modifications

The purity of the primary beam bombarding the sample surface is of utmost importance in trace SIMS analysis. Ionized metallic and gaseous impurities, present in the discharge plasma of the ion source, are extracted with the source gas ions and are subsequently implanted and sputtered from the sample surface.

Beam impurities commonly include Al, Cr, Mn, Fe, Ni and Cu and using a pure GaAs sample and a cold cathode source [2], this contamination gives rise to apparent concentrations in the range 10^{15} to 10^{17} atoms cm^{-3}. It should be noted that this type of contamination also occurs with the duoplasmatron ion source. Mass filtration of the primary beam with a simple Wien filter [3] leads to a reduction in contamination by at least two orders of magnitude (Table 1). The Si^+ ion which may not be completely separated from O_2^+, can be eliminated if a high purity (99.99 %) Ni cathode is used.

Contamination of the sample surface is also caused by the sputtering of surfaces in close proximity to the sample. All metal components, originally of stainless steel, which are in line of slight of the sample, have been replaced with Ta or Ti. This has led to the detection limit for Fe being reduced from about 5×10^{14} atoms cm^{-3} to 7×10^{13} atoms cm^{-3} (see Table 1).

Table 1 Detection limits in a depth profiling mode for elements in GaAs

Element	Detected ion	[ppma]	[atoms cm^{-3}]
Be	$^{9}Be^+$	0.002	8×10^{13}
B	$^{11}B^+$	0.01	5×10^{14}
Al	$^{27}Al^+$	0.01	4×10^{14}
Si	$^{28}Si^+$	0.01	4×10^{14}
Cr	$^{52}Cr^+$	0.001	4×10^{13}
Mn	$^{55}Mn^+$	0.001	4×10^{13}
Fe	$^{56}Fe^+$	0.002	7×10^{13}
Cu	^{63}Cu	0.02	1×10^{15}
Zn	$^{64}Zn^+$	0.1	5×10^{15}
Ge	$^{74}Ge^+$	0.2	1×10^{16}
Sn	$^{120}Sn^+$	0.1	5×10^{15}

Experimental conditions: primary beam O_2^+ at 12 keV with 2 μA target current, raster scan width 0.6x0.6 mm, sputter rate ~4 μm h^{-1}. Energy discrimination of molecular ions used with target voltage off-set [1]. Screening electrode in operation.

Improved transport of secondary ions from the sample surface to the entrance of the energy analyser has been obtained by the addition of a screening electrode to the entrance side of the energy analyser (Fig. 1). This electroce which is independently biassed, gives a more localized extraction field and leads to an increase in ion transmission by a factor of about 3 [4]. A further improvement in transmission has been obtained by reversing the polarity of the field applied to the quadrupole rods. In this mode of operation, the acceptance characteristics of the quadrupole more closely match the output of the energy

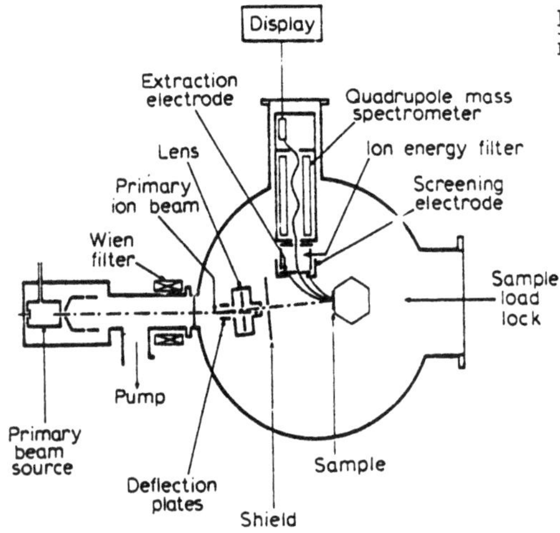

Fig. 1 Schematic diagram of modified SIMS system

analyser [5] in both ion energy and angular distribution. Taken overall the transmission efficiency of the analysis system has been improved by a factor of about 6. This can be judged from the results obtained from a bulk GaAs sample, uniformly doped with Cr at $1x10^{17}$ atoms cm^{-3}. With the standard system, the count rate at mass 52 was $2x10^3$ c/s per μA (O_2^+) compared with the present value of $1.2x10^4$ c/s. As the measured sputter yield for GaAs is 2.8 atoms per O_2^+ ion [6], the useful ion yield (ions detected per atom sputtered) is $\sim 5x10^{-4}$.

Analytical Performance

Broadening of the concentration profile due to re-location of the target atoms during sputter erosion is of particular interest in a SIMS analysis. It has been shown that the mass, angle of incidence and energy of the bombarding ion, have controlling effects on the shape of the depth profile [7,8]. We have investigated SIMS profile broadening as a function of beam energy using ion implanted Si samples. Figure 2 shows Sb profiles measured with energies of 1.5 and 12 keV (O_2^+) and it can be seen that significant broadening occurs at 12 keV. Extrapolation to 'zero energy', of the profile width as a function of ion energy (1.5 to 12 keV range) enables an estimate of the true shape of the profile to be obtained [2]. This corrected profile is shown in Fig. 2 and it can be seen that extrapolation leads to a reduction of about 10 % in the relative width of the 1-5 keV profile.

A near surface peak at 10 nm is a dominant feature of the 12 keV Sb profile shown in Fig. 2. With the lower beam energy, this peak is only just detectable at ~3 nm. Both surface peaks, together with the rapid rise in concentration in the regions 0 to 3 nm and 0 to 10 nm, are associated with the built up of oxygen in the target surface during the initial stages of sputtering. This process gives rise to an increase in ion yields until equilibrium is established after these regions have been sputtered away. The 1.5 keV profile is therefore unreliable in the depth range 0 to 3 nm.

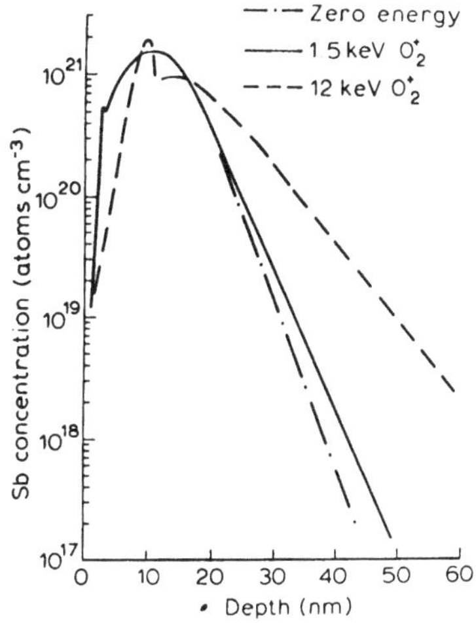

Fig. 2 Depth profiles of ^{121}Sb (10^{15} cm^{-2} at 10 keV) implanted in crystalline Si. Two SIMS primary beam energies used to demonstrate beam induced profile broadening. The estimated "true" distribution of Sb is given by the zero energy profile

The observed SIMS profile broadening is most probably due to a combination of atomic mixing effects which occur during the ion bombardment of the sample. These include cascade and recoil mixing and enhanced diffusion due to the creation of defects in the near surface region [9]. It may well be that enhanced diffusion is the dominant process giving rise to SIMS profile broadening.

The elemental detection limits of our system, with the screening electrode in operation, are given in Table 1. These limits were determined by comparing peak signal intensities (I_s) from carefully prepared GaAs samples (ion implanted, bulk and epitaxial layers) containing known concentration of dopants with the corresponding intensities (I_u) from selected pure GaAs samples (Bridgeman, LEC and MBE). The limits (C_{MIN}) were calculated from the realtionship, $C_{MIN} = I_u.C_s/I_s$, where C_s is the element concentration as measured by atomic absorption [10] or deduced from the ion implanted dose [11]. It should be noted that I_u contains a contribution due to system noise measured at non-unit mass (~ 1 c/s per μA O_2^+ at mass 57.5). Lower limits are obtained if allowance is made for this background. The limits given in Table 1 not only demonstrate the high detection capability of our system but also indicate the inherent purity of the GaAs samples used in this study.

With this analytical performance, we have been able to investigate the incorporation of important acceptor elements (Cr, Fe and Mn) at concentrations below 10^{14} atoms cm^{-3} in a variety of GaAs samples [12]. This work is illustrated by the SIMS depth profile of Cr in a GaAs layer grown by the molecular beam technique (Fig. 3). It can be seen that the intensity at mass 52 which is essentially constant throughout the layer, is significantly greater than the background signal. After allowance is made for background, the estimated Cr concentration in the layer is about $5x10^{13}$ atoms cm^{-3} or 1 ppba. The profile also shows a rapid rise at the interface with the Cr doped substrate.

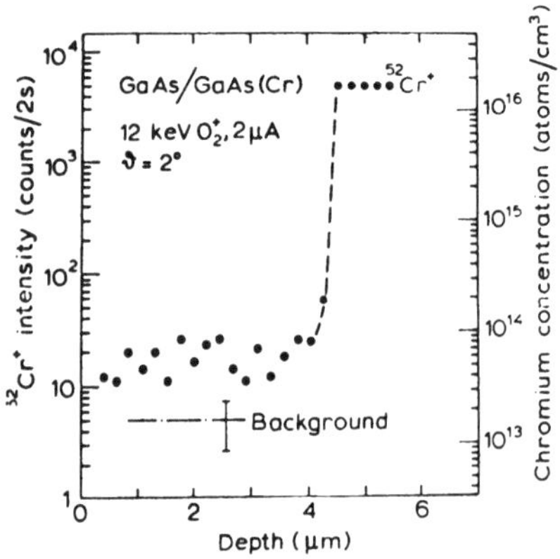

Fig. 3 Depth profile of Cr in a GaAs layer (4.4 μm), grown by the molecular beam technique, on a Cr doped GaAs substrate. The background intensity measured at 57.5 amu. Screening electrode in operation

Other SIMS measurements [13,14] on MBE GaAs have indicated high levels of Cr in undoped layers ($4x10^{14}$ to 10^{16} atoms cm^{-3}) with the frequent occurrence of Cr out-diffusion from the substrate. Our SIMS measurements on MBE material show that out-diffusion is an infrequent phenomenon and that the residual Cr levels in the layer are generally in the mid 10^{13} atoms cm^{-3} region [15].

Summary and Conclusions

The limits of detection in a SIMS system are critically dependent on detailed features of the instrument design and application. Unless these features are carefully controlled measurements of impurity concentrations at the low levels of importance in semiconductor characterisation can be meaningless.

The limits of detection of a number of elements of interest in semiconductors have been measured at 10^{13} atoms cm^{-3} (similar results can be expected for electronegative ions sputtered using Cs ion bombardment [16] and the dynamic range and precision attainable in depth profiles has been confirmed experimentally. Semiconductor analysis is a challenging task and development of the SIMS technique must continue in this direction if the needs of materials scientists and device engineers are to be satisfied.

Acknowledgements

The author would like to thank Dr K. Wittmaack of Gesellschaft für Strahlen und Umweltforschung, MBH, Neuherberg, Germany, and colleagues at Atomika Ltd. for their contributions to the improved system design. The semiconductor samples and associated data were kindly supplied by Drs J. Stagg, T. Foxon, J.M. Shannon and Mr J. Neave of Philips Research Laboratories (PRL). Useful comments and advice were given by Mr E.J. Millett (PRL).

References

1. J.B. Clegg, Surf. Inter. Anal. 2, 91 (1980)
2. K. Wittmaack, in D.R. Beaman et al. (eds),"Proc. Eight Int. Congress on X-ray Optics and Microanalysis", Pendell Publ. Comp. Midland, 1980, p. 31

3. K. Wittmaack, J. Vac. Sci. Technol. 16, 1027 (1979)
4. K. Wittmaack, J.B. Clegg and M.G. Dowsett, to be published
5. K. Wittmaack, private communication
6. A.E. Morgan and J.B. Clegg, Spectrochimica Acta 35B, 281 (1980)
7. H.H. Andersen, Appl. Phys. 18, 131 (1979)
8. J.A. McHugh, Rad. Effects 21, 209 (1974)
9. V. Littmark and W.O. Hofer, Nucl. Insts. Meths 168, 329 (1980)
10. M.R. Brozel, J.B. Clegg and R.C. Newmann, J. Phys. D. Appl. Phys. 11, 1331 (1978)
11. T.W. Sigmon, in Secondary Ion Mass Spectrometry SIMS-II, ed. by A. Benninghoven, C.A. Evans, Jr., R.A. Powell, R. Shimizu, H.A. Storms, Springer Series in Chemical Physics, Vol. 9 (Springer, Berlin, Heidelberg, New York 1979) p. 80
12. J.B. Clegg, G.B. Scott, J. Hallais and A. Mircea-Roussel, J. Appl. Phys. 52, 1110 (1981)
13. H. Morkoc, C. Hopkins, C.A. Evans and A.Y. Cho, J. Appl. Phys. 51, 5986 (1980)
14. A.M. Huber, Eight Int. Symp. on GaAs and Related Compounds (1980) to be published
15. J.B. Clegg and T. Foxon, to be published
16. R.G. Wilson and V.R. Deline, Appl. Phys. Lett. 37, 793 (1980)

Impurity Redistribution in GaAS Epilayers

A.M. Huber, G. Morillot, P. Merenda[1], M. Bonnet, and G. Bessonneau[1]
THOMSON-CSF - L.C.R. - BP 10
F-91401 Orsay, France

Introduction

It is generally recognised that secondary ion mass spectrometry (SIMS) is one of the most powerful characterization techniques available in the electronics industry. Several problems have been solved thanks to SIMS techniques. In particular it contributes to the improvement of the expensive and difficult preparation and treatment of gallium arsenide, one of the basic materials of new semiconductor technology [1-6].

This paper presents a sensitive and quantitative determination of Mg, Cr, Mn, Fe in GaAs epilayers by this method. SIMS has already been successfully applied to the determination of Cr and Fe (deep acceptor level elements) [7,8]. However the distribution of Mg (shallow acceptor) and of Mn (generally believed to create deep acceptor level) in GaAs epilayers has not yet been reported in conjunction with Cr and Fe profiles.

The possible origins of these impurities in GaAs epilayers will be discussed. In the context of these SIMS results we will also consider the redistribution of silicon and sulphur (shallow donors) in some epilayers, in an attempt to present a more complete material characterization. Furthermore, it must be remembered that a comprehensive assessment should include quantitative oxygen and carbon measurement. Unfortunately, the sensitivity of present analytical techniques is limited to $3 \cdot 10^{16}$ atom cm^{-3} and $1 \cdot 10^{17}$ atom cm^{-3} for C and O respectively. The use of a ion pump close to the target chamber offers a possible improvement to this situation.

The epilayers studied here have been grown by chloride [9,10] or by a metalorganic vapour phase epitaxy (MOVPE) system [11,12] for the production of metal Schottky gate field effect transistor (FET). In recent years the development of MOVPE techniques has been found to offer the possibility of growing large area epilayers for FET requirements. For FET applications, the majority of producers still use material prepared by the $Ga/AsCl_3/H_2$ and $Ga/AsH_3/HCl/H_2$ "halogen" growth processes. In this paper the impurity redistributions in chloride VPE and MOVPE layers will be compared.

---

[1]THOMSON-CSF - D.C.M. - BP 10, F-91401 Orsay, France

Experimental

The analysis was performed using a CAMECA IMS 3F instrument. The target chamber was evacuated by cryopumping to a pressure of $1 \cdot 10^{-9}$ Torr. The surface of the sample was scanned with a focussed oxygen primary ion beam (1.5 μA, O^+ at 12.5 keV), while the intensities of secondary ion species: $^{24}Mg^+$, $^{52}Cr^+$, $^{55}Mn^+$, $^{56}Fe^+$, with $^{75}As^+$ as internal reference element were simultaneously recorded. A mass resolution of 1100 M/ΔM was used to avoid possible interference with ionised hydrocarbons. For $^{28}Si^-$ and $^{32}S^-$ analysis, a caesium primary ion beam (1 μA, 14.3 keV) was used, the internal reference element being $^{69}Ga^-$. The scanned area was 250 x 250 μm for all the working conditions and the analysed region was 60 μm in diameter. Counting time per measurement point was 2 s for each element. The erosion rates were 25 $Å \cdot s^{-1}$ and 100 $Å \cdot s^{-1}$ respectively, as determined by Talysurf measurements with an estimated depth precision of ±10 %.

The secondary ion signals in GaAs were converted to concentration using two calibration methods: either
(i) by standard samples previously analysed by calibrated spark source mass spectrometry or by nuclear activation methods, or
(ii) by the use of implantation to prepare Mn, Mg, Fe, Cr, Si and S doped GaAs samples. Implantation doses for each element varied from 10^{12} to 10^{14} atom cm^{-2} at an energy of 300 keV.

Taking into account the uncertainty of standard calibration and of the SIMS measurements, the precision of impurity determination is ±20 % above $5 \cdot 10^{15}$ atom cm^{-3}. Below this value the quantum noise on signal and background reduces the signal to noise ratio in the measurement and becomes predominant, reducing the precision to ±50 % at $1 \cdot 10^{14}$ atom cm^{-3}, and this results in an associated detection limit of $1\text{-}5 \cdot 10^{14}$ atom cm^{-3}. To simplify the simultaneous presentation of multiple element profiles in the same layer, smoothed curves are presented, but the magnitude of the fluctuations is indicated by error bars.

The sample were analysed in two different areas about 10 mm apart. Generally, analysis of 2 clean areas gives reproducible and representative results of the material. Careful observation of the analysed areas was carried out, using Normarsky interference microscopy, following analysis, to check the absence of hillocks or dust in craters. Ten chloride VPE and twenty five MOVPE layers have been analysed, and we will present the typical profiles showing general aspects of both epitaxial techniques. The growth conditions of the chloride and the metalorganic epilayers used in this study are listed in Table 1. In the case of a few layers we looked for a correlation between SIMS results and device performance.

To facilite the explanation of the results a schematic cross section of an FET structure is shown in Fig. 1. The epitaxial layers used for these devices, have been grown on a 450 μm thick semi-insulating GaAs substrate, Cr doped or undoped. In the chloride VPE process a 3-5 μm high resistivity buffer layer

Table 1

Method of VPE	Preheating	In-situ etching	Growth temp.
Chloride $Ga/AsCl_3/H_2$	YES at 750°C 20 min	YES	750°C
MOVPE $TMG/AsH_3/H_2$	YES at 600°C 10 min	NO	600°C

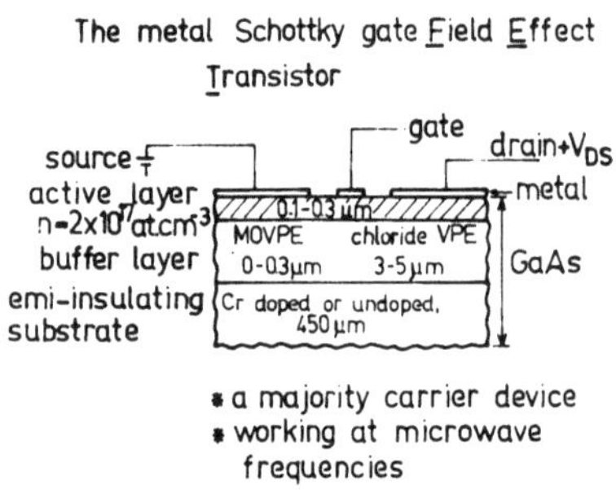

Fig. 1 Schematic cross section of an FET structure

is always incorporated, but in the MOVPE system, due to the relatively high background donor level, either no buffer or at best a thin 0.3-0.5 μm thick layer is grown. The active layer consists of a layer 0.2-0.3 μm thick, doped with Si or S at $n = 1\text{-}2 \cdot 10^{17}$ atom cm^{-3}. It should be noted here that the performance of a GaAs FET is significantly affected by trapping impurity centers at the interface of active layer and low power gain and large noise figure have been reported to originate from these centers. However, the identity of the centers which cause these problems has not previously been determined (with the exception of Cr).

Results and Discussion

Before reporting the results, we point out that one of the important differences between the chloride VPE and MOVPE is the absence of an in-situ vapour-etch in the case of the latter. In the chloride process a minimum of 5 μm of material is removed during the in-situ etch prior to layer growth. By using HCl or $AsCl_3$ as an etching agent in MOVPE, it was not possible to achieve simultaneously good morphology, low carrier concentration and high mobility for any given set of reactor conditions [13].

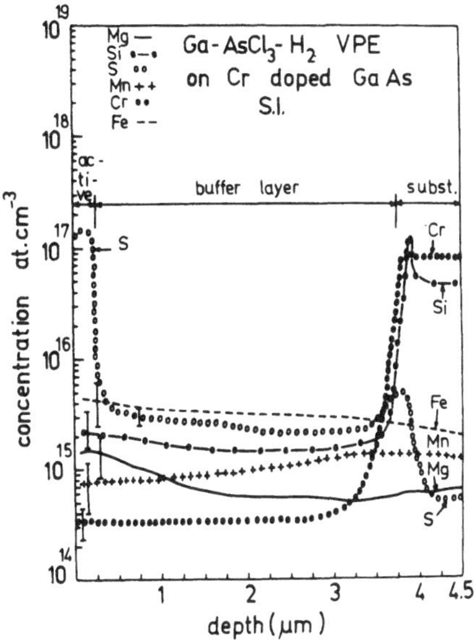

Fig. 2 S, Si, Mg, Mn, Cr and Fe redistribution measured by SIMS in chloride VPE layer. (FET output power of this sample is 1 W for 2400 μm gate width, 1.5 μm length at 7 GHz, gain 7 dB Vds = 9 V)

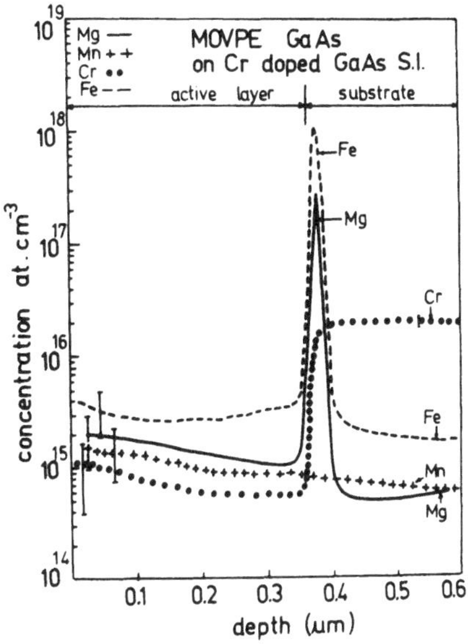

Fig. 3 Mg, Mn, Cr and Fe redistribution measured by SIMS in a MOVPE layer grown on Cr doped semi-insulating substrate (output power 0.4 W same geometry as in Fig. 2)

Figure 2 shows typical Mg, Mn, Cr and Fe distributions in a chloride VPE layer. One can observe that the Mg, Mn and Cr levels lie between $5 \cdot 10^{14}$-$1 \cdot 10^{15}$ atom cm^{-3}. Fe, with $3.5 \cdot 10^{15}$ atom cm^{-3}, has the highest deep level element concentration in this layer. The presentation together with the shallow donor level element S (doping element of the active layer) and Si (residual element) gives a more complete image of the material. (These elements were analysed by caesium primary ion bombardment). Since the doping level is $1 \cdot 10^{17}$ atom cm^{-3} in the active layer, Fe does not noticeably disturb the electron conduction, however it can contribute to the compensation of the residual shallow donors (S and Si) in the buffer layer. This deep acceptor behaviour of Fe can be exploited to produce high resistivity buffer layers by controlled compensation. This layer and the others with similar material quality have produced good power FET results (output power 1 W for 2400 μm gate width, 1.5 μm length at 7 GHz, gain 7 dB, Vds = 9 V).

A more complex profile situation is observed in MOVPE layers. Fig. 3 shows typical profiles obtained on a layer grown on a Cr doped substrate. One can observe considerable accumulation of Mg, Fe at the interface (Mg: $3 \cdot 10^{17}$ atom cm^{-3}, Fe: $1 \cdot 10^{18}$ atom cm^{-3}). It is interesting to see that noticeable Mn accumulation at the interface was not detected. Again Fe is the predominant impurity in the layer with $3 \cdot 10^{15}$ atom cm^{-3}, but the Cr level is somewhat higher than in a VPE chloride layer. The high Mg acceptor concentration at the interface can generate a space charge which results in a so-called "substrate effect" and output power reduction. The power FETs, fabricated from this epilayer produced poor output power (0.4 W with 2400 μm gate width, 1.5 μm gate length at 7 GHz, gain 7 db, Vds = 9 V). When the layers are deposited on undoped semi-insulating substrates, the results are generally better [14] and good output power has been measured on a number of FETs made from these layers, with or without a buffer (output power 1 W, 7 GHz, same geometry).

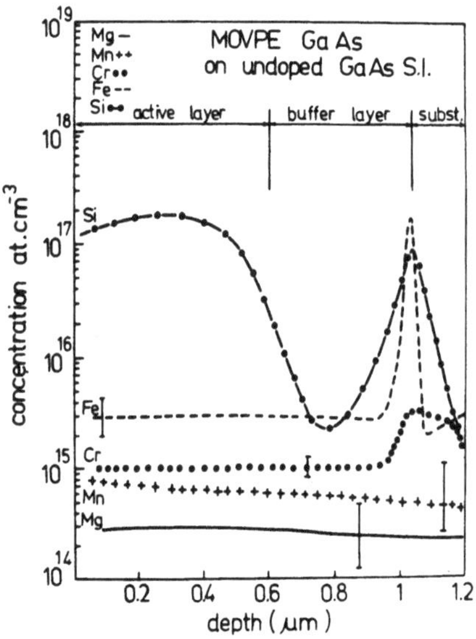

Fig. 4 Si, Mg, Mn, Cr and Fe redistribution measured by SIMS in a MOVPE layer grown on undoped semi-insulating substrate (output power 1 W same geometry as in Fig. 2)

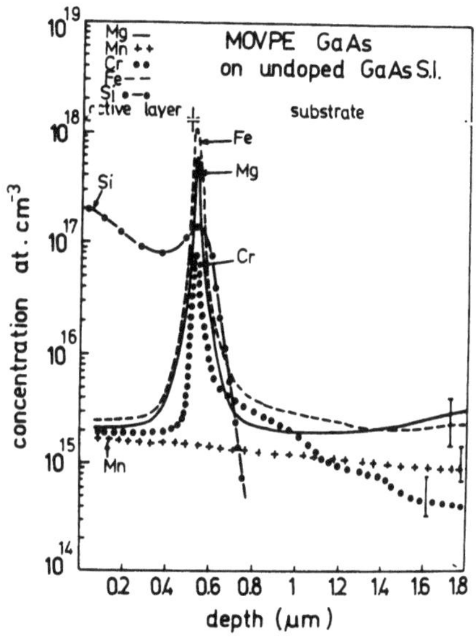

Fig. 5 Si, Mg, Mn, Cr and Fe redistributed measured by SIMS in a MOVPE layer grown on undoped semi-insulating substrate (output power 0.5 W same geometry as in Fig. 2)

Figure 4 shows the impurity and Si doped profiles obtained on one such layer. It can be seen that Fe at $1.6 \cdot 10^{17}$ atom cm^{-3} and Si $9 \cdot 10^{16}$ atom cm^{-3} accumulate at the interface at doses smaller than in the Cr doped substrate, and that the Mg level is very low: $3 \cdot 10^{14}$ atom cm^{-3}.

Unfortunately, this behaviour is not always observed; probably due to substrate inhomogeneity and poor FET data can also be obtained on undoped substrates corresponding again to considerable Mg and Fe accumulation at the interface (see Fig. 5). The origin of these impurities at the interface and in the epilayers is partly related to the substrate. Cr, Mn, Fe and Mg outdiffusion during capped or capless annealing from semi-insulating GaAs has been mentioned by several authors [1,15], due to stress and vacancies. Since the epilayer deposition begins with a preheating treatment in both chloride VPE and in MOVPE processes at 750°C and 600°C respectively, during

10 minutes minimum, the possibility of outdiffusion exists. Although in the case of chloride VPE the in-situ etch eliminates the accumulated impurities, in MOVPE in-situ etchings are still not possible.

Figure 6 presents a Fe profile in a MOVPE layer; the depleted region in the substrate offers confirmation of the substrate outdiffusion process. The dose of the accumulated element indicates that they can originate from depths as great as 20 μm. The measured diffusion coefficient in this case at 600°C is 10^{-9} cm^{-2} s^{-1}. Nevertheless, it must be acknowledged that some impurities may also originate from the vapour phase and surface contamination.

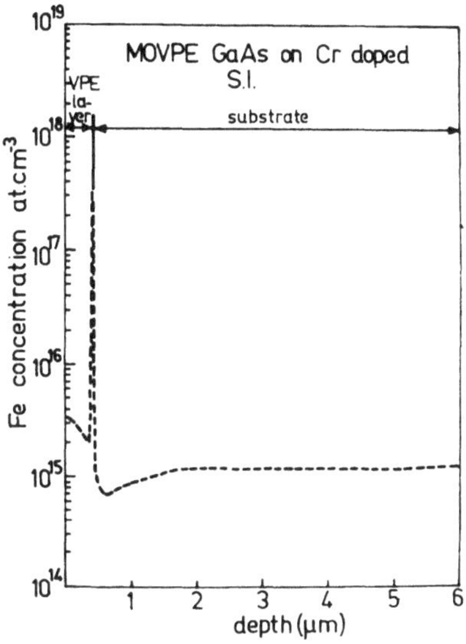

Fig. 6 Fe redistribution measured by SIMS in a MOVPE layer grown on Cr doped substrate

Analysis of a larger number of epilayers will be necessary for any given set of growth conditions, to specify the contribution of each impurity source in MOVPE layer growth. Mg, Fe, Mn and Cr redistribution was determined simultaneously, in chloride VPE and in MOVPE GaAs layers by SIMS analysis. The chloride process produces a maximum of $4 \cdot 10^{15}$ atom cm^{-3} impurity content of Fe, Mg, Cr and Mn in the layers and there is no apparent impurity accumulation at the layer-substrate interface. The presence of Fe about $3 \cdot 10^{15}$ atom cm^{-3} can create deep acceptor level which results in a high resistivity buffer layer by controlled compensation. Improtant impurity accumulation, particularly for Mg and Fe, and variable impurity content were encountered in the MOVPE layers grown on Cr doped substrates. On undoped semi-insulating substrates the situation is better but not yet reproducible, probably as a result of substrate inhomogeneities.

There is no doubt that systematic assessment of material by SIMS will be a usefull tool for the optimization of MOVPE growth conditions, choice and preparation of substrate and deposition process.

Acknowledgement

We would like to express our thanks to Dr. P.R. Jay and H.N. Migeon for helpful discussions.

References

1. A.M. Huber, G. Morillot, N.T. Linh, P.N. Favennec, B. Deveaud and B. Toulouse, Appl. Phys. Lett. 34, 858 (1979)
2. C.A. Evans, Jr., V.R. Deline, T.W. Sigmond and A. Lidow, Appl. Phys. Lett. 35, 291 (1979)
3. J. Kasahara and N. Watanabe, Jpn J. Appl. Phys. 19, L151 (1980)
4. P.K. Vasudev and R.G. Wilson, Appl. Phys. Lett. 37, 308 (1980)
5. M.H. Badawi, B.J. Sealy and J.B. Clegg, Electron Lett. 16, 554 (1980)
6. R.D. Fairman, R.T. Chen, J.R. Oliver and D.R. Ch'en, IEE Trans. Electron. Devices vol. ED-28, 135 (1981)
7. A.M. Huber, G. Morillot, P. Merenda, N.T. Linh, in Secondary Ion Mass Spectrometry SIMS-II, ed. by A. Benninghoven, C.A. Evans, Jr., R.A. Powell, R. Shimizu, H.A. Storms, Springer Series in Chemical Physics, Vol. 9 (Springer, Berlin, Heidelberg, New York 1979) pp. 91-94
8. R.G. Wilson, P.K. Vasudev, D.M. Jamba, C.A. Evans, Jr., and V.R. Deline, Appl. Phys. Lett. 36, 215 (1980)
9. J.R. Knight, D. Effer and P.R. Evans, Solid-State Electron 8, 178 (1965)
10. L. Hollan, Acta Electronica 21, 117 (1978)
11. S.J. Bass, J. Cryst. Growth, 31, 172 (1978)
12. J.P. Duchemin, J. Vac. Sci. Technol. 18 (3) 753 (1981)
13. Rajaram Bhat and Sorab K. Ghandi, J. Electrochem. Soc. 125, 771 (1978)
14. M. Bonnet, N. Visentin, G. Bessonneau and J.P. Duchemin, Conference AJACCIO (1981) to be published in J. Cryst. Growth (1981)
15. A.M. Huber, G. Morillot, P. Merenda, J. Perrocheau, J.L. Debrun, M. Valladon and D. Koemmerer, in "Gallium Arsenide and Related Compounds" (Vienna) 56, 579 (1980)
16. P.B. Klein, P.E.R. Nordquist and P.F. Siebenmann, J. Appl. Phys. 51, 4861 (1980)

Quantitative Distribution Analysis of B, As and P in Si for Process Simulation

M. Grasserbauer, G. Stingeder, E. Guerrero[1], H. Pötzl[1], R. Tielert[2], and H. Ryssel[3]

Institute for Analytical Chemistry, Technical University Vienna, A - Wien, Austria

1. Definition of Problem

For the development and improvement of process simulation accurate information about the (depth) distribution of dopant elements (electrically active fraction and total elemental concentration) as a function of process parameters has to be obtained. The analytical requirements for distribution analysis as applied for process simulation are stringent:

i) high precision and accuracy

ii) high detection power and large dynamic range (concentration range of interest: 10^{14}-$5 \cdot 10^{21}$ at/cm^3)

iii) high depth resolution (shallow p/n-structures in modern devices)

2. Survey of Analytical Methods

For distribution analysis of the electrically active fraction of the dopant elements measurement of electrical properties (four-point, two-point resistance and C/V measurements) is used. Depth resolution is either obtained by measurement of angle lapped specimens or through combination with chemical removal of thin surface layers [1]. The analytical figures of merit vary for the individual methods: 2-point-spreading resistance measurements are very sensitive (DL ~ 10^{14} at/cm^3), but difficult to perform. 4-point resistance measurements are less sensitive (DL ~ 10^{17}-10^{18} at/cm^3), easier to apply but very time consuming. These electrical techniques show generally sufficient depth resolution (10-40 nm) and high accuracy of analysis. The analytical error in concentration of electrically active elements is in the order of several percent.
C/V measurements [2] are very useful for the determination of the distribution of low dopant concentrations. For high concentrations ($\gtrsim 10^{19}$ at/cm^3) this method is not applicable, how-

[1]Department of Physical Electronics, Institute of General Electrical Engineering and Electronics, Technical University Vienna, and Ludwig-Bolzmann-Institute for Solid State Physics, A - Wien, Austria

[2]Siemens AG, Central Research Laboratories, D-8000 München, Fed. Rep. of Germany

[3]Institute for Solid State Technology, Fraunhofer Gesellschaft, D-8000 München, Fed. Rep. of Germany

ever, because high voltages would be necessary to establish a sufficient depth of the space charge region. Furthermore at high impurity density the depletion approximation becomes questionable. Therefore C/V-measurements have not been used for this work. For elemental distribution analysis of B, As and P in Si mainly neutron activation analysis (NAA) [2] and SIMS [3-7] can be applied.

3. Practical Analytical Performance of SIMS

Due to the requirements of high detection power and accuracy for generation of suitable data for improvement of process simulation only the most modern in instrumental technology can be applied. In this work the CAMECA IMS-3f ion probe analyzer was used. All figures of merit given below refer to this instrument and the use of O_2^+ as primary ions.

3.1 Practical Limits of Detection

i) B: The detection limit was found to be in the order of 10^{14} at/cm^3 ($\triangleq$ 1 ng/g) [3,7,8]. A dynamic range of 10^7 can be achieved if sputtering of deposited material on the immersion lens is reduced by switching to small rasters during measurement of the profile [9]. The possibility of redistribution at high primary current densities [10] has to be investigated however.

ii) As: Elimination of the Si_2O-interference can be performed by energy filtering [8,11-15] or application of high mass resolution (ΔM = 0.02345 amu, $M/\Delta M \geq 3000$). With both techniques a detection limit of $2 \cdot 10^{16}$ at/cm^3 ($\triangleq$ 1 μg/g) can be obtained. The dynamic range is about 10^5.

iii) P: Highly accurate distribution analysis of P can only be carried out with high mass resolution ($M/\Delta M \geq 4000$) due to the SiH/P-interference (ΔM = 0.00782 amu). In order to achieve the necessary stability of the mass spectrometer implementation of a special measurement cycle to reduce hysteresis effects was necessary [16]. With this technique a detection limit of $5 \cdot 10^{15}$ at/cm^3 and a dynamic range of 10^6 could be obtained. H (as SiH) and P can be measured simultaneously (Fig. 1). For routine analysis of P a magnetic stability of $5 \cdot 10^{-6}$ during the measurement is desirable. This value can be obtained using a temperature stabilized Hall probe. A further increase in detection power for As and P is possible by use of Cs^+-primary ions.

3.2 Practical Limits of Determination

The practical limit of determination defines the lower limit of quantitative evaluation of the depth profile. Due to an increased probability of occurrence of systematic errors near the detection limit (redeposition, stability of magnetic field or sample potential) the practical limit of determination proved to be about a factor of ten lower than the detection limit (B ~ 10^{15} at/cm^3, As ~ 10^{17} at/cm^3, P ~ $5 \cdot 10^{16}$ at/cm^3).

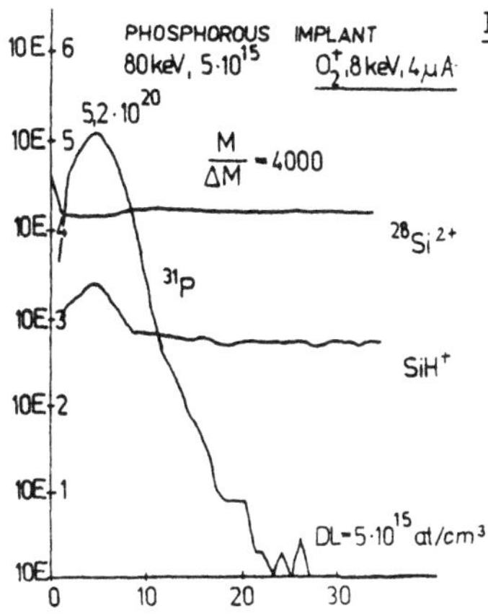

Fig. 1 SIMS-profiles of P and SiH in Si

3.3 Precision

Since precision of distribution analysis at concentrations above the limit of determination is mainly influenced by instrumental fluctuations (e.g. drift) it was determined by measuring the changes in the relative sensitivity factor in short-time intervals (during one day) and within a period of several months. The results are that for all elements the long-range reproducibility of the relative sensitivity factors is low (changes of a factor between 1.4 [B] and 3 [As] occur). Short time precision is however sufficient for accurate distribution analysis: e.g. for Boron the relative standard deviation of the calibration factor during one day is in the range of 1-5 %, for As typically in the order of 10 %. For P measurements the precision has not been determined yet, because the temperature stabilized Hall probe has not been available. The reproducibility of the sputter rate was found to be not high enough for quantification of the depth scale by a calibration factor. This has the consequence that the depth of each crater has to be determined individually.

3.4 Accuracy

The accuracy of distribution analysis of SIMS was determined by using NAA as a reference method and evaluation of systematic and statistical errors. For calibration of the concentration scale of SIMS profiles external standards of sufficient homogeneity or the integration method (calculation of relative sensitivity factor from implantation dose) were used. In absence of systematic errors the external standard method yields typically an accuracy of 10 % for B and As at peak concentrations of about 10^{20}-10^{21} at/cm^3. For P systematic investigations have not been performed yet.

The accuracy of the integration method is influenced by several statistical and systematic errors, which have to be accounted for:

- accuracy of the absolute implantation dose ~2-5 %
  lateral homogeneity of implantation <1 % rel.

- matrix effect for oxidized samples: The concentration of dopant elements in the oxide layer (thickness about 50 nm in the samples investigated) has to be determined accurately, otherwise analytical errors may be encountered (enrichment of B in oxide during annealing). This systematic error could be eliminated by use of oxygen flooding during analysis. An oxygen pressure of $5\cdot10^{-5}$ mbar proved to be optimal. A decrease in sputter rate by a factor 3-5 is however encountered (0.2-0.5 nm/s). Therefore the means of elimination of this systematic error by determination of the relative sensitivity factors of these dopant elements in an SiO_2-matrix will be investigated (preparation of suitable standards and development of accurate reference methods).

For high accuracy of quantitative distribution analysis the integration method is combined with the external standard technique, which is able to achieve an accuracy of the determination of B and As of 5-10 % at peak maxima. Considering the complexity of SIMS and the concentration for which these figures apply (10^{20} at/cm^3 = 0.2 at%) these values are absolutely satisfactory.

For calibration of the depth scale mechanical measurement of each crater with a stylus technique combined with interference microscopy and ellipsometry (to determine the position of the SiO_2/Si-interface) was used. The accuracy of depth calibration with this technique is in the order of 10 %. For implantation profiles calibration through the mean projected range $R_p \pm \Delta R_p$ was applied. The accuracy of this method is estimated to be in the order of 10 %. Generally a depth resolution of 10-30 nm can be obtained with SIMS.

4. Combination of Methods for Quantitative Distribution Analysis

The complexity of SIMS and its large number of possibilities for systematic errors generate the necessity of combining it with other analytical techniques to obtain maximum accuracy. As reference methods analytical techniques can be used which exhibit suitable characteristics (like sufficient detection power and high accuracy) and *different* systematic errors than SIMS. Evaluation of these aspects for various quantitative trace analytical methods led to the conclusion that NAA is at the present state the most suitable reference method for distribution analysis of B, As and P in Si. The basic analytical characteristics and systematic errors of NAA were therefore investigated and compared with SIMS (Table 1) [2].

The results which were obtained by analysis of identical samples can be summarized in the following items:

i) SIMS as a direct method of analysis is much less time consuming.
ii) The detection power of SIMS is higher for As and P.
iii) Measurement of P is rather difficult due to interference with the Si-line at 1710 keV and necessity of β measurement.

Table 1 Comparison of SIMS and NAA for quantitative distribution analysis of dopant elements (B: flux 10^9 n/cm^2/s subthermal, analyzed area 1.5 cm^2, As, P: flux 10^{13} n/cm^2/s, analyzed area 0.5 cm^2)

	SIMS				NAA			
ELEMENT	SI	DYN. RANGE	DL(AT/CM^3)	ACC.(%REL)	REACTION	DYN.RANGE	DL(AT/CM^3)	ACC.(%REL)
B	$^{10}B^+$, $^{11}B^+$	10^7	10^{14}	5-10	$^{10}B(n,\alpha)^7Li$	10^3	10^{14}	1
As	$^{75}As^+$	10^5	2.10^{16}	10	$^{75}As(n,\gamma)^{76}As$	5.10^4	10^{17}	10
P	$^{31}P^+$	10^6	5.10^{15}	10	$^{31}P(n,\gamma)^{32}P$ β-DECAY	5.10^4	10^{17}	10

iv) After correction or elimination of systematic errors NAA provides high accuracy for distribution analysis and is therefore well suited as a reference method to SIMS (Fig. 2).

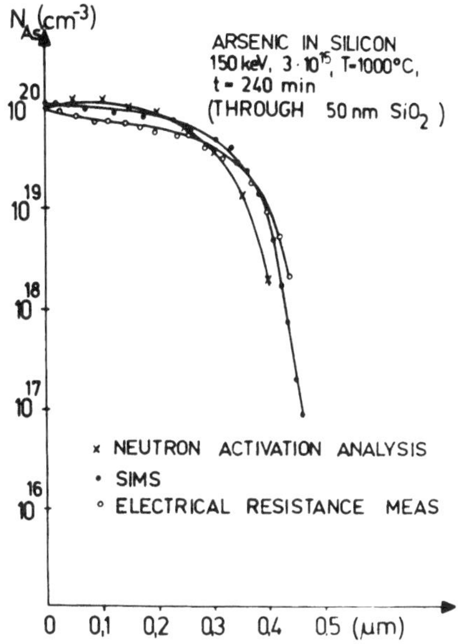

Fig. 2 Comparison of methods for quantitative distribution analysis - As in Si: total elemental concentration: NAA/SIMS, electrically active dopant concentration: 4-point resistance measurements

The conclusion from this methodological evaluation is that SIMS has some advantages for quantitative distribution analysis for reasons speed, cost, dynamic range (B) and sensitivity (P and As), but it should be combined with NAA to obtain maximum accuracy.

5. Application for Process Simulation

For the simulation of 1- or 2-dimensional diffusion of dopant elements the following complex interaction phenomena have to be taken into account and described mathematically [17,18]:

i) ion implantation - dependence of implantation profile on energy and dose
ii) redistribution during annealing:
  α) dependence of diffusion coefficient on concentration
  β) influence of partial activation (e.g. clustering)
  γ) mutual influence of different dopant elements on diffusion (cooperative diffusion)
  δ) oxide influence (oxide enhanced diffusion)

Quantitative distribution analysis with SIMS provides input parameters for diffusion models and serves as the reference method for calculated profiles (test of accuracy of calculations). Among the phenomena listed above the following items have been studied so far:

5.1 Input Parameters for Process Simulation

5.1.1 Implantation Profile

The distribution of the dopant elements after implantation may be calculated for a given substrate from energy and dose using the Pearson IV [17] calculation or two half Gaussian profiles. Measurement of B, As and P (Fig. 4) showed that a deviation from simulation was encountered for all three elements due to channeling effects (implantation through oxide in <100> direction). This deviation can be described by an additional ex-

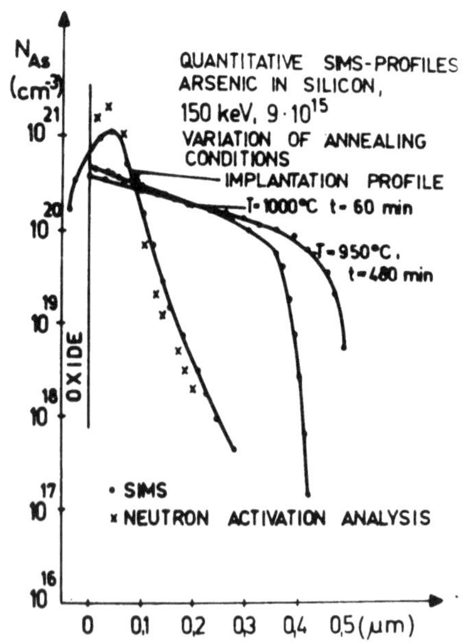

Fig. 3 Quantitative distribution analysis of As in Si: implantation profile - comparison NAA/SIMS, diffusion profile - influence of annealing conditions (SIMS)

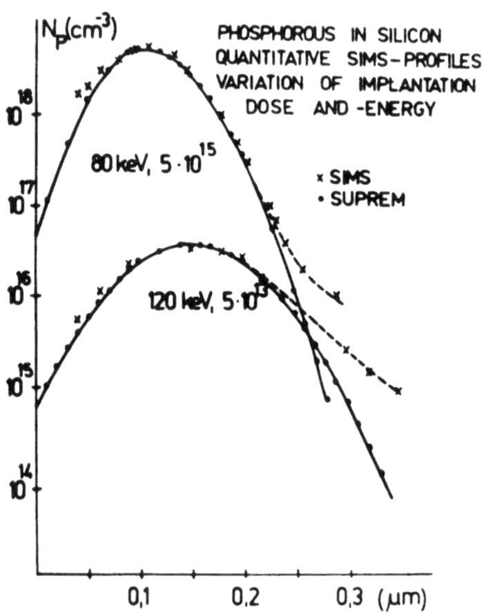

Fig. 4 Comparison of quantitative SIMS-profiles of P-implantations in Si with calculated profiles (Pearson IV - distribution in SUPREM II-05 process simulation program)

ponential term [19,20,21] and has to be taken into account for process simulation because these channelling effects may have a significant influence on the properties of a transistor (e.g. MOS channel implantation).

5.1.2 Redistribution during Annealing

α) Diffusion Profile

For the accurate determination of the concentration dependence of the diffusion coefficient the distribution of the dopant elements after annealing as a function of process parameters like temperature, time and atmosphere has to be measured with maximum precision and accuracy. Fig. 3 shows as an example the redistribution of As during annealing and under specific conditions.

β) Partial Activation

The fraction of electrically inactive dopant elements, which exhibit a strong influence on diffusion properties, can be determined as a function of total dopant concentration from electrical and SIMS measurements (see Fig. 2). Fig. 5 shows the equilibrium between the electrically active and the inactive fraction for As at an annealing temperature of 1000 °C (independent on implantation energy and annealing time). Statistical evaluation of a large number of samples hints that this equlibrium is best described mathematically by a higher-order function, rather than by a linear relationship which is commonly used [17,21]. From the equilibrium "active solubility limit" could be determined as $3 \cdot 10^{20}$ at/cm^3.

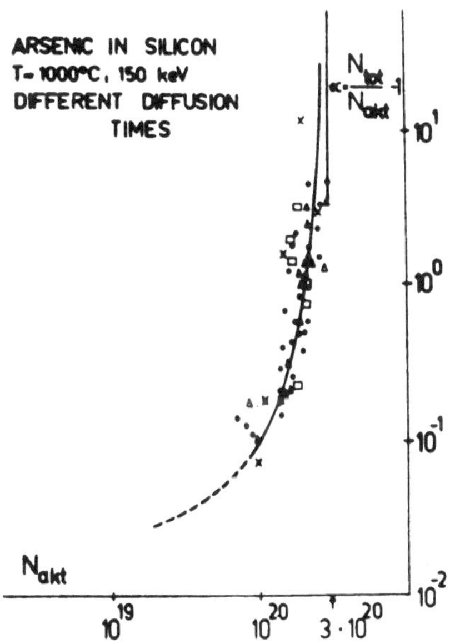

Fig. 5 Characterization of partial activation of As in Si. Equilibrium between electrically active dopant concentration (N_{akt}) and ratio inactive/active concentration (α) determined from SIMS profiles and electrical measurements

5.2 SIMS as a Reference Method for Calculated Profiles

Based on its high accuracy and sensitivity SIMS enables the check of calculated profiles. Fig. 6 shows the calculation of the redistribution of B in Si with SUPREM II-05 [17] and the elemental distribution obtained with SIMS. Generally there is good agreement between the calculated and the measured distribution. Small differences observed (near the surface and at lower concentrations) hint at possibilities of improvement of the process simulation model used, but still have to be evaluated in detail. Large discrepancies between calculated and measured profiles are usually encountered for As and P.

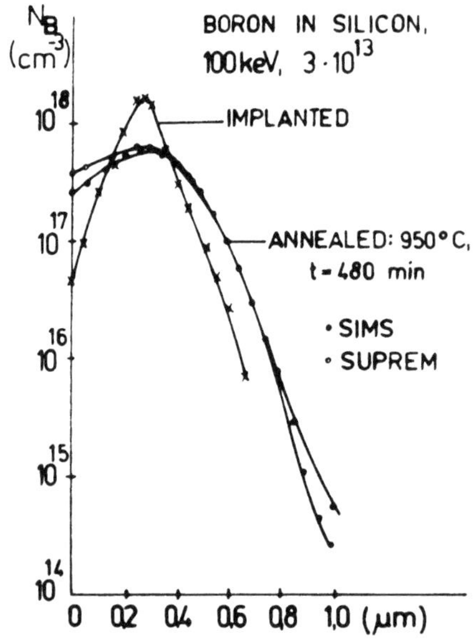

Fig. 6 Comparison of quantitative SIMS profiles of B in Si after annealing with results of process simulation program SUPREM II-05

References

1. E. Guerrero, G. Dobolka and A. Baghai, Thin Solid Films 76, 237 (1981)
2. H. Ryssel and I. Ruge, "Ionenimplantation", Teubner, Stuttgart, 1978
3. T.W. Sigmon, in Secondary Ion Mass Spectrometry SIMS-II, ed. by C.A. Benninghoven, C.A. Evans, R.A. Powell, R. Shimizu, H.A. Storms, Springer Series in Chemical Physics, Vol. 9 (Springer, Berlin, Heidelberg, New York 1979)
4. H.W. Werner, Mikrochim. Acta, Suppl. 8, 25 (1975)
5. E. Zinner, Scanning 3, 57 (1980)
6. R.E. Honig, Thin Solid Films 31, 89 (1976)
7. C.A. Evans, R.J. Blattner, Ann. Rev. Mater. Sci. 8, 181 (1978)
8. M. Grasserbauer, G. Stingeder, E. Guerrero and F. Fallmann, Mikrochim. Acta (in press)
9. H.N. Migeon, private communication

10. C.R. Helms, J. Rouse, R.W. Barton and S.A. Schwarz, Integrated Circuits Laboratory, Stanford Electronics Laboratories, Report 15. Feb. 1980, p. 246
11. K. Wittmaack, Appl. Phys. Lett. 29, 552 (1976)
12. P. Williams and C.A. Evans, Appl. Phys. Lett. 30, 559 (1977)
13. C.W. Magee, J. Electrochem. Soc. 126, 660 (1979)
14. P. Williams, R.K. Lewis, C.A. Evans and P.R. Hanley, Anal. Chem. 49, 1399 (1977)
15. D.P. Leta and G.H. Morrison, Anal. Chem. 52, 514 (1980)
16. E. Zinner, private communication
17. D.A. Antonaidis, S.E. Hansen and R.W. Dutton, Integrated Circuits Laboratory, Stanford Electronics Laboratories, Computer Program - SUPREM II-05
18. R. Tielert, IEEE Trans. on Electron Devices ED-27, 1497 (1980)
19. D.A. Antoniadis and R.W. Dutton, IEEE Trans. on Electron Devices 26, 490 (1979)
20. A. Desalvo, R. Galloni, R. Rosa and F. Zignani, J. Appl. Phys. 51, 1994 (1980)
21. E. Guerrero, H. Pötzl, R. Tielert, M. Grasserbauer, and G. Stingeder, paper submitted to J. Electrochem. Soc.

High Spatial Resolution SIMS Depth Profiling of Cr Dopant in CdSe Thin Film Transistors

J.D. Brown, F.R. Shepherd[1], and W.D. Westwood[1]

Faculty of Engineering Science and Centre for Interdisciplinary Studies in Chemical Physics, The University of Western Ontario London, Canada

Introduction

CdSe thin film transistors (TFTs) whose structure is illustrated in Fig. 1, have been fabricated with Cr source and drain electrodes having separations varying between 5 and 50 μm [1]. The thermal annealing process employed to activate the devices has been shown to cause diffusion of Cr from the electrodes into the CdSe layer in the gap, doping the film [1,2]. Auger electron spectroscopy [2] was unable to determine the detailed Cr distribution in the device. Since the conductivity behaviour of CdSe and other polycrystalline semiconductor TFTs depends critically on the doping level [3,4], an attempt was made to determine the Cr distribution across the source drain gap and throughout the depth of the CdSe film in annealed, operational devices.

In the devices studied the source-drain electrodes were 30 μm wide by 600 μm long (over the gate) and the source-drain gap width was 40 μm (Fig. 1). The CdSe film on top of an Al_2O_3 insulating layer was about 85 nm. Since the Cr concentrations are in the low ppm atomic range, the measurement of the Cr distribution requires high sensitivity. High lateral and depth resolution are required to characterize the spatial distribution in the CdSe layer. This combined requirement of high sensitivity with spatial constraints is frequently encountered in device analysis.

Procedure

The ion microprobe analyses were carried out using an Applied Research Laboratories ion microprobe mass analyser (IMMA). In the particular instrument used in this study, a 600 ℓ/s ion pump is located directly opposite the outlet from the duoplasmatron ion source ensuring excellent vacuum conditions in the vicinity of the sample. Typically, the base pressure of the instrument after overnight pumping is $1\cdot10^{-9}$ Torr and during analysis using either nitrogen or oxygen in the duoplasmatron, the pressure rises to not more than $2\cdot10^{-7}$ Torr.

CdSe sputters very rapidly under ion bombardment. A major problem in analysis of CdSe films is therefore the limitation of sputter rate so that a sufficient number of measurements can

[1]Bell-Northern Research, Ottawa, Canada

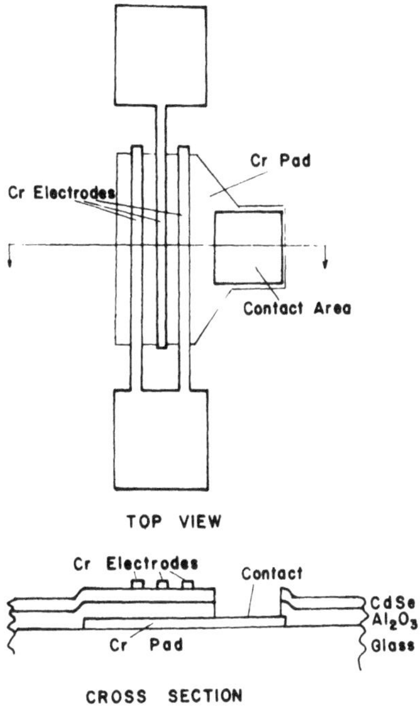

Fig. 1 Thin Film Transistor structure. The thickness of the films in the cross section is greatly exaggerated. The Cr electrodes are 30 μm by 600 μm

be made through the thickness of the film to define the concentration gradients. The usual methods of limiting sputter rate by rastering the primary beam over a large area, say 100 x 100 μm or by defocussing the primary bean cannot be used if lateral resolution is to be maintained. Fortunately in the electrode gap of the transistor structure the concentrations would be e ected to vary only with depth in the film and with distance from the electrodes. During analysis, the primary ion beam was scanned along a 400 μm line parallel to the electrode structure by using the primary beam deflection plates. To further decrease the sputter rate, while at the same time giving good lateral resolution, the primary beam of ${}^{32}O_2^+$ ions was reduced to the minimum practical current of 0.1 nA and to the lowest possible energy of 10 keV. The resultant beam diameter was about 2 μm (actually slightly elliptical with the major axis of the ellipse parallel to the direction of the line scan). With this primary beam current and a linear scan of 400 μm, about 15 minutes was required to sputter through the 100 nm CdSe film giving a sputter rate of just over 0.1 nm/s.

Four mass species were measured sequentially under computer control, ${}^{27}Al^+$, ${}^{52}Cr^+$, ${}^{80}Se^+$ and ${}^{114}Cd^+$. Each cycle of measurements required approximately 60 seconds so that each element was measured 15 times during the sputtering through the film. However, with the very low primary beam current several adjustments had to be made to the secondary ion mass spectrometer to increase sensitivity so that good measurements particularly of the Cr^+ intensity could be made in a reasonable time. The α and β slits in the instrument were opened so that the measured secondary ion intensities increased by a factor of about 10 over the normal settings. At the same time, the mass resolution was still sufficient to separate adjacent masses up to at least mass 150.

After setting the primary beam and performing the necessary computer set-up of the secondary spectrometer controls, a test depth profile was run on an area of the specimen near the electrodes but not critical to the desired gradient measurements. When such a test run was deemed satisfactory, the analysis involving 7 or 8 depth profiles in the 40 μm transistor gap was carried out completely under computer control including the necessary stage movement between profiles. The total elapsed time for the complete measurements was about 2 to 3 hours. During that period the primary beam intensity remained stable to better than 5 %.

To establish the absolute levels of Cr measured during the profiles, a CdSe film implanted with ^{52}Cr at energy of 50 keV to a total dose of $3 \cdot 10^{15}$ atoms/cm^2 was depth profiled under identical conditions. The area under the profile curve is proportional to the total dose. Estimates of the Cr concentrations are based on this standard.

Results and Discussion

The integrated area under the peak for the ion-implanted ^{52}Cr standard was $2.6 \cdot 10^7$ counts during a depth profile which required 700 seconds. Typical count rates through the peak of the distribution were $4 \cdot 10^4$/s. This measurement permits the calculation of the sensitivity of the IMMA for the Cr analysis in CdSe on the assumption that a crater 400 x 1.5 μm was sputtered by the primary ion beam. The total number of Cr atoms in that area of film with total implanted dose of $3 \cdot 10^{15}$ a/cm^2 is $1.8 \cdot 10^{10}$ atoms. Hence, roughly 1.4 Cr atom in 1000 sputtered by the ion beam is actually detected in the secondary mass spectrometer. This sensitivity will be used in the calculations to obtain measured concentrations in the transistor structures.

A part of the craters left by the ion beam are shown in Fig. 2. The micrograph shows approximately one-third of the total length of the craters. Centre-to-centre distance of the craters is 4.6 μm and the width of the craters is about 1.5 μm; this width is difficult to measure exactly because of the different focus of the microscope on the top and bottom of the craters. Optical interference microscopy indicated that the crater depths were approximately 90 nm. At the bottom of Fig.2 is a surface profilometer trace across a set of craters (unfortunately not the same craters as in the photomicrograph). the stylus tip had a radius of ~1.5 μm. This trace indicates that the ridges between the craters have been slightly rounded and that the depth of CdSe removed is about 10-12 nm. Such an effect will smear the measured depth distribution to a small extent. Note that because the tip diameter exceeds the crater

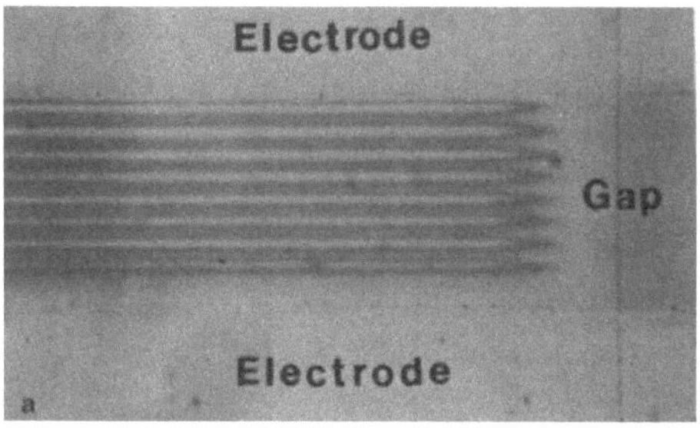

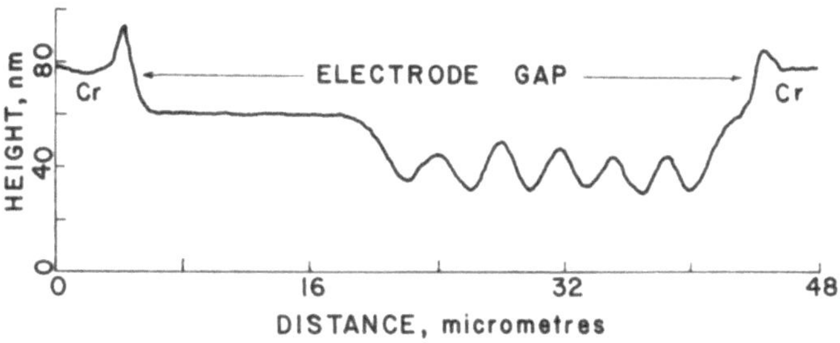

(b)

Fig. 2 (a) Optical micrograph of part of ion beam craters. The gap is 40 μm. (b) Profilometer trace across craters from a different transistor than (a)

width, the full depth of the craters is not seen by the profilometer. The rounding indicated by the profilometer is not observed in the interference microscope - perhaps because the 10 nm is close to the limits of measurement of fringle shifts of the green line of mercury used.

A depth profile for the third crater from the top of Fig. 2 is reproduced in Fig. 3. Because of a noisy preamplifier, the normal background count rate of less than 1 count/s had increased to 7.5 ± 0.8 counts/s. This background has been subtracted from the data of Fig. 3. Note that the Cr count rate in this region is still 2-4 times above this higher background. Note also that the count rates in the device are three orders of magnitude less than those measured in the ion implanted specimen. The integrated count through the thickness of the film of $1.2 \cdot 10^4$ indicates a total Cr concentration of $1.4 \cdot 10^{12}$ atoms/cm^2. In terms of bulk concentration, assuming that concentration in a CdSe film thickness of 100 nm, the Cr concentration is four parts per million atomic with a limit of detection based on a 1:1 peak to background ratio of about one part per million atomic. These figures represent sensitivity at 0.1 nA primary beam current and a beam size of less than 2 μm. These low Cr concentrations in the device indicate that the thermally diffused Cr atoms are much more effective in such transistors than ion implanted Cr where many of the Cr atoms may be required to neutralize traps introduced in the implantation step before electrically active impurities result.

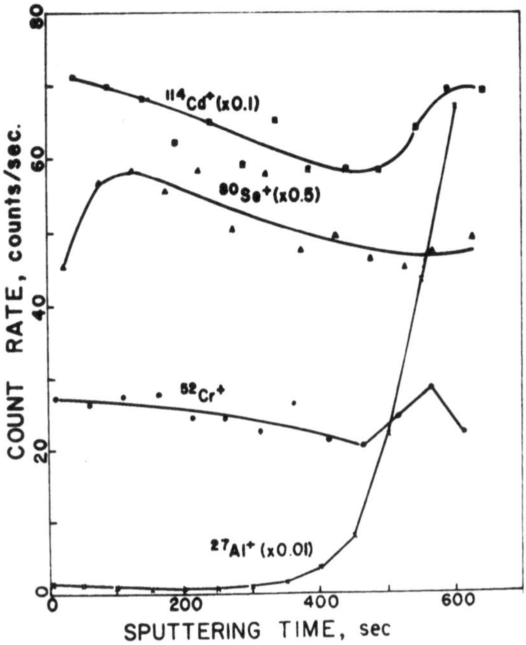

Fig. 3 Depth profile corresponding to the third crater from top of Fig. 2 (a)

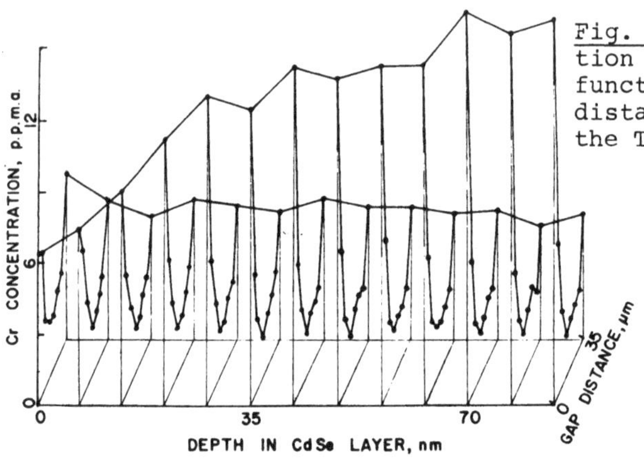

Fig. 4 Chromium concentration profile measured as a function of depth and distance across the gap for the TFT of Fig. 2(a)

The Al intensity of Fig. 3 is also typical of the profiles for Al seen in these linear scans - a small concentration at the surface presumably due to a small contamination falling to a low value halfway through the CdSe film, then rising rapidly toward the back surface perhaps due either to diffusion of the Al from the thin Al_2O_3 layer under the CdSe or due to ions sputtered through microcracks in the remaining CdSe film. However, a careful SEM and optical microscopic examination of the craters has never revealed such cracks. The Cd and Se intensities are roughly constant throughout the film, with some enhancement of the Cd intensity at the back interface probably due to an enhanced secondary ion yield caused by the increased O concentration at the Al_2O_3 layer.

A composite plot of the Cr concentrations for the eight depth profiles in a transistor gap is shown in Fig. 4. The Cr concentration tends to be high on the surface near the electrodes, falling to lower values near the centre of the gap. Measurable concentrations above background are found however through the entire three dimensional structure.

Conclusions

The Cr concentration has been measured with a lateral resolution of a few μm and a depth resolution of about 10 nm in the 40 μm gap of a CdSe thin film transistor. The limit of detection is about one part per million atomic using a primary beam of 0.1 nA of O_2^+ with a beam diameter of less than 2 μm. Measurements with this degree of lateral resolution and sensitivity to low concentrations are applicable to impurity distributions in many types of solid state devices.

Acknowledgements

The authors wish to thank the Metaalinstituut TNO, Apeldoorn, The Netherlands for their cooperation in making the ion microprobe measurements possible and I.V. Mitchell, Chalk River Nuclear Laboratories, Atomic Energy of Canada Ltd., for preparing the ion implant standard.

References

1. F.R. Shepherd, H. Nentwich, W.D. Westwood and S.J. Ingrey, J. Vac. Sci. Technol. 17, 485 (1980)
2. S.J. Ingrey, F.R. Shepherd and W.D. Westwood, J. Vac. Sci. Technol. 17, 481 (1980)
3. F.R. Shepherd, W.D. Westwood, P.J. Scanlon, J. Levinson, I.V. Mitchell and H. Plattner, J. Vac. Sci. Technol. 18, 899 (1981)
4. J. Levinson, F.R. Shepherd, P.J. Scanlon, W.D. Westwood, G. Este and M. Rider, to appear in J. Appl. Phys.

SIMS Investigation of p-n Junction Quality in Ion Implanted cw Laser Annealed Silicon

M. Maier, D. Bimberg, H. Baumgart[1,2], and F. Phillipp[1]

Basislabor of SFB 56 "Festkörperelektronik", Technical University, D-5100 Aachen, Fed. Rep. of Germany

Introduction

It is well established that silicon samples amorphized by ion implantation can be recrystallized by cw laser annealing [1,2]. The damaged surface layer is reconstructed by laser induced solid phase epitaxy [2,3,4]. In contrast to conventional furnace annealing and pulsed laser annealing the depth distribution of the implanted dopant remains unaltered during cw laser annealing [2,5], a distinct advantage for many device applications. Depending on laser power and dwell time, either optically perfect surface morphologies or surfaces showing slip lines are obtained [6,7]. Slip lines are observed at high laser power or long dwell time and indicate the generation of slip dislocations due to the thermally induced stresses [6,8]. At first sight samples with optically perfect surface morphology might be preferred for device fabrication. In contrast, the SIMS profiling studies of implanted and cw laser annealed diode structures reported here give clear evidence of much better quality of the p-n junctions in terms of lower reverse bias current for samples with the "slipped" surface morphology. TEM analysis reveals that the crystal quality of the p-n junction region (and not of the surface) is indeed decisive for the electrical behaviour of the diode.

Experimental

(100)-oriented p-type silicon specimens ($N \simeq 2 \cdot 10^{15}$ cm^{-3}) were amorphized by implantation of 60 keV ions at a dose of $2 \cdot 10^{15}$ cm^{-2}. The specimens were subsequently irradiated by a cw argon ion laser with an output power of approx. 8 W. Various irradiation times were achieved by scanning the specimen with respect to the laser beam at controlled velocities. SIMS analyses were performed with an ATOMIKA instrument using both O_2^+ and Ar^+ primary ions of 10 keV energy. The focussed and scanned 1 μA primary ion beam formed a quadratic spot of 0.75 mm x 0.75 mm. The quadrupole mass filter of the SIMS

[1] Max-Planck-Institut für Festkörperforschung, Max-Planck-Institut für Metallforschung, D-7000 Stuttgart, Fed. Rep. of Germany

[2] present address: Bell Laboratories, Murray Hill, N.J. 07974, USA

instrument was tuned to the Si_2^+ matrix molecular secondary ions. The dependence of the Si_2^+ secondary ion intensity on sputtering time was determined for the differently annealed samples. TEM studies were performed in an AEI-EM7 high voltage electron microscope operating at 1000 keV. The investigation of plan-view as well as cross-section specimens permitted determination of nature and depth distribution of the crystal defects.

Results and Discussion

The laser induced surface morphologies obtained at different exposure times are shown in the Nomarski optical micrograph of Fig. 1. The exposure times are 200 ms (Fig. 1a) and 50 ms (Fig. 1b). The slip line pattern in Fig. 1a (type A sample) is known to be generated by dislocation glide across the central region of the momentary laser spot [8]. Using a laser dwell time of 50 ms suppressed generation of slip dislocations and yielded samples with a nearly perfect surface morphology (Fig. 1b, type B sample).

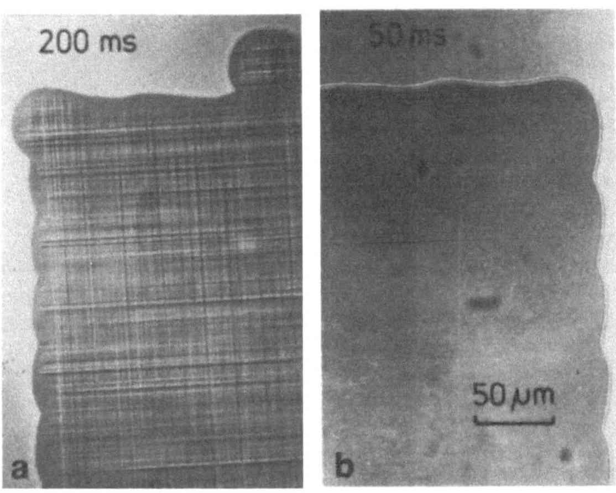

Fig. 1 Nomarski optical micrograph of solid phase epitaxially annealed silicon showing
a) slip lines
b) perfect surface morphology

The depth profiles of the intensity of the Si_2^+ secondary ions were found to be constant in the case of Ar^+ primary ions, while a considerable variation was obtained with an O_2^+ primary ion beam as shown in Fig. 1. For sake of clarity the depth profile of the implanted arsenic is also shown in Fig. 2. The location of the p-n junction was determined by equating the implanted arsenic concentration with the boron contents of the substrate material.

The variation of the Si_2^+ signal with depth is caused by a corresponding variation of surface charging. The SIMS instrument is equipped with an energy filter which allows only ions with a predetermined kinetic energy of small bandwidth to enter the quadrupole mass analyser. If the sample bias is varied ions with a correspondingly changed initial energy pass the energy filter. Thus energy distribution spectra can be determined via a variation of the sample bias. Fig. 3 shows such

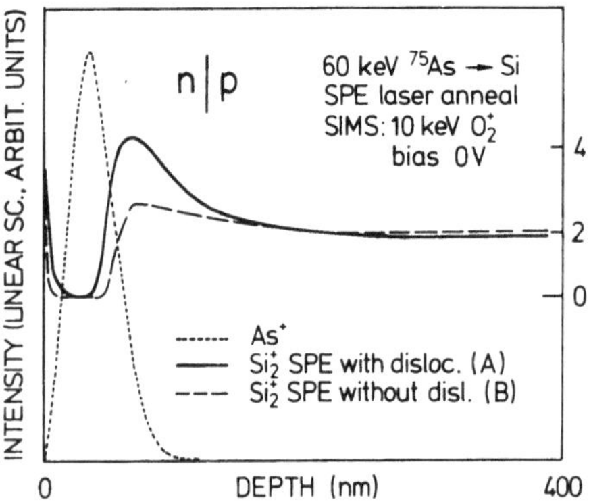

Fig.2 Secondary ion signal intensity and relative surface potential vs depth of implanted and cw laser annealed silicon analysed by oxygen primary ions

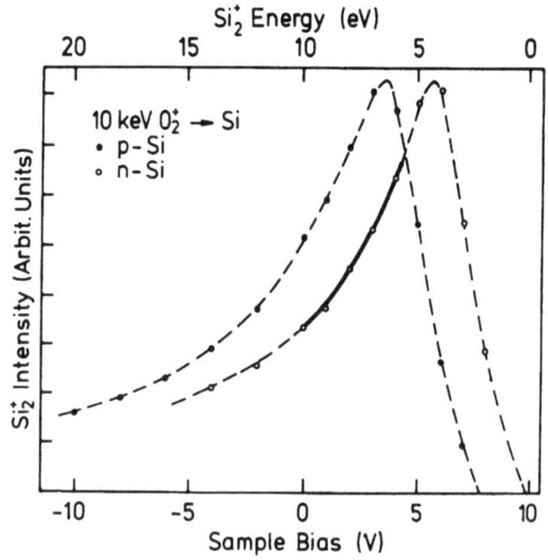

Fig.3 Intensity vs sample bias as Si_2^+ secondary ions sputtered from n-type and p-type silicon by oxygen primary ions (Si_2^+ energy spectrum) solid line: working curve (see text)

energy spectra of Si_2^+ ions for the n-type layer and the p-type substrate for bombardment with O_2^+ primary ions. The peaks in both spectra are of equal height but their respective positions are shifted by approx. 2 V. The surface of the p-type substrate during sputtering is apparently more positively charged by 2 V than the surface of the n-type layer.

A fixed sample bias of 0 V was used for the determination of the profiles shown in Fig. 2. The differences in intensities of the Si_2^+ signal of the p-type substrate and the n-type layer can now be explained by a shift of the energy spectrum due to different surface charging. Fig. 3 shows that different charging causes the energy spectrum only to shift but not to alter its shape and peak intensity. Thus the energy spectra of Fig. 3 can serve as a calibration curve for the transformation of the Si_2^+ signal intensity scale into a relative surface potential scale. The solid line in Fig. 3 (n-type spectrum) shows the intensity changes which are found in Fig. 2 and the corresponding surface charge changes. Apart from the differences of the secondary ion signal in the n- and p-type regions Fig. 2 displays peaks of different heights for type A and B samples in the depletion region of the junction. Using Fig. 3 it is found that the surface charge in the depletion region is up to 2 V larger than the surface charge of the p-type substrate.

How can we understand these large differences in surface charging? Sputtering of silicon with O_2^+ primary ions at normal incidence produces an almost perfect SiO_2 layer [9]. The thickness of this layer is theoretically given by the projected range of the implanted oxygen ions [10] and amounts to approx.

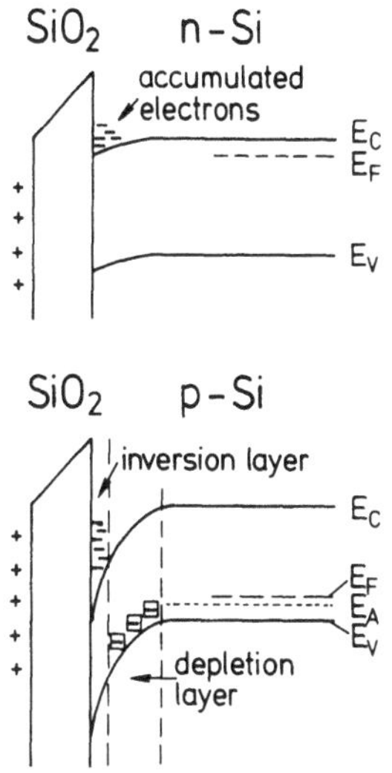

Fig. 4 Schematic energy band diagrams for metal oxide semiconductor structures for both n-type and p-type silicon for a positively charged surface

10 nm. Oxide layers in MOS transistors have typically the same thickness. Since sputtering with Ar^+ primary ions does not yield any surface potential variation it is this SiO_2 layer which is essential for the understanding of our results. From the physics of MOS devices [11] it is known that positive biasing of a metal contact causes accumulation of electrons at the interface between the SiO_2 layer and the n-type silicon. Positive bias creates a depletion layer in the case of p-type silicon and an inversion layer at increased positive bias. In Fig. 4 both cases are depicted schematically replacing the positively biased electrode by fixed positive charges caused by the positive ion bombardment. Under stationary conditions of constant primary current at the surface of the SiO_2 layer a positive charge is built up and a tunnel current I_t through the SiO_2 layer is induced by the resulting electric field being large enough to prevent further charging. The size of this current is $I_t = I_p + I_s = 2\ \mu A$ where I_p is the primary ion current and I_s is the current flowing away from the surface (secondary electrons and ions). The magnitude of the surface charge is given by the thickness of the insulating layer and the magnitude of the current which has to flow. Going from n-type to p-type silicon the thickness of the insulating barrier varies. Already a small surface charge leads to an accumulation of electrons at the SiO_2/n-Si interface. Surface charging of the SiO_2 leads to a depletion layer of high resistivity at the SiO_2/p-Si interface. This depletion layer is an additional barrier where electrons have to tunnel through. The extra 2 V of surface potential found for SiO_2/p-Si is a consequence of the increased tunnel barrier thickness.

The current flowing must be kept constant. The surface potential peaks when the sputtering beam reaches the beginning of the depletion layer of the p-n junction. Then the insulating layer has a maximum thickness. Further sputtering leads to a decrease of the thickness of this layer and results in a decreasing surface potential. At the actual position of the junction the surface potential has almost completely decreased to the bulk p-Si value as proved by Fig. 2. It is important to note that the given location of the junction is derived independently as mentioned above.

The different surface potential at the p-n junction of type A and B samples under exactly identical sputter conditions (and current flowing) gives clear evidence that the insulating properties of the junction under reverse bias conditions of the type B samples are much inferior. A higher reverse bias current flows than in the type A samples at the same backward voltage. In view of Fig. 1 this result sounds surprising. It is obvious now that an inspection of the defect properties at the surface does not allow any conclusions on the quality of the junction.

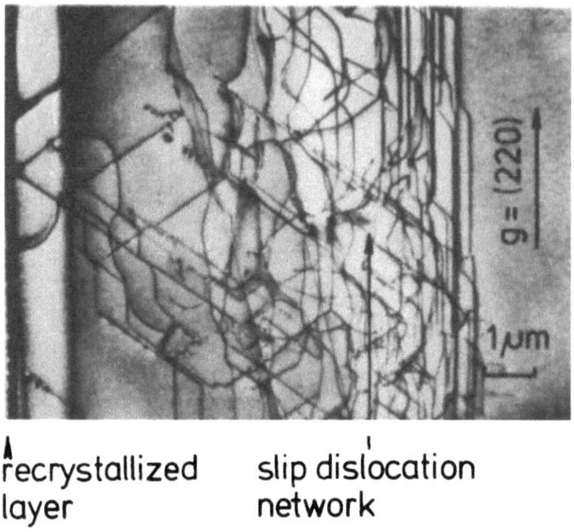

Fig. 5 Cross section TEM micrograph showing slip dislocation network and recrystallized layer containing small dislocation loops

Cross-sectional electron micrographs, however, reveal the depth dependence of the defects and supply an independent verification of the above conclusions. The slip dislocations in the type A samples are mainly located deep below the specimen surface (Fig. 5) whereas the crystalline quality of the p-n junction region (i.e. within the recrystallized layer) is essentially determined by the presence of small dislocation loops that are found in both types of samples. The density of slip dislocations penetrating the p-n junction is less than a factor of 50 compared with the dislocation density corresponding to the loops. Thus it can be assumed that the influence of the slip dislocations on the electrical behaviour of the p-n junctions is negligible. As releaved by plan-view specimens the density of dislocation loops is less by a factor of three in the type A samples as compared to the samples of type B. Apparently this decrease of the loop density due to the longer annealing is responsible for the better p-n junction quality of the type A samples.

It is thus shown that the implanted and cw laser annealed diode structure with the slipped surface morphology has the better electrical quality. This surprising result was found by SIMS thus demonstrating that SIMS being applied only for chemical analysis is also an extremely interesting method for the assessment of the electrical properties and quality of implanted and annealed p-n junctions. A particular advantage of SIMS is that no electrical contacts need to be made on top of the thin annealed layer.

We are indebted to D.R. Young for helpful discussions and to R. Uebbing for technical support.

References

1. A. Gat, J.F. Gibbons, T.J. Magee, J. Peng, V.R. Deline, P. Williams and C.A. Evans, Jr., Appl. Phys. Lett. 32, 276 (1978)
2. J.S. Williams, W.L. Brown, H.J. Leamy, J.M. Poate, J.W. Rogers, D. Rousseau, G.A. Rozgonyi, J.A. Schelnutt and T.T. Sheng, Appl. Phys. Lett. 33, 542 (1978)
3. D.H. Auston, J.A. Golovchenko, P.R. Smith, C.M. Surko and T.N.C. Venkatesan, Appl. Phys. Lett. 33, 539 (1978)
4. A. Gat, A. Lietoila and J.F. Gibbons, J. Appl. Phys. 50, 2926 (1979)
5. A. Gat, J.F. Gibbons, T.J. Magee, J. Peng, P. Williams and C.A. Evans, Jr., Appl. Phys. Lett. 33, 389 (1978)
6. G.A. Rozgonyi and H. Baumgart, J. de Physique, Suppl. au 41, (5), C 4-85 (1980)
7. K. Ishida, H. Okabayashi and M. Yoshida, Appl. Phys. Lett. 37, 175 (1980)
8. H. Baumgart, F. Phillipp, G.A. Rozgonyi and V. Gösele, Appl. Phys. Lett. 38, 95 (1981)
9. W. Reuter and K. Wittmaack, Appl. Surf. Sc. 5, 221 (1980)
10. J.C.C. Tsai and J.M. Morabito, Surf. Sc. 44, 247 (1974)
11. S.M. Sze, "Physics of Semiconductor Devices" Wiley, Interscience, New York, 1969, p. 425

Profiles of Implanted or Diffuses Dopants (Be, Zn, Cr, Se) in Indium Phosphide

M. Gauneau, A. Rupert, and P.N. Favennec

LAB.ICM/MPA - CNET
F-22301 Lannion, France

Introduction

Indium phosphide is of technological interest since it has been shown that optical fiber communications systems in the 1.1-1.6 μm range could be based on the InP/InGaAsP material. It is of primary importance to develop appropriate device fabricating techniques both for InP itself and alloys grown on it. The paper to be presented is concerned with the SIMS technique applied to the determination of some dopant profiles, i.e. Be, Zn as p-type, Se as n-type and Cr in semi-insulating substrates.

Previous work on this subject have been described by D.P. LETA et al. [1-2], W.J. DEWLIN et al. [3], J.E. BAKER et al. [4], but this study presents some new aspects: 1) Be, Cr, Se are implanted at high energies (400-800 keV); 2) Zn profiles, obtained by diffusion under $PH_3 + H_2$ atmosphere, and Se profiles are tracked using a primary cesium ion source; 3) much attention is paid to the precise measurement of projected ranges and standard deviations; 4) the comportment of InP is compared to that of GaAs.

Experimental

The samples were <100> crystals. They were implanted at doses and energies listed in Table 1. Some impurity/substrate couples have been added (Cl/InP, B/GaAs, O/GaAs) to better demonstrate the discrepancy between experimental and theoretical values of Rp and ΔRp. Implantation has been carried out with a Van de Graaff accelerator; the single crystals were tilted at an angle of 7° to the incoming ion beam to prevent any channeling.

Two secondary ion microanalyzers have been used, either the IM 20 AEI apparatus, or the IMS-3F manufactured by CAMECA. Zn diffusion profiles and Se implanted profiles have been obtained by using a primary cesium ion source fitted on the IMS-3F microanalyzer [5] and monitoring $ZnCs^+$ secondary ions according to H.A. STORMS et al. [6], or Se-H secondary ions for selenium implant. After SIMS analysis, the depth of each sputtered crater was measured using a TALYSTEP stylus type surface profiler; with such an apparatus it has been shown that an accuracy better than 10 % was achieved when depth calibration was performed with depth standards.

Table 1 Measured projected range (Rp) and standard deviation (ΔRp) for some dopants in InP or GaAs, comparison with theoretical values.

Implant / Substrate	Doses [atom cm^{-2}]	Energy [keV]	Reduced energy [$\sqrt{\varepsilon}$]	Measured Rp [Å]	Theoretical Rp [Å]	Theo.Rp / Meas.Rp	Measured ΔRp [Å]	Theoretical ΔRp [Å]	Theo.ΔRp / Meas.ΔRp
Be/InP	10^{14}	400	5.05	11.440	10.680	0.93			
	10^{15}						1.830	2.340	1.28
Be/GaAs	10^{14}	400	5.05	11.250	12.000	1.07	1.860	2.630	1.41
B/GaAs	$3.6\cdot10^{14}$	200	3.13	5.460	5.000	0.92	1.490	1.700	1.14
	$4.7\cdot10^{14}$	400	4.42	9.400	9.500	1.01	1.650	2.340	1.42
	$1\cdot10^{14}$	500	4.94	10.820	11.500	1.06	1.662	2.550	1.53
	$5.9\cdot10^{14}$	1.000	6.99	17.500	20.650	1.18	1.925	3.235	1.68
O/GaAs	$6\cdot10^{14}$	1.000	5.26	14.750	13.400	0.91			1.09
							2.360	2.580	
Cl/InP	$3\cdot10^{15}$	500	2.22	4.650	3.650	0.70			0.79
							1.530	1.220	
Cr/InP	10^{13}	800	2.14	5.680	3.605	0.63			
	10^{14}						1.940	1.140	0.58
	10^{15}								
Cr/GaAs	10^{13}	400	1.52	2.498	2.050	0.82	1.220	780	0.64
	10^{13}	800	2.14	5.530	4.050	0.73	2.070	1.280	0.62
	10^{14}								
	10^{15}								
	10^{13}	1.000	2.40	7.245	5.200	0.72	2.200	1.470	0.67
Se/InP	$9\cdot10^{13}$	400	1.12	1.510	1.230	0.81	860	490	0.57

Results and Discussion

All dopant substrate combinations studied here are listed in Table 1. The SIMS technique allows to measure

- Rp, ΔRp, projected range and standard deviation of the atomic distribution;
- C max, C min, maximum and minimum profile concentrations, and so, to get a standardization of the analysis;
- Y +/-, the useful yield of the secondary ion emission process.

Experimental values of Rp and ΔRp, comparison with LSS theory

As an example, Fig. 1 shows a berillium profile corresponding to 10^{15} atoms cm^{-2} implanted at 400 keV in InP. This profile is well fitted by a theoretical curve drawn according to a Gram - charlier development taking into account the first moments: Rp, ΔRp, skewness and kurtosis, and presents the expected skewing towards the surface. Rp and ΔRp have been measured for each dopant substrate couple quoted in Table 1 and compared with theoretical values taken from the ion-implantation range statistics compiled by GIBBONS et al. [7]. The calculations assume the projected range and standard deviation values for GaAs and Ge to be identical, and, for InP, to be 0.89 times these values. Extrapolation has been made for missing species. The ratios (theoretical Rp/measured Rp), and (theoretical ΔRp/measured ΔRp) are reported in Table 1. The deviation, for Rp, is within the experimental errors for Be and B (about 10 % as mentioned above due to the inaccuracies in the depth measurements and slight non-uniformities in the stability of the sputter rates during analysis). But the discrepancy reaches 30-40 % for chlorine, chromium, selenium, and the difference is always bigger for the standard deviation ΔRp.

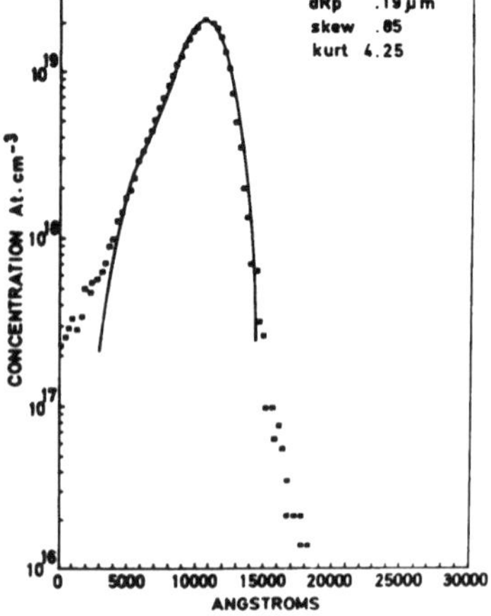

Fig. 1 Be profiled in InP for a 10^{15} atoms cm^{-2} dose implaneted at 400 keV

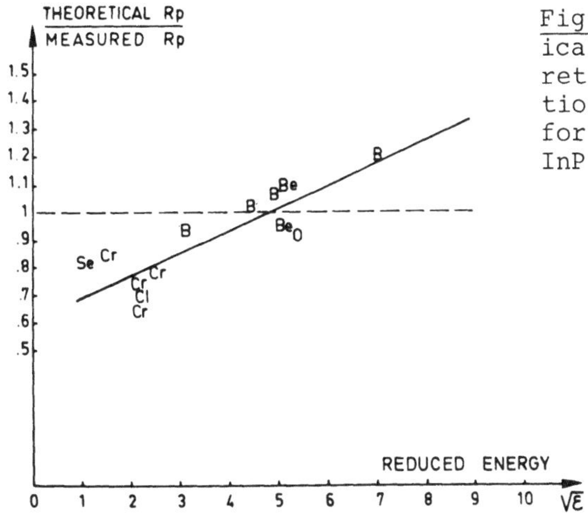

Fig. 2 Variation of (theoretical Rp/measured Rp) and (theoretical ΔRp/measured ΔRp) ratios against reduced energy for Be, B, O, Cl, Cr, Se in InP and GaAs

This effect depends on the implanted ions and target masses, and on the energy. Fig. 2 depicts the variation of the theo.Rp/meas.Rp ratio against reduced energy according to H.E. SCHIOTT calculations [8]. It is seen that the deviation depends on $\sqrt{\varepsilon}$. D.P. LETA et al. [2] found, as a general trend, that experimental values of Rp and ΔRp were higher than theoretical ones at energies ranging from 100 to 300 keV. That is confirmed here at higher energies. This experimental fact cannot be related to: (i) the knock-on effect as it has been demonstrated by D.P. LETA, the well defined deep side of berillium profile on Fig.1, in agreement with LSS theory, can be considered as an added proof; (ii) the channeling of implanted ions since theoretical overestimation of Rp and ΔRp are observed for Be and B.

As a conclusion, it may be assumed that the electronic stopping power used to compute the Rp and ΔRp values is slightly overestimated for heavy elements.

Useful yields in InP, detection limits of some elements, comparison with GaAs:

Depth profiling of implanted elements allows to determine C max and C min, the maximum and the minimum concentration of the distribution [1]. The useful yield (y +/- in the text) defined as the ratio of the ions collected, for the implant under investigation, with respect to the atoms sputtered from the actual sampling area [9] can also be deduced. This useful yield reveals parameters on which the detection limit depends on:

i) apparatus parameters (transmission coefficient, ratio of the signal to background);
ii) material parameters (ionization probabilities which depend on the element under consideration and on the matrix);
iii) secondary effects (residual impurities, interference with matrix molecular ions or vacuum contaminants, contribution of secondary ions coming from other parts than the center of the craters - edges, extractive lens, neutral sputtering).

Results obtained for Be, Cl, Cr and Se in InP or GaAs are listed in Table 2. To provide self-consistent data base the secondary ion signal have been normalized:

- in a same matrix and different microanalyzer, for the same analysed volume of material during each second;
- in different matrix, but the same microanalyzer, for the same experimental conditions (primary beam intensity, analysed area, matrix species ion intensities).

Table 2 Measured C max, C min, and useful secondary ion yields for Be/InP, Cl/InP, Cr/InP, Se/InP; comparison with GaAs for Be and Cr

Element Sub-strate	Primary beam / Ion micro-analyzer	Analysed ions	Dose [atoms cm^{-2}]	Present results C max [atoms cm^{-3}]	C min [atoms cm^{-3}]	Y +/- useful yield
Be/InP	$^{32}O_2^+$ 15 keV	$^{9}Be^+$	10^{14}	$2\cdot10^{18}$	$6\cdot10^{16}$	$8\cdot10^{-7}$
	AEI IM 20		10^{15}	$2\cdot10^{19}$	-	-
Be/GaAs	$^{32}O_2^+$ 15 keV	$^{9}Be^+$	10^{14}	$2\cdot10^{18}$	$3\cdot10^{16}$	$5\cdot10^{-6}$
	AEI IM 20					
Be/InP	$^{40}A^+$ 8 keV	$^{9}Be^+$	10^{14}	$2\cdot10^{18}$	$4\cdot10^{15}$	$9\cdot10^{-6}$
	CAMECA IMS-3F		10^{15}	$2\cdot10^{19}$	-	-
Cl/InP	$^{40}A^+$ 25 keV	$^{35}Cl^-$	$3\cdot10^{15}$	$8\cdot10^{19}$	$1\cdot10^{18}$	$3\cdot10^{-5}$
	AEI IM 20					
Cr/InP	$^{32}O_2^+$ 15 keV	$^{52}Cr^+$	10^{13}	$2\cdot10^{17}$	$2\cdot10^{15}$	$2\cdot10^{-4}$
	AEI		10^{14}	$2\cdot10^{18}$	-	-
	IM 20		10^{15}	$2\cdot10^{19}$		
Cr/GaAs	$^{32}O_2^+$ 15 keV	$^{52}Cr^+$	10^{13}	$2\cdot10^{17}$	$1\cdot10^{16}$	$8\cdot10^{-4}$
	AEI		10^{14}	$2\cdot10^{18}$		
	IM 20		10^{15}	$2\cdot10^{19}$		
Se/InP	$^{133}Cs^+$ 10 keV	$^{81}SeH^-$	$9\cdot10^{13}$	$4\cdot10^{18}$	$2\cdot10^{17}$	$1\cdot10^{-3}$
	CAMECA IMS-3F					

Apparatus effect

If comparison of useful yields is made between two ion micro-analyzers: AEI-IM 20 and IMS-3F, for Be inP (see Table 2), as an example, it is observed that the useful yield is ten times bigger for the IMS-3F and higher dynamic ranges are obtained on profiles. But these experiments with the IMS-3F microanalyzer are certainly not optimized because they have been undertaken with argon primary beam and oxygen backfilling; it is well known that ionization probabilities depend on the primary ion species, and higher yields should have been obtained with oxygen primary beam.

Matrix effect

Careful examination of Table 2 shows that useful ion yield for Be and Cr are higher in GaAs than in InP (X 4-6). This fact is visualized on Fig. 3 where it is seen that for the same dopant concentration in each substrate higher count rates are obtained in GaAs. This result is in agreement with V.R. DELINE et al.'s measurements [9] described a general matrix effect and interpreted it as being proportional to $1/\dot{Z}$ ($\dot{Z}$ = linear erosion rate); here the erosion rates of GaAs and InP are respectively 436 and 1200 Å/min. That is to say the sensitivity should not be as high in InP as in GaAs.

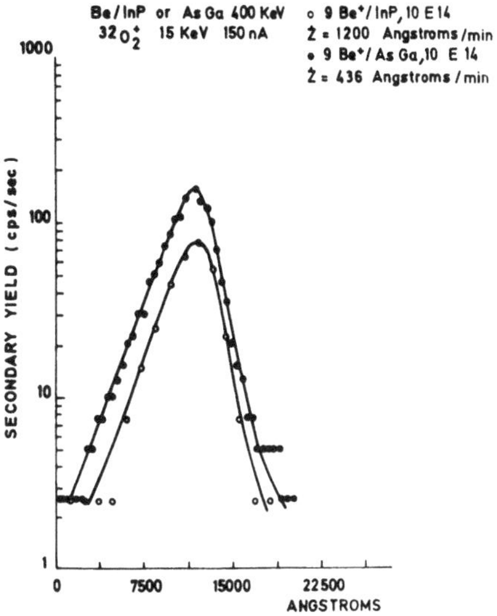

Fig. 3 Matrix effect on useful secondary ion yields of Be in InP or GaAs

Ionization probabilities of the element, choice of the primary beam

Figure 4 illustrates two profiles: Se and Zn in InP, tracked with $^{133}Cs^+$ primary ions. Zinc had been diffused in n-type substrate with residual carrier concentration equal to $2\cdot10^{16}$ $e\cdot cm^{-3}$. Zn and Se have been respectively profiled by monitoring either $ZnCs^+$ or SeH^-. Such a comportment of zinc has been described by H.A. STORMS et al. [6]. While a high background at 10^{18} $atoms\cdot cm^{-3}$ was observed with oxygen primary beam, here a dynamic range of three order of magnitude, from $2\cdot10^{19}$ to 10^{16} $atoms\cdot cm^{-3}$, is achieved. For 400 keV Se-implanted InP a very poor sensitivity was obtained by recording the $^{80}Se^-$ secondary ions. This comportment of selenium in InP is not yet well understood, it is supposed that his dopant has been implanted as the SeH^+ species. These two examples are not direct evidence of the improvement, with cesium primary ions, of the senisitvities for element having high electron affinities, but they show the complexity of the secondary ion emission process and the need of various experimental conditions.

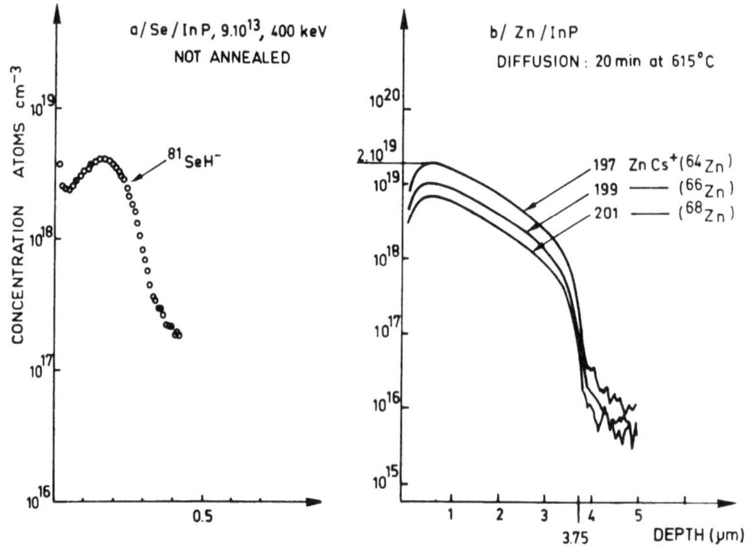

Fig. 4 Se (a) and Zn (b) profiles in InP determined with a $^{133}Cs^+$ primary beam

Secondary effects

Higher useful yield does not always mean that lower detection limits are reached. At least, three secondary effects can be encountered:

i) residual impurity in the substrate. As an example, we can quote chlorine whose useful yield is equal to $3 \cdot 10^{-5}$, but whose detection limit is only $1 \cdot 10^{18}$ atoms$\cdot$cm^{-3};

ii) interferences with matrix molecular ions or vacuum contaminants if high mass resolution or energy selection is not performed;

iii) contribution of secondary ions coming from other parts than the center of the craters (edge effects, neutral sputtering, secondary ion emission from the extractive lens). In general, a higher background is recorded for the higher implanted doses. The best way to attenuate such an effect is to increase the signal/background ratio by increasing the erosion rate.

As a conclusion, the detection limits obtained in InP, for Be, Cr, Se, Zn, are summarized in Table 3. So far, if we regard p, n dopants and residual impurities in InP, the best sensitivities and the lower detection limits should be obtained:

- with $^{32}O_2^+$ primary beam for Be, Mg, B, Al, Fe, Cr;
- with $^{133}Cs^+$ primary beam for Zn, Cd, Hg, Si, Ge, Sn, S, Se, Te, Cu, O, C.

Table 3 Detection limits of some dopants in InP, comparison with GaAs

Elements	AsGa [atoms·cm^{-3}]	InP (literature) [atoms·cm^{-3}]	InP (this work [atoms·cm^{-3}]
p-type			
Be	$2\cdot10^{16}$ [1]	$7\cdot10^{15}$ [1]	$4\cdot10^{15}$
	$8\cdot10^{13}$ [10]		
Zn	$4\cdot10^{17}$ [11]		$1\cdot10^{16}$
	$1\cdot10^{16}$ [10]		
n-type			
Se	$8\cdot10^{16}$ [1]		$2\cdot10^{17}$
	$9\cdot10^{17}$ [11]		
deep level impurity			
Cr	$8\cdot10^{13}$ [10]		
	$2\cdot10^{14}$ [11]		$6\cdot10^{14}$
	$2.5\cdot10^{15}$ [12]		

Conclusion

In this study concentrating on implanted or diffused dopants in InP and GaAs, the following main results can be pointed out: i) discrepancy observed at low energies between measured and theoretical Rp, or ΔRp, is confirmed at higher energies. The electronic stopping power used to compute the LSS theoretical values is not well estimated; ii) SIMS detection limits of dopants in semiconductors depend on many factors: apparatus, ionization probabilities, matrix, residual impurities, interferences, etc..., but due to a matrix effect these detection limits should not be so low in InP as in GaAs; some values are given for Be, Zn, Se, Cr in InP.

This work results from the collaboration with F. DEGREVE of Aolunium Pechiney Laboratory, with H.N. MIGEON of the CAMECA. The authors wish also to thank M. SALVI, L. HENRY and H. L'HARIDON for performing the implantation and the diffusion.

References

1. D.P. Leta and G.H. Morrison, Anal. Chem. 52, 514 (1980)
2. D.P. Leta and G.H. Morrison, Int. J. Mass Spectrom.and Ion Physics 34, 147 (1980)
3. W.J. Dewlin, D.P. Leta, L.F. Eastman, G.H. Morrison and J. Comas, Proc. Int. Symp. on AsGa and Related Compounds, 1978, p. 510

4. J.E. Baker, P. Williams, D.J. Worford, J.F. Oberstar, B.G. Steeman, in Secondary Ion Mass Spectrometry SIMS-II, ed. by A. Benninghoven, C.A. Evans, Jr., R.A. Powell, R. Shimizu, H.A. Storms, Springer Series in Chemical Physics, Vol. 9 (Springer, Berlin, Heidelberg, New York 1979) p. 106
5. M. Gauneau, A. Rupert, P.N. Favennec, L. Henry, H. L'Haridon and H.N. Migeon, J. Micros. Spectros. Electron. 6, 213 (1981)
6. H.A. Storms, K.F. Brown and J.D. Stein, Anal. Chem. 49, 2023 (1977)
7. J.F. Gibbons, W.J. Johnson and S. Mylroie, "Projected Range Statistics", distributed by Halsted Press, 1975
8. H.E. Schiott, Rad. Effects 6, 107 (1970)
9. V.R. Deline, W. Katz, C.A. Evans and P. Williams, Appl. Phys. Lett. 33(9), 832 (1978)
10. J.B. Clegg, Surf. and Interf. Anal. 2, 91 (1980)
11. A. Huber, G. Morillot and P. Merenda, J. Micros. Spectrom. Electron. 4, 493 (1979)
12. R.G. Wilson, P.K. Vasudev, D.M. Jamba, C.A. Evans and V.R. Deline, Appl. Phys. Lett. 36(3), 215 (1980)

Applications of SIMS in Studies of Slow Diffusion and Isotope Diffusion

A. Lodding, H. Odelius, and U. Södervall

Physics Department and Materials Science Centre,
Chalmers University of Technology
S-41296 Göteborg, Sweden

Quantitative measurement of tracer diffusion coefficients by means of SIMS raises strict requirements of detection sensitivity, in-depth resolution and spectral purity [1]. During recent years, accurate measurements have been carried out on single-crystal metals and semiconductors [2-6], involving the determination of *very low diffusion coefficients* (as an example, see Fig. 1, from [3]) as well as the detection and identification of *very dilute impurities* (see e.g. Fig. 2, from [6]).

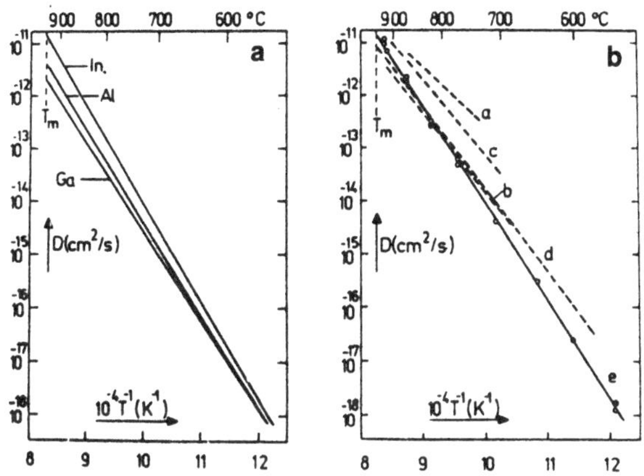

Fig. 1 Arrhenius representation of tracer diffusion in germanium, measured by SIMS between the melting point and ca 2/3 T_m [3]. (a) Diffusion coefficients of Al, Ga and In in Ge. The sputtered profile depth for the lowest D values is of the order of 2000 Å. The tracer diffused from a surface layer ca 50 Å in thickness. (b) Experimental points; comparison with other investigations for In in Ge

Among the precautions required to control the numerous artifacts were the following:

a) The homogeneity and smoothness of the sample surface before SIMS-analysis was studied by microscope. Etching or deposition of surface coating was employed to minimize crater-edge effects.

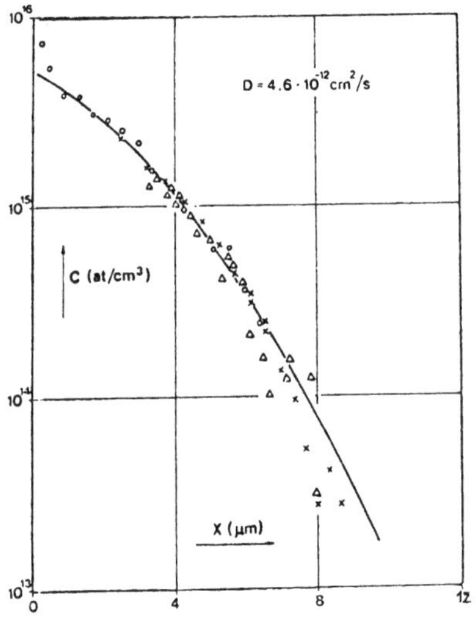

Fig. 2 Diffusion profile of Se in Si (p^+n diode), obtained by SIMS (rings), CCS (constant capacitance spectroscopy, triangles), and DLTS (deep level transient spectroscopy, crosses) [6]. The CCS and DLTS data are shifted upward by ca. a factor 8, to fit the SIMS concentration calibration. The detection limit of Se in Si by SIMS, using Cs^+ primary ions, is of the order of 1 atom ppb

b) The primary beam was rastered over an area at least four times that of the spot to be analyzed. The smoothness of the crater bottom and the conicity of the crater were controlled by optical interference fringes or by "talysurf" surface level monitor. Samples exhibiting faceting at particular angles between primary beam and crystallographic axes were discarded.

c) The ionic yields were enhanced and crystallographic effects lowered by reactive sputtering species (O_2^+ or Cs^+), or by oxygen leak.

d) Spectral contaminations were controlled by checks with high mass resolution and, wherever practicalbe, by "offset" in secondary ion energy. A low contaminant background was accepted and subtracted only if found constant at deep penetrations.

e) Corrections were made for the counting dead time of the amplifier.

f) The magnet field corresponding to the top of each mass peak was repeatedly checked during profiling. Flat-topped peaks were consistently used in isotope work.

g) The constancy of sputter rate and of emitting parameters was checked by sequential recording of one or several intrinsic matrix peaks.

h) When studying very shallow diffusion profiles, requiring relatively feeble primary ion density and wide raster, contamination from residual atmosphere could become serious enough to impair accuracy, posing a limit to the possibilities of measuring very low diffusion coefficients. To keep low the ratio [7] of the impact frequency from residual atom species to the arrival frequency of primary ions, cryogenic pumping ("cold-finger") was established in the sample vicinity. Routinely, the current density of primary ions was maintained as high as allowed by the shallowness of profile, the necessity of gating, and the limitations of counting deadtime.

j) Computer software was developed to allow evaluation of diffusion coefficients even from penetration profiles exhibiting surface artifacts due to contamination or degassing.

For practical reasons (i.e. to utilize as intense a signal and as wide a dynamic range as possible), most of the tracer-matrix systems selected for study were characterized by high ionizability (i.e. low ionization potential, or high electron affinity, or high yield of molecular species) of the tracer; and by relatively low sputtering yield [8] or high oxygen sticking efficiency of the matrix.

To determine the *isotope effect* of diffusion, the demands on high spectral purity and wide dynamic range of in-depth profiling are particularly stringent. Although SIMS in principle offers a most attractive alternative to radioactive tracer techniques for such measurements, only very few results have hitherto been published [1,9,10]. In the following, the results of recent studies of the diffusion of $^{69-71}$Ga in silicon will be reported.

The isotope effects is defined (see e.g. [9]) as

$$E = (-2)(\Delta D/D)/(\Delta M/M), \qquad (1)$$

where ΔD is the difference in diffusivity between two isotopes of mass difference ΔM; D and M denoting the resp. mean values. The physical interpretation (see e.g. [11]) is

$$E = f\cdot\Delta K, \qquad (2)$$

where f is the Bardeen-Herring correlation factor and ΔK a factor denoting the degree of sharing of kinetic energy in the saddle-point of diffusion jump.

Experimentally, the isotope effect is conveniently obtained if the tracer isotope ratio c_β/c_α (varying with depth, as the light isotope penetrates faster than the heavy one) is plotted in a log-log diagram against the concentration c of the tracer. It may be derived [12] that

$$E = A\cdot[d\ln(c_\beta/c_\alpha)/d\ln c], \qquad (3)$$

where A = 1 for in-diffusion from a thin surface film of tracer, and A lies between 0.5 (at sample surface) and unity (at deep penetrations) for diffusion under the conditions of constant surface concentration. In the diffusion annealed under present discussion, the latter geometry was the case; the tracer was diffused into the sample from saturated Ga vapor in a sealed quartz ampoule, and the surface concentration corresponded to the solubility limit of Ga in Si (of the order of 0.1 atom% at the temperatures of the annealling).

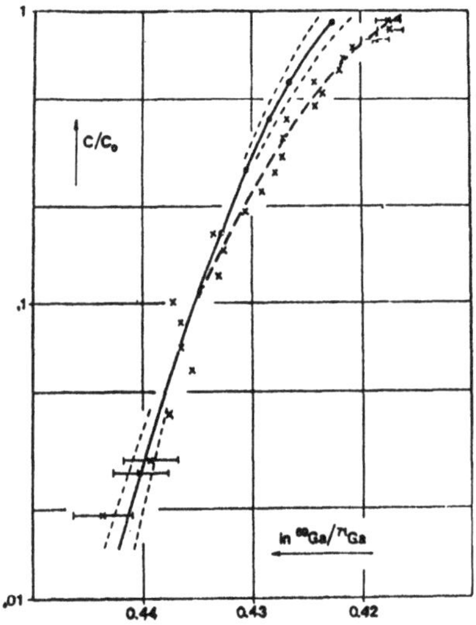

Fig. 3 Logarithmic plot of the gallium isotope ratio as a function of gallium concentration, after diffusion into silicon from saturated vapor. The isotope effect of diffusion may be obtained from the gradient of the dead-time corrected curve (whole-drawn) via Eq. (3)

In the SIMS evaluation of the diffusion profiles and isotope ratios, the single crystal silicon surfaces were sputtered with O_2^+ primary ions in a Cameca IMS-3f instrument. Typically, the Ga concentration decreased by two to three orders of magnitude within ca. 8 μm from the surface. Deeper profiles were considered as less reliable because of insufficient flatness of crater bottoms. With a 1 μA primary beam, ca. 100 μm in diameter, with a raster of ca. 300 μm, and utilizing an analyzed area of 60 μm, the 8 μm in-depth profile took about one hour of sputtering. Sequentially the 69 and 71 Ga^+ isotope peaks were measured, alternating with the 75 peak $(^{29}Si^{30}SiO)^+$ as matrix indication.

In all samples the variation of c_{Ga} with depth was found to exhibit, to a satisfactory approximation, the error-function type behaviour expected for this type of diffusion geometry, and could be represented by a straight line in a probability diagram. The diffusion coefficients at three temperatures, obtained [13] from the slopes of such lines, are listed in Table 1. They may be described by $D = D_o\exp(-Q/kT)$, with $D_o = 54\ cm^2/s$ and $Q = (3.81 \pm 0.25)$ eV. This agrees very well with earlier results obtained by other techniques [14], viz. 60 cm^2/s and 3.89 eV, respectively.

The measured isotope ratios at different diffusion depths are shown in Fig. 3, plotted logarithmically vs. the Ga concentration. As only relative changes in $^{69}Ga/^{71}Ga$ are of relevance here, no correction has been made for the mass discrimination typical of SIMS [7,15]. At deep penetrations (*low* c_{Ga}) the relatively wide uncertainty margins are given by the counting statistics; ca $5 \cdot 10^5$ counts collected for each isotope, corresponding to an accuracy of the order of 0.2 % in the isotope ratio. At *high* counting intensities (i.e. for $c/c_o \gtrsim 0.2$, the surface concentration c_o yielding about 10^5 counts/s for the 71 isotope), the dominant source of error is the counting dead time of the emplifier, in the custom-built SIMS equipment as high as 60 to 80 ns. Under these circumstances, at the very moderate counting rate of 10^5 counts/s, the dead time causes the ion current ratio $^{69}Ga^+/^{71}Ga^+$ to deviate from the "correct" isotope ratio by as much as 0.6 %. This may be compared to the total observed diffusive shift of the isotope ratio, i.e. only ca 1.5 %. The magnitude of the necessary dead-time correction can be clearly seen in Fig. 3. Obviously, if higher isotope ratios than those of Ga (ca 1.5) were to be measured, the dead-time induced error would be so great as to jeopardize the meaningfullness of the results; for such measurements, amplifiers with much shorter, and continuously checked [16], dead-time appearance is needed.

The shape of the dead-time-corrected curve in Fig. 3 is in satisfactory agreement with theoretical expectations [12]. From the gradients and from tabled values [12] of the factor A at different c/c_o, the isotope effect E may be derived via Eq.(3). The results at three temperatures are listed in Table 1.

Table 1 Tracer diffusion coefficients D and the isotope effect E of Ga in Si

T [^oC]	1100	1160	1200
$D \cdot 10^{12}$ (cm^2/s)	0.38 ± 0.03	1.60 ± 0.11	3.30 ± 0.40
E	0.31 ± 0.03	0.27 ± 0.03	0.27 ± 0.04

Although a certain temperature dependence of E may possibly be masked by the margins of error, one may at the present stage of the investigation conclude that the isotope effect for Ga diffusion in silicon lies within some 10 % from 0.28 in the temperature range 1100-1200 ^oC.

From measurements of the isotope effect of self-diffusion in a related matrix [17] (germanium), and the knowledge that for self-diffusion by the vacancy mechanism in the diamond-type matrix f = 0.5 (see, e.g. [11]), it has been deduced that the energy-sharing factor ΔK in Eq.(2) is likely to lie between ca 0.50 and 0.62. Accordingly, the present results suggest $f_{Ga-Si} \simeq 0.50$, i.e. a very similar behaviour as in self-diffusion.

Continuing studies are hoped to yield the isotope effect with an accuracy sufficient to distinguish also the temperature dependence of E and to yield information on the differences between tracer atoms and matrix atoms in regard of vacancy jump frequencies. The SIMS technique also appears to offer a means of direct determination of ΔK in silicon, if the enriched 28-29 Si isotopes are employed as tracers in an isotope effect study.

Acknowledgements

This work has been financially supported by the Swedish Board of Technical Development and the Swedish Natural Science Research Council. We acknowledge the valuable aid of N.H.Migeon, Courbevoie, in several stages of the experimental work.

References

1. A. Lodding, Adv. in Mass Spec. 8, 471 (1980)
2. W. Gust, M.B. Hintz, A. Lodding and H. Odelius, phys. stat. sol. (1981) (in press)
3. P. Dorner, Thesis, Univ. of Stuttgart (1980)
4. P. Dorner, W. Gust, M.B. Hintz, A. Lodding, H. Odelius and B. Predel, Acta Metall. 28, 291 (1980)
5. L. Kalinowski and R. Seguin, Appl. Phys. Lett. 35, 211 (1979)
6. H.G. Grimmeiss, E. Janzen, B. Skarstam and A. Lodding, J. Appl. Phys. 51, 6238 (1980)
7. A. Lodding, Proc. Euroanalysis-IV, Rev. Anal. Chem. (1981) (in press)
8. V.R. Deline, in Secondary Ion Mass Spectrometry SIMS-II, ed. by A. Benninghoven, C.A. Evans, Jr., R.A. Powell, R. Shimizu, H.A. Storms, Springer Series in Chemical Physics, Vol. 9 (Springer, Berlin, Heidelberg, New York 1979) p. 48
9. Th. Hehenkamp, A. Lodding, H. Odelius and V. Schlett, Acta Metall. 27, 827 (1979)
10. J. Verlinden and R. Gijbels, Adv. in Mass Spec. 8, 485 (1980)
11. J.R. Manning, "Diffusion Kinetics of Atoms in Crystal", v.Nostrand, Princeton (1968)
12. A. Lodding and U. Södervall, to be published in Z. Naturforsch. (1981)
13. U. Södervall, Diploma work, Chalmers Univ. Tech., Gothenburg (1979)
14. D. Shaw (Ed.), "Atomic Diffusion in Semiconductors", Plenum Press, London (1973)
15. J.C. Lorin, A. Havette and G. Slodzian, these Proceedings
16. E. Zinner and M. Grasserbauer, these Proceedings
17. D. Campbell, Phys. Rev. B12, 2318 (1975)

Rapid Diffusion and Gettering Studies of Bulk Oxygen in Silicon by Cs/SIMS

C.A. Evans, Jr., B.K. Furman, and T.J. Magee[1]

Charles Evans & Associates
San Mateo, CA 94402, USA

The analysis of the light elements in materials has always been of ultimate importance due to their ubiquitous nature in nature and the environment. However, the analysis also is confounded or made more difficult by the presence of these species in the residual vacuum or sample environment of an analytical system. Modern instrumentation, in particular, high performance secondary ion mass spectrometers, employ a variety of experimental conditions which have made the analysis of these species more attainable. In this paper we would like to address the characterization of oxygen unintentionally incorporated in silicon during the CZOCHRALSKI (Cz) silicon growth method with the oxygen coming from the silica quartz crucible used to contain the melt. Although this oxygen is unintentionally incorporated in the silicon, it is known to have beneficial qualities on the resultant material. In particular, the oxygen supposedly adds structural strength to the silicon and through a variety of processing and thermal cycling the oxygen can be made to precipitate into regions which allow the stress resulting from this precipitation to getter impurities unintentionally incorporated during device fabrication.

The classical method for determining the chemical concentration of oxygen in silicon has been Fourier Transform infrared spectrometry. Although this technique can be made very quantitative, it does suffer from two limitations. The first is that the oxygen measured by this technique is only that which is contained at interstitial sites and second, the technique is not microanalytical, thereby not allowing lateral distributions or in-depth profiles to be obtained. Secondary ion mass spectrometry operating in the high performance mode does allow for the three-dimensional characterization of the oxygen contained in the material with approximately 100 Å depth resolution and submicrometer lateral resolution. Unfortunately, this analysis is complicated by the presence of oxygen in the residual vacuum, thereby necessitating specialized operating conditions in order to reduce the influence of the residual vacuum oxygen. For these reasons, only recently has the direct measurement of oxygen in silicon become possible. A variety of papers have been published from Japanese workers who have used secondary ion mass spectrometry to observe oxygen in silicon [1-3].

[1]ARACOR, Sunnyvale, CA 94086, USA

Experimental

These studies have examined a variety of silicon materials grown by the Cz method and processed in the manner discussed with each experiment. The secondary ion mass spectrometer employed was the Cameca IMS-3f Ion Microanalyzer equipped with a cesium primary ion source and a recently developed pumping system which consists of a turbo-molecular pump on the introduction lock in order to reduce the pressures during introduction of the samples and a circulating helium cryogenic pump reentrant into the sample chamber. This cryogenic pump provides 800 liters per second of pumping speed at the sample surface. Although the ion gauge has not been well calibrated, and the vacuum does degrade due to the high thru-put of samples employed in our laboratory, we believe the vacuum to be in the height 10^{-9} Torr during these experiments.

Quantitative conversion of the ion intensity into atomic densities were obtained by a quantitative ion implant of oxygen into FZ silicon. The contribution of the residual vacuum in the sample chamber to the analytical signal was determined by the analysis of a piece of float zone silicon known to contain approximately 10^{16} atoms·cm^{-3} of oxygen. The background was thus determined to be of the order of $2\text{-}3\cdot10^{17}$ atoms·cm^{-3} under the conditions employed, a concentration far below that of the 10^{18} or higher concentrations found to be in the Cz silicon.

Experimental Results

The results obtained to date on the study of oxygen in silicon have almost been in a series of publications [4-7]. These previous publications contain not only the data but a discussion of the results with respect to the impact on the material and materials processing technology. In this paper we would like to discuss the results in order to demonstrate the efficacy of such an analysis and its potential contribution to an understanding of silicon materials processing. As with many electrically inactive impurities unintentionally incorporated into semiconductor crystals, oxygen diffuses very rapidly and will getter into regions containing residual crystalline defects. Moreover, the oxygen will precipitate in these regions of residual damage as SiO_x species are thereby increasing the stress and strain in the crystal and causing more oxygen gettering to occur. Unlike other species such as fluorine in silicon [8], and chromium in GaAs [9], the damage cannot be subsequently removed, since the SiO_2 cannot be redissolved under normal processing conditions.

The gettering of oxygen in silicon and the residual damage can be illustrated as we have done in previous studies of fluorine in silicon and chromium in GaAs. In this instance, a boron implant is made into the silicon and thermally processed so as to remove the majority of the damage introduced by the implantation process. This analysis shows the collection of oxygen in that region between the heavily amorphized silicon and the undamaged silicon. During the annealing to remove residual damage, the oxygen is moving into the region containing the residual damage and causing a build-up above the normal background concentration. If we examine a situation in which there is more damage during the implantion, thereby providing more residual damage such as is found with an arsenic ion implant into silicon, we find that there is significant amount of oxygen movement so as to not only increase the oxygen concentration in the residual damage region to above 10^{19} atom·cm^{3}, but the depletion of oxygen can also be seen due to the diffusion length of the oxygen in this material.

There is a possible second source of oxygen for diffusion in the silicon from through-oxide implantation which is commonly practiced in the semiconductor industry. In this case, the concern is that as the implanting ions traverse an overlaying SiO_2 film the oxygen from this layer will be recoiled and implanted into the underlaying silicon. We have observed such an effect for selenium and krypton implantation through a silicon nitride encapsulant into the underlying GaAs. We studied samples which had had an overlayer of silicon dioxide and had been implanted with arsenic ions whose energy was designed so as to have the arsenic come to rest in the near surface region of the silicon just under the SiO_2. The SiO_2 layer was removed subsequent to implantation so as to reduce the effects of stress during thermal processing as well as to allow for better depth profiling during the SIMS analysis. There was found to be a significant amount of oxygen recoiled into the sample. After thermal processing, we were able to observe arsenic diffusion and the redistribution of the oxygen to the residual damage. In this case, the amount of oxygen gettered into damage is significantly greater since there is more oxygen present in the region around the damage due to the recoil implantation process adding to the amount of oxygen present in the bulk of the silicon.

We were able to use both the profiling and imaging modes of the IMS-3f to detect the presence of SiO_x precipitates in the silicon. The depth profiling mode detected the inhomogeneous presence of oxygen in precipitates due to an increase in the statistical fluctuations of the oxygen ion intensities as one enters into the region containing the precipitation. The alternative approach used the imaging mode of the ion microscope so as to observe the precipitates present in the internal regions of the material as flashes of oxygen ion intensities. If experimental conditions allow, one can use isotopic labelling of the oxygen with ^{18}O in order to track a given process as well as reduce the impact of residual vacuum contamination. Since by using ^{18}O enriched oxygen in the experiments, one can bring the oxygen content far above the ^{18}O natural background level. The results of such experiments have been described in a study by MICHELSON [10] and the reader is referred to these publications for a detailed discussions. In summary, we conclude that the analysis of oxygen on silicon is an exacting but rewarding experiment for conditions must be employed which reduce the contribution of vacuum to the analytical signal and instrumentation with ion optics necessary to reject the crater edge effects and neutral beam sputtering of the surface regions of the material which all contribute to the limitation of the analysis via contributions to background.

References

1. H. Koyama, J. Appl. Phys. 51, 3202 (1980)
2. T. Hirao, G. Fuse, K. Inoue, S. Takayanagi, Y. Yaegashi and S. Ichikawa, J. Appl. Phys. 50, 5251 (1979)
3. K. Hoh, H. Koyama, K. Uda, and Y. Miura, Jap. J. of Appl. Phys. 19, L375 (1980)
4. T.J. Magee, C. Leung, H. Kawayoshi, B.K. Furman and C.A. Evans, Jr., Appl. Phys. Lett. 38, 891 (1981)
5. ibid 39, 260 (1981)
6. ibid 39, 1413 (1981)
7. T.J. Magee, C. Leung, H. Kawayoshi, B.K. Furman, C. Hopkins and C.A. Evans, Jr., J. Appl. Phys. 51, 3202 (1980)
8. M.Y. Tsai, D.S. Day, B.G. Streetman, P. Williams and C.A. Evans, Jr., J. Appl. Phys. 50, 188 (1979)
9. T.J. Magee, J. Hung, V.R. Deline and C.A. Evans, Jr., Appl. Phys. Lett. 37, 53 (1980)
10. J.R. Michelson, Appl. Phys. Lett. (in print)

Water Diffusion in Fused Silica and Iron-Making Slag

M. Kobayashi, K.S. Goto, and M. Someno[1]

Department of Metallurgy, Tokyo Inst. Technology
Tokyo, Japan

1. Introduction

Secondary ion mass spectrometry is a powerful tool for analysis of hydrogen in small parts of a specimen. Purpose of this study is to measure diffusion coefficients of hydrogen in iron- and steel-making slags at high temperatures. As the first step, a preliminary investigation of heavy "water" diffusion in a silica glass was made. Here "water" diffusion means diffusion of hydrogen-bonding species which exist in a slag (or a silicate glass) through a reaction of water vapor with the slag on the surface.

2. Experimental

Silica glass as received was made by vacuum-melting of quartz powder. Infrared absorption spectrum of this glass indicated water content under the detection limit. Specimens with their size of 20 mm x 10 mm x 1-3 mm were mechanically polished and then chemically etched in a 10 % HF solution to remove surface cracks. Diffusion-annealing was done in a mixed gas of nitrogen and heavy water vapor (P_{D_2O} = 355 mbar). After the annealing, six steps were made for depth profiling on the specimen surface by changing etching time of the specimen in the HF solution.

A Hitachi ion microanalyzer (IMA-2) was used with 2 keV energy, current of $5 \cdot 10^{-7}$ A and beam spot size of 1 mm diameter. In order to eliminate charge accumulation on the specimen surface during analysis, gold was deposited on the surface with its thickness of about 50 nm and electrons with 3 keV were sprayed on the surface with the same current density as the primary ion.

3. Results and Application

3.1 Charging-up

Variation of the energy distribution of secondary positive ions of D^+ and Si^+ was observed with the electron spray current, the energy distribution shifted towards lower energy region, which means decrease in positive charge on the specimen surface.

[1]Department of Mechanics, Technological Univ. Nagaoka, Nagaoka, Japan

This positive charge could be minimized by the electron spray when the spray current density on the surface was the same as the current density of the primary ion.

The energy distribution of D^+ was quite different from the distribution of Si^+; intensity ratio of D^+/Si^+ was dependent on the energy. Therefore it is important to set the energy window of the spectrometer and the electron spray current, respectively, at a fixed value.

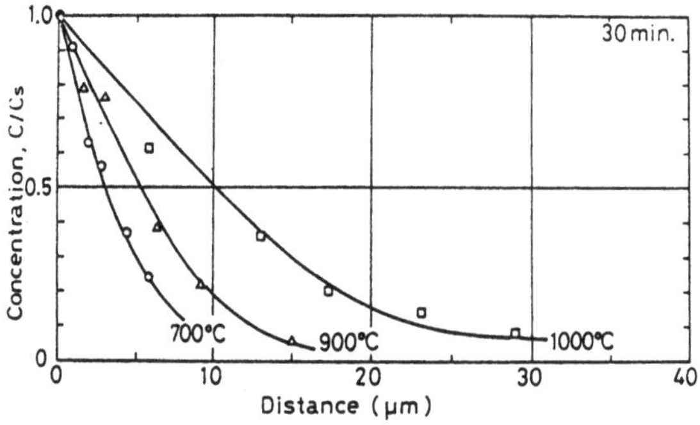

Fig. 1 Depth profiles of deuterium in silica glass specimens diffusion-annealed at 700 °C, 900 °C and 1000 °C for 30 minutes

3.2 Depth Profiling

Figure 1 shows an example of depth profiles of deuterium in silica glass specimens. The normalized concentration was calculated from the normalized intensity ratio of D^+/Si^+, considering no matrix effect in silicate materials [1]. The surface concentration of deuterium was determined as number of oxygen-deuterium bondings in the SiO_2 matrix. This concentration was estimated from the depth profile and the infrared absorption spectrum which gives overall concentration of OD bonding in a specimen. From the concentration and the intensity ratio, relative secondary ion yield of deuterium to silicon was calculated and tabulated in Table 1. The average of $1.3\cdot10^{-2}$ was used to estimate the surface concentration of deuterium shown in Fig. 2.

Table 1 Relative secondary positive ion yield of deuterium in fused silica (S_{D^+}/S_{Si^+})

No.	T [°C]	$I_s(D^+)/I_s(^{28}Si^+)$	$C_s(-OD/SiO_2)$ [ppmw]	$K = S_{D^+}/S_{Si^+}$
1	800	$3.4\cdot10^{-5}$	520	$1.8\cdot10^{-2}$
2	900	$1.4\cdot10^{-5}$	320	$1.2\cdot10^{-2}$
3	1000	$1.6\cdot10^{-5}$	320	$1.4\cdot10^{-2}$
4	1100	$9.8\cdot10^{-6}$	320	$8.4\cdot10^{-3}$

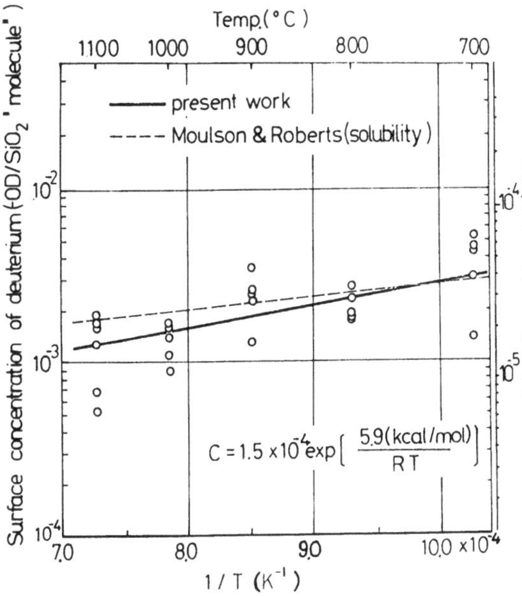

Fig. 2 Surface concentration of deuterium calculated from the intensity ratio of D^+/Si^+ (relative ion yield = $1.3 \cdot 10^{-2}$)

Comparison of the estimation with the solubility of water, which is shown in the figure by a dashed line, revealed equilibrium of the surface reaction of water vapor with the silica glass: the diffusion coefficient could be estimated in case of negligible effect of the surface reaction.

3.3 Diffusion Coefficient

The diffusion coefficient of water in the silica glass is shown in Fig. 3 as a function of reciprocal absolute temperature. The least square fit of the data is written as,

$$D(cm^2/s) = 2.3 \cdot 10^{-6} \exp(-88(kJ/mol)/RT) .$$

Results showed good agreement with those reported by the different techniques of infrared absorption spectroscopy [2] and in-depth measurement of tritium β-ray emission [3].

3.4 Application

Figure 4 shows an application to a 40 % SiO_2 - 40 % CaO - 20 % Al_2O_3 synthetic slag. Diffusion coefficients were determined below the temperature at which crystallization and viscous flow of the speciment occured. It is noted that the diffusion coefficient, and the activation energy of diffusion of water in this slag are somewhat greater than in the silica glass.

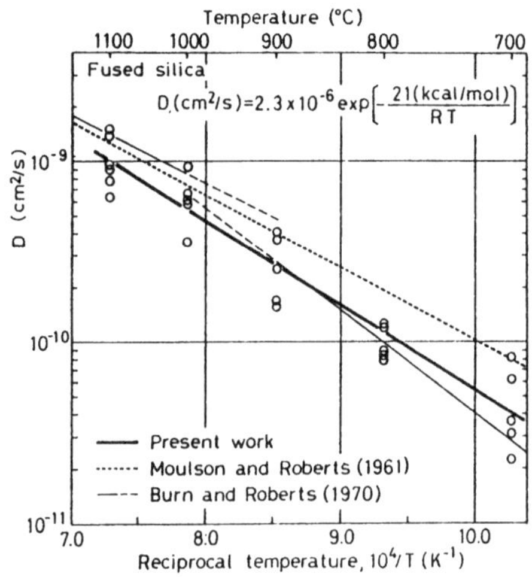

Fig. 3

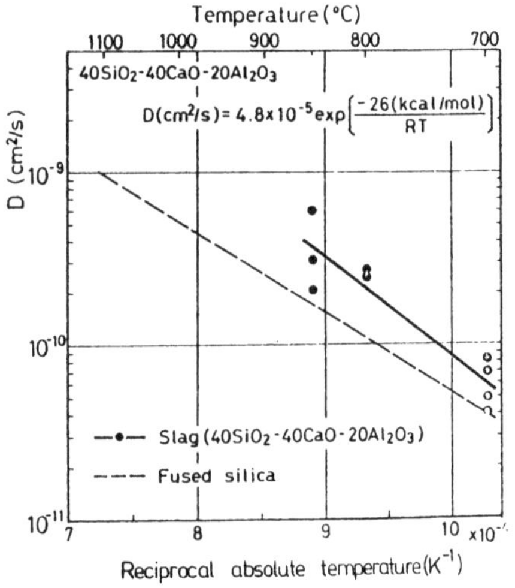

Fig. 4

Fig. 3 Temperature dependence of estimated diffusion coefficient of deuterium in a silica glass

Fig. 4 Temperature dependence of diffusion coefficient of deuterium in a 40 % SiO_2 -40 % CaO - 20 % Al_2O_3 slag

References

1. J.R. Hinthorne and C.A. Anderson, Amer. Mineralogist
2. A.J. Moulson and J.P. Roberts, Trans. Faraday Soc. 57, 1208 (1961)
3. I. Burn and J.P. Roberts, Phys. Chem. Glasses 11, 106 (1979)

Combined SIMS-AES-XPS Investigation of the Composition and Interface Structure of Anodic Oxide Layers on $Cd_{0.2}Hg_{0.8}Te$ (CMT)

U. Kaiser, O. Ganschow, J. Neelsen, H.M. Nitz, L. Wiedmann, and A. Benninghoven

Physikalisches Institut, Universität Münster, D-4400 Münster, Fed. Rep. of Germany

1. Introduction

Because of its small band gap, cadmium mercury telluride (CMT) is widely used as a material for IR detectors. The constancy and long-time stability of its composition, as well as a low impurity concentration are the crucial requirements for successful applications of this semiconductor. Anodic oxide layers are frequently used for passivation of the CMT surface. In the present investigation, the surface and in-depth composition of anodic oxide layers with different thickness from 20 to 230 nm on $Cd_{0.2}Hg_{0.8}Te$ has been characterized by quasi simultaneous secondary ion mass spectrometry (SIMS), Auger electron spectroscopy (AES), and X-ray photoelectron spectroscopy (XPS).

Bare CMT surface shows appreciable changes of elemental concentrations with time, which are due to mercury evaporation [1]. During sputter depth profiling at room temperature, the simultaneous occurrence of preferential sputtering and mercury diffusion [1] results in a depletion profile of this element below the actual surface, which in sputter equilibrium extends to a depth of about 1000 nm. At 180 K, however, diffusion is negligible, and the region changed by preferential sputtering is confined to the uppermost three atomic layers. Therefore our experiments have been performed in the 170-180 K range. Apart from the specimen temperature, the experimental arrangement was as described in [1].

We limit our discussion to the example of the 230 nm layer, as the composition of the thinner layers is quite similar. A full account of our results will be published elsewhere.

2. Preferential Sputtering and Depth Profiling

The elemental concentrations as determined from the electron spectroscopic data are weighted mean values across the information depth of the method used, which is larger than the inhomogeneous surface layer induced by preferential sputtering. As the escape of the emitted photoelectrons (O 1s, Te 3d, Cd 3d and Hg 4f) is larger than of the Auger electrons (O KLL, Te MNN, Cd MNN and Hg NVV), the XPS data give the true elemental concentrations with greater accuracy than the AES data. From the differences between XPS and AES we can estimate the

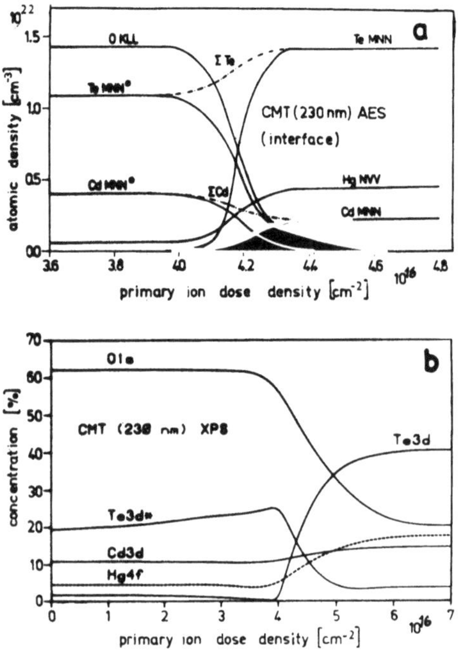

Fig. 1 Atomic densities of elements as determined from the electron spectroscopic signals as a function of primary ion dose density for the 230 nm layer. (a) from AES, (b) from XPS. *denotes the shifted lines from the oxide layer

concentration gradients at the surface and the partial sputtering yields.

Figure 1 shows the AES and XPS data for the 230 nm layer. The AES signals are constant for a primary ion dose density $D < 3.6 \cdot 10^{16}$ cm^{-2} and therefore not shown in the figure. As compared to the XPS data, the concentrations determined by AES are smaller for Hg and O and larger for Cd. In view of the smaller information depth of the AES transitions, these differences indicate concentration gradients of these elements due to preferential sputtering in such a way, that for the total sputter yields $S_{Hg} < S_{Te} < S_{Cd}$ must hold, as has been found for the bare CMT [1]. From the same argument it can be concluded that oxygen is sputtered preferentially as well. This can additionally be seen from the appearance of an unshifted Te 3d doublet due to non-oxidized Te soon after the onset of sputtering.

The mean erosion rate of the oxide has been determined from the interface position in the AES data as $(5.3 \pm 0.3) \cdot 10^{-22}$ cm^3 per incident primary ion (Ar^+, 5 keV), similar to the value of $(5.8 \pm 0.3) \cdot 10^{-22}$ cm^3 per primary ion for the bare substrate at room temperature [1].

The slow change of the XPS signals from the interface to the matrix is a crater edge effect. Because of the small specimen diameter, the area which could be homogeneously rastered by the ion beam had to be chosen smaller than the acceptance area of XPS. In AES, these effects could be excluded. Therefore quantitative depth data were derived from the AES measurements.

3. Chemical Information

From various XPS investigations of anodic oxides on CMT [2-4], chemical shifts of the Cd MNN Auger transition and the Te 3d photoelectron lines with respect to their energy in the CMT matrix are known. In this study, we additionally observed shifts of the Te MNN, Hg NVV, Te 4d, and Cd 4d lines. The most prominent features are the shifted Te 3d doublet and the shifted Te MNN and Cd MNN lines, which are shown together with the unshifted contribution from the matrix in Fig. 1.

More detailed information can be obtained by SIMS. The secondary ions emitted from the samples consisted of atomic and molecular species characteristic of the matrix and the oxide such as $TeO_m^{\pm}$, $Te_2O_m^{\pm}$ (m = 0,1,2...), $Cd^{\pm}$, O^- etc. as well as a number of contaminations like C^-, C_2^-, C_2H^-, $C_2H_2^-$, F^-, Cl^-, OH^-, K^+ and Na^+. The depth distribution of the most prominent of these species is given in Fig. 1 for the 230 nm layer. Within the experimental detection limit, no $CdO_m^{\pm}$ or $HgO_m^{\pm}$ has been detected. In investigating these types of molecular ions, the problem of interference of different secondary ion species (e.g. TeO_m^-, CdO_{m+1}^- and $CdO_{m+1}H^-$, or $Te_2O_m^-$, $CdTeO_{m+1}$ and $CdTeO_{m+1}H^-$) has been solved by numerical separation of the intensities of the constituent molecules according to their natural isotopic abundances. This analysis yields an upper limit for CdO_m^- and CdO_mH^- of 1 % of the TeO_{m-1}^- signal, so that there is no evidence for Cd oxidation from these data.

4. The Structure of the Oxide Layer and the Interface

In contrast to the AES and XPS results, which suggest a rather homogeneous oxide layer, the SIMS data indicate an appreciable inhomogeneity of the layer. The increase of the TeO_m^- ions in the inner part of the oxide layer (see Fig. 2b) must primarily be attributed to an inhomogenity with respect to the oxidation number of the elements and the ionicity of chemical bonds, which influence the ionization probability very strongly.

An estimate of the changes in the oxidation state can be made by the valence model of PLOG, WIEDMANN and BENNINCHOVEN [5]. Although there is no evidence for direct Cd or Hg oxidation, the "lattice valency" G^- of Te can be evaluated from the group of TeO_m^- ions. G^- changes from 1.4 at the surface to a maximum value of 2.5 in the inner part of the oxide layer, thus indicating a higher oxidation state of Te in this region. This evidence is supported by increase of the Te 3d* signal in XPS near the interface, which corresponds to the maximum signal of the oxide specific secondary ions. Note that due to the preferential sputtering of oxygen a decrease of the apparent

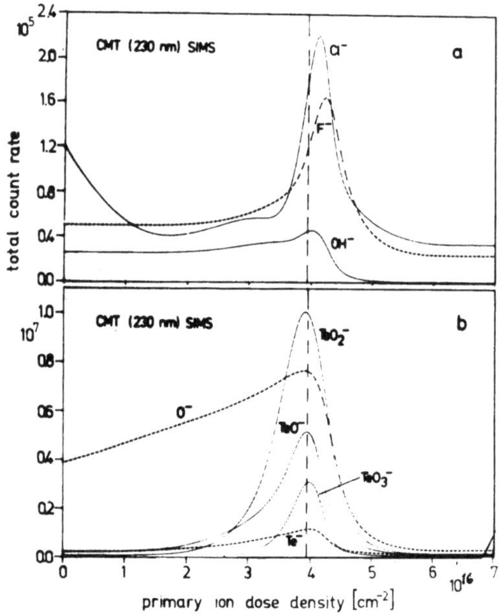

Fig. 2 Integrated count rates of characteristic secondary ions as a function of the primary ion dose density for the 230 nm layer. (a) impurity ions, (b) oxide-specific ions

oxidation number would rather be expected during depth profiling.

As mentioned above, a number of contaminations has been detected by SIMS. The intensities of these contaminations are peaked at the interface (Fig. 2a). The total amount of the contaminations accumulated in the interface region is estimated to be in the order of $3 \cdot 10^{14}$ atoms/cm^2, independent of the layer thickness within a factor of 2. We therefore suggest that these contaminations are present at the surface at the very beginning of anodization and are not continuously incorporated into the layer during this process. Residues from the wet etching procedure preceding anodization have been detected in the interface region only, which further confirms this argument. Because of their large electronegativity, they may well represent the reason for the accumulation layer observed in CMT MIS structures [2]. It should be noted that the maxima of the oxide-specific ions and the impurities do not coincide. Thus we exclude that the increase of secondary ion intensity might be merely caused by the increased impurity amount in this region.

Further details of the interface structure can be deduced from the AES data (Fig. 1a). The structure of the interface and the oxide layer as a whole is summarized in Fig. 3. It is interesting that the Te and Cd "interface" (as indicated by the transition from shifted to unshifted Auger lines) are

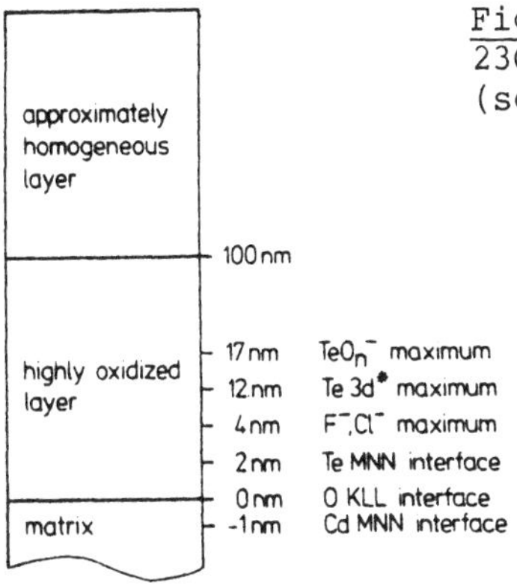

Fig. 3 Composition and structure of the 230 nm layer as a function of depth (schematically)

separated by about 3 nm, thus indicating a further inhomogeneity. A similar interface separation is known for Ga and As at the interface between GaAs and its anodic oxide [6].

5. Comparison of CMT Anodic Oxides with Tellurium Compounds of Known Stoichiometry

In order to elucidate the stoichiometry of the anodic oxide layers, the TeO_m^- region of the SIMS spectra of both the low and high oxidation state region have been compared with a number of tellurium oxides, oxyacids and their salts including $CdTeO_3$, TeO_2, Te_2O_5, TeO_3, H_2TeO_4, and H_6TeO_6. None of their fingerprint spectra resembles the low oxidation state region, whereas the high oxidation state is matched to a certain extent by H_6TeO_6, thus indicating an appreciable amount of Te^{6+} in this region. In contrast to these results, the chemical shifts in XPS show no significant differences between the investigated Te compounds, and seem therefore to be unsuitable for deciding between these possibilities.

Acknowledgement

We thank the Fraunhofer-Gesellschaft for financial support.

References

1. H.M. Nitz, O. Ganschow, U. Kaiser, L. Wiedmann and A. Benninghoven, Surface Sci. 104, 365 (1981)
2. H.J. Richter, U. Solzbach, M. Seelmann-Eggebert, H. Brendecke, H. Maier, J. Ziegler, and R. Krüger, in Insulating Films on Semiconductors, ed. by M. Schulz and G. Pensl, Springer Series in Electrophysics, Vol. 7 (Springer, Berlin, Heidelberg, New York 1981) p. 298
3. U. Solzbach and H.J. Richter, Surface Sci. 97, 191 (1980)
4. T.S. Sun, S.P. Bucher and N.E. Byer, J. Vac. Sci. Technol. 17, 1067 (1980)
5. C. Plog, L. Wiedmann and A. Benninghoven, Surface Sci. 67, 565 (1977)
6. P.A. Breeze, H.L. Hartnagel and P.M.A. Sherwood, J. Electrochem. Soc. 27, 454 (1980)

Part VI

Application II. Surface Studies, Ion Microscopy

The Chemical Compositon of Oxide Films on Aluminium and Its Influence on Surface Properties Studied by SIMS, XPS and AES

M. Textor

Swiss Aluminium Ltd., Research and Development,
CH-8212 Neuhausen am Rheinfall, Switzerland

The chemical composition of natural oxide films on aluminium foil and thin strip of technical purity has been studied using SIMS depth profiling, photoelectron spectroscopy (XPS) and Auger electron spectroscopy (AES). Special emphasis was placed on understanding diffusion processes leading to surface enrichment of low concentration impurities, in particular alkali and alkaline earth metals in technically pure aluminium.

A semi-quantification of SIMS intensities has been carried out for lithium and magnesium in aluminium oxide films partly by using bulk aluminium standards and recording the SIMS in-

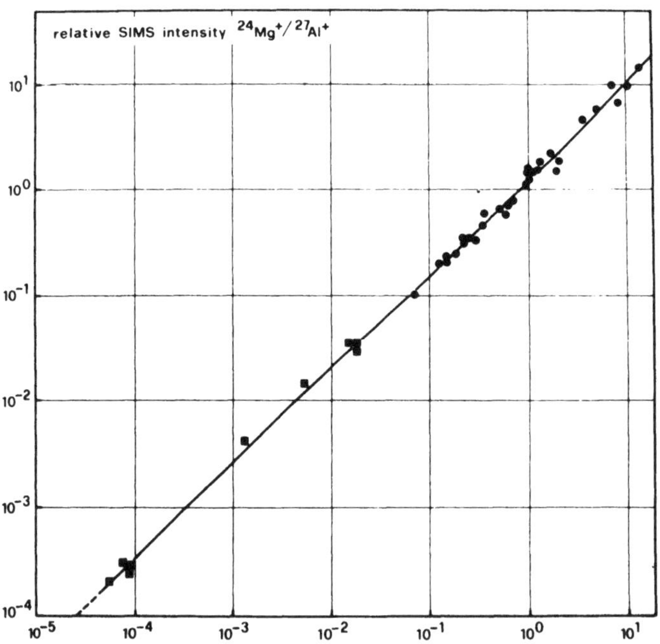

Fig. 1 Relationship between the SIMS intensity ratio $^{24}Mg^+/^{27}Al^+$ and the atomic ratio Mg/Al. The latter was obtained through either quantitative XPS measurements (●) or chemical analysis of the bulk metal (■). Primary ion: 3 kV O_2^+

tensities of $^{27}Al^+$, $^{24}Mg^+$ and $^7Li^+$ under oxidatively saturated surface conditions (bombardment by O_2^+ or Ar^+ with an oxygen background pressure of $\approx 1 \cdot 10^{-5}$ mbar) and partly by comparison of SIMS intensities and quantitative XPS and AES results. Combined SIMS/XPS measurements on Al standards and on a large number of commercially produced foil and thin strip samples has led to a calibration curve (Fig. 1) which allows a semiquantitative (within a factor of two) determination of magnesium in aluminium oxide films over a concentration range of almost six orders of magnitude. A similar calibration curve has been obtained for lithium in aluminium oxide layers.

The strong surface enrichment of alkali and alkaline earth impurity elements in aluminium occurs in practice during heat-treatment processes and has been studied for foil and strip samples at 280 °C, a temperature typical of the final annealing treatment of foil. Fig. 2 shows the SIMS intensities of $^7Li^+$, $^{23}Na^+$ and $^{24}Mg^+$ relative to $^{27}Al^+$ recorded on the surface of electropolished samples of technically pure aluminium as a function of the annealing time at 280 °C. There is a pronounced increase of the surface concentration of these elements within the first few hours of the heat-treatment.

A typical example of a SIMS depth profile of Li, Na and Mg through the oxide layer of an annealed 5 µm foil sample is shown in Fig. 3. There is a basic difference in the surface

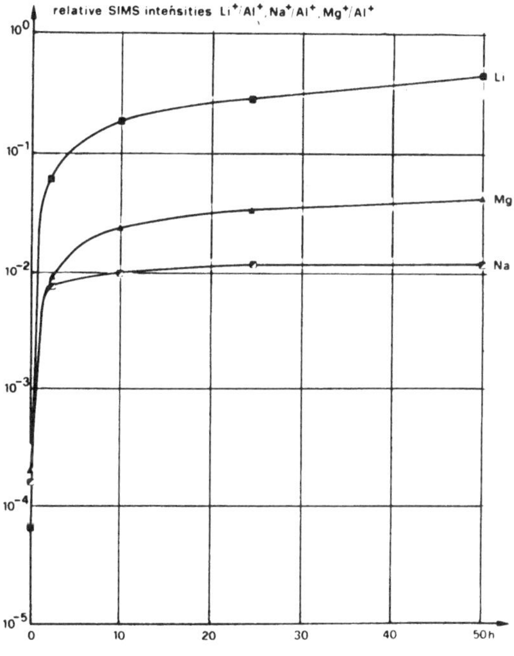

Fig. 2 Relative SIMS intensities of $^7Li^+$, $^{23}Na^+$ and $^{24}Mg^+$ measured on electropolished aluminium samples of 99.2 % purity (metal analysis: 3 ppm Li, 8 ppm Na, 12 ppm Mg) as a function of annealing time at 280 °C in oxygen. Primary ion: 3 kV O_2^+

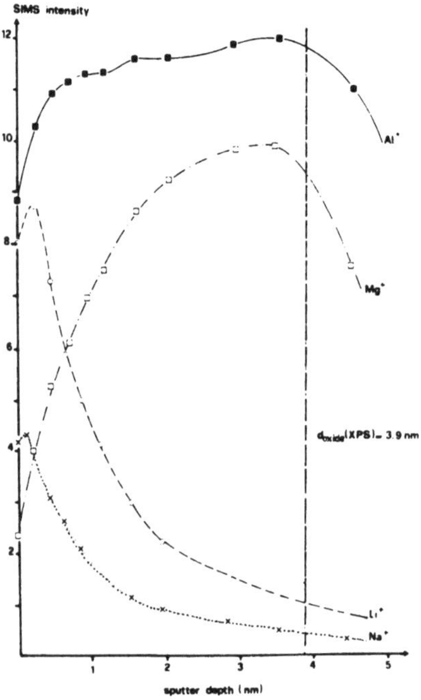

Fig. 3 SIMS intensities of $^{7}Li^{+}$, $^{23}Na^{+}$, $^{24}Mg^{+}$ and $^{27}Al^{+}$ as a function of sputter depth for an annealed (280 °C) aluminium foil of 99.2 % purity (metal analysis: 3 ppm Li, 8 ppm Na, 12 ppm Mg). The SIMS intensities have been multipled by $2\cdot10^{-5}$ for Al^{+}, $1\cdot10^{-3}$ for Mg^{+}, $5\cdot10^{-4}$ for Na^{+} and $1\cdot10^{-4}$ for Li^{+}. The oxide film thickness as determined from XPS measurements is 3.9 nm. Primary ion: 3 kV Ar^{+}, 68 nA

enrichment mechanism of the alkali and alkaline earth elements: the alkali elements Li and Na are enriched in a very thin (≈1 nm) layer at the immediate phase boundary oxide/air. This is taken as an indication for a mechanism where Li and Na diffuse at higher temperature through the oxide film and are oxidized at the outermost surface by a direct reaction with oxygen. On the other hand for magnesium (as well as beryllium and calcium) at low bulk metal concentration (<200 ppm) the thermal enrichment leads to a concentration maximum near the oxide/metal interphase. This is most probably due to reaction where Mg reduces Al_2O_3 according to

$$3\ Mg + Al_2O_3 \rightarrow 3\ MgO + 2\ Al\ ,$$

the reduced metallic aluminium being oxidized again by oxygen. Magnesium oxide enrichment on aluminium has been shown [1] to occur under UHV condition at higher temperature according to the above equation.

Based on the above mentioned quantification of SIMS data, surface enrichment factors (i.e. the ratio of surface to bulk metal concentration) have been determined for technically pure aluminium foil and strip samples. The dependence of surface on bulk metal concentration for lithium (metal content 0.1-15 ppm) and magnesium (3 ppm - 3 %) is shown in Fig. 4 and 5, respectively (annealing: 24 h at 280 °C). The surface to bulk enrichment factors are typically 5000 for lithium and 1000 for magnesium. There is a strong dependence of the surface accumulation of these elements on the annealing temperature. The surface concentration of both lithium and magnesium increases

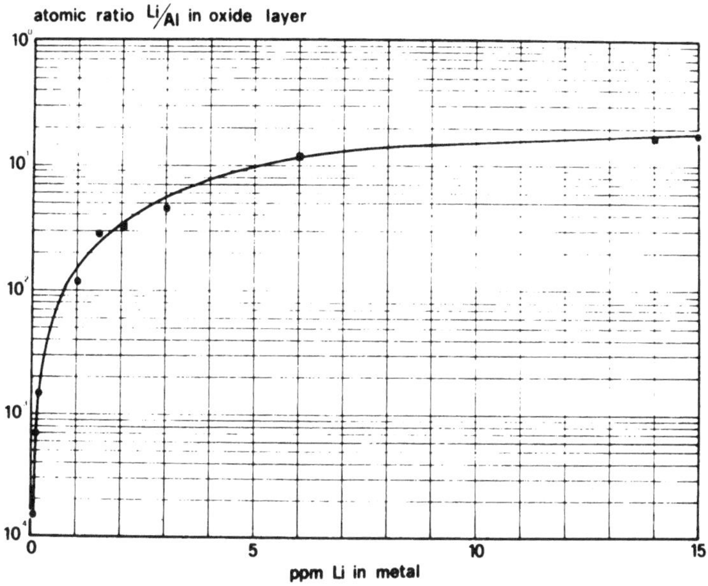

Fig. 4

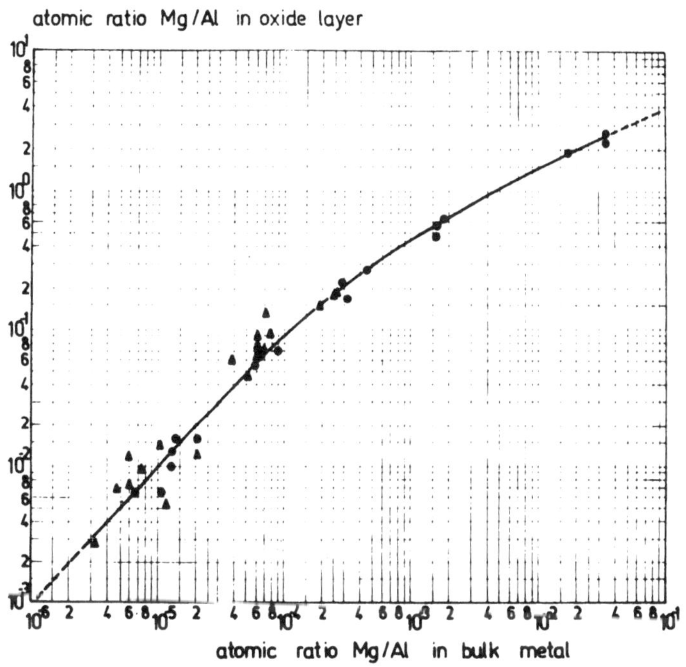

Fig. 5

Fig. 4 Surface enrichment of Li as a function of Li metal content for Al samples annealed for 24 h at 280 °C in oxygen. The plotted atomic ratios Li/Al correspond to the maxima in the Li depth profiles (see Fig. 3)

Fig. 5 Surface enrichment of magnesium as a function of the Mg metal content for a variety of aluminium samples (base metal purity 99.2 %) with different Mg metal concentrations, annealed in the laboratory (24 h at 280 °C, oxygen) (●). Included are also various measurements on commercial foils of 99.2-99.5 quality (annealing condition: 270-290 °C, 20-100 h, air) (▲).

by a factor of abt. 2 for every 30 ^{o}C temperature rise in the range 250-350 ^{o}C (annealing time: 24 h).

The discussed enrichment processes have significant consequences as far as the general behaviour of foil and thin strip is concerned. Lithium in bulk metal concentrations as low as 2-3 ppm can significantly reduce the protective properties of the oxide film on annealed foil leading to an enhanced susceptibility to surface corrosion particularly in the presence of high humidity. Magnesium on the other hand has a strong influence on the adhesion strength of lacquers in converted foil; in particular it may cause lacquer to foil delamination when the lacquered foil is in contact with an acid environment (e.g. packing foil for fruit juices) due to the low chemical resistance of magnesium oxide under these conditions. Both phenomena can only be understood on the basis of the massive surface enrichment of the two elements occuring already at the relatively low annealing temperature of 280 ^{o}C. The quantification of surface-analytical data discussed above has led to a more detailed understanding of the surface enrichment processes and, from a practical point of view, to the establishment of upper concentration limits for these impurities in the production of aluminium foil and thin strip.

Reference

1. B. Goldstein and J. Dresdner, Surface Science 71, 15 (1978)

Study of the Adsorption of Water on Titania by Secondary Ion Mass Spectrometry

J. Marien[1] and E. De Pauw

Institute of Chemistry, University of Liège, Liège, Belgium

Introduction

Titania is known to dissociate water photocatalytically [1] but many aspects of the process are not fully understood. Complete agreement does not exist with respect to the character (associative, dissociative) of the water adsorption and the nature of the sites. In this paper, we report preliminary SIMS results concerning the adsorption of water on clean titania powders whose surface stoichiometry is varied and checked by a combination of Auger and SIMS. The main purpose of the present work is then to investigate the potentialities of SIMS as regarding the adsorption of water on oxides. Particularly, we are interested to see if the degree of coverage and the various phases of adsorption can be deduced from static SIMS.

Experimental

The RIBER SIMS equipment previously described [2] includes a large rod quadrupole (Q 156) preceded by an energy filter and a differentially pumped ion gun (CI 50 RB). Spectra of clean and water covered surfaces are recorded in the conditions of static SIMS (Ar^+; 3 keV; 10^{-9} $A \cdot cm^{-2}$) [3]. An electron floodgun is used during the SIMS analysis of insulating surfaces such as stoichiometric TiO_2 and ice films. The Auger spectra were got with a high sensitivity 10 keV RIBER Auger [4] allowing analysis with low primary electron excitation (2 keV, $4 \cdot 10^{-8}$ A). The TiO_2 (99 % pure) powder is pressed into pellets; they are mounted on a UHV manipulator which can be heated (up to 1000 K) and cooled (down to 120 K). The temperatures indicated here are true surface temperatures calibrated with thermocouples pasted on both sides of the pellet.

Deionized water degassed by repeated freeze-thaw cycles was introduced into the UHV chamber just in front of the sample by means of a leak valve followed by a capillary. High exposures to water (10^{10} Langmuirs) have also been performed in a UHV specimen preparation chamber attached to the analysis cell.

---

[1]Research Associate of the National Fund for Scientific Research, Belgium

Results and Discussion

Clean samples and surface stoichiometries

The TiO_2 pellets which are initially white and insulating are carefully cleaned by heating at 800 K in UHV followed by intense argon ion sputtering (1.5 keV, $2 \cdot 10^{16}$ ions/cm^2). After those reducing treatments, the sample is blue and conducting. We have thus evidence that a surface layer rich in Ti_2O_3 has been built up. Plenty of Ti^{3+} centers related to oxygen vacancies are created by preferential sputtering. This has been demonstrated previously by UPS and ELS [5,6]. The positive and negative SIMS spectra of the reduced clean surface are shown in Fig. 1.

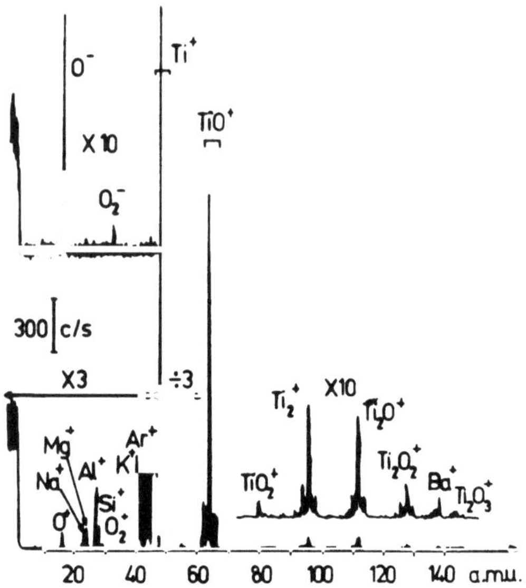

Fig. 1 + and - SIMS spectra of clean sputtered TiO_2 (Ar^+; 3 keV; 10^{-9} A·cm^{-2})

Besides characteristic peaks of titanium oxide; one finds the traces of impurities (Al^+, Mg^+, Mn^+, Si^+) present in the bulk and the low level common surface impurities (Na^+, K^+, Ar^+). The peaks around 96, 112, 128, 144 amu have been identified at higher primary ion current density and resolution to discriminate between $(Ti_xO_y)^+$ and $(Ti_{x-1}O_{y+3})^+$. The intensities of the characteristic oxide ions relative to $^{48}Ti^+$ are collected in Table 1 (row A) for this sputtered surface.

The clean surface after intense ion bombardment has also been examined by Auger spectrometry. From the peak-to-peak intensities in the dN/dE Auger spectrum a ratio O (508 eV)/Ti (381 eV) of 1.85 is got. The latter when corrected by the relative sensitivity factors of O (KLL) and Ti (LMM) [7] amounts to 1.60. Taking into consideration the facts that a matrix effect is likely and that the Auger signal is averaged over several layers, the 1.6 value is highly indicative of a Ti_2O_3 superficial structure. It is characterized by the low TiO^+/Ti^+ value (0.18). Besides the high Ti_2^+ intensity is likely to be related with the Ti^{3+} pairing 6 in Ti_2O_3. To change the superficial oxygen content, the sputtered sample has been heated in UHV (700 K, 240 minutes). This treatment called B in Table 1, keeps the surface clean as Auger and SIMS show. The relative ion intensities collected in row B of Table 1 clearly indicate that the surface is enriched in oxygen. In particular, the relative ion intensity of $Ti^{48}O^+$ is doubled and the parent ion $Ti^{48}O_2^+$ is tripled. Contrary to what is observed for the sputtered sample, TiO_2^+ is now more intense than Ti_2^+. The ratio of the Auger oxygen line (508 eV) to the titanium one (381 eV) amounts to 2.15 for the annealed surface. It means that after heating the oxygen to titanium surface concentration is 1.90 and thus, owing to the oxygen diffusion from the bulk, the TiO_2 surface stoichiometry is almost recovered. Intermediate surface stoichiometries have been prepared by varying the annealing time. They are characterized in the rows C, D, E of Table 1. As Fig. 2 shows, a plot of the SIMS ratio $^{48}TiO^+/^{48}Ti^+$ versus the Auger ratio O (508 eV)/Ti(381 eV) exhibits a linear relationship.

Table 1

Surface preparation	$\frac{O^+}{^{48}Ti^+}$	$\frac{^{48}TiO^+}{^{48}Ti^+}$	$\frac{^{48}TiO_2^+}{^{48}Ti^+}$	$\frac{^{48}Ti_2^+}{^{48}Ti^+}$	$\frac{^{48}Ti_2O^+}{^{48}Ti^+}$	$\frac{^{48}Ti_2O_2^+}{^{48}Ti^+}$	$\frac{^{48}Ti_2O_3^+}{^{48}Ti^+}$	Auger $\frac{O(508\ eV)}{Ti(381\ eV)}$
A	$2.6\cdot10^{-3}$	0.18	$6\cdot10^{-4}$	$7\cdot10^{-3}$	$6\cdot10^{-3}$	$2\cdot10^{-3}$	$3.5\cdot10^{-4}$	1.60
B	$2.7\cdot10^{-3}$	0.34	$1.9\cdot10^{-3}$	$1.4\cdot10^{-3}$	$2.4\cdot10^{-3}$	$2.2\cdot10^{-3}$	$6\cdot10^{-4}$	1.90
C	$2.4\cdot10^{-3}$	0.23	$9\cdot10^{-4}$	$5.5\cdot10^{-3}$	$4.8\cdot10^{-3}$	$2.6\cdot10^{-3}$	$4.2\cdot10^{-4}$	1.67
D	$2.2\cdot10^{-3}$	0.27	$1.1\cdot10^{-3}$	$4.8\cdot10^{-3}$	$4\cdot10^{-3}$	$1.9\cdot10^{-3}$	$5.1\cdot10^{-4}$	1.73
E	$2.6\cdot10^{-3}$	0.30	$1.5\cdot10^{-3}$	$2.2\cdot10^{-3}$	$3\cdot10^{-3}$	$2.4\cdot10^{-3}$	$4.3\cdot10^{-4}$	1.80

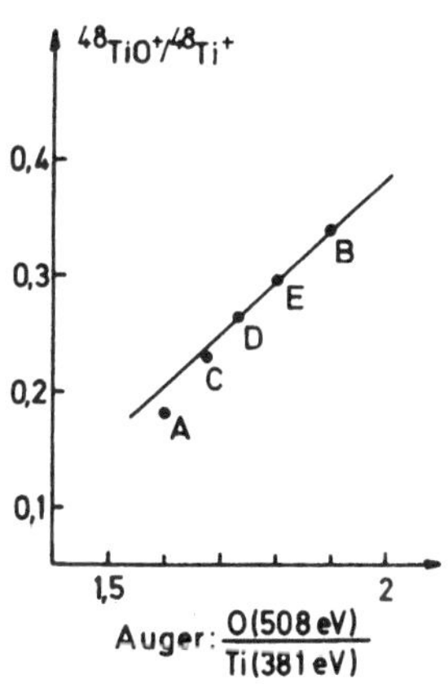

Fig. 2 Ratio of the SIMS peaks $^{48}TiO^+/^{48}Ti^+$ versus the ratio of Auger peak-to-peak amplitudes O(508 eV)/Ti(381 eV)

Point A in Fig. 2 deviates from the line but this can be explained just as above for the 1.6 value. Table 1 shows also a monotonical increase of TiO_2^+ and a monotonical decrease of Ti_2^+ as the surface oxygen increases. After this Auger calibration, it is thus allowed to estimate the surface stoichiometry from SIMS measurements only.

As shown in Fig. 3 after water adsorption at room temperature, the sputtered TiO_2 surface exhibits the following main features. Intense new peaks have appeared (OH^+, OH^-, TiH^+, TiO_2H^+); traces of (H_2O^+, H_3O^+, $Ti(H_2O)^+$, $Ti(H_3O)^+$) are hardly detected but present. The relative intensities in the TiO^+ series are strongly modified by the contribution of $Ti(OH)^+$. We also note that, in contrast to the clean state the parent peak $(TiO_2H)^+$ is higher than Ti_2^+; there is a promotion effect of the parent peak due to protonation. The intensity of the highly oxygenated ions species also increases when compared to the clean state. It is especially convenient to follow the water adsorption by recording the ratio R between the 65 amu line ($^{48}TiOH$ + ^{49}TiO) and the 62 amu one (^{46}TiO).

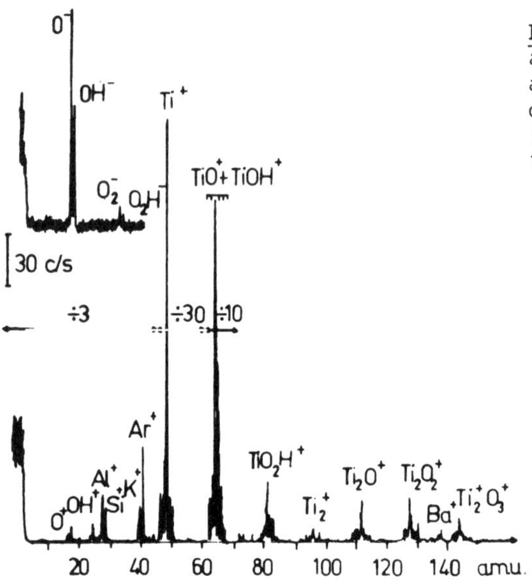

Fig. 3 SIMS spectrum after H_2O adsorption at 300 K (10 Langmuirs) on sputtered TiO_2 (Ar^+; 3 keV; 10^{-9} $A{\cdot}cm^{-2}$)

Fig. 4 Relation between the amount of adsoprtion (300 K on sputtered TiO_2) and the water exposure

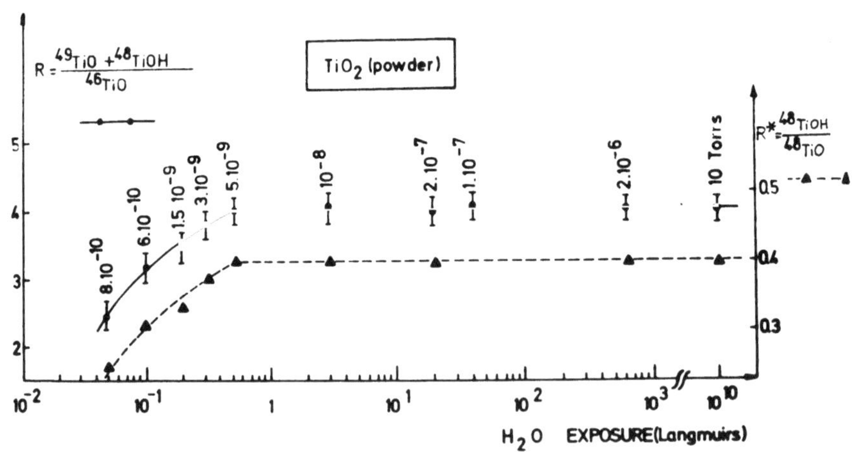

After isotopic correction, the ratio R^* ($^{48}TiOH^+/^{48}TiO^+$) is deduced. In Fig. 4 the irreversible adsorption has been measured; it is seen that saturation is got at 300 K for an H_2O exposure as low as 0.6 Langmuir. No reversible contribution has been found. We observed a high sputtering coefficient (about 30 at 3 keV) so that the ion dose must be minimized to get reliable R values at low H_2O equilibrium pressures.

Adsorption of H_2O in relation to the TiO_2 surface stoichiometry

Room temperature adsorption of water has been performed on TiO_2 surfaces whose stoichiometries correspond to B and C of Table 1. The SIMS R^* values recorded at saturation are linearly related to the superficial stoichiometry. The amount of adsorbed water increases with the density of oxygen vacancies. This is in agreement with other results showing that Ti^{3+} could be the active center for strong water adsorption [8] and that a sputtered surface mainly chemisorbs water dissociatively.

Adsorption of H_2O at 120 K on sputtered TiO_2

The sputtered surface kept at 120 K has been exposed permanently to water. For exposures lower than one Langmuir (L), the SIMS spectrum looks roughly like the one got for room temperature adsorption in Fig. 3 except that R^* has rised to 1.1 and then levels out. In particular, H_2O^+ and H_3O^+ are still lower than OH^+ and TiO_2H^+ predominates in the series around 80 amu. As ions comprising molecular water are quite small, it is likely that such a spectrum is characteristic of chemisorbed hydroxyls and the 1.1 R^* value could be related to a full monolayer. During the water dose increase from a few tens to 200 Langmuirs, the 18 and 19 amu are overcoming the 17 amu; $Ti(H_2O)^+$ gets higher; new lines at 82, 83, 84, 99, 100, 101, 117, 118 amu appear and keep growing. This can be seen in Fig. 5 which shows the spectrum after an exposure of 150 L when $H(H_2O)_n{}^+$ ions begin to appear but are not yet characteristic of adsorbed multilayers. We believe that a spectrum like Fig. 5 indicates

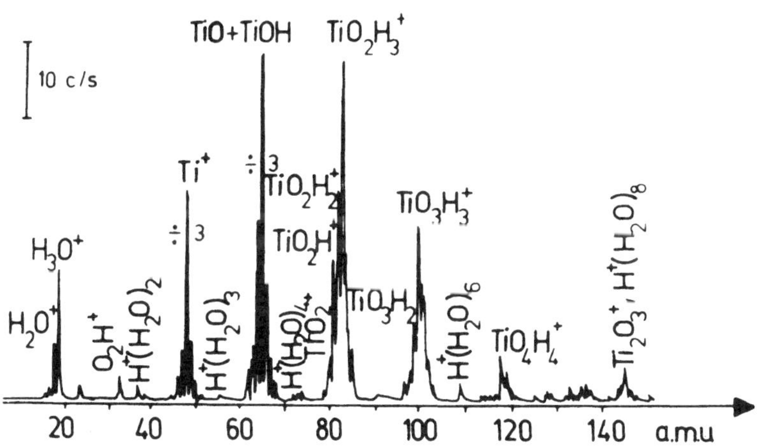

Fig. 5 Adsorption of 150 L of water on sputtered TiO_2(120 K) (Ar^+; 3 keV; 10^{-9} A·cm^{-2})

a physisorbed water phase, perhaps one monolayer on top of the hydroxylated surface. The new cluster ions could be interpreted as being $Ti(H_2O)_n^+$, $TiO(H_2O)_m^+$, $TiOH(H_2O)_p^+$ and $TiO_2H(H_2O)^+$. One finds thus the same ions as in room temperature adsorption but they are associated with water molecules. For exposures higher than 300 L the cluster ions $H(H_2O)^+$ predominate but Ti^+ and TiO^+ are still present. A thick insulating ice film is built up at 600 L; the spectrum then consists only of the water polymers $(H_2O)^+$ and $H(H_2O)_n^+$ as it has been showed previously in the case of alkaly halides [9] and metals [10].

References

1. H. Gerischer, in Solar Energy Conversion, ed. by B.O. Seraphin, Topics in Applied Physics, Vol. 31 (Springer, Berlin, Heidelberg, New York 1979) p. 115£ J.M. Lehn, J.P. Sauvage and R. Ziessel, Nouv. J. Chim. 5, 291 (1981)
2. J. Marien and E. De Pauw, Bull. Soc. Chim. Belg. 88, 115 (1979)
3. A. Benninghoven, Surface Sci. 35, 427 (1973)
4. *We thank the RIBER Company who placed the Auger spectrometer at our disposal*
5. Y.W. Chung, W.J. Lo and G.A. Somorjai, Surface Sci. 64, 588 (1977)
6. V.E. Henrich, G. Dresselhaus and H.J. Zeigez, Phys. Rev. Lett. 36, 1335 (1976)
7. L.E. Davis, N.C. McDonald, P.W. Palmberg, C.E. Riach and R.E. Weber,"Handbook of Auger Electron Spectroscopy" (Physical Elect., Eden Prairie, Minn. 1978)
8. W.J. Lo, Y.W. Chung and G.A. Somorjai, Surface Sci. 71, 49 (1978)
9. J. Estel, H. Hoinkes, H. Kaarmann, H. Nahr and H. Wilsch, Surface Sci. 54, 393 (1976)
10. G.M. Lancaster, F. Honda, Y. Fukuda and J.W. Rabalais, J. Am. Chem. Soc. 101, 1951 (1979)

SIMS Studies on Oxygen Adsorption on Aluminium and Its Alloys

D. Marton, Á. Csanády[1]

Institute of Physics, Technical University Budapest, H-1111 Budapest, Hungary

1. Introduction

SIMS has been used widely for surface chemistry studies, i.e. to try to identify species formed on metal or semiconductor surfaces during their reaction with the gaseous ambience, the latter being mostly oxygen. These studies can provide reliable information if the SIMS method is combined with other surface sensitive methods as TDMS [1], IRS [2], etc. One of us has developed a new method of using SIMS to measure oxygen adsorption on silicon surface [3]. The present paper deals with this method for the analysis of oxygen adsorption on aluminium and some aluminium alloys.

2. Experimental and Results

Oxidation studies have been performed on 5 different probes 1) High purity monocrystalline Al(99.999 %) cut along a high index plane and thus having a polycrystalline-like surface after the ion bombardment; 2) Al thin layer deposited on glass substrate (in $p < 10^{-3}$ Pa background pressure), and Al alloys, containing 9 % Mg; 1.5 % Zn; 2 % Mg and 1.5 % Zn(samples 3, 4, 5 respectively). All of the samples were cleaned in the SIMS chamber before adsorption measurements by means of ion bombardment (3 keV Ar^+ ions).

The measurements were carried out essentially as it has been described in earlier papers [3,4]. A Balzers SIMS equipment [5] has been used with Ar^+ primary ions of 3 keV energy and incident angle of 60° to the surface normal. The crater wall effect was excluded by a diaphragm in the way of the secondary beam.

The sample surfaces were exposed to oxygen doses varying from $1.7 \cdot 10^{-3}$ Pa·s to 5.3 Pa·s. After each exposure oxygen was pumped out of the chamber before starting the SIMS measurement. The background pressure was $p < 10^{-7}$ Pa. The positive and negative secondary ion yields were obtained in separate experiments. The ion yields (I_1) were measured sequentially (Al^+, $AlO^{\pm}$, Al_2^+, O^-). The time consumed for one measurement was 1.25 s, so the second measurement of the same species was made Δt = 2.5-5 s later (I_2). Thus ion yields $I_1(L)$ and $I_2(L)$ have been obtained. We define

---

[1]ALUTERV-FKI, H-1389 Budapest, Hungary

$$\frac{dI(L)}{dt} \equiv \frac{I_2(L)-I_1(L)}{t} \tag{1}$$

and

$$\frac{dI(L)}{dL} \equiv \frac{I_1(L'')-I_1(L')}{L''-L'} \tag{2}$$

where L" and L' note two oxygen doses one of them (L') being smaller and the other (L") higher than the one (L) for which the derivate is being calculated. (For extreme L values L' or L" is taken to be equal with L.) Using definitions (1,2) we can calculate the function

$$\alpha(L) \equiv \frac{dI}{dL}/\frac{dI}{dt} \tag{3}$$

and the experimental results can be characterized with $\alpha(L)$ curves (Fig. 1).

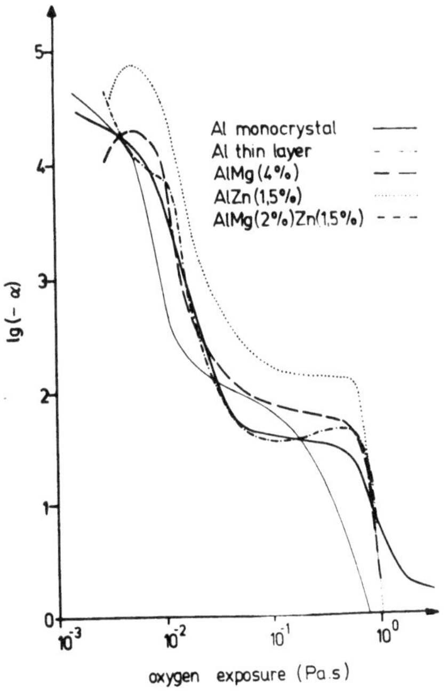

Fig. 1 (dI)/dL)/(dI/dt) values versus oxygen exposure. The curves are calculated on the base of Eqs.(1-2) and using fitted polynoms of the 5th order

Discussion

The ion yield variation during the SIMS measurement after each exposure is due to the change of the coverage of the surface. This can be described by the equation

$$\frac{dI}{dt} = \frac{dI}{d\vartheta} \cdot \frac{d\vartheta}{dt} = - \frac{dI}{d\vartheta} \cdot S \frac{\vartheta \nu}{N_o} \tag{4}$$

where the variation of the coverage according to BENNINGHOVEN [6] is

$$\frac{d\vartheta}{dt} = -S\,\frac{\vartheta\nu}{N_o}\ . \tag{5}$$

Here S is the sputtering coefficient (atoms/ions), ν is the primary ion current density (ions/cm^2s) and N_o is the number of the adsorption sites per unit area on the surface. From Eqs.(3) and (4) a differential equation can be obtained:

$$\frac{d\vartheta}{\vartheta} = -\,\frac{S\nu}{N_o}\,\alpha dL\ . \tag{6}$$

The integral of this equation

$$\ln\frac{\vartheta_2}{\vartheta_1} = \frac{S\nu}{N_o}\int_{L_1}^{L_2}(-\alpha)dL \tag{7}$$

provides a possibility to determine ϑ_2 for any L_2 if one knows ϑ_1 for some L_1 and the $\alpha(L)$ function. So, generally speaking, SIMS alone does not provide unambigous coverage data, but it can be useful when some relevant data from other measurements are available.

The most detailed oxygen adsorption measurements have been carried out by KRUEGER and POLLACK using a gravimetric method on Al thin films [7]. So we try to base our calculations on their results, namely that $\vartheta = 1$ was obtained at about $L = 1.3\cdot10^{-2}$ Pa·s and $N_o = 1.13\cdot10^{15}\ cm^{-2}$ [7]. The adsorption isotherms obtained using this reference point for our thin film sample is shown in Fig. 2. It is obvious that the curve obtained grows too rapidly at low doses and is too flat at

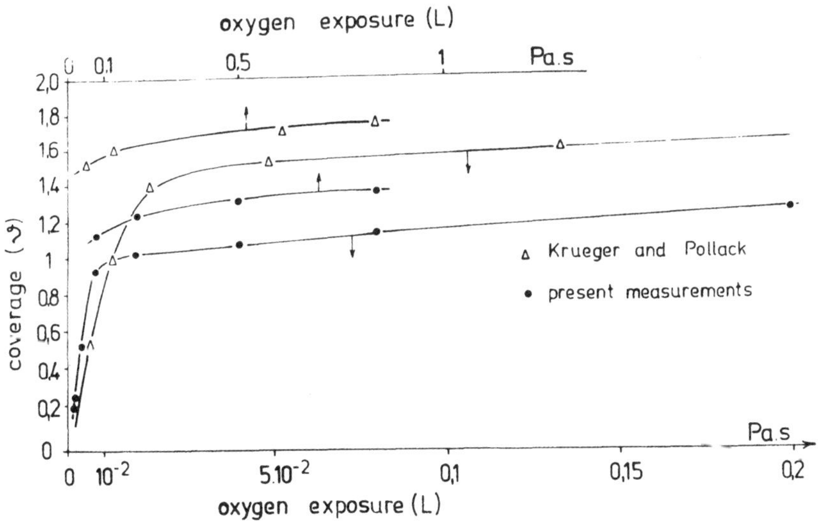

Fig. 2 Adsorption isotherms (at room temperature) of oxygen on Al thin films, measured by KRUEGER and POLLACK [7] and by SIMS, the latter measurement was fitted to the first one at $\vartheta = 1$

higher ones. On the other hand, the fitting at a certain exposure is unjustified because the measurement of exposures is very uncertain (for the comparison of results obtained in two different equipments). So it seems reasonable to try other points of the $\alpha(L)$ curve to see whether they belong to $\vartheta = 1$.

For such a point we can use Eq.(5):

$$\left.\frac{d\vartheta}{dt}\right|_{\vartheta=1} = -\frac{S\nu}{N_o} \,. \tag{8}$$

On the other hand we can suppose that for $\vartheta \approx 1$

$$\vartheta = kI \,, \tag{9}$$

i.e. the measured ion yields are proportional to the coverage. Then

$$\left.\frac{d\vartheta}{dt}\right|_{\vartheta\approx 1} = \frac{1}{I_{10}} \cdot \frac{dI}{dt} \tag{10}$$

where I_{10} is the ion yield obtained in the first measurement at the exposure closest to the one producing $\vartheta = 1$. So we obtain from Eqs.(8) and (10)

$$\frac{S\nu}{N_o} = -\frac{1}{I_{10}} \frac{dI}{dt} \,. \tag{11}$$

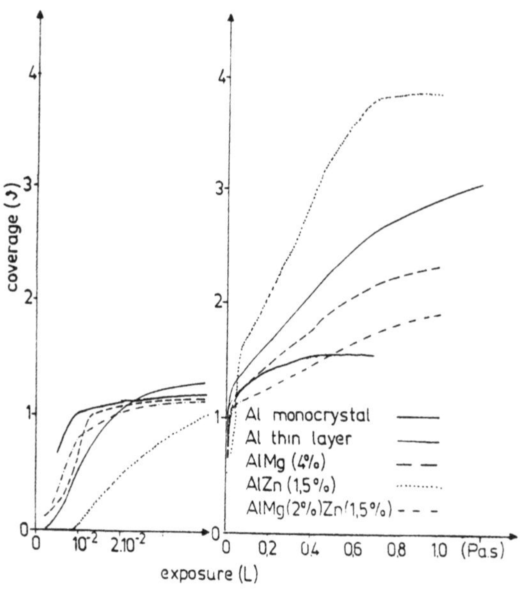

Fig. 3 Adsorption isotherms of oxygen on different Al and Al alloy samples. The curves are calculated supposing that $\vartheta = 1$ is produced by the L exposure at which the change of α in Fig. 1 is most abrupt

Thus we can rewrite Eq.(7) as

$$\ln\vartheta = -\frac{1}{I_{10}}\frac{dI}{dt}\int_{L_o}^{L}(-\alpha)dL \ . \qquad (12)$$

Supposing that exposure L_o results in $\vartheta = 1$, we can now calculate the full $\vartheta(L)$ isotherm . From a series of such isotherms we can select the one being closest to the model one of KRUEGER and POLLACK.

We obtained this way the isotherma for the thin Al film as shown in Fig. 3. The exposure corresponding to $\vartheta = 1$ we found likely to be $L = 8\cdot10^{-3}$ Pa·s. This exposure lies in the middle of the abruptly decreasing part of the $-\alpha(L)$ curve. This feature is found to be decisive for the shape of the isotherm . Thus we supposed this to be a general rule and so we calculated the other $\vartheta(L)$ curves of the Fig. 3.

These results should not be interpreted as quantitative ones, but show that SIMS measurements can be very informative in adsorption studies especially if supplemented with at least one other surface sensitive method. Using SIMS this way would be especially fruitful for the analysis of extremely low coverages and would avoid many problems described in context of the use of SIMS for oxidation studies [8].

References

1. P. Beckmann, K.H. Müller, M. Schemmer and A. Benninghoven, in Secondary Ion Mass Spectrometry SIMS II, ed. by A. Benninghoven, C.A. Evans, Jr., R.A. Powell, R. Shimizu, H.A. Storms, Springer Series in Chemical Physics, Vol. 9 (Springer, Berlin, Heidelberg, New York 1979) p. 137
2. E. Janssen, ibid. p. 171
3. D. Marton, Appl. in Surf. Sci. 5, 65 (1980)
4. D. Marton, V.K. Josepovits and Á. Csanák, Surf. and Interface Anal. 1, 132 (1979)
5. F. Pavlyák, L. Bori, J. Giber and R. Buhl, Japan J. Appl. Phys. 16, 335 (1977)
6. A. Benninghoven, Surf. Sci. 28, 541 (1971)
6. W.H. Krueger and S.R. Pollack, Surf. Sci. 30, 263 (1972)
8. K. Wittmaack, Surf. Sci. 68, 118 (1977)

Oxidation and Segregation at the Surfaces of Different Aluminium Foils and Sheets

Á. Csanády, D. Marton[1], T. Turmezey
ALUTERV-FKI
H-1389 Budapest, Hungary

Introduction

Only a few SIMS investigations concerning bulk aluminium materials have been published in the literature. At the surface of differently heat treated Al-0.4 w% Fe and Al-0.4 w% Fe--0.06 w% Mg foils produced from 99.99 aluminium, Be, B and Mg impurities were observed by DORSEY [1] but their origin was not clearly understood. Aluminium sheets and foils were investigated by JANSSEN [2]. Deleterious effects on surface properties such as heat-seal adherence where explained by impurities (mainly magnesium) enriched in the surface oxide formed during annealing. DEGREVE et al. [3] have shown by an IMMA of extremely good mass resolution that Be enriches between amorphous oxid and metal in a heat treated Al-Zn-Mg alloy. To ensure advantageous surface properties i.e. corrosion behaviour of aluminium and aluminium alloy semiproducts, a detailed knowledge and understanding of the mechanism of surface processes are essential.

The aim of our work was to study the surface composition changes of aluminium with different impurities and aluminium alloys by SIMS. These observations were completed by the results of other, morphological and structural methods like electron microscopy and X-ray analysis.

Experiments and Results

The SIMS experiments were carried out in a modified Balzers SIMS equipment. The mean parameters were the following: background pressure $5 \cdot 10^{-7}$ Pa, Ar pressure $6 \cdot 10^{-4}$ Pa, Ar^+ primary ions with 3 keV energy, $I_p = 5 \cdot 10^{-8} - 1 \cdot 10^{-6}$ A ion current, sputtered area $A_1 = 0.1$ cm^2, investigated area $A_2 = 0.015$ cm^2. The secondary ions were analysed by a quadrupole system furnished with a special modification of ion optics [4]. The mass resolution of the spectrometer was $M/\Delta M = 300$. In the course of the measurements, a series of spectra and depth profiles of the observed elements were taken up to 100-150 nm depth. The sputtering speed was examined by talystep measurements and was found $s' = 1.7$ nm/min in the case of $I_p = 1$ μA. The depth data were drawn on the basis of these measurements. Since the aim of our work was to analyse the amorphous oxide layer at the surface, very low primary ion currents were used at the beginning of the profile measurements.

---

[1]Technical University, H-1502 Budapest, Hungary

To eliminate the difficulties of evaluation due to the ion current changes and matrix effects, ion yield ratios were calculated. Instead of the ion yield $I'_x = I_x/I_{Al}$ ion yield ratios (x = Li, Mg, Cu and Zn) are given.

In Table 1 the mean parameters (chemical composition and heat treatment) of the investigated samples are listed. The compositional data are based on the bulk analysis of the materials. (Atomic adsorption and wet chemical analysis.) To 99.99% Al very small quantities of Li, Mg and Cu additives were given (No. 1-3), I'_{Li}, I'_{Mg} and I'_{Cu} data are shown versus depth on Fig. 1. The second group of samples (No. 4-7) are Al-Mg and Al-Zn-Mg alloys with Cu and Li additives, respectively. In the case of these samples the surfaces were measured without and with heat treatment. The data of the as surface measurements are summarized in Table 2. For illustration in Fig. 2 the enrichments of Mg due to heat treatment are shown. The samples No. 8-10 are aluminium (99.5) foils with different chemical compositions. They contain the above listed alloying elements in different impurity quantities. The I'_{Li} and I'_{Mg} data versus depth are shown in Fig. 3.

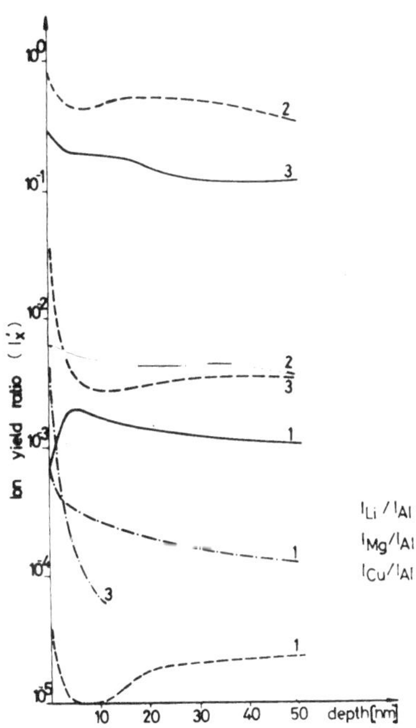

Fig. 1 I'_{Li}, I'_{Mg} and I'_{Cu} ion yield ratios versus depth for samples 1-3

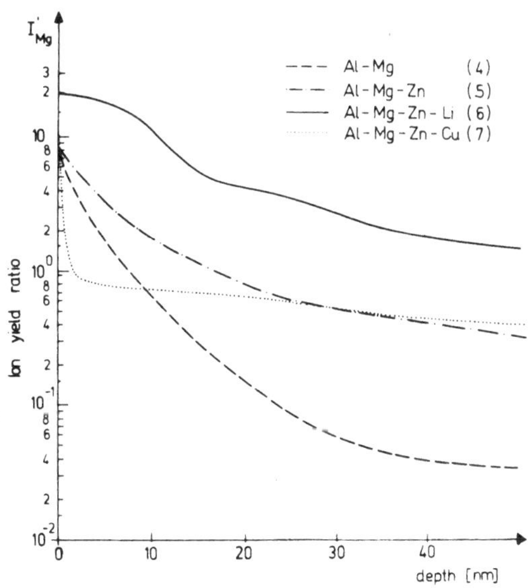

Fig. 2 I'_{Mg} ion yield ratios versus depth for samples 4-7

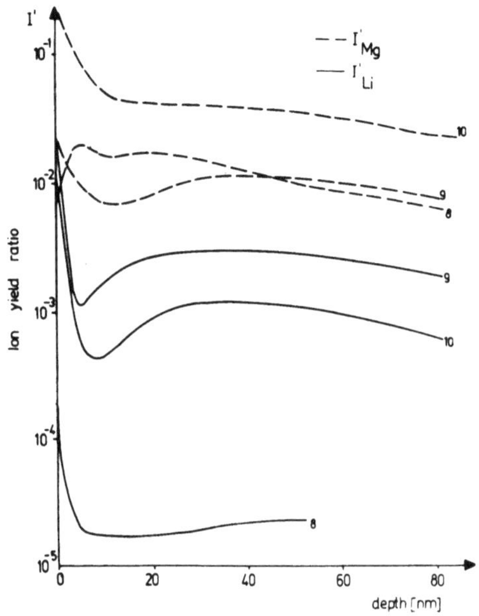

Fig. 3 I'_{Li} and I'_{Mg} ion yield ratios versus depth for samples 8-10

Table 1

Sample	Matrix	Chemical composition w% $^{x}\mu g/g$							Heat treatment			
		Li	Na	Mg	Zn	Cu	Fe	Si	°C	h	°C	h
1) Al(+Cu)	99.99	0.2^{x}	5^{x}	3.5^{x}	-	440^{x}	-	-	400	2	-	-
2) Al(+Li)	99.99	85^{x}	3^{x}	4.0^{x}	-	14^{x}	-	-	400	2	-	-
3) Al(+Mg)	99.99	0.2^{x}	5^{x}	920.0^{x}	-	31^{x}	-	-	400	2	-	-
4) Al-Mg	99.99	-	-	2.1	5	-	-	-	350	1	-	-
5) Al-Zn-Mg	99.99	-	-	2.1	5	-	-	-	350	1	-	-
6) Al-Zn-Mg-Li	99.99	0.2	-	2.1	5	-	-	-	350	1	-	-
7) Al-Zn-Mg-Cu	99.99	-	-	2.1	5	-	-	-	350	1	-	-
8) Al	99.5	0.18^{x}	1^{x}	17^{x}	40^{x}	220^{x}	0.32	0.15	240	5	330	$3^{\ominus}$
9) Al	99.5	1.6^{x}	8^{x}	10^{x}	160^{x}	20^{x}	0.31	0.1	280	8	330	$8^{\ominus}$
10) Al	99.5	0.63^{x}	3^{x}	70^{x}	80^{x}	10^{x}	0.34	0.12	280	8	330	$8^{\ominus}$

$^{\ominus}$Industrial heat treatments

Table 2

Sample	Element	$I' = I_x/I_{Al}$ at the surface: before heat treatment	$I'' = I_x/I_{Al}$ at the surface: after heat treatment	I''/I'
Al-Mg (4)	Mg	$1.0\cdot10^{-2}$	8	$8.0\cdot10^{2}$
Al-Zn-Mg (5)	Mg	$2.5\cdot10^{-2}$	8.3	$3.3\cdot10^{2}$
	Zn	$2.0\cdot10^{-4}$	$7.3\cdot10^{-3}$	$3.6\cdot10$
Al-Zn-Mg-Li (6)	Mg	$1.2\cdot10^{-2}$	$2.2\cdot10$	$1.8\cdot10$
	Zn	$1.8\cdot10^{-4}$	$2.0\cdot10^{-2}$	$1.1\cdot10^{2}$
	Li	$2.2\cdot10^{-1}$	$6.3\cdot10^{2}$	$2.9\cdot10^{3}$
Al-Zn-Mg-Cu (7)	Mg	$7.2\cdot10^{-3}$	1.2	$1.6\cdot10^{2}$
	Zn	$4.5\cdot10^{-4}$	$2.5\cdot10^{-3}$	$5.5\cdot10$
	Cu	$5.0\cdot10^{-3}$	$5.0\cdot10^{-2}$	10

Discussion

The amorphous oxide layers on aluminium and aluminium alloys are so called very thin oxide films (< 5 nm). They are characterized by very high metal/oxygen ratio. Discussion concerning the form of metal excess [5] has not been finished. The γAl_2O_3 pseudo spinel structure exhibits a poorly developed long range order. In the crystal structure of γAl_2O_3 there are 32 oxygen ions arranged in an FCC spinel type lattice and 24 accompanying cation sites. In stoichiometric γAl_2O_3 on the average, only 21.3, i.e. less than the available cation sites, are occupied by aluminium. It follows that γAl_2O_3 has a lot of vacant cation sites [6]. Impurities (cations) can be easily taken by these vacant sites but the diffusion through the lattice is influenced by the cations built in. The following were established:

- Zn and Cu are built in the oxide layer (Fig. 1, Table 1), Cu is enriched to a greater extent than Zn below the amorphous oxide.

- Oxide phases of Zn and Cu are not formed neither in presence of Mg nor without it even at higher temperatures.

- The aluminium is continuously depleted in Li and Mg independently of their original amounts.

- Oxide phases of Li and Mg are formed already below 400 $^{\circ}C$ between amorphous oxide and metal (Li_2O, MgO and $Mg_2Al_2O_4$ etc.), they do not inhibit the transformation of the outer surface to lithium-aluminium-oxide and magnesium oxide [7]. The situation is contrary to the case of Be. Due the oxide phase formed by Be between the amorphous oxide and metal [8] the further diffusion of Be to the surface is inhibited, moreover that of Mg also [9].

- Li is diffusing in the metal and also in the oxide faster than Mg [10]. Oxide is formed easier than in case of Mg. In the presence of both elements - at high concentration or low concentration - at the outer surface Li is the dominating element (Table 1, Fig. 3).

- The corrosion properties of the materials are hardly influenced by this sequence of oxidation and diffusion. The diffusion of Mg can be slightly prevented by Zn, more by Cu (Table 1, Fig. 1); similar inhibiting effect in connection with Li could not be observed.

On the surface of heat treated samples containing only impurity amounts of Li and Mg, lithium oxides and magnesium oxide could be verified by SIMS spectra, secondary electron pictures and by X-ray diffraction, too [7]. The corrosion of surfaces with oxide crystallites is different from pure aluminium oxide. The chemical stabilities of lithium oxides and magnesium oxides differ from that of aluminium oxide [11].

Various types of hydroxide crystals formed on the top of such layers can be explained by the presence of lithium and magnesium oxides (Fig. 4). The degree of corrosion is influenced by these oxides too (Fig. 5).

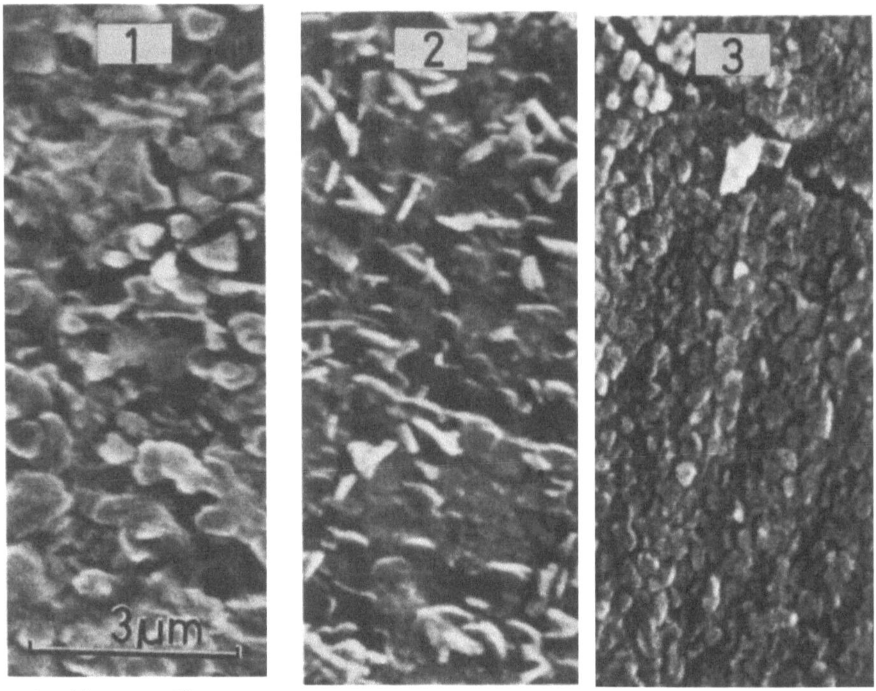

Fig.4▲ Fig.5▼ Captions see opposite page

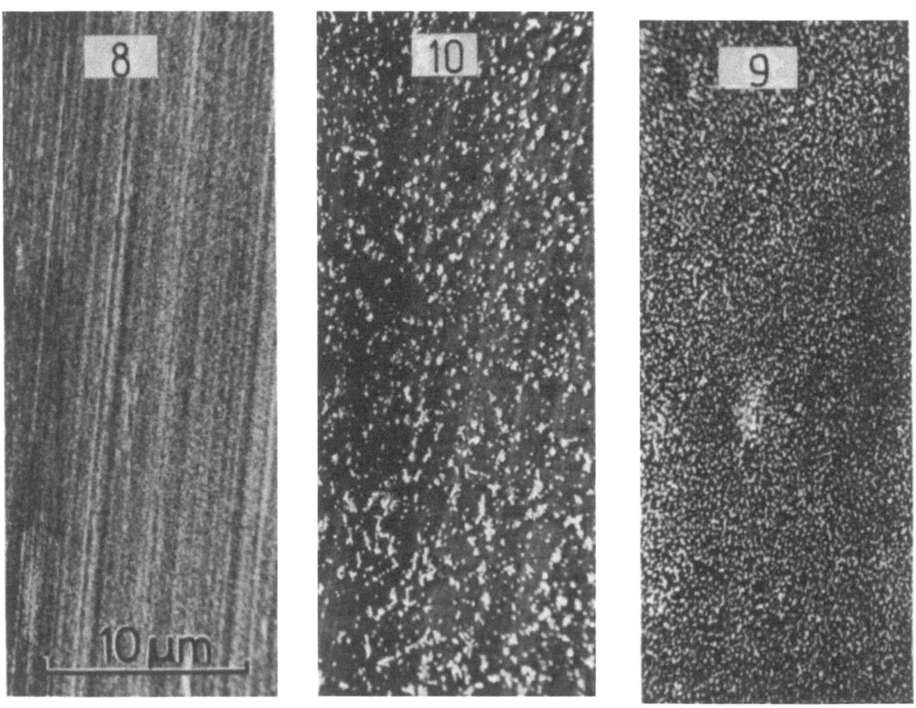

References

1. G.A. Dorsey, Plating 60, 35 (1973)
2. E. Janssen, Mat. Sci. and Eng. 42, 309 (1980)
3. F. Degreve, R. Figaret and P. Laty, Int. Journ. of Mass Spectr. and Ion Phys. 29, 351 (1979)
4. F. Pavlyák, L. Bori, J. Giber and R. Buhl, Jap. J. Appl. Phys. 16, 335 (1977)
5. M. Pryor, Oxidation of Metals 3, 523 (1971)
6. B.C. Lippens, J.J. Steggerda, in "Physical and Chemical Aspects of Adsorbents and Catalysts" (ed. G.B. Linsen) Academic Press, New York, 1970, p. 171
7. Á. Csanády et al., to be published (1981)
8. P. Schmoker, Dissertation, Zürich (1974)
9. H. Ginsberg and P.K. Datta, Aluminium 42, 681 (1966)
10. M. Bishop and K.E. Fletcher, Int. Met. Rev. 17, 203 (1972)
11. M. Pourbaix, "Atlas d'equilibres électrochimiques", 1963, Paris, Gauthier - Villars, 139, 169

Fig.4 Scanning electron micrographs of aluminium hydroxide crystals formed at the surface of samples 1-3, during 16 day corrosion test at 40 °C in wet (90% rel. hum.) atmosphere

Fig.5 Scanning electron micrographs of aluminium hydroxide crystals formed at the surface of samples 8-10, during 10 day corrosion test at 40 °C in wet (90% rel.hum.) atmosphere

Oxygen Adsorption on Polycrystalline PT_3Pb at Elevated Temperatures. A SIMS Study

W. Unger[1], L. Bori, and D. Marton

Institute of Physics, Technical University Budapest
H-1111 Budapest, Hungary

Introduction

The surface of solid platinum is an important subject of the modern methods of surface science caused by the special demands of research in catalysis [1]. Especially some work has been done to examine the oxygen adsorption on platinum single crystal surfaces at different temperatures reviewed by NIEHUS et al. recently [2]. The main feature is the experimental evidence for a chemisorption state of oxygen at temperatures below ~800 K and for a second so-called "oxide" state above ~800 K. It was our aim to utilize the known advantages of SIMS to describe the temperature dependence of the oxygen adsorption phenomena at the surface of a binary alloy of platinum, polycrystalline Pt_3Pb.

Experimental

Experiments were carried out with a Balzers SIMS equipment using Ar^+ primary ions, as described elsewhere [3]. A basic pressure of about $1.3 \cdot 10^{-8}$ Pa was achieved in the target chamber. During the measurements the Ar partial pressure rose to $8 \cdot 10^{-4}$ Pa, but the essential partial pressures of CO and O_2 were lower than 10^{-7} Pa. The target was pressed Pt_3Pb powder. The identity of the Pt_3Pb powder was confirmed in X-ray diffraction measurements. The sample cleaning procedure in UHV included several cycles of Ar^+ ion sputtering and exposure to oxygen at 800 K. The target was estimated to be "clean" when the SIMS signal of carbon disappeared and the oxygen signal remained constant. The main impurities resulting from the target preparation are Cl, O, Al, Ca and Si.

The oxidation studies were performed by dosewise exposure of the sample to oxygen without ion bombardment. The ion yield data were taken at the onset of the corresponding sputter curves at every dose. The doses are ranging from $L = 6.7 \cdot 10^{-3}$ Pas to $L = 5.3 \cdot 10^{-1}$ Pas. Using different time and pressure values for the $6.7 \cdot 10^{-3}$ Pas dose no dependence on oxygen pressure ($1.3 \cdot 10^{-4}$ Pa ... $2.7 \cdot 10^{-3}$ Pa) was found within the scatter of experimental data. Before every new adsorption experiment the target was cleaned again by sputtering. The target temperature was measured by a Ni/NiCr thermocouple placed at the reverse of the sample.

[1]Permanent address: Zentralinstitut für Physikalische Chemie, AdW der DDR GDR-1199 Berlin

Results and Discussion

Independent on the temperature variation the spectra show particularly for platinum some mixed clusters resulting from the main impurities, e.g. $PtCa^+$ and $PtCl^-$.

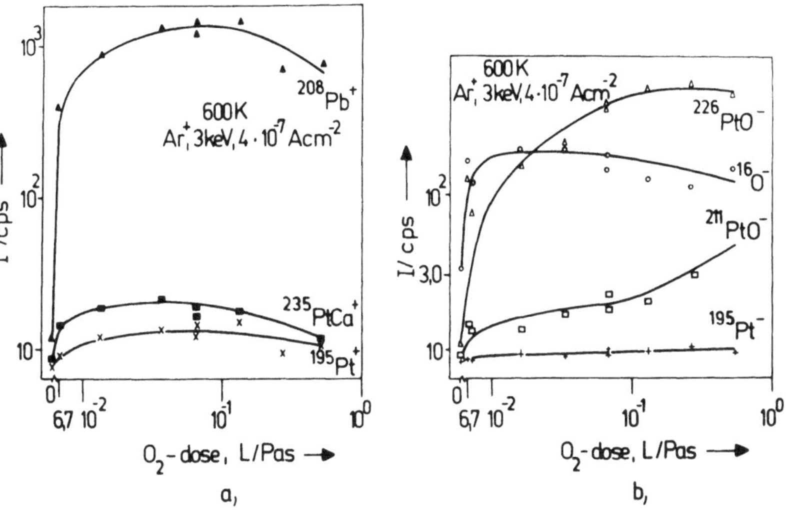

Fig. 1 Influence of oxygen adsorption on polycrystalline Pt_3Pb on characteristical secondary ion yields at 600 K

O_2 adsorption at 600 K (Fig. 1a, 1b) results in a strong enhancement of the Pb^+ yield in contrast to the Pt^+ yield and in the formation of PtO^- ions and, more efficiently, of PtO_2^- ions. Pt^+ and Pt^- yields are nearly unaffected. Saturation appears at ~10^{-1} Pas O_2 but not concerning the PtO^- yield. O_2 exposure at 790 K (Fig. 2a, 2b, 2c) results in a decreasing Pb^+ yield down to a constant value already after a small oxygen dose and in the formation of Pt^- ions as well as negative oxid ions of Pt and Pb. The Pt^+ yield remains unaffected. The negative lead oxid ions are formed immediately at small O_2 doses, PbO_2^- being more efficient than PbO^-. A substantial enhancement of Pt^-, PtO^- and PtO_2^- yields is visible first at an O_2 dose of ~$3\cdot10^{-2}$ Pas. Applying the same dose range as for the O_2 adsorption at 600 K for Pt^- and PtO_2^- yields no saturation was reached. At 790 K the bulk solved Cl becomes active and influences the surface chemistry. The formation of $MeCl^-$ and $MeClO^-$ cluster ions is due to this effect.

O_2 exposure at 980 K (Fig. 3) results only in a weak enhancement of Pt^- yield and in a decreasing Pb^+ yield. Metal oxide ions are not visible at all. Saturation occurs at a dose of ~10^{-1} Pas. H_2 exposure of the oxygen treated surface ($2.7\cdot10^{-1}$ Pas O_2) of Pt_3Pb at 600 K as well as 790 K results in a rapid decreasing of all enhanced MeO_n^- yields and Pb^+ or Pt^-, respectively. Supplying $1\cdot10^{-4}$ Pas H_2 only one tenth of the original oxid yields remains and they disappear after a dose of $1\cdot10^{-2}$ Pas H_2.

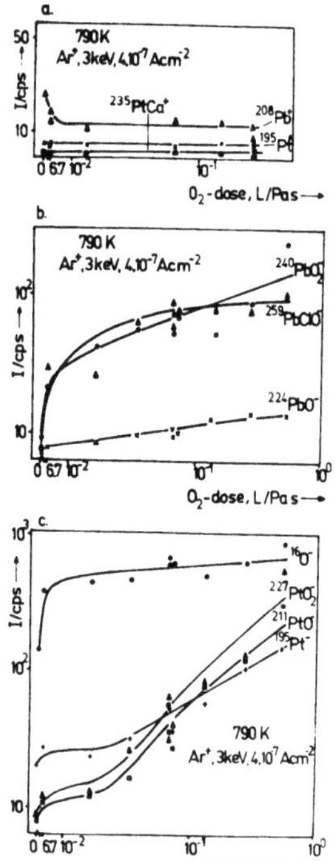

Fig. 2 Influence of oxygen adsorption on polycrystalline Pt_3Pb on characteristical secondary ion yields at 790 K

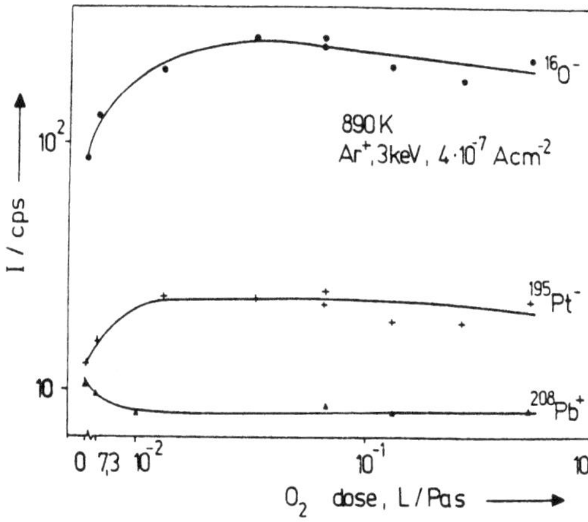

Fig. 3 Influence of oxygen adsorption on polycrystalline Pt_3Pb on characteristical secondary ion yields at 890 K (Note the lack of oxid ions as well as Pt^+ and $PtCa^+$, too)

Some principal features of the dependence of typical secondary ion yields on oxygen adsorption are outlined by WILLIAMS [4]. In the given picture the strong enhancement of Pb^+ yields supports the idea of oxygen chemisorption on Pt_3Pb at 600 K, too. The lack of significant enhancement of Pt^+ yield is due to the fact, that the presence of Ca prevents Pt^+ emission and promotes the emission of $PtCa_n{}^+$ clusters, e.g. $PtCa^+$ [4]. Moreover, oxygen is more strongly bound to Pb. The decreasing Pb^+ yield at higher O_2 doses ($L \gtrsim 10^{-1}$ Pas) indicates, that oxygen incorporation occurs. For the oxygen adsorption at 790 K the immediately decreasing Pb^+ yield displays subsurface oxygen even at low O_2 doses resulting in a strong enhancement of negative species typical for oxidized metal surfaces. In contrast to that no oxide ion was formed at 890 K. Obviously, at this temperature no stable surface oxide can be generated. Only the Pt^- formation process remains, but with diminished yields.

In the discussion of the rapid disappearence of oxid ions after H_2 coadsorption one has to be careful. Firstly the effect could be due to a real removal of oxygen from the exposed Pt_3Pb surface. There is experimental evidence for this possibility, however, on Pt single crystal and at room temperature [5]. Secondly one has to take into account that according to the surface polarization model adsorbed hydrogen can change the electronical properties of the exposed oxidized surface unfavourably for the emission of the species showing decreasing yields. There is also experimental evidence for a high sticking coefficient for H_2 on oxidized Pt surfaces (at 375 K) [6].

Finally we refer to the fact of coincidence of the conclusions following from the surface polarization model applied to the Pt_3Pb-O_2 adsorption system with our experience regarding the special catalytical behaviour of this alloy in general. Interpretation of SIMS results on such kind of adsorption systems is nearly impossible without information from completing methods.

Acknowledgements

The authors thank Dr. G. Lietz for preparing the alloy powder. They appreciate the critical discussion by the group of Dr. J. Völter.

References

1. G.A. Somorjai, Surf. Sci. 89, 496 (1979)
2. H. Niehus and G. Comsa, Surf. Sci. 93, 147 (1980)
3. I. Bårsony, D. Marton and J. Giber, Thin Solid Fils 51, 275 (1978)
4. P. Williams, Surf. Sci. 90, 588 (1979)
5. R. Ducros and R.P. Merrill, Surf. Sci. 66, 227 (1976)
6. R.W. McCabe and L.D. Schmidt, Surf. Sci. 65, 189 (1977)

SIMS Investigation on TiFe Nitrided by NH_3-Pretreatment

G. Kirch and H. Züchner

Institut für Physikalische Chemie der Universität Münster, D-4400 Münster, Fed. Rep. of Germany

Introduction

TiFe has been found in the last years to be one of the best materials for hydrogen storage and purification [1]. The idea to use hydrogen storage materials as hydrogenation catalysts [2] has been picked up to investigate the catalytic activity of TiFe specially pretreated for the ammonia synthesis [3]. The apparent activation energy has been found to be only 62 kJ/mol for TiFe compared to values of 150 kJ/mol for pure Fe and 90 kJ/mol for technical contacts. The special pretreatment is a cycling hydrogenation and dehydrogenation followed by a nitridation with ammonia at temperatures of about 450°C. The total nitridation leads to a compound of a stoichiometry $TiFeN_{1.8}$. X-ray Debye-Scherrer patterns from this material only show reflexes which can be appointed to TiN, no iron containing compounds can be identified and no X-ray reflexes of TiFe occur [3]. Thus it has to be assumed, that a reconstruction of the original TiFe lattice into a new phase, probably nitrided Ti and finely dispersed Fe, has taken place. The Fe shows the high catalytic activity. These results are confirmed by magnetic and Mössbauer measurements which clearly show a disappearance of the original TiFe [3,4].

The purpose of the present work is to verify the decomposition of the original intermetallic phase TiFe by the ammonia treatment applying the method of secondary ion mass spectrometry and to study surface segregation effects in TiFe which seem to play an important role in hydrogen uptake behaviour [5,6] and the catalytic activity of these materials [7]. Results from X-ray and magnetic investigations and especially from AES studies are used to interpret the secondary ion emission.

Experimental

The above mentioned activation procedure, cycling hydrogenation-dehydrogenation, usually leads to a pulverization of the TiFe. To prevent this the pretreatment was changed in such a way, that the TiFe disks, cut from a regulus with a polished plane surface of about 1 cm^2 only, were treated with NH_3 gas at 450°C for about 100 h. This procedure does not pulverize the material, the bulk remains undestroyed. Nevertheless the surface reacts with the ammonia up to a depth of approximately 1 μm even without preceding hydrogen cycling.

Samples prepared in this way were investigated together with disks of untreated TiFe by means of SIMS and AES in a combined SIMS-AES equipment (BALZERS + PHI). The ratemeter output of the SIMS and the multiplexer output of the AES equipment can be connected to a computer (TEKTRONIX 4051) to store the intensities of preselected secondary ions or the peak-to-peak amplitudes of Auger peaks during the measurement of depth profiles. This combination allows an easy standardization and in connection with a plotter an easy plot of the measured and standardized SIMS and AES depth profiles. They are recorded while bombarding the specimens with Ar^+ primary ion densities greater than $1 \cdot 10^{-6}$ A cm^{-2} in order to get a complete depth profile in a few hours allowing a sufficient view into the interesting reaction region. No special attention is paid to exact determination of the profile depth in this investigation.

Results and Discussion

Figure 1 presents the depth profiles of the peak-to-peak amplitudes of the Auger analysis on an untreated TiFe sample, which can be interpreted in good approximation as concentration profiles for the respective elements. After sputtering for only 5 s the observed Auger peaks (Ti (387 and 418 eV), Fe (703 eV), O (510 eV)) reach values which do not change any more during further bombardment.

The corresponding Auger concentration profiles of the $TiFe/NH_3$ sample (Fig. 2) look quite different. The Fe concentration indicates two maxima and minima. The first minimum and the Ti concentration show a distinct shoulder. The second broad and flat minimum appears in the maximum nitrogen concentration range. Here the Fe and Ti concentrations are only 60 % and 71 %, respectively, of the values for the untreated

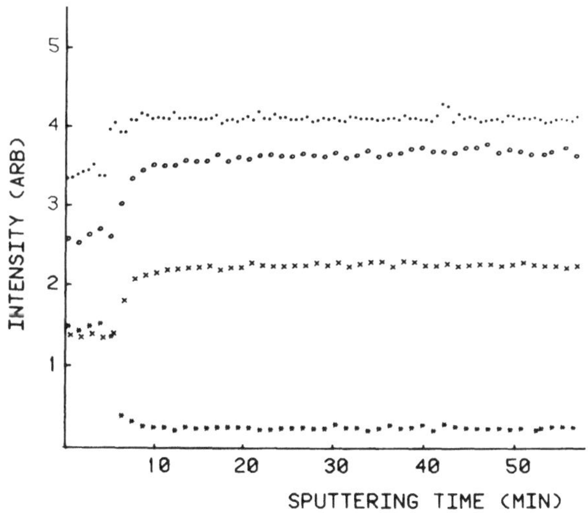

Fig. 1 Auger profiles of untreated TiFe; (o Ti (387 eV), x Ti (418 eV), * O (510 eV), • Fe (703 eV))

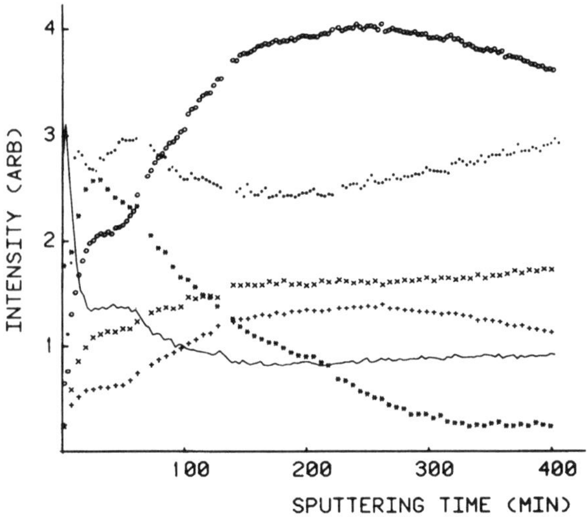

Fig. 2 Auger profiles of TiFe pretreated with ammonia; (o (Ti (387 eV) + N (381 eV)) x 0.5, x Ti (418 eV), • Fe (703 eV), * O (510 eV), + N (381 eV) x 0.25, —— Fe (703 eV)/Ti (418 eV)). The N profile (+) has been derived from the measured curve (o) by subtracting the Ti (387 eV) profile calculated from the Ti (418 eV) curve using the ratio Ti (387 eV)/Ti (418 eV) from Fig. 1

sample indicating the dilution effect caused by incorporated nitrogen. According to this explanation the Ti and Fe concentrations increase with increasing sample depth after passing the maximum N concentration while N concentration decreases. Moreover, the Fe/Ti ratio standardized to the corresponding ratio of the untreated sample and plotted in Fig. 2, too, demonstrates a Ti enrichment in the region of high N concentration. In the near-surface region (sputter time <100 min) a distinct enrichment of Fe can be observed (Fe/Ti > 1) despite the fact that the oxygen concentration[1)] shows a maximum. By means of the inner nitridation of the titanium Fe particles segregate to the surface forming there the active centres for catalysis.

Figures 3 and 4 present the secondary ion depth profiles of Fe^+ and Ti^+, respectively, for the untreated and the $TiFe/NH_3$ sample. For the untreated sample the decrease of the intensities of Fe^+ as well as of Ti^+ with increasing profile depth shows the decreasing oxygen content. A more detailed view makes clear that the Fe^+ intensity already decreases whilst the Ti^+ inten-

1) An incorporation of oxygen into TiFe normally causes a Ti enrichment in the oxidized region and a squeezing out of Fe to the surface, too [6]. In this case the N content is much higher than the O content unpurposely inserted during the preparation. Therefore the influence of the oxygen is overcompensated, but remains visible in the shoulder of the Fe/Ti run.

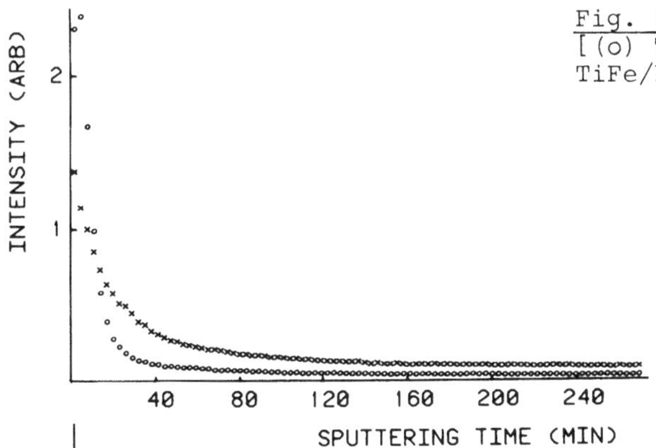

Fig. 3 Fe^+ profile of TiFe [(o) TiFe untreated, (x) TiFe/NH_3]

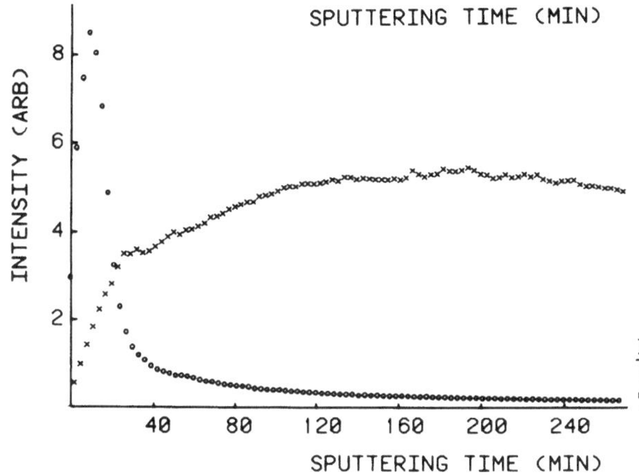

Fig. 4 Ti^+ profile of TiFe [(o) TiFe untreated, (x) TiFe/NH_3]

sity still aims towards a maximum. This seems to be an indication for a Fe segregation to the surface and an enrichment of Ti just below the surface relative to the concentration distribution in the bulk. In the sample pretreated with NH_3 the Ti^+ and Fe^+ profiles look at least in the initial shape like the corresponding Auger profiles in Fig. 2: the Ti^+ intensity increases. According to the interpretation of the Auger profiles the Ti^+ and Fe^+ curves first seem to be determined by the decrease of the Fe enrichment. After passing the maximum nitrogen concentration[2)] the chemical surrounding of the Ti changes with

2) A nitrogen profile cannot be observed by SIMS and is seen only in the Auger study. The samll N^+ emission out of the pretreated sample does not give sufficient information about the N concentration change. A few pretreated samples show enhanced mass peaks at m/e = 28 [7]. They have to be appointed to a N_2^+ emission from weakly bound N confirming the annealing experiments on such samples where only weakly bound nitrogen escapes [3].
It cannot be decided whether the mass peak at m/e = 110 (Fig. 5) has to be ascribed to a Ti_2N^+ emission - which seems to be probable - or to a $(^{54}Fe + ^{56}Fe)^+$ emission. Moreover, it is conceivable that the mass peak m/e = 112 of the pretreated sample is not only caused by Fe_2^+ and Ti_2O^+ but also by $Ti_2NH_2^+$. A TiN^+ emission has not been observed.

decreasing N content in such a way that the Ti^+ intensity decreases (Fig. 4) even though the Ti concentration increases (Fig. 2). The matrix effect predominates the concentration influence.

From an interpretation of the absolute and standardized intensities of the secondary ions including molecular ions which is based on the results from AES but also magnetic and X-ray measurements further information about the nitridation effect is available. Fig. 5 shows two mass groups from the positive secondary ion mass spectra of a Ti, Fe and TiFe sample recorded after sputtering for a considerable time until obtaining constant intensities and of the TiFe sample pretreated with NH_3 out of a profile depth near the nitrogen concentration maximum. All spectra have been recorded under the same conditions ($j_P = 2 \cdot 10^{-6}$ A cm^{-2}). Average intensity values for some characteristic secondary ions taken from ten successively obtained spectra for each sample are listed in Table 1. Additionally they are standardized to the Ti^+ or/and Fe^+ intensity.

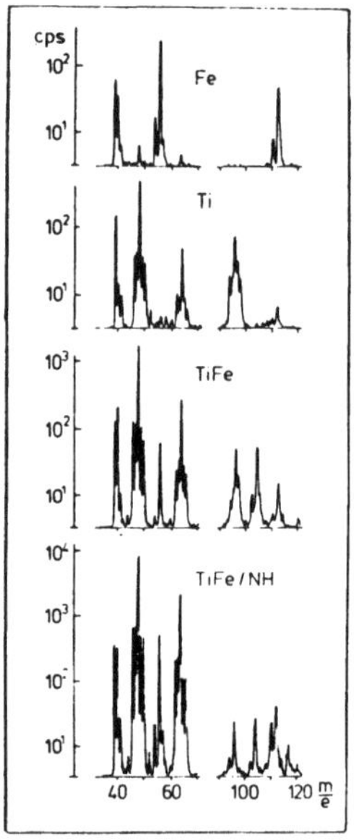

Fig. 5 Comparison of positive SI-spectra ($j_P = 2 \cdot 10^{-6}$ A cm^{-2})

The Ti^+ intensity increases from the titanium to the pure TiFe and again to the TiFe/NH_3 even though the Ti concentration in the TiFe sample is reduced compared to that in pure Ti. This behaviour indicates the change in the matrix and at least in the binding state of the Ti from the titanium to the intermetallic phase TiFe and to the nitrided TiFe. This may be due partly to the influence of oxygen. - The increase of the oxygen concentration is expressed by the increase of the TiO^+ emission as well as by the oxygen Auger profile. - The smaller Ti_2^+ fraction of the TiFe sample compared to that of pure Ti should be attributed no more to the oxygen influence but rather to the reduced probability to find Ti-Ti-pairs because of the compound formation. The effect is simplified by the nitridation, so the

Table 1 Comparison of secondary ion emissions ($j_p = 2\cdot10^{-6}$ A cm^{-2})

	m/e	Ti		Fe		TiFe			TiFe/NH_3		
		cps	/Ti^+	cps	/Fe^+	cps	/Ti^+	/Fe^+	cps	/Ti^+	/Fe^+
Ti^+	48	590	1	-	-	1950	1	30	9780	1	16.6
Fe^+	56	-	-	270	1	65	0.03	1	590	0.06	1
TiO^+	64	56	0.09	-	-	350	0.18	5.4	2350	0.24	4
Ti_2^+	96	71	0.12	-	-	50	0.026	0.77	27	0.0028	0.046
$TiFe^+$	104	-	-	-	-	50	0.026	0.77	31	0.0032	0.053
$Fe_2^+ + Ti_2O^+$	112	7.3	0.012	54	0.2	14.7	0.0075	0.23	49	0.005	0.083

Ti_2^+/Ti^+ ratio is reduced again by one order of magnitude. The Fe^+ emission seems to be less affected by oxygen. Its reduction from Fe to TiFe is to be attributed to the compound formation with a smaller Fe concentration fraction. The surprisingly high Fe^+ emission in the pretreated sample compared to the pure TiFe could be ascribed to the nitridation whereby Fe particles are set free and pressed to relatively unstable places in the lattice (weakly bound Fe). Because the ratio Fe_2^+/Fe^+ does not increase compared to pure TiFe it can be assumed that no distinct iron clustering takes place[3].

The SIMS results (Table 1) do not show a dilution effect by incorporated nitrogen. On the contrary, the N incorporation effects an increase of the Ti^+ and Fe^+ emission. Moreover, the ratio Fe^+/Ti^+ increases from pure TiFe to the pretreated sample despite the fact, that Ti should be enriched in the nitrided region. The apparent inversion of the enrichment effect must be attributed to the fact, that the iron remains weakly bound when nitriding the Ti in the TiFe. The destruction of the TiFe by the nitridation process becomes more obvious when regarding at the $TiFe^+$ emission. It decreases from the pure TiFe sample to the pretreated one. The ratios $TiFe^+/Ti^+$ and $TiFe^+/Fe^+$ are reduced by a factor of ten.

Conclusion

The results obtained by SIMS at TiFe pretreated with ammonia allow, together with results from AES, magnetic and X-ray measurements, a consistent interpretation: the reaction of TiFe with NH_3 leads to an inner nitridation and destruction of the TiFe under precipitation of nitrided Ti and weakly bound Fe. Ti is enriched in the nitride region compared to unreacted TiFe, while iron segregates to the surface forming there the catalytic active centres.

3) A Ti_2O^+ and perhaps a $Ti_2NH_2^+$ emission have to be considered besides the Fe_2^+ emission in the mass peak m/e = 112. Therefore the ratio Fe_2^+/Fe^+ should even be smaller than 0.083.

Nevertheless, it has to be emphasized that the AES investigations are indispensable for the interpretation of the SIMS results. Based on the Auger results yielding information about element concentrations, the SIMS measurements give information about matrix effects and changes in binding behaviour.

References

1. J.J. Reilly and R.H. Wiswall, Inorg. Chem. 13, 218 (1974)
2. K. Soga, H. Imamura and S. Ikeda, J. Catal. 56, 119 (1979)
3. E. Schwab and E. Wicke, Z. Phys. Chem. N.F. 122, 217 (1980)
4. E. Schwab, private communication
5. L. Schlapbach, A. Seiler, F. Stucki and P. Zürcher, Proc. Int. Conf. Hydrogen in Metals, Münster 1979, Akadem. Verlagsgesellschaft, Wiesbaden, eds. E.Wicke and H.Züchner
6. H. Züchner and G. Kirch, to be published
7. G. Kirch, R. Hempelmann, E. Schwab and H. Züchner, Z. Phys. Chem. N.F. (1981)...

SIMS/TDMS Studies of Hydrocarbon Interaction with Nickel

M. Schemmer, P. Beckmann, D. Greifendorf, and A. Benninghoven

Physikalisches Institut der Universität Münster,
D-4400 Münster, Fed. Rep. of Germany

Introduction

As compared with other surface analytical techniques, static secondary ion mass spectrometry offers the following advantages:

- hydrogen sensitivity
- isotope sensitivity
- high detection sensitivity (< 10^{-6} ML)
- compound detection by molecular ions

The latter feature is of special interest in conjunction with the investigation of surface reactions. It is, however, also connected with serious problems. Firstly, there is no straightfoward relationship between the composition of secondary ions and that of corresponding surface complexes. Secondly, an adsorbate induced change of ionization probability hinders the quantitative interpretation of SIMS results.

It is the aim of this work to demonstrate with the aid of the nickel-ethylene interaction that an unambigous identification of surface compounds, and, in addition, the quantitative determination of their concentration is possible by consideration of secondary ions Ni_mA^+ (A: adsorbate, Me: metal) and intensity ratios Ni_mA^+/Ni_m^+, respectively. Based on these relations, a description of complex surface reactions, like the Ni-C_2H_4 interaction, e.g. is possible, and it is in good agreement with previous results obtained with other techniques. Furthermore, the SIMS investigation supplies additional information on the process of the nickel-ethylene interaction.

Results and Discussion

1. *Characteristic Secondary Ion Emission from Uniform Adsorbate Layers*

If we consider the positive SIMS spectrum of an ethylene covered Ni surface (Fig. 1), a great number of secondary ions are changing their intensity during C_2H_4 exposure. Our investigation is based on only some of these observed secondary ions.

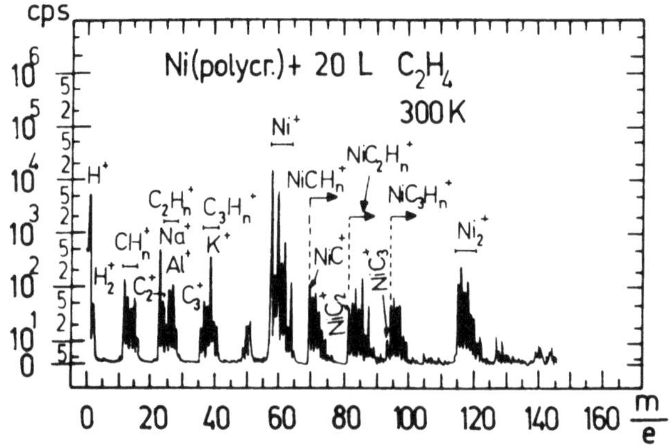

Fig.1 Positive SIMS spectrum after 20 L C_2H_4 exposure at 300 K. Primary ion current: $3 \cdot 10^{-9}$ A, Ar^+, 3 keV, bombarded target area: 0.1 cm^2

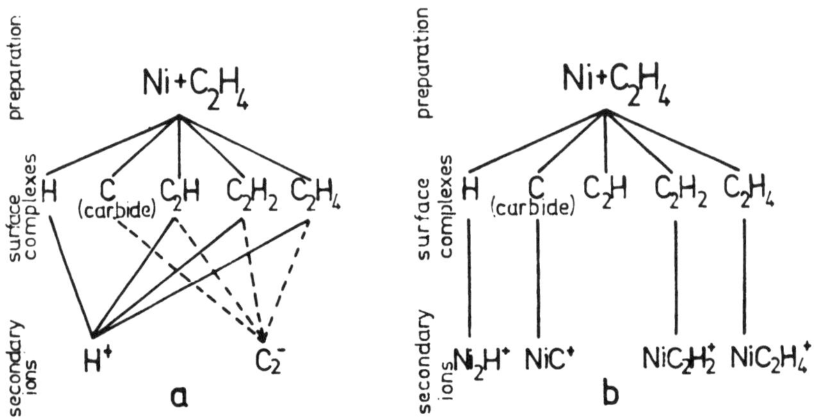

Fig. 2 (a) Surface complexes after 300 K ethylene adsorption that may emit H^+ and C_2^-, respectively;
(b) Correlation between surface complexes and secondary ions NiA^+

The most intense positive and negative ions, H^+ and C_2^-, respectively, cannot be used for this purpose, because they may be emitted from more than one of those surface compounds, that have to be assumed to be present on the surface after 300 K C_2H_4 exposure (Fig. 2a).

From earlier investigations of simple adsorbate layers on various metal substrates, it resulted that one type of secondary ions was especially related to an adsorbate A on a metal surface Me: the ions Me_mA^+. Assuming this type of characteristic secondary ions also for surface complexes present on the surface after ethylene exposure at 300 K, an unambigous identification of these surface complexes by their related secondary ion emission would be made possible, as shown in Fig. 2b.

To establish the relation assumed above, the secondary ion emission from surfaces covered virtually only by one compound has to be investigated. Table 1 shows the investigated surface compounds, the technique applied to their preparation, and the "characteristic" secondary ion species Ni_mA^+.

Table 1 Surface complexes that are formed after 300 K C_2H_4 exposure together with method of their isolated preparation and "characteristic" secondary ion speices

Surface complex	Preparation	"Characteristic" secondary ion NiA^+, Ni_2A^+
Ni-H	H_2-exposure at 150 K	Ni_2H^+
Ni-C	C_2H_4- or C_2H_2-exposure at 300 K, followed by heating to 450 K	NiC^+
$Ni-C_2H_2$	C_2H_2-exposure at 80 K	$NiC_2H_2^+$
$Ni-C_2H_4$	C_2H_4-exposure at 80 K	$NiC_2H_4^+$

2. *Quantitative Description of Surface Compound Concentration*

During ethylene exposure, not only the intensity of hydrocarbon or hydrocarbon fragment ions is enhanced, but also that of Ni^+ and Ni_2^+. This is due to an increase in ionization probability, induced by the adsorbed species. A similar increase is observed during oxygen adsorption. Such an increase can also be assumed for all other secondary ions. We were able to eliminate this influence of the ionization probability on secondary ion intensity by considerating the intensity ratios.

First hints on this method were given by SOMENO et al. [1] and by EVANS et al. [2] some years ago. They investigated hydrogen bulk concentration in various metals and showed, that in many cases a linear relationship held between MeH^+/Me^+ and the hydrogen bulk concentration. The same method was used for a determination of hydrogen surface concentration by BENNINGHOVEN et al. [3] and later by WINOGRAD et al. [4], BORDOLI et al. [5], and BARBER et al. [6] with respect to CO and H_2 surface concentration. A shortcoming of all of the above measurements, however, was the lack of an independent, simultaneously applied analytical technique in order to determine the surface concentration. Such measurements have been carried out first by BENNINGHOVEN et al., when they investigated the $Ni-H_2$ interaction [7]. In those experiments, combined SIMS and TDMS measurements allowed to relate SIMS intensity ratios Ni_2H^+/Ni_2^+ to relative hydrogen coverage, which was measured by the TDMS peak areas.

Similar experiments have been carried out recently with respect to ethylene and acetylene surface concentration [8]. Also in these cases, a linear relationship could be established. Due to a lack of electron spectroscopy facilities in the apparatus used for these investigations, no independent method of carbon coverage measurement exists. It seems, however, to be justified, to assume the NiC^+/Ni^+ ratio to be proportional to the carbon concentration in a carbidic surface compound.

3. *Application of Intensity Ratios to the Investigation of the Nickel-Ethylene Interaction*

3.1 Room Temperature Interaction

The $Ni-C_2H_4$ interaction has been studied by many other groups with various analytical techniques [9]. Our SIMS investigation, applying the method of intensity ratios, leads to results that are in good agreement with the literature. During ethylene exposure, initial dissociation which leaves carbon and hydrogen on the surface, is followed by adsorption of an acetylenic species and undissociated ethylene molecules [9,10]. Nickel-hydrogen bonds were, up to our investigations, only assumed to be present on the surface from the observation of H_2 thermal desorption.

3.2 Carbide Precoverage

A carbide covered Ni-surface can be prepared by heating the surface up to 450 K after hydrocarbon adsorption. The results of ethylene exposure of such a carbide covered surface is an immediate adsorption of undissociated C_2H_4 if the carbide concentration has already reached its maximum (Fig. 3a). If the carbide concentration is below its saturation value, the new C_2H_4 exposure leads to dissociation up to NiC^+ saturation, before full molecular adsorption occurs (Fig. 3b).

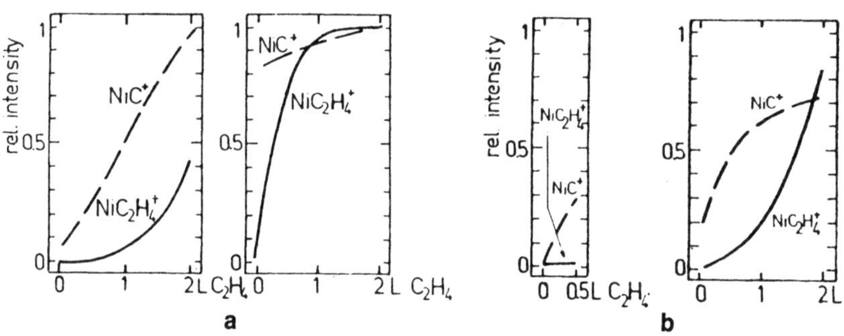

Fig.3. (a) Behaviour of secondary ion intensities NiC^+ and $NiC_2H_4^+$ during 2 L C_2H_4 (left part) and subsequent 2 L C_2H_4 exposure (right part). Between both exposure experiments, the surface had been heated up to 450 K. (b) Same as (a), but first exposure dose is only 0.5 L

3.3 Thermal Decomposition of an 80 K Ethylene Adsorption Layer

The 80 K C_2H_4 exposure of a nickel surface mainly results in the adsorption of undissociated C_2H_4 molecules (Fig. 4). The $NiC_2H_2^+$ and Ni_2H^+ emission that is observed at this stage may be due either to fragmentation upon ion formation or to a small amount of dissociation even at this temperature.

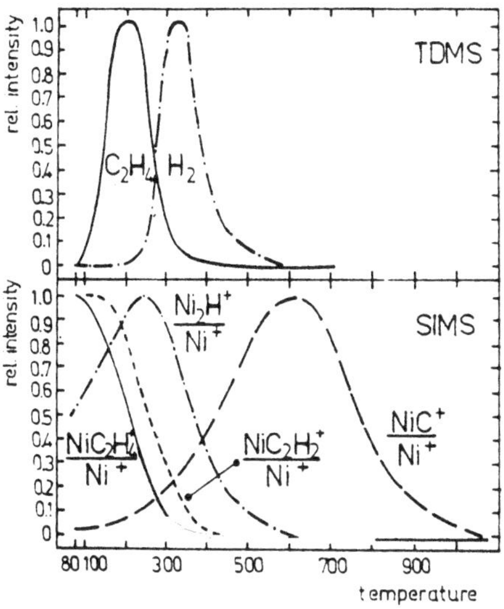

Fig. 4 Secondary ion emission and TDMS signals during heating of an 80 K C_2H_4 saturated Ni surface [8]

Application of intensity ratios to determine the surface complex concentration leads to the following conclusions with respect to the nickel-ethylene interaction: two different reaction paths can be distinguished for ethylene molecules physisorbed at 80 K. One leads to thermal desorption of the main part of ethylene at approx. 200 K; the other results in a series of decomposition steps. Both reactions are reflected by the immediate decrease of $NiC_2H_4^+/Ni^+$. The decomposition results in nickel-hydrogen bond formation, indicated by an increase of Ni_2H^+/Ni_2^+, and in the formation of an acetylenic species. This can be concluded from the delay in the decrease of the $NiC_2H_2^+/Ni^+$ ratio, which otherwise would diminish together with its main parent compound, $NiC_2H_4^+$. During further heating, also a decomposition of the acetylenic species takes place, resulting in an additional increase and maximum of Ni_2H^+/Ni_2^+. The subsequent decrease of this ratio is accompanied by thermal desorption of H_2 at approx. 330 K. At this temperature, all other hydrogen containing signals also dissapear from the surface. This depletion of hydrogen complexes results in a formation of a nickel-carbide layer, indicated by the increase of NiC^+/Ni^+. The decrease of this ratio at approx. 700 K corresponds to carbon bulk solution.

References

1. M. Someno, H. Saito and M. Kobayashi, Trans. JIM 26, 305 (1975)
2. P. Williams, C.A. Evans, Jr., M.L. Grossbeck and H.K. Birnbaum, Anal. Chem. 48, 969 (1976)
3. A. Benninghoven, K.-H. Müller and M. Schemmer, 7th Int. Mass Spectrometry Conf., Florence, 1976
4. T. Fleisch, G.L. Ott, W.N. Delgass and N. Winograd, Surface Sci. 81, 1 (1979)
5. R.S. Borodli, J.C. Vickerman and J. Wolstenholme, Surface Sci. 85, 244 (1979)

6. M. Barber, R.S. Bordoli, J.C. Vickerman and J. Wolstenholme, "Proc. 7th Int. Vac. Congr. & 3rd Int. Conf. Solid Surfaces", Vienna (1977) p. 983
7. A. Benninghoven, P. Beckmann, D. Greifendorf, K.-H. Müller and M. Schemmer, Surface Sci. 107, 148 (1981)
8. M. Schemmer, Ph.D. Thesis, Münster (1981);
P. Beckmann, D. Greifendorf, M. Schemmer and A. Benninghoven, Proc. of ECOSS4 (Münster) 1981, to be published in Surf. Sci.;
M. Schemmer, P. Beckmann, D. Greifendorf and A. Benninghoven, to be published in Surface Sci.
9. See for example
J.E. Demuth, Surface Sic. 76, L 603 (1978)
J.C. Bertolini and J. Rousseau, Surface Sci. 83, 531 (1979)
S. Lehwald and H. Ibach, Surface Sci. 89, 425 (1979)
and references therein
10. A. Benninghoven, P. Beckmann, D. Greifendorf and M. Schemmer to be published in Surface Sci. Lett.

SIMS Investigation of Adsorption and Chemical Modification of C_2H_4 and C_2H_2 on a Polycrystalline Ni-Surface

H. Kaarmann, B. Leidenberger, H. Hoinkes, and H. Wisch

Physikalisches Institut der Universität Erlangen-Nürnberg, D-8520 Erlangen, Fed. Rep. of Germany

Introduction

Nickel is an important catalyst in the hydrogenation and dehydrogenation of hydrocarbons. As with other transition metals, this feature is connected with the capability of the catalytic active surface to adsorb H_2 dissociatively. In the study of these catalytic processes the chemical changes in ethylene have often served as a relatively simple example and many investigations on this system with different methods are reported in the literature [1-7]. For surface temperatures below 150 K, molecular adsorption of C_2H_4 is observed [1] whereas for temperatures between 200 K and 400 K hydrogenation as well as dehydrogenation could be detected [1-3]. Some authors were able to demonstrate the simultaneous existence of both corresponding species. Furthermore, also polymerization and cracking could be found [4,5]. For temperatures above 400 K most commonly a complete dehydrogenation of C_2H_4 is observed with the formation of a carbon layer on the Ni-surface [3]. With increasing temperature this carbon layer diminishes and disappears at temperatures above 700 K [3]. Some interest is also concentrated on the influence of preadsorption of hydrogen [4,5]. Very similar findings are reported for acetylene on Ni [6]. At a temperature of 150 K, molecular adsorption and partial dehydrogenation coexist. For temperatures near and above 300 K cracking and formation of a CH-species are quoted. Hydrogenation and polymerization were found at temperatures above 400 K, also the formation of a graphite layer. A special feature seems to be the fact that the CH-species only looses hydrogen at the substantially higher temperature of 670 K (as compared to the case of ethylene on Ni).

It was the aim of the work presented here to get detailed knowledge on the formation of the different surface species by means of SIMS. As in SIMS changes of one atomic unit as well as atomic hydrogen itself can be detected unambigously, it was intented to identify all surface species rather directly, an idea successfully applied earlier in the case of the decomposition of NH_3 on a Fe(110)-surface [8].

Experimental

The experimental measurements were carried out in a stainless steel UHV-apparatus equipped with SIMS, LEED and AES, which has been described in detail elsewhere [9]. Typical operating conditions were vacuum in the 10^{-8} Pa range, primary ions: Ar^+ of about 10 nA/cm^2 current density at 500 eV energy in the static SIMS mode and 0.2 $\mu A/cm^2$ at 1.5 keV in the sputtering (cleaning) mode, respectively. The gases were introduced through a movable capillary tube system mounted on a manipulator and ending in front of the Ni sample. This device was chosen in order to avoid contamination and memory effects of the UHV-chamber. It was adjusted to achieve saturation of the surface within about 10 min at a surface temperature of 130 K. The sample was of high purity Ni, obtained from Goodfellow Metals. Standard cleaning procedures were applied in UHV.

Results and Discussion

The sample was cooled to 130 K whereby no observable changes in the SIMS-spectra occured. Then the gas inlet valve was opened and C_2H_4 and C_2H_2 admitted to the Ni-surface, respectively. The variation of the secondary ion signals and their dependence on the surface temperature were recorded. In the present investigation especially the behaviour of secondary ion signals like CH_n^+ and $C_2H_n^+$ with n = 0, 1, 2, ... was of interest in order to get information about corresponding surface species. The emission process of secondary ions, however, by itself is capable of producing fragments from parent molecules at the surface. Consequently a direct correlation of observed secondary ions to a surface species does not necessarily exist. In addition to that, the secondary ion yield may not be directly proportional to the surface concentration of the species in question and may perhaps depend on the surface temperature. These facts of course introduce considerable difficulties into the interpretation of measurements. The investigation of all observed positive ion signals however, together with the comparison to the temperature dependence of detectable negative ion signals during the same heating process can be interpreted within a reasonable reaction scheme.

Ethylene Exposure

In this short report only some examples for typical measurements can be given. Fig. 1 demonstrates the increase of the H^+-signal from the Ni-surface as soon as ethylene has been adsorbed at 130 K. The presence of hydrogen at low temperatures and the diminution of hydrogen with increasing temperatures is quite obvious. Indeed, the H^+-ion can have its origin in atomic hydrogen on the surface as well as in all surface species containing hydrogen. Likewise the temperature behaviour of other positive secondary ion signals is very complex as demonstrated in Fig. 2. Fig. 3 shows two negative ion yields as a function of surface temperature (which is very similar to the case of acetylene on Ni). This behaviour is compatible with the following interpretation: a) With adsorption of C_2H_4 on Ni at T_{SF} = 130 K, dehydrogenated species are immediately created (this statement of course has to correlate with the interpretation of Fig. 1); b) Above 300 K further successive dehydrogenation occurs which is completed at a surface temperature of 550 K to 600 K. Similar interpretation of the observed secondary ion signals within consistent intercorrela-

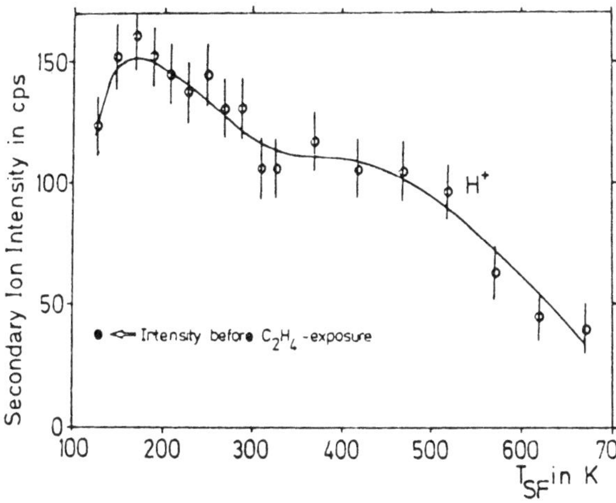

Fig. 1 H^+ signal in the SIMS-analysis of C_2H_4 on Ni for relevant temperatures

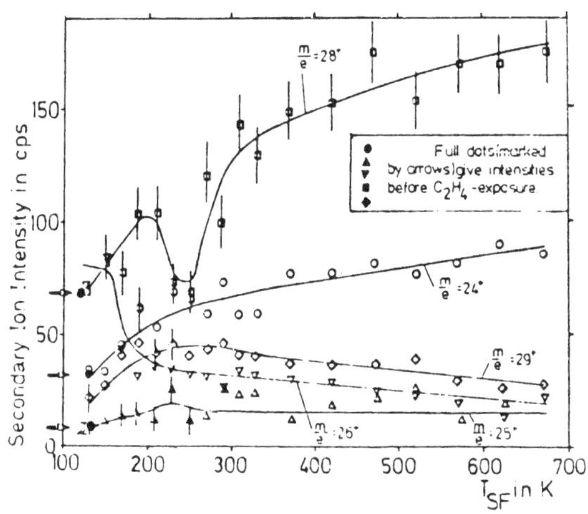

Fig. 2 Temperature dependence of some positive secondary ion signals during heating of a Ni-surface with ethylene adsorbed on it

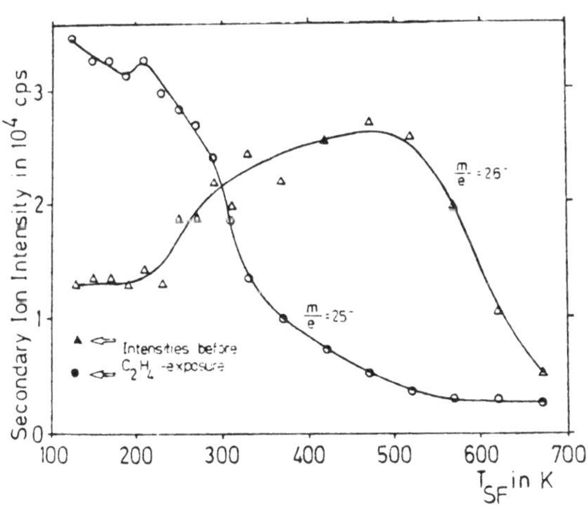

Fig. 3 Same as Fig. 2 for two selected negative ion signals

tions as described above has led us to the following possibilities for the interaction of ethylene with a polycrystalline Ni-surface:

i) C_2H_4 adsorbs on a polycrystalline Ni-surface at T_{SF}=130 K predominantly dissociatively by splitting off two hydrogen atoms which remain adsorbed in addition to $(C_2H_2)_{ads}$.
ii) The surface species $(C_2H_2)_{ads}$ at 130 K shows hydrogenation as well as dehydrogenation and cracking, resulting in surface species as $(C_2H_5)_{ads}$, $(C_2H)_{ads}$, $(C_2)_{ads}$, and $(CH_n)_{ads}$ with n = 0, 1, 2.
iii) Upon heating hydrogenation is observed for temperatures between 130 K and 200 K: $(C_2H_n)_{ads}$ with n = 0, 1, 2 gives $(C_2H_5)_{ads}$.
iv) Hydrogenation of the cracking products occurs, too: $(CH_n)_{ads}$ with n = 0, 1, 2 gives $(CH_3)_{ads}$.
v) At temperatures above 300 K the species $(C_2H_5)_{ads}$ is decomposed by dehydrogenation.
vi) Dehydrogenation of the species $(CH_3)_{ads}$ is observed at temperatures above 400 K.
vii) Dehydrogenation of the surface species leads to the formation of a carbon layer on the nickel surface. The disappearance of carbon at surface temperatures above 700 K is represented in Fig. 4. This investigation is performed by means of SIMS and Atomic Beam Scattering. Further experimental details will be published elsewhere [10].

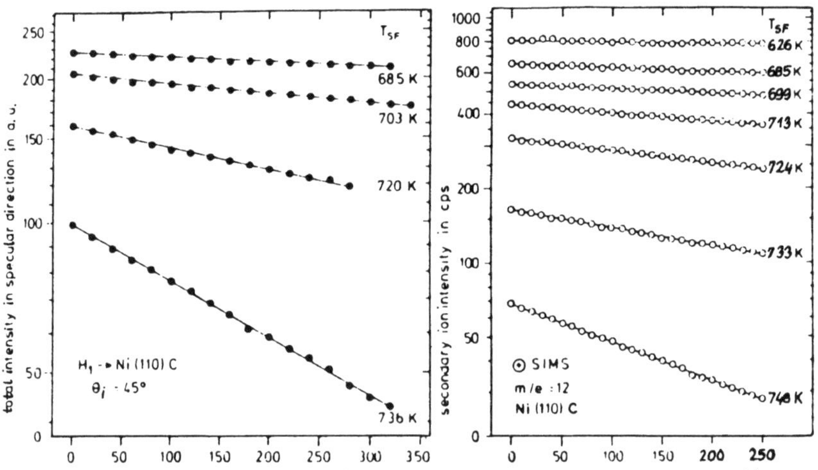

Fig. 4 Disappearance of a carbon layer from a Ni(110)-surface. The carbon layer was produced by ethylene adsorption at T_{SF} = 500 K. The secondary ion yield of C^+ as well as the signal from Atomic Beam Scattering proved to be proportional to the relative surface coverage with carbon

Acetylene Exposure

Acetylene exposure of the nickel target yielded, in part, very similar results as described above with the following differences:

i) C_2H_2 adsorbs on polycrystalline nickel at T_{SF} = 130 K molecularly, which led to $(C_2H_2)_{ads}$.
ii) Step iv) in the case of ethylene on nickel, i.e. the hydrogenation of cracking products, could not be observed. This behaviour can be referred to the smaller amount of adsorbed atomic hydrogen in the case of acetylene on nickel.

Conclusions

Some or other of these surface reactions and processes have been found by different authors [1-7] with the whole variety of surface analytical methods. The present SIMS-investigation, which was restricted to the analysis of the temperature behaviour of secondary ions like $CH_n^{\pm}$ and $C_2H_n^{\pm}$ supports many of these earlier results within an extended reaction scheme. However, it must be confirmed that an unambigous interpretation of the rather extensive and complex SIMS-results remains difficult and could not have been achieved without the positive knowledge of many details, reported before in the literature. It is evident, that a reliable interpretation of secondary ion signals still requires further exploration. The formation of ratios of SI-counting rates like $NiCH_n^+/Ni^+$ and $NiC_2H_n^+/Ni^+$ as utilized by BENNINGHOVEN and coworkers [7] could be a successful approach.

References

1. J.E. Demuth, H. Ibach and S. Lehwald, Phys. Rev. Lett. 40, 1044 (1978)
2. J. Erkelens and Th.J. Liefkens, J. Catalysis 8, 36 (1967)
3. R.A. Zuhr and J.B. Hudson, Surface Sci. 66, 405 (1977)
4. G.A. Martin and B. Imelik, Surface Sci. 42, 157 (1974)
5. O. Beck, Disc. Farad. Soc. 8, 118 (1950)
6. S. Lehwald and H. Ibach, Surface Sci. 89, 425 (1979)
7. A. Benninghoven, P. Beckmann, D. Greifendorf and M. Schemmer, Appl. Surface Sci. 6, 288 (1980)
8. M. Drechsler, H. Hoinkes, H. Kaarmann, H. Wilsch, G. Ertl and M. Weiss, Appl. Surface Sci. 3, 217 (1979)
9. J. Estel, H. Hoinkes, H. Kaarmann, H. Nahr and H. Wilsch, Surface Sci. 54, 393 (1976)
10. H. Hoinkes, H. Kaarmann and H. Wilsch, submitted for publication in Surface Sci. Lett.

Secondary Ion Emission from UHV-Deposited Amino Acid Overlayers on Clean Metal Surfaces

W. Lange, M. Jirikowsky, D. Holtkamp, and A. Benninghoven

Physikalisches Institut der Universität Münster,
D-4400 Münster, Fed. Rep. of Germany

1. Introduction

In SIMS work on organic materials samples are usually prepared under atmospheric conditions using aqueous or alcoholic solutions (e.g. dipping technique [1], electrospray [2], micropipette [3]). Major drawbacks of these methods are possible chemical reactions occuring between the target material and some components of the solution, and contaminations with e.g. hydrocarbons. Furthermore, it is not possible to produce well-defined substrate surfaces. Therefore, when investigating the influence of surface chemistry on secondary ion formation, it is necessary to perform sample preparation under UHV-conditions.

Most amino acids can be sublimed in appreciable amounts at temperatures below 450 K [4], where thermal decomposition can be neglected. Thus, overlayers can be produced in situ by exposing a substrate surface to a molecular beam. Prior to the evaporation of the sample material the surface can be cleaned or chemically changed in a well-defined manner.

For this purpose an apparatus was developed which simultaneously allows target exposure to a molecular beam, ion and electron bombradment, and SIMS-analysis of the surface. Both the ion and the electron gun are supplied with a scanning device (scanning area more than 1 cm^2). By gating the pulse counting device crater effects during enhancend ion or electron bombardment are avoided. The target can be cooled and heated in the temperature range between 170 K and approx. 1000 K. Calibration of the molecular beam source is accomplished by means of a quartz crystal microbalance. Auger experiments are carried out with a modified single pass CMA. Registration of spectra using a pulse counting technique permits a reduction of the primary electron current into the 10^{-10} A range [5]. Thus, beam damages during analysis are virtually avoided (static AES). Further details of the apparatus are described elsewhere [6].

We examined the secondary ion emission of some amino acids deposited on various substrates. For all combinations under investigation we found a quite similar behaviour of characteristic secondary ions during layer growth [6]. The intensity of the parent ions $(M\pm H)^{\pm}$ appears to be strongly dependent on the chemical state of the substrate surface. A particularly high parent ion yield is observed for noble metal surfaces (Au, Pt) with amino acid coverages in the monolayer range. However, remarkable differences between the $(M+H)^+$ and $(M-H)^-$ ion emission were found for all amino acids. As a typical example results for glycine are presented in this paper.

2. Experimental Results

2.1 Vapor Pressure and Layer Growth

By means of a quartz crystal microbalance (sensitivity: 10^{-8} g) the rate of layer growth as a function of the evaporator temperature was determined. Furthermore, reevaporation of amino acid from the heated quartz crystal was examined. Within the temperature range from 300 K to 450 K the resulting vapor pressures are in good agreement with an Arrhenius law: $\log p = A/T + B$. The constants A and B of some amino acids are given in Table 1.

Table 1 Vapor pressure constants of some amino acids (p in mbar, T in K)

amino acids	A/1000	B
methionine	-6.22	11.94
phenylalanine	-6.67	13.07
leucine	-6.97	14.54
glycine	-6.52	13.10

It is generally noticed that reevaporation cannot be neglected at room temperature (e.g. glycine: $p = 10^{-8}$ mbar at 310 K). Therefore the target has to be cooled in order to get stable multilayers. On an uncooled Cu surface (T = 310 K) layer growth stops after a total exposure density of $5 \cdot 10^{14}$ molecules per cm^2, as measured with AES (Fig. 1). This amount corresponds to a monolayer coverage (see for example [7]). Strong binding forces between substrate and adsorbate prevent reevaporation in the mono- and submonolayer range. After occupation of all

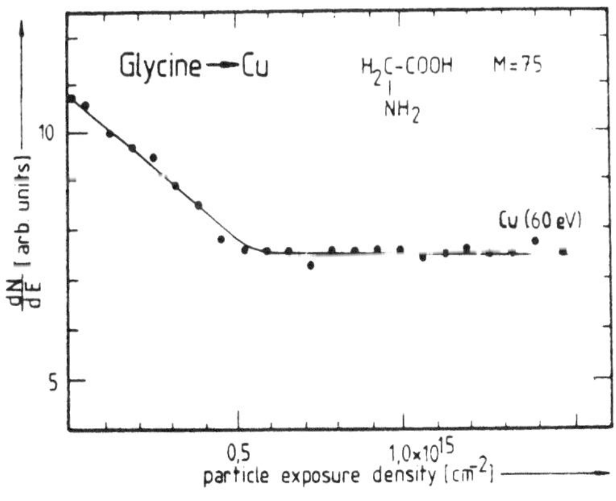

Fig. 1 Cu MVV peak-to-peak height during layer growth of glycine (target temperature: 310 K)

adsorption sites on the metal surface further adsorption takes place on top of an amino acid layer. Because of weaker bonds thermal desorption occurs, governed by the equilibrium vapor pressure of the amino acid. Therefore, multilayers are only formed if the particle flux exceeds the reevaporation rate. Earlier SIMS measurements also indicate a limitation of coverage to about 1 monolayer [6].

2.2 Secondary Ion Emission

The intensities of the protonated and deprotonated molecular ions were measured as a function of the particle exposure density (Fig. 2). There are great differences regarding various substrates. However, the influence of the substrate material is much lower for the deprotonated species. On reactive metals as Cu or Ni the increase of the protonated ion intensity starts at a particle exposure density which corresponds to the completion of the first monolayer (as measured by AES). On noble metal surfaces, however, both ions show a very similar behaviour. Furthermore, on Au or Pt a $(M-H)^+$ ion is observed with a yield comparable to $(M+H)^+$.

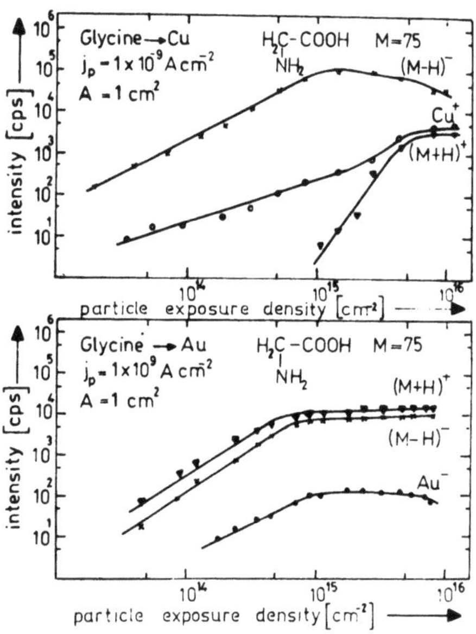

Fig. 2 Secondary ion intensities during layer growth of glycine on Cu and Au

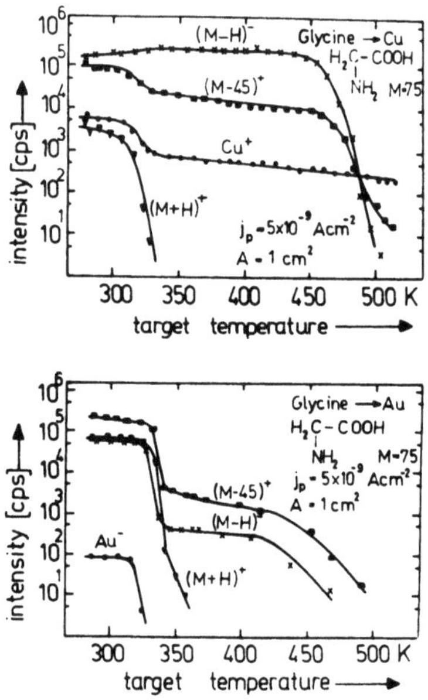

Fig. 3 Changes of secondary ion intensities during heating of glycine covered Cu and Au targets

During heating (1 K/s typically) of a Cu target covered by approximately one monolayer of glycine two changes in secondary ion emission can be distinguished (Fig. 3): at about 320 K the intensity of the protonated ion disappears whereas the $(M-H)^-$ intensity remains nearly constant. The $(M-COOH)^+$ intensity drops to a lower level and then remains constant. At about

450 K the $(M-H)^-$ emission disappears, too. In this temperature range there is a re-increase of the substrate AES signals to a level corresponding to the clean metal surface. A different behaviour of the secondary ion signals is observed during heating of an Au target (Fig. 3). Again, the $(M+H)^+$ intensity decreases sharply at about 320 K. In contrast to the Cu target, however, the $(M-H)^-$ intensity is also reduced by two orders of magnitude.

In the submonolayer region beam damages due to ion (Ar^+; 4.5 keV) or electron (1 keV) bombardment were measured (Fig. 4). The parent peaks show an exponential decrease, as expected from monolayer sputtering [8]. From the slope of the straight line (in a logarithmic diagram) cross sections for desorption and destruction can be calculated. Compared to electron bombardment the cross sections for ion bombardment are greater by a factor of 100 to 1000 (the values resulting from Fig. 4 are: $7.9 \cdot 10^{-17}$ cm^2 and $1.9 \cdot 10^{-14}$ cm^2 for $(M+H)^+$; $1.1 \cdot 10^{-17}$ cm^2 and $7.3 \cdot 10^{-15}$ cm^2 for $(M-H)^-$). Again, remarkable differences are observed for protonated and deprotonated ions. For both ion and electron bombardment the cross sections for the $(M+H)^+$ ion are substantially higher than those of the $(M-H)^-$ ion.

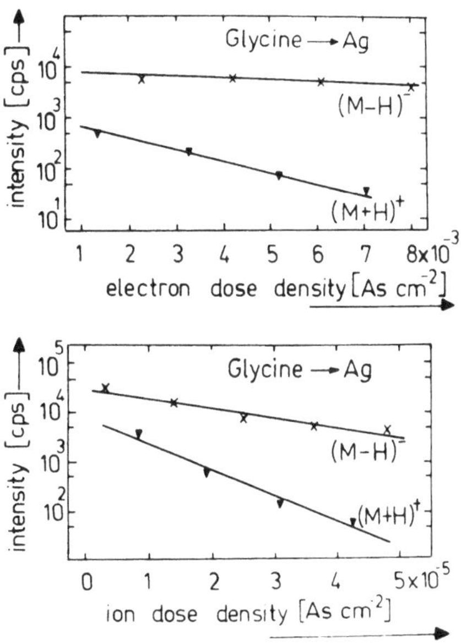

Fig. 4 Changes in the parent ion intensities by electron or ion bombardment (submonolayer coverage on clean Ag)

3. Discussion

In the submonolayer and monolayer range secondary ion emission from amino acid overlayers is governed by the chemical nature of the substrate. The different behaviour of $(M+H)^+$ and $(M-H)^-$ ions can be explained by assuming different binding mechanisms between amino acid molecules and the target surface.

On relative metals chemisorption takes place resulting in the formation of stable complexes as [(M-H)—Me]. The emission of protonated molecular ions from this surface compound is a rather unprobable process. The deprotonated ion, however, can easily be formed by cleavage of the complex. After completion of the first amino acid layer physisorption occurs which is characterized by weak interactions between adjacent amino acid molecules. Now both protonated and deprotonated ions can be formed by proton exchange processes between the molecules. Regarding submonolayers on noble metal surfaces, the similar behaviour of the protonated and deprotonated species indicates that on these metals bond formation occurs without loss of protons (physisorption), thus leading to comparable probabilities for the formation of $(M+H)^+$ and $(M-H)^-$ ions.

The results obtained during sample heating and enhanced ion or electron bombardment support the assumption that $(M+H)^+$ emission is possible from weakly bound amino acid molecules only. The disappearance of the $(M+H)^+$ emission slightly above room temperature is the result of thermal desorption of physisorbed molecules. The simultaneous decrease of fragment ion intensities (e.g. $(M-COOH)^+$) indicates a depletion of the total coverage of the surface. Due to the strong bonds of the chemisorbed phase, the $(M-H)^-$ emission disappears at higher temperatures. The decrease of the $(M-H)^-$ intensity from Au targets in the low temperature range by more than two orders of magnitude strongly supports the assumption that amino acid molecules are physisorbed on noble metals.

References

1. A. Benninghoven, D. Jaspers and W. Sichtermann, Appl. Phys. 11, 35 (1975)
2. E. Brunix and G. Rudstam, Nucl. Inst. Meth. 13, 131 (1961)
3. A. Benninghoven and W. Sichtermann, Anal. Chem. 50, 1180 (1978)
4. H.J. Svec and D.D. Clyde, J. Chem. Eng. Data 10, 151 (1965)
5. A. Benninghoven, O. Ganschow, P. Steffens and L. Wiedmann, J. Electron Spectrosc. Rel. Phenom. 44, 19 (1978)
6. A. Benninghoven, M. Jirikowsky and W. Lange, to be published
7. L.L. Atanasoska, J.C. Cuchholz and G.A. Somorjai, Surface Sci. 72, 189 (1978)
8. A. Benninghoven, Surface Sci. 35, 427 (1973)

SIMS Investigation of Adsorption of O_2, H_2O, CO, CO_2, CH_2O, and CH_3OH and Coadsorption of O_2 with CH_2O and CH_3OH on Polycrystalline Silver Surfaces

L. Wiedmann, N.L. Wang[1], R. Jede, L.D. An[2], O. Ganschow, and A. Benninghoven

Physikalisches Institut der Universität Münster,
D-4400 Münster, Fed. Rep. of Germany

Introduction

In the past few years, the importance of a surface analytical approach to the study of heterogeneous catalysis has become evident [1]. In order to ensure the relevance of the surface physical results obtained in ultra-high vacuum for the catalytic processes under technical conditions (usually high pressure and temprature), we have set up an instrument which is a combination of a technical microreactor and a surface analytical system incorporating SIMS, AES, XPS, ISS and TDMS. The details of this combined system will be published elsewhere. It should, however, be pointed out that three kinds of experiments can be conducted in this instrument, which are essential for the understanding of catalytic processes:

(1) well-defined reactions on model surfaces under "clean" conditions,
(2) surface analytical characterization of catalysts as drawn from the plant,
(3) catalytic reactions in the microreactor and subsequent surface characterization of the catalyst after transferring it into the surface analysis section in vacuum and within a few minutes.

One technically important catalytic reaction which is being investigated with this combined apparatus is the oxidation or dehydrogenation of methanol to formaldehyde on silver. Whereas the experiments of type (2) and (3) on this system will be published elsewhere, it is the purpose of this paper to report some type (1) of experiments, namely the adsorption of various gases on clean and preoxidized polycrystalline silver surfaces under UHV conditions.

In recent TDMS investigations of the adsorption of methanol [2] and formaldehyde [3] on preoxidized silver surfaces, intermediate species such as CH_3O [2] and H_2COO [3] have been postulated to occur in these surface reactions. The study of adsorption of possible intermediate or desorption products

[1]Guest scientist from Shanghai Institute of Metallurgy, Academia Sinica, China

[2]Guest scientist from Lanchow Institute of Chemical Physics, Academia Sinica, China

like O_2, CO, CO_2, H_2O, and the reactants CH_2O and CH_3OH themselves on clean silver is likely to give some characteristic "fingerprint spectra", which may be compared to those obtained in the case of adsorption on preoxidized surfaces. Besides, the results are of direct relevance for the corresponding type (2) and (3) experiments mentioned above.

As SIMS offers low detection limits and a great ability for detecting and distinguishing different surface compounds [4], this method has been chosen for monitoring the adsorption and coadsorption experiments. Because of UHV considerations, the adsorbate pressure did not exceed 10^{-5} mbar, and the total adsorbate exposition was limited to about 10^4 L. All measurements were taken at room temperature. The experiments will be extended to higher gas expositions and different temperatures in a separate preparation chamber, where adsorbate pressures up to 1 bar are possible.

Results

1. Adsorption of Oxygen on Clean Silver

Oxygen adsorbs very slowly on clean silver surfaces. Fig. 1a shows the intensity of the oxide-specific secondary ions AgO^-, $Ag_2O^{\pm}$, and $Ag_2O_2^-$ as a function of the oxygen dose. Significant intensities emerge at about 10-100 L, and saturation is not reached at 10^4 L, the maximum dose applied. Besides the silver ions $Ag^{\pm}$ and $Ag_2^{\pm}$ themselves $O^{\pm}$ and O_2^- are observed, which behave similarly to the ions shown in Fig. 1a. No AgO^+ could be detected. This dose dependence is consistent with estimates of the initial sticking coefficient of oxygen on Ag, which is the order of 10^{-3} [5-6].

2. Exposure of Clean Silver to CO, CO_2, H_2O and CH_3OH

Figure 1b shows the intensity of characteristic ions during methanol exposition. In contrast to oxygen adsorption, only very little fragment ion intensity (CH_3^+ and CH_3O^+) emerges at exposures greater than 10^3 L. The Ag^+ and Ag^- intensities remain approximately constant throughout the exposure range, thus excluding significant changes of both ionization probability and overlayer coverage. Furthermore, compound ions of the type $AgR^{\pm}$ or $Ag_2R^{\pm}$, where R is a fragment of the parent molecule CH_3OH, are completely absent. From these results we conclude that methanol does not adsorb on clean silver to a significant amount in the investigated exposure range. Similar results (not shown in the figures) have been obtained for CO, CO_2, and H_2O exposure. None of these gases is found to adsorb on clean silver at doses up to 10^4 L.

3. Adsorption of Methanol on Preoxidized Silver

The results of methanol exposition to a silver surface preoxidized with 10^3 L O_2 is shown in Fig. 2. Intensity of two characteristic ion species, $AgCO^{\pm}$, is observed emerging at 0.1-0.3 L and passing through a maximum at 2 L. At its maximum intensity, the oxide-specific ions Ag_2O^+ disappear.

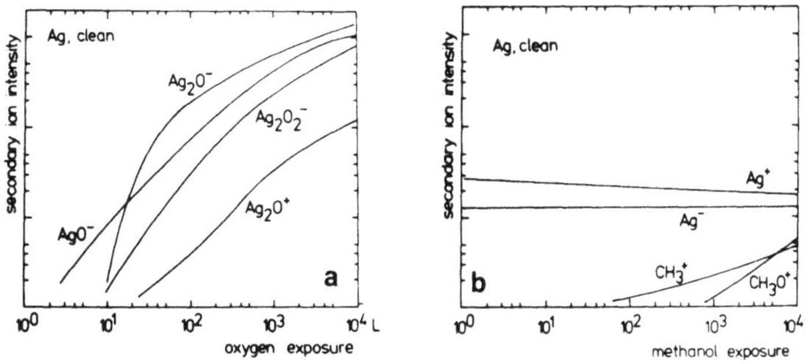

Fig. 1 Intensity of oxide-specific secondary ions as a function of gas exposure. (a) oxygen clean silver, (b) methanol on clean silver

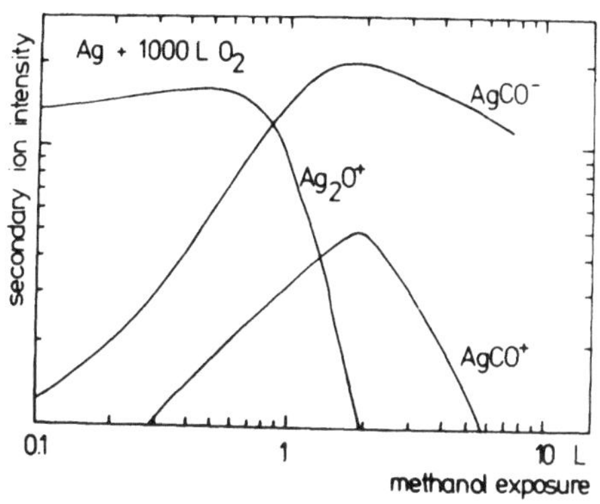

Fig. 2 Intensity of characteristic secondary ions as a function of methanol exposure to a silver surface preoxidized with 1000 L oxygen

These features can be explained by a surface reaction between the preadsorbed oxygen and the methanol accomodating at the surface. The observed $AgCO^{\pm}$ emission is characteristic for a surface species formed during this reaction, whereas the surface oxygen characterized by Ag_2O^+ is completely consumed. As we have shown in the previous section, the emitted species $AgCO^{\pm}$ cannot be attributed to adsorbed CO or CO_2. Hence, the adsorbed methanol cannot have been oxidized to one of these final products, but must be oxidized or dehydrogenated to one of its intermediate oxidation products, methoxide or formaldehyde. One reasonable explanation of the observed surface reaction is the formation of surface methoxide, $CH_3O_{(a)}$, as postulated by WACHS and MADIX [2]. During adsorption of methanol on clean Pd(100) [7] and Ni(111) [8] surfaces, $CH_3O_{(a)}$ is also concluded to be the adsorbed species. If this assumption is true, the preadsorbed oxygen must be consumed by uptake of the excess hydrogen and subsequent desorption as H_2O, because this species does not adsorb on clean silver. The possibility of formaldehyde formation will be discussed in the following section.

4. Adsorption of Formaldehyde on Clean and Preoxidized Silver

In order to relate the fingerprint spectra of formaldehyde to the methanol-oxygen surface reaction on silver, the adsorption of formaldehyde on clean and preoxidized silver has been studied. Fig. 3 shows the intensity of characteristic positive

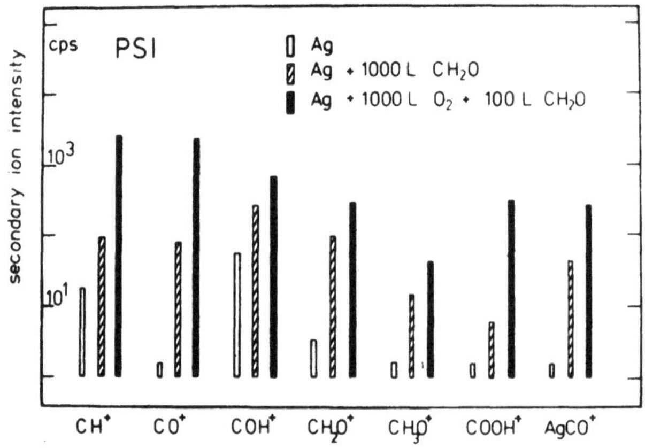

Fig. 3 Intensity of characteristic positive secondary ions for (a) clean silver, (b) 1000 L formaldehyde adsorbed on clean silver, (c) 100 L formaldehyde adsorbed on silver preoxidized with 1000 L oxygen

secondary ions for clean silver and the Ag + CH_2O and Ag + O_2 + CH_2O systems. In contrast to CH_3OH, CH_2O does adsorb on clean silver. After 1000 L exposure, the intensities of CO^+, CH_2O^+ and $AgCO^+$ have increased by more than an order of magnitude as compared to clean silver. A further order-of-magnitude increase of CH^+, CO^+ and $COOH^+$ is observed for 100 L CH_2O adsorption on preoxidized silver as compared to adsorption on clean silver. Specifically, this strong increase of $COOH^+$ indicates further oxidation of the formaldehyde by preadsorbed oxygen. The formation of surface species of the type $H_2COO_{(a)}$ as postulated by WACHS et al. [3] seems to be a possible explanation. It should, however, be noted that $AgCO^+$ and (not shown) $AgCO^-$ are present in both the Ag + CH_2O and Ag + O_2 + CH_2O systems. For CH_2O adsorption on clean silver, the $AgCO^{\pm}$ ions are even the most prominent species of the type $AgR^{\pm}$. As $AgCO^{\pm}$ are also the most intense species in the Ag + O_2 + CH_3OH system, the appearance of these ions is therefore not indicative of whether surface methoxide or formaldehyde is formed by this surface reaction.

Further experiments at higher exposures and different temperatures with both simultaneous and successive exposition to oxygen and methanol or fomaldehyde, will have to elucidate the nature of the surface species formed by methanol oyidation and the details of the reaction mechanism.

Acknowledgement

We wish to thank the Bundesminister für Forschung und Technologie for financial support.

References

1. H.P. Bonzel, Surface Sci. 68, 236 (1977)
2. I.E. Wachs and R.J. Madix, Surface Sci. 76, 531 (1978)
3. I.E. Wachs and R.J. Madix, Appl. Surface Sci. 5, 426 (1980)
4. A. Benninghoven, Surface Sci. 53, 596 (1975)
5. H.A. Engelhardt and D. Menzel, Surface Sci. 57, 591 (1976)
6. H. Albers, W.J.J. van der Waal, O.L.J. Gijzeman and G.A. Bootsma, Surface Sci. 77, 1 (1978)
7. K. Christmann and J.E. Demuth, submitted to J. Chem. Phys.
8. J.E. Demuth and H. Ibach, Chem. Phys. Letters 60, 395 (1979)

Distribution of Ni, Co, Ga, and Cu in Iron Meteorites

J. Okano and H. Nishimura

Institute of Geological Sciences, College of General Education, Osaka University, Osaka 560, Japan

Introduction

Distribution of nickel in taenite veins of iron meteorites have been thoroughly investigated by J.A. WOOD [1], J.I. GOLDSTEIN and J.M. SHORT [2] using electron probe microanalyzers (EPMA) in connection with the problem of cooling rates of meteorite parent bodies. M. SHIMA et al. [3] reported the results of the line analyses by EPMA for iron, nickel and cobalt on the four iron meteorites. Studies on the distributions of minor components such as gallium, germanium and copper in taenite veins have been limited due to a lack of microanalytical techniques with high sensitivities. Recently, ion microprobe mass analyzer (IMMA) has been developed and proved to have a much higher sensitivity compared with EPMA. In 1979, S.J.B. REED et al. [4] reported the results on line analyses using IMMA for gallium in iron meteorites and they found M type profiles that were strongly correlated with the profiles for nickel.

The authors have carried out IMMA line analyses for nickel, cobalt, copper and gallium across the taenite veins in two iron meteorites, Odessa and Canyon Diablo. Besides, the isotopic ratio of $^{58}Ni/^{60}Ni$ has been measured for kamacite phase and for taenite phase. The experimental results and the brief discussion will be described in the following.

Experimental

Specimens used in this work are Odessa and Canyon Diablo. Both are coarse octahedrites belonging to group I. The bulk compositions are shown in Table 1a [5] and 1b [6]. The specimens were cut and the surfaces were polished with alumina powder. They were slightly etched by dilute nitric acid and cleaned ultrasonically.

Hitachi IMA 2A ion microprobe mass analyzer has been used for this work. Primary ions were 17 keV O_2^+. The specimens were mounted on the sample holder which was movable three dimensionally with a micrometer device. The specimens can be observed with a zoomed optical microscope during the analyses. The total secondary ion image or the specific secondary ion image can be displayed on the CRT screen.

Table 1a Chemical composition of Odessa [5]

[%]					[ppm]		
Ni	Co	P	C	S	Ga	Ge	Cu
7.35	0.38	0.25	0.2	0.5	75	285	200

Table 1b Chemical composition of Canyon Diablo [6]

[%]					[ppm]		
Ni	Co	P	C	S	Ga	Ge	Cu
7.10	0.46	0.26	1.0	1.0	80	320	140

Line Analyses for nickel, cobalt, copper and gallium

The line analyses were made along the probed lines, across the taenite veins, by moving the samples step by step. The diameter of the ion probe was 10-20 μm. The secondary ion intensities at mass 59, 60, 63 and 69 were used as the measures of the local concentrations of cobalt, nickel, copper and gallium, respectively. The intensities were normalized to the $^{56}Fe^+$ intensity in order to compensate the topographycal effect of the sample surface. The contributions of interference ionic species at these mass peaks were evaluated to be negligibly small. Measurements for germanium were found to be difficult due to the interference of FeO^+.

The concentration profiles along the probed lines (bb', ff'; aa', bb') are shown in Fig. 1 and Fig. 2 for Odessa and in Fig. 3 and Fig. 4 for Canyon Diablo. From these results, it is obvious that the profiles for nickel, copper and gallium are strongly correlated to each other. Copper and gallium are remarkably enriched near the border of the taenite veins. It seems that the profiles for copper and gallium give very sensitive evidence of the thermal history of the meteorites.

Isotopic Analyses for Ni

Isotopic abundance ratio, $^{58}Ni/^{60}Ni$ has been measured for the center part of taenite phase and for arbitrary selected portions of kamacite phase which are located around the taenite phase. The $^{58}Ni^+/^{60}Ni^+$ ratio obtained for a pure nickel metal of four nines was used to determine the instrumental mass discrimination factor. FeH^+/Fe^+ ratio has been determined from peak height ratio of mass 57 to mass 56 with every run. The hydride formation ratio was 0.1 ~ 0.2 percent in usual cases. $^{58}Ni^+$ intensity was determined by subtracting the contributions of $^{58}Fe^+$ and $^{57}FeH^+$ from the intensity of mass 58 peak. $^{60}Ni^+$ intensity was determined by subtracting the contribution of $^{59}CoH^+$ from the intensity of mass 60 peak. The contribution of $^{59}CoH^+$ was usually less than 0.1 percent and negligibly small in most cases.

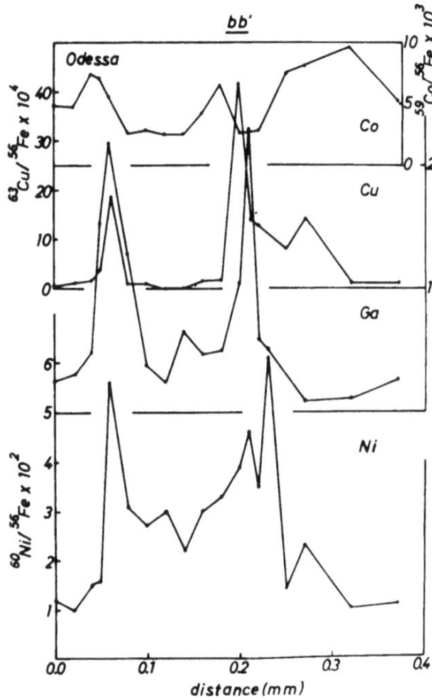

Fig. 1 Profiles for Ni, Ga, Cu and Co along the probed line bb'

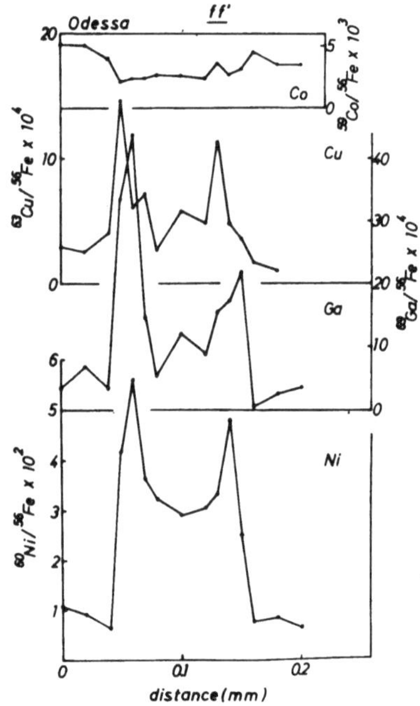

Fig. 2 Profiles for Ni, Ga, Cu and Co along the probed line ff'

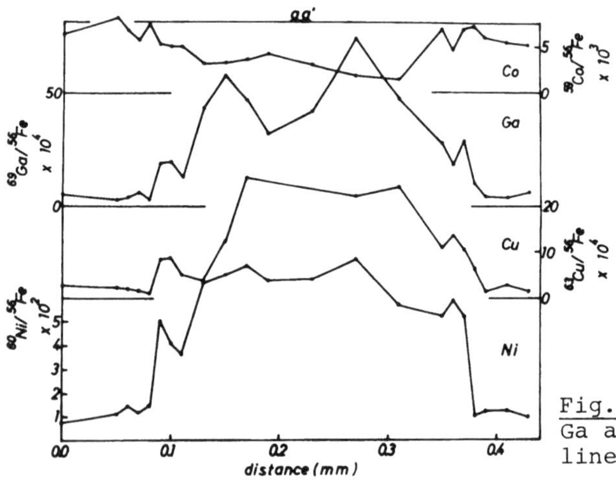

Fig. 3 Profiles for Ni, Cu, Ga and Co along the probed line aa'

The results are shown in Fig. 5. The $^{58}Ni^+/^{60}Ni^+$ ratios for the taenite phase are clearly large compared with those for the kamacite phase. This means that the light isotope, ^{58}Ni, has been enriched in taenite phase during the cooling process of the meteorite parent body. The differences of the $^{58}Ni/^{60}Ni^+$ ratio between two phases are about 1.5 percent for Odessa and about 1.0 percent for Canyon Diablo.

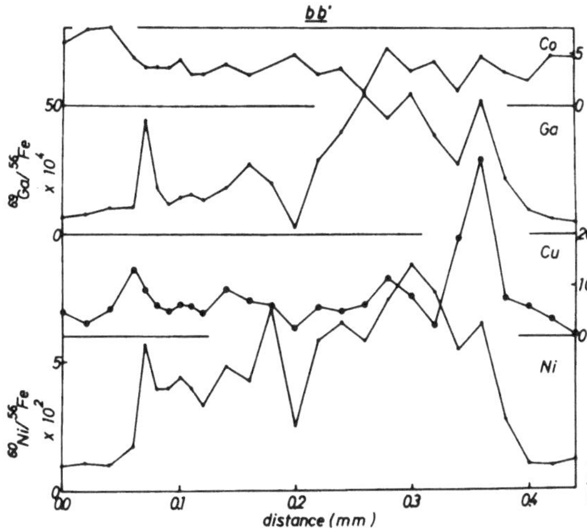

Fig. 4 Profiles for Ni, Cu, Ga and Co along the probed line bb'

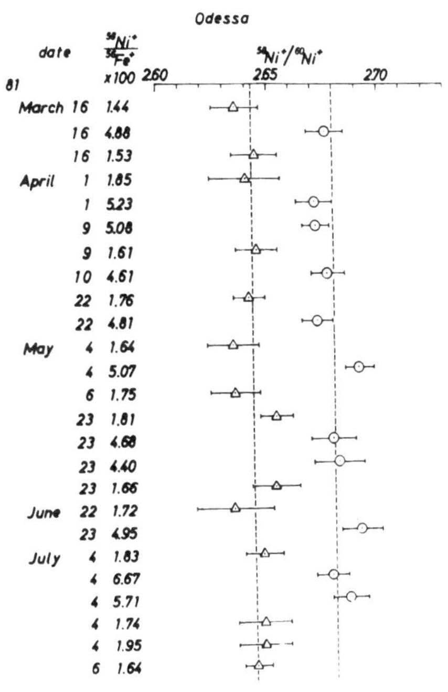

Fig. 5 Measured isotopic ratio, $^{58}Ni/^{60}Ni$ for kamacite phase (triangle) and for taenite phase (circle)

Assuming that the diffusion velocity is inversely proportional to the square root of the isotopic mass, the calculation by computer simulations has been carried out under the following conditions: bulk nickel concentration, 7.5 wt percent; kamacite band width, 1.6 mm; cooling rate at 500 °C, 10 °C/10^6 y. The result of the calculation showed that the difference in the isotopic ratio between two phases was 0.56 percent. This value is somewhat smaller than experimentally obtained values. The reason for the discrepancy is not known but the idealization used for the calculation may be one of the sources.

References

1. J.A. Wood, Icarus 3, 429 (1964)
2. J.I. Goldstein and R.E. Ogilvie, Geochim. Cosmochim. Acta, 29, 893 (1965) and 31, 893 (1965)
3. M. Shima et al., Bulletin Nat. Sci. Museum, Series E, 1, 1 (1978)
4. S.J.B. Reed et al., Meteoritics 14, 523 (1979)
5. V.F. Buchwald, "Handbook of Iron Meteorites", Vol. 2, 381 (1975)
6. ibid. vol. 3, 937

Metallurgical Applications of Ionic Microscopy

E. Darque-Ceretti, R. Dennebouy[1], J.C. Pivin[2], and C. Roques-Carmes[3]

Ecole des Mines, CEMEF
F-06560 Valbonne, France

Mechanical and physico-chemical properties of metallic materials are greatly affected by the distribution of impurities or minor additions in the matrix structure on a microscopic scale. Most information on the composition of precipitates or diffusion gradients can be obtained by EMA. But light elements such as B, Be, C, N are important constituents of metallic materials, of which detection is almost impossible by EMA. The study of sub-microscopic precipitations and atomic segregations at interfaces are also out of the field of this microanalytical method. Auger scanning microscopy is the most appropriate non-destructive method to study segregations at interfaces and submicroscopic precipitations, because of its high spatial resolution and of its better sensibility to light elements (for local contents over 1 %).

Ionic microscopy proves to be a complementary method for the microanalysis of light elements and elements with low contents, because of:

a) its high sensibility,
b) the possibility to detect all elements and isotopes,
c) the information on the chemical state of elements afforded by polyatomic ions and chemical effects on emission yields,
d) the 0.5 μm spatial resolution of the CAMECA ionic microscope.

The heterogeneity of metallic materials affects also the data of analytical technics, and a first paragraph will be devoted to give the evidence of the usefulness of ionic microscopy to characterize alloys before using them as standards. Two examples of applications to metallurgical studies will be presented in a second paragraph. They were selected among research developed by a team of metallurgists in collaboration with Professor SLODZIAN, as specific applications of ionic microscopy with respect to the data of other methods. They illustrate simultaneously several possibilities (points a, b, c, d) of the method.

---

[1]Laboratoire de Physique des Solides, Université Paris-Sud
F-91405 Orsay Cedex, France

[2]Laboratoire de Métallurgie Physique, Université Paris-Sud
F-91405 Orsay Cedex, France

[3]Laboratoire de Microanalyse des Surfaces, E.N.S.M.M.
F-25030 Besancon Cedex, France

Characterization of the Purity and Microstructure of Metallic Standards

Materials prepared by classical methods of metallurgy generally contain impurities and heterogeneities on a microscopic scale. They are unsuitable for accurate SIMS calibrations because of the high sensibility of ion emission yields to chemical effects and of possible mass superpositions. However areas free of major heterogeneities can be selected by ionic microscopy for approximate calibrations.

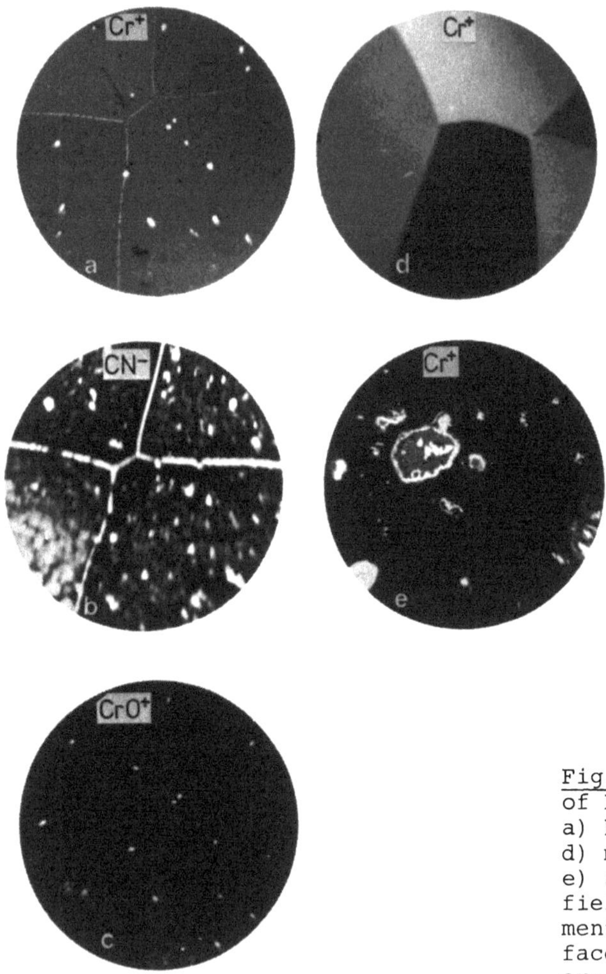

Fig. 1 Ionic micrographs of Fe-15 % Cr alloys - a) b) c) molten in air, d) molten in Ar/H_2 plasma, e) sintered (viewing field = 250 μm, bombardment = 5.5 keV Ar^+, surface free of chemisorbed oxygen)

As an example Fig. 1 shows the structures of Fe-Cr alloys prepared by melting in air, sintering or melting in a Ar/H_2 plasma. In spite of the choice of pure Fe, Cr rods or powders as base materials and the carefulness of the preparations:

- alloys prepared by melting in air (or even in primary vacuum) contain chromium oxides and carbonitrides along grain boundaries;
- sintered alloys contain grains of pure chromium, which were not dissolved because they were protected by a thin chromia film.

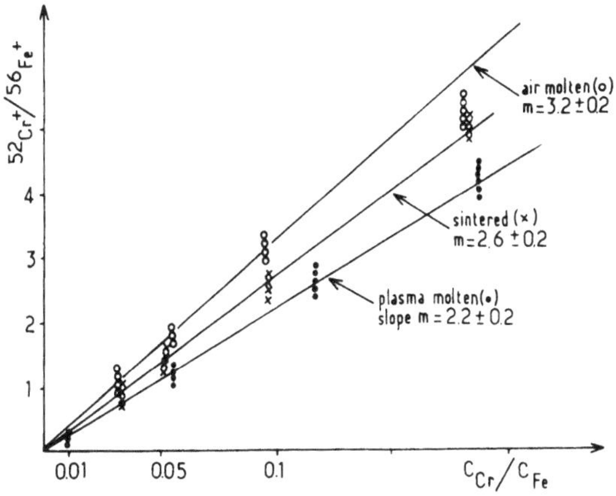

Fig. 2 Calibration curves for Fe-Cr alloys of different purities (Ar^+, surface saturated with oxygen)

The average emission yield of chromium ions varies of a factor 10 with the purity of the standard. Even on selected areas (free of particles observable by ionic microscopy) the slope m of the calibration curve $ICr^+/IFe^+ = f(C_{Cr}/C_{Fe})$ varies from 2.2 to 3.2 (Fig. 2). On the other hand calibration curves are sometimes established on two-phase systems in order to determine the amount of precpitates in a matrix. In such cases ionic microscopy permits to eliminate standards with an heterogeneous distribution of precipitates (their density may vary from one grain to another or in the vicinity of grain boundaries). Moreover the differential sputtering of precipitates and the matrix induces generally a modification of their superficial density, which is different according to the crystallographic orientation of the analysed grain. Such effects were observed on Fe-C [1] and Cu-Be [2,3] alloys, using ionic microscopy.

Applications of Ionic Microscopy to Studies of Segregations, Precipitations and Diffusion Gradients in Metallic Materials

Structure of protective coating for gas turbines

Metallic materials are often protected by coatings in order to conciliate specific mechanical properties of the bulk material and a high resistance to wear or corrosion. For instance alloys for turbine blades are generally NiCr or CoCr matrix hardened with small additions of Ti, V, Mo, Al, B, N and C. Their surface is protected against corrosion by coatings enriched in aluminium [5]. Their adherence and mechanical properties are ensured by an intricate structure due to additions in the coating (Ta, Y ...) and interdiffusion processes with the matrix. Ionic microscopy permitted establishing (in relatively short time) the maps of distribution of all elements and polyatomic ions (MO^+, MC^+, CN^- ...), whatever their mass and content. New phases were identified which could not be analysed by EMA, because of their small size (about 1 µm) and of their composition (carbonitrides, boron)) [4].

Mechanism of internal oxidation of Cu-Be alloys

The precipitation of submicroscopic particles in a matrix is commonly used to increase the hardness of materials. For instance the wear resistance of electric contacts is improved by finely distributed BeO or Al_2O_3 precipitates in the copper matrix. For that purpose the Cu-Be (or Cu-Al, Cu-Si) dilute alloy is oxidized under well defined conditions, so that an oxide film cannot form on the surface but oxygen diffuses in the matrix and reacts with the nearly moveless oxidizable constituent to form precipitates of diameter about 100 to 500 Å. The most schematic model of the conditions of oxidation [8] neglects the contribution of the dopant diffusion to the kinetics of oxidation. The front of internal oxidation moves inwards at a rate depending only of the diffusion coefficient of oxygen. But in the case of Cu-Be alloys the diffusion of Be must be taken into account for fitting the kinetics [8]. Evidence of this diffusion was first obtained by ionic microscopy. It confirmed also the influence of grain boundaries on the oxidation kinetics and on the structure of the oxidized layer.

Figure 3 shows ionic images of a cross section of a cylindrical Cu-1.7 Be sample. Of course the BeO precipitates cannot be separated because of their size (100 Å) and distance from each other but the chemical enhancement of ionic emissions due to oxygen permits to observe the oxidized zone on Be^+ or even Cu^+ images (the Cu^+ emission is enhanced at matrix-pre-

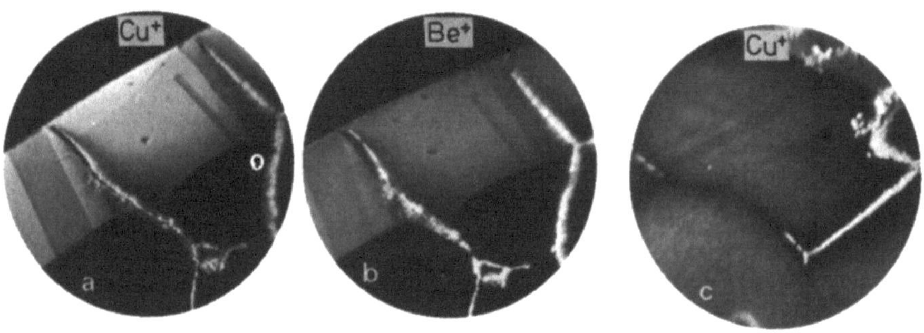

Fig.3 Ionic micrographs of a Cu-1.7% Be polycrystalline alloy oxidized in the pack of RHINES [6]. (a,b) in the oxidized zone (250 µm, Ar^+, surface free of oxygen), (c) in the underlying alloy near the oxidation front, Be pumping along a grain boundary (250 µm,Ar^+, surface saturated with oxygen)

cipitates interfaces). Crystallographic contrast on these images are due to channeling effects but also to variations in the superficial density of precipitates after ion bombardment (confirmed by experiments on monocrystalline samples rotating under the bombardment [3]). Grain boundaries are more emissive in the underlying matrix because of a larger diffusion coefficient of oxygen than in grain. In the oxidized zone grain boundaries are surrounded by less emissive zones. This effect was also attributed to the difference of diffusion coefficients within grains and at grain boundaries: Be is pumped in the vicinity of grain boundaries since oxygen enters more quickly in deeper layers upon them.

Ionic images of the matrix near the oxidation front permitted to detect and quantify a diffusion gradients of Be. For these images, the surface was saturated with chemisorbed oxygen under Ar^+ bombardment or bombarded with O_2^+ ions. In such conditions both emission yields of Be^+ or Cu^+ are related to the Be local content because of the chemical effect of BeO clusters in the vicinity of Cu atoms. On basis of a previously developed model [13] the ionization probabilities may be written:

$$P^{Be^+} = P_{Cu}^{Be^+} \cdot C_{Cu} + P_{Be}^{Be^+} \cdot C_{Be}$$

$$P^{Cu^+} = P_{Cu}^{Cu^+} \cdot C_{Cu} + P_{Be}^{Cu^+} \cdot C_{Be}$$

so that, using proper simplifications, I_{Be^+} and I_{Cu^+} vary linearly with C_{Be}. The ionization probabilities were calibrated on standard solid solutions and the optical density of images was measured in order to determine the diffusion gradient of Be (Fig. 4). This gradient was fitted by calculated ones with different values of the diffusion coefficient D_{Be}. It proved to be an interesting method to measure a diffusion coefficient with an accuracy better than a factor 1.5.

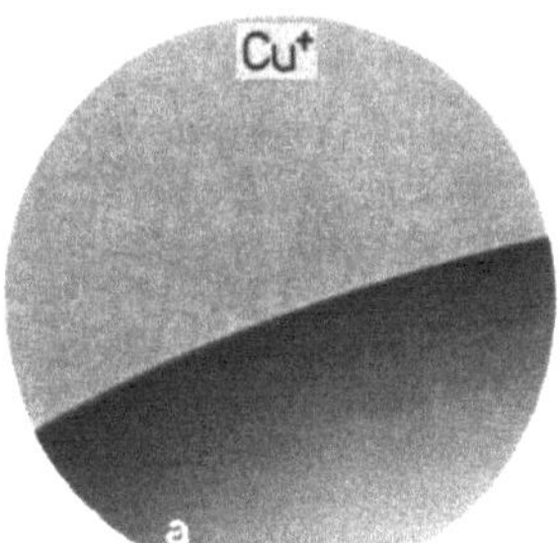

Fig. 4a Cu^+ micrograph straight upon the front of internal oxidation on a Cu-1.7 % Be monocrystal (field diameter = 250 μm, bombardment Ar^+, surface saturated with oxygen)

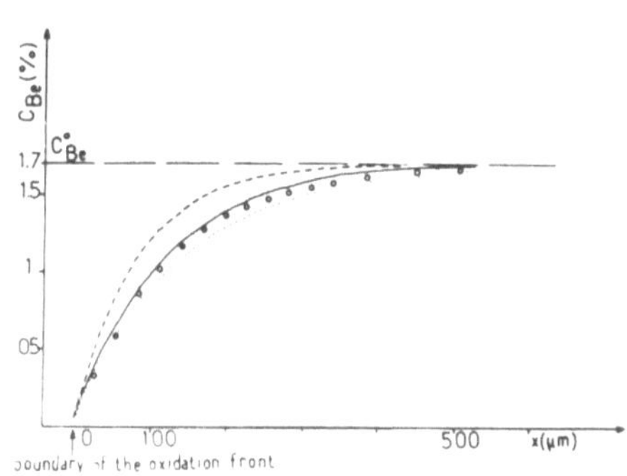

Fig. 4b Be diffusion gradient deduced from micrograph 4a (open circles) and calculated gradients with
--- $D_{Be} = 2 \cdot 10^{-9}\ cm^2 \cdot s^{-1}$
—— $D_{Be} = 3 \cdot 10^{-9}\ cm^2 \cdot s^{-1}$
··· $D_{Be} = 3.5 \cdot 10^{-9}\ cm^2 \cdot s^{-1}$

Conclusion

Other applications - developed in our laboratory [9,10] or by other authors [11,12] - could have been presented. All of them would have demonstrated the usefulness of ionic microscopy to determine the distribution of light elements or elements in trace amounts, in small precipitates or at interfaces. Nevertheless difficulties may be encountered such as:

a) charge accumulations on insulating layers or precipitates,
b) the quick removal of small precipitates when these precipitates are only scattered in the studied sample,
c) artefactual contrasts at interfaces between two phases due to a sputtering relief of charges,
d) other artefacts and chemical effects specific to SIMS.

However one must first of all have a good knowledge of basic principles of ionic microscopy (relations between spatial resolution and detection, collection efficiency, sensibility of films) [14,15,16] to use this method optimally.

References

1. E. Darque, thesis Orsay (1976)
2. O. Lyon thesis Orsay (1975)
3. G. Blaise, O. Lyon and C. Roques-Carmes, Surf. Sci. 71, 630 (1978)
4. G. Thauvin, C. Roquies-Carmes, G. Blaise, G. Slodzian and R. Pichoir, La Recherche Aerospatiale, O.N.E.R.A. 4, 217 (1976)
5. Galmiche, Materiaux et Techniques, Avril-Mai 181 (1973)
6. F.N. Rhines, W.A. Johnsen and W.A. Anderson, Trans. AIME 147, 205 (1942)
7. M.C. Montlouis, thesis Orsay (1979)
8. J.L. Meijering, Advance in Materials Research 5 (1971); Trans. AIME 218, 963 (1960)
9. J.C. Pivin, C. Roques-Carmes and P. Lacombe, Mem. Sci. Rev. Met. 74, n°3, 161 (1977); Revues on High Temperatues Materials, to be published in 1981
10. C. Ducroqo, C. Roques-Carmes and G. Beranger, J. of Nucl. Mat., to be published in 1981
11. V. Leroy, J.P. Servais and L. Habraken, internal report C.R.M. n° 35 (1973), Centre de Recherche Metallurgique, Liège, Belgium
12. P. Maitrepierre, R. Namdar-Irani, J. Rofes-Vernis, B.J. Thomas and G. Henri, Proceedings of the 11th Conf. of Microbeam Analysis Soc. (1976), Miami- and internal reports, IRSID, St Germain en Laye, France
13. J.C. Pivin, C. Roques-Carmes and G. Slodzian, Int. J. Mass Spect. and Ion Phys. 26, 219 (1978); J. Appl. Phys. 51, 4158 (1980); in Secondary Ion Mass Spectrometry SIMS II, ed. by A. Benninghoven, C.A. Evans, Jr., R.A. Powell, R. Shimizu, H. Storms, Springer Series in Chemical Physics, Vol. 9 (Springer, Berlin, Heidelberg, New York 1979) p. 160
14. G. Slodzian, thesis Paris (1963); Anal. Phys. 9, 591 (1964)

15. G. Slodzian, NBS Spec. Publ. 427, "Proc. of the Workshop on SIMS and IMMA", Gaithersburg (1974) p. 33
16. M. Chaintreau, thesis Orsay (1978)

Secondary Ion Mass Spectrometry of Organic Compounds

A. Benninghoven

Physikalisches Institut der Universität Münster,
D-4400 Münster, Fed. Rep. of Germany

Parent-like secondary ions of the general composition $(M+H)^+$ and $(M-H)^-$, where M stands for an organic molecule, are emitted with very high absolute yields from corresponding molecular layers on solid surfaces during ion bombardment. That was the surprising result of two systematic SIMS investigations of amino acids (1), and further groups of organic compounds (2,3) published in 1976 and 1977. Besides these and other parent-like ions as $(M+cation)^+$ e.g. (4), many characteristic fragment ions are emitted too (Fig. 1).

This emission of undestroyed large organic molecules was a completely unexpected result at this time. Sputtering was regarded to be highly destructive for large molecules and only fragments were expected to be emitted from the bombarded surface. In contrast to this, very often parent-like secondary ions represent the strongest line in an organic SIMS spectrum.

Meanwhile many groups have entered this new field of organic mass spectrometry (4-9). The main reason for this is the analytical capacity of this new development in SIMS: Submonolayers of even non-volatile organic compounds can be detected and identified with rather simple SIMS equipment.

From the large body of experimental facts that is available today, some general features governing the secondary ion emission from organic molecules can be extracted:

- For all organic compounds that have been investigated so far, the emission of only slightly changed charged molecules can be accomplished by an appropriate sample preparation. Among these so-called parent-like secondary ions are positively charged protonated molecules $(M+H)^+$, negatively charged deprotonated molecules $(M-H)^-$, positively charged cationized molecules $(M+Cat.)^+$, such as $(M+Na)^+$ or $(M+Ag)^+$, positively charged double cationized molecules $(M+Cat.+Cat.-H)^+$ in which the additional cation replaces a hydrogen atom (10), and negatively charged anionized molecules $(M+An.)^-$, such as $(M+Cl)^-$.
- In addition to these parent-like secondary ions containing only one molecule M, secondary ions containing two molecules, e.g., $(2M+H)^+$, $(2M-H)^-$ or $(2M+Cat.)^+$, are observed.

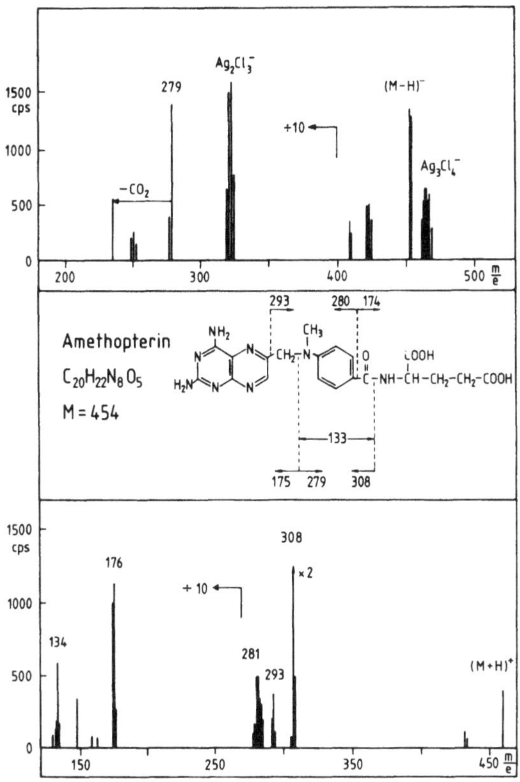

Fig. 1 Positive and negative secondary ion spectrum of amethopterine, an involatile folic acid derivate. 10^{-9} g amethopterine had been deposited on 0.1 cm^2 of a Ag substrate by drying 1 µl of a 10^{-3}-mol. solution in 0.1n HCl. Primary ions: Ar^+, 3 keV, $3 \cdot 10^{-10}$ A on 0.1 cm^2. Besides the parent-like secondary ions $(M+H)^+$ and $(M-H)^-$, characteristic positive and negative fragment ions are emitted. The strong emission of $Ag_2Cl_3^-$ and $Ag_3Cl_4^-$ originates from the HCl component in the sample solution.

- The absolute as well as the relative yields of all these parent-like secondary ions completely depend on the chemical surface composition of the target. This composition is determined by the substrate material (at least in the case of monolayer deposition), the solvent (in the case of wet target preparation), and the concentration of additional atomic or molecular species in this solvent or on the sample surface.

- By an appropriate sample preparation (surface chemistry) extremely high yields (up to some 10 %) for one or several of these parent-like ions species can be achieved for many of the molecules M studied so far.

- In additional to the parent-like secondary ions, characteristic fragment ions, positive as well as negative, are emitted (Fig. 1). Some of them have the general composition $(M-A)^{\pm}$,

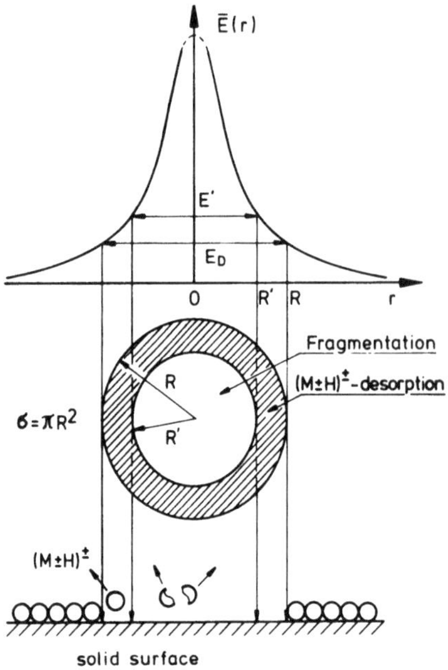

Fig. 2 Parent-like ion formation during sputtering of large organic molecules M from a solid surface. The energy distribution $\bar{E}(r)$ represents the average energy which is trasferred by an impact cascade to a substrate surface atom at a distance r from the point of primary ion impact (r = o). The rapid ($< 10^{-12}$s) energy transfer to the molecules M results in a disappearance of M for $r \leq R$. Only for $R' \leq r \leq R$ does this disappearance result in the emission of parent-like ions. Note that this scheme does not describe one single impact process, but averages a large number of such events !

where A is a small stable complex such as CO_2 or H_2O. But also other complexes, sometimes in a cationized state, are observed.

- Disappearance cross sections σ (see Fig. 2), for ion as well as for electron bombardment, and disappearance temperatures may be very different for different parent-like secondary ions, e.g. $(M+H)^+$ and $(M-H)^-$ of the same molecule M (11).

- By isotope experiments it could be demonstrated for amino acids, that the proton transfer, resulting in the formation of $(M+H)^+$- and $(M-H)^-$-ions, occurs in the same way as in a solution, i.e. only the hydrogen atoms in the CNH_2-COOH group contribute to this transfer (11).

Considering these and other general features of organic secondary ion emission, a tentative, qualitative emission model can be established, that assumes the formation of a precursor at the solid surface before the excitation process, followed by a rapid evaporation-like desorption of this precursor. During this ejection the charge sign of the precursor shows a tendency to remain conserved. This fast evaporation model (12) may be summarized as follows:

1. Precursor Formation

Prior to any ion bombardment a precursor of the finally emitted secondary ion exists at the target surface. This concerns the atomic composition as well as the average charge state of the finally emitted secondary ions.

2. Fast Evaporation

The low energy tail of the impact cascade, that is generated near the surface as a result of the ion impact, is responsible for the fast evaporation ($< 10^{-12}$s) of the unfragmented precursors (Fig. 2).

3. Charge Sign Conservation

A tendency exists to maintain the average charge sign of the precursor during its transformation in the corresponding secondary ion.

4. Fragmentation

The high energy part of the surface cascade $\bar{E}(r)$ produces fragments of the surface precursors, including atomic species. Additional fragment formation may occur during the decomposition of larger excited molecular secondary species after their separation from the surface.

Organic SIMS has an extraordinary analytical power. It enables the detection and identification of even non-volatile organic compounds in the subnanogram range, even with the existing instruments. Organic SIMS is increasingly applied in organic mass spectrometry with double focussing magnetic instruments, where for technical reasons neutral atoms are used as bombarding primary particles (13). There is no doubt that possible improvements in sample preparation and instrument transmission will lower the detection limits by orders of magnitude. Even the present sensitivity in the subnanogram range makes organic SIMS an extremely useful detector in liquid (14) and thin layer (15) chromatography, in drug detection (16), for the trace detection and analysis of organic compounds in body fluids (17) etc. Another important field of general SIMS application- surface studies of catalysts (18)- will become increasingly important for organic SIMS.

References

1 A.Benninghoven and W.Sichtermann, Appl.Phys. 11 (1976) 35.
2 A.Benninghoven and W.Sichtermann, Org.Mass Spectrom. 12 (1977) 595.
3 A.Benninghoven and W.Sichtermann, Anal. Chem. 50 (1978) 1180.
4 H.Grade, N.Winograd and R.G.Cooks, J.Am.Chem.Soc. 99 (1977) 7725.
5 R.J.Colton, J.S.Murday, J.R.Wyatt, and J.J.DeCorpo, Surf. Sci. 84 (1979) 235.
6 R.J.Day, S.E.Unger, and R.G.Cooks, Anal.Chem. 52 (1980) 557A.
7 K.D.Klöppel and W.Seidel, Int.J.Mass Spectrom.Ion Phys. 31 (1979) 151.
8 H.Kambara and S.Hishida, Org.Mass Spectrom. 16 (1981) 167.
9 J.A.Gardella and D.M.Hercules, Anal.Chem. 52 (1980) 226.

10 W.Sichtermann and A.Benninghoven , Int.J.Mass Spectrom. Ion Phys. 40 (1981) 177.
11 W.Lange, M.Jirikowsky and A.Benninghoven, to be published in Surf.Sci.
12 A.Benninghoven, to be published.
13 M.Barber, R.S.Bordoli, R.D.Sedywick,and A.N.Tyler, J.C.S. Chem.Comm (1981) 325.
14 A.Benninghoven et al., Org.Mass Spectrom. 15 (1980) 459.
15 S.E.Unger, A.Vince, R.G.Cooks, R.Chrisman,and E.D.Rothman, Anal.Chem. 53 (1981) 976.
16 A.Benninghoven and W.Sichtermann, Int.J.Mass Spectrom.Ion Phys. 38 (1981) 351.
17 W.Sichtermann, M.Junack, A.Eicke, and A.Benninghoven, Fresenius Z.Anal.Chem. 301 (1980) 115.
18 A.Benninghoven, P.Beckmann, D.Greifendorf, and M.Schemmer, Surf.Sci., in press

Index of Contributors

An, L.D. 421
Antal, J. 297

Barcz, A. 134
Bauer, E. 128
Baumgart, H. 336
Beckmann, P. 405
Bedrich, W. 81
Benninghoven, A. 66,365, 405,416,421,438
Bentz, B.L. 30
Bernheim, M. 151
Bessonneau, B. 314
Bimberg, D. 336
Bonnet, M. 314
Bori, L. 269,394
Brandstaetter, E. 97,238
Brown, J.D. 72,330

Chaumont, J. 274
Cherepin, V.T. 49,77,233
Clegg, J.B. 308
Csanády, Á. 383,388

Darque-Ceretti, E. 431
Dennebouy, R. 431
De Pauw, E. 377
Didenko, P.J. 227
Dittmann, J. 61
Domanski, M. 134
Dubinsky, I.N. 49
Düsterhöft, H. 124
Dyad'kin, Ya.Ya. 49

Evans, C.A., Jr. 88,172, 357

Favennec, P.N. 342
Flentje, G. 165
Frenzel, H. 94
Furman, B.K. 88,357

Ganschow, O. 66,365,421
Gauneau, M. 342
Gerlach, R.L. 22
Giber, J. 216,269
Gnaser, H. 43,282
Goto, K.S. 361
Gras, D.J. 72
Grasserbauer, M. 292,321
Greifendorf, D. 405
Griepentrog, M. 302
Guenther, K.H. 97,238
Guerrero, E. 321
Gyulai, J. 269

Hauffe, W. 206
Hauser, E. 97,238
Havette, A. 140,288
Hennequin, J.-F. 57
Higatsberger, M.J. 38,43
Hobi, G. 97,238
Hofer, W.O. 165,201
Hofmann, S. 186
Hoinkes, H. 411
Holtkamp, D. 416
Honig, R.E. 172
Huber, A.M. 274,314

Inglebert, R.-L. 57

Jede, R. 66,421
Jirikowsky, M. 416
Jurela, Z. 257

Kaarmann, H. 411
Kaiser, U. 365
Kerkow, H. 302
Kirch, G. 398
Klöppel, K.D. 159
Klose, H. 302
Kobayashi, M. 361
Koch, B. 81
Konishi, M. 211
Kosyachov, A.K. 233
Krasyuk, A.D. 233
Kubicki, D. 102
Kugler, S. 297

Lange, W. 416
László, J. 216
Le Goux, J.J. 52
Leidenberger, B. 411
Liebl, H. 30
Linke, U. 165
Litovchenko, V.G. 227
Littmark, U. 201
Lodding, A. 351
Loison, D. 274
Lorin, J.C. 140

Magee, C.W. 172
Magee, T.J. 357
Mai, H. 81
Maier, M. 336
Marchenko, R.I. 227
Marien, J. 377
Marton, D. 216,282,383, 388,394
Marton, J. 282
Maul, J.L. 94
Merenda, P. 314
Migeon, H.N. 52
Mizsei, I. 216
Morillot, G. 274,314
Morrison, G.H. 244
Müller-Jahreis, U. 124, 302

Neelsen, J. 365
Nishimura, H. 426
Nitz, H.M. 365

Odelius, H. 351
Oechsner, H. 106
Ohashi, Y. 211
Okano, J. 426
Ol'khovsky, V.L. 77

Pahlke, C. 124
Phillipp, F. 336
Pivin, J.C. 274,431
Pötzl, H. 321
Pollinger, P. 38,43
Prigge, S. 128

Riedel, M. 282,297
Romanova, G.Ph. 227
Roques-Carmes, C. 274, 431
Rüdenauer, F.G. 2,38,43, 282
Rupert, A. 342
Ryssel, H. 321

Schemmer, M. 405

Schou, J. 165
Seidenkranz, U. 81
Shepherd, F.R. 330
Simondet, F. 102
Slodzian, G. 115,140,151, 288
Södervall, U. 351
Sólyom, A. 269
Someno, M. 361
Speckmann, W. 128
Steiger, W. 43
Stingeder, G. 321
Stopka, A. 264
Studnicka, H. 38,43

Suba, J. 264
Suzuki, T. 211
Syhre, H. 81

Textor, M. 372
Tielert, R. 321
Tsunoayama, K. 211
Turmezey, T. 388

Unger, W. 394

Vasiliev, M.A. 233
Voigtmann, R. 81

Wang, N.L. 421
Westwood, W.D. 330
Wiedmann, L. 365,421
Wierer, P. 97
Wilsch, H. 411
Wirth, Th. 222
Wojtowicz-Natanson, B. 134

Yefremov, A.A. 227

Zinner, E. 292
Züchner, H. 398

MIX
Papier aus verantwortungsvollen Quellen
Paper from responsible sources
FSC® C105338

If you have any concerns about our products,
you can contact us on
ProductSafety@springernature.com

In case Publisher is established outside the EU,
the EU authorized representative is:
Springer Nature Customer Service Center GmbH
Europaplatz 3, 69115 Heidelberg, Germany

Printed by Libri Plureos GmbH
in Hamburg, Germany